1-22-03
se

D0060910

Practical TV & Video Systems Repair

Online Services

Delmar Online
To access a wide variety of Delmar products and services on the World Wide Web,
point your browser to:
        http://www.delmar.com/delmar.html
        or email: info@delmar.com

thomson.com
To access International Thomson Publishing's
home site for information on more than 34 publishers
and 20,000 products, point your browser to:
        http://www.thomson.com
        or email: findit@kiosk.thomson.com

A service of I(T)P®

Practical TV & Video Systems Repair

John Ross

Fort Hays State University

Delmar
Thomson Learning™

Africa • Australia • Canada • Denmark • Japan • Mexico
New Zealand • Philippines • Puerto Rico • Singapore
Spain • United Kingdom • United States

DISCARD

LIBRARY
FORSYTH TECHNICAL COMMUNITY COLLEGE
2100 SILAS CREEK PARKWAY
WINSTON-SALEM, NC 27103-5197

NOTICE TO THE READER

Publisher does not warrant or guarantee any of the products described herein or perform any independent analysis in connection with any of the product information contained herein. Publisher does not assume, and expressly disclaims, any obligation to obtain and include information other than that provided to it by the manufacturer.

The reader is expressly warned to consider and adopt all safety precautions that might be indicated by the activities herein and to avoid all potential hazards. By following the instructions contained herein, the reader willingly assumes all risks in connection with such instructions.

The Publisher makes no representation or warranties of any kind, including but not limited to, the warranties of fitness for particular purpose or merchantability, nor are any such representations implied with respect to the material set forth herein, and the publisher takes no responsibility with respect to such material. The publisher shall not be liable for any special, consequential, or exemplary damages resulting, in whole or part, from the readers' use of, or reliance upon, this material.

Delmar Staff:

Business Unit Director: Alar Elken
Executive Editor: Sandy Clark
Acquisitions Editor: Gregory L. Clayton
Developmental Editor: Michelle Ruelos Cannistraci
Editorial Assistant: Jennifer Thompson
Executive Marketing Manager: Maura Theriault
Channel Manager: Mona Caron

Marketing Coordinator: Paula Collins
Executive Production Manager: Mary Ellen Black
Production Manager: Larry Main
Senior Project Editor: Christopher Chien
Art Director: Nicole Reamer
Technology Project Manager: Tom Smith

COPYRIGHT © 2000
Delmar is a division of Thomson Learning. The Thomson Learning logo is a registered trademark used herein under license.

Printed in the United States of America

1 2 3 4 5 6 7 8 9 10   XXX   05 04 03 02 01 00

For more information, contact Delmar, 3 Columbia Circle, PO Box 15015, Albany, NY 12212-0515;
or find us on the World Wide Web at http://www.delmar.com

Asia
Thomson Learning
60 Albert Street, #15-01
Albert Complex
Singapore 189969

Australia/New Zealand
Nelson/Thomson Learning
102 Dodds Street
South Melbourne, Victoria 3205
Australia

Canada
Nelson/Thomson Learning
1120 Birchmont Road
Scarborough, Ontario
Canada M1K 5G4

International Headquarters
Thomson Learning
International Division
290 Harbor Drive, 2nd Floor
Stamford, CT 06902-7477
USA

Japan
Thomson Learning
Palaceside Building 5F
1-1-1 Hitotsubashi, Chiyoda-ku
Tokyo 100 0003 Japan

Latin America
Thomson Learning
Seneca, 53
Colonia Polanco
11560 Mexico D.F. Mexico

Spain
Thomson Learning
Calle Magallanes, 25
28015-Madrid
Espana

UK/Europe/Middle East
Thomson Learning
Berkshire House
168-173 High Holborn
London
WC1V 7AA United Kingdom

Thomas Nelson & Sons Ltd.
Nelson House
Mayfield Road
Walton-on-Thames
KT 12 5PL United Kingdom

ALL RIGHTS RESERVED. No part of this work covered by the copyright hereon may be reproduced or used in any form or by any means—graphic, electronics or mechanical, including photocopying, recording, taping or information storage and retrieval systems—without the written permission of the publisher.

You can request permission to use material from this text through the following phone and fax numbers.
Phone: 1-800-730-2214; Fax 1-800-730-2215; or visit our Web site at http://www.thomsonrights.com

Library of Congress Cataloging-in-Publication Data

Ross, John A. (John Allan), 1955–
    Practical TV & video systems repair / John Ross.
      p. cm.
    Includes index.
    ISBN 0-8273-8547-1
    1. Television—Repairing. 2. Television circuits—Maintenance and repair. 3. Video tape recorder—Maintenance and repair. 4. Home video systems—Maintenance and repair.
  I. Title.

TK6653 .R67 1999
621.388'87—dc21
                                    99-046635

CONTENTS

Preface   ix

PART 1   BUILDING A FOUNDATION

CHAPTER 1   Fundamentals **3**

    Introduction   4
1-1   Electronic Systems   4
1-2   Energy, Signals, and Frequencies   6
1-3   Radio-Frequency Signals   9
1-4   Circuit Functions   11
1-5   Manipulating Signals for Communication   16
1-6   Troubleshooting Methods   20
    Summary   23
    Review Questions   23

CHAPTER 2   Digital Fundamentals **27**

    Introduction   28
2-1   Boolean Algebra   29
2-2   Digital Number Systems   30
2-3   Digital Building Blocks   31
2-4   Multivibrators, Schmitt Triggers, and Flip-Flops   32
2-5   Control   34
2-6   Timing and Comparison   35
2-7   Conversion   37
2-8   Major Integrated Circuit Families   37
2-9   Using Digital Circuits to Manipulate Frequencies   38
2-10   Microprocessors   40
2-11   Troubleshooting Digital Circuits   42
    Summary   43
    Review Questions   43

PART 2   POWER SUPPLIES AND DEFLECTION CIRCUITS

CHAPTER 3   Linear Power Supplies **49**

    Introduction   50
3-1   Power Supply Basics   50
3-2   Voltage Multipliers   64
3-3   Overvoltage and Overcurrent Protection   65
3-4   Linear Power Supply Circuits   67
3-5   Troubleshooting Power Supply Circuits   70
    Summary   78
    Review Questions   78

CHAPTER 4   Switched-Mode Power Supplies **83**

    Introduction   84
4-1   SMPS Basics   84
4-2   Scan-Derived Power Supplies   88
4-3   Switched-Mode Power Supply Circuits   94
4-4   Troubleshooting Switched-Mode Power Supply Problems   99
    Summary   110
    Review Questions   110

CHAPTER 5   Sync Signals and Horizontal Deflection **115**

    Introduction   116
5-1   Sync Separation   116
5-2   Horizontal Deflection Basics   121
5-3   Horizontal AFC Circuits   123
5-4   Horizontal Oscillator Circuits   129
5-5   Horizontal Driver and Output Circuits   130
5-6   Complete Horizontal Deflection Systems   134
5-7   IC-Based Horizontal Scan Systems   135

5-8   Troubleshooting Sync and Horizontal
        Circuits   137
      Summary   147
      Review Questions   147

**CHAPTER 6   Sync Signals and Vertical
            Deflection** **151**

      Introduction   151
6-1   Vertical Sync Signals   152

6-2   Vertical Deflection System Basics   155
6-3   Vertical Oscillator Circuits   160
6-4   Vertical Driver and Output Amplifier
        Circuits   164
6-5   IC-Based Vertical Scanning Circuits   166
6-6   Troubleshooting Vertical Deflection
        Circuits   171
      Summary   183
      Review Questions   183

PART 3   PROCESSING SIGNALS

**CHAPTER 7   Tuners and Tuning Control
            Systems** **191**

      Introduction   192
7-1   Tuner Basics   192
7-2   R-F Amplifier Stage Operation   193
7-3   Oscillator Stage Operation   195
7-4   Mixer Stage Operation   196
7-5   Varactor Diode Tuning   198
7-6   Microprocessor Control Systems   199
7-7   Troubleshooting Tuner Systems   209
      Summary   213
      Review Questions   213

**CHAPTER 8   Intermediate Frequency, AGC,
            and AFT Circuits** **217**

      Introduction   218
8-1   Processing Intermediate Frequency
        Signals   218
8-2   Selectivity and Attenuation   222
8-3   I-F Amplifier Stage Gain, Coupling, and
        Neutralization   225
8-4   Automatic Gain Control   229
8-5   I-F Amplifier Circuit Characteristics   232
8-6   Detection of the Picture and Sound
        Carriers   238
8-7   I-F Section Troubleshooting   242
      Summary   245
      Review Questions   246

CHAPTER 9   Processing the Sound Signal ... **251**

      Introduction   251
9-1   Producing and Shaping the Sound
        Signals   253

9-2   Demodulating the Sound Signal   256
9-3   Audio Amplifiers   260
9-4   Monaural Sound Systems   266
9-5   Troubleshooting Sound Circuits   273
      Summary   280
      Review Questions   281

**CHAPTER 10   Luminance Signals and
             Video Amplifiers** **285**

      Introduction   285
10-1   Luminance Signal Basics   286
10-2   Amplifying the Luminance Signal   288
10-3   Video Signal Gain Requirements   294
10-4   Luminance Processing Applications   301
10-5   Troubleshooting Video Amplifier
         Circuits   304
      Summary   309
      Review Questions   309

**CHAPTER 11   Chrominance Signals and
             Chroma Amplifiers** **313**

      Introduction   314
11-1   Where do Chrominance Signals Come
         From?   314
11-2   Separating the Chrominance and Composite
         Video Signals   319
11-3   Revisiting the Luminance/Chrominance
         Integrated Circuit   326
11-4   RGB Driver and Video Output Stages   335
11-5   Troubleshooting Color-Processing
         Circuits   339
      Summary   346
      Review Questions   346

PART 4   REPRODUCING THE IMAGE

**CHAPTER 12   CRTs and Projection
             Television Systems** **355**

      Introduction   356
12-1   Cathode-Ray Tube Basics   356

12-2   Display Adjustments   361
12-3   Projection Television Display
         Methods   369
12-4   Troubleshooting Picture-Tube Faults   376

Summary   378
Review Questions   378

CHAPTER 13 HDTV and Computer Monitors 381

Introduction   382
13-1   Interlaced and Noninterlaced
          Scanning   382

13-2   High-Definition Television   383
13-3   Computer Monitors   386
13-4   Computer Monitor Operation   392
13-5   Computer Monitor Circuits   395
13-6   Troubleshooting Computer Monitors   401
         Summary   406
         Review Questions   406

PART 5   FOCUSING ON THE CUSTOMER

CHAPTER 14 Basic Approaches to Service and Professional Development 413

Introduction   413
14-1   Finding Success Through People
          Skills   414

14-2   Gaining Versatility   416
14-3   Succeeding Through Business Skills   420
14-4   Working with Parts Suppliers   426
         Summary   427
         Review Questions   428

Appendix—Schematics   429

Glossary   439

Index   451

Foldouts

PREFACE

Rapid changes in consumer electronics technology have prompted corresponding changes in technology education. Throughout the past decade, the emphasis in technology has moved away from the use of analog circuits to more efficient digital circuitry. Circuits once housed on removable, replaceable modules are now found encapsulated within integrated circuit packages. Because of these primary and complex changes, the quantity of knowledge required to service and maintain modern electronic products has grown substantially. As a result, technology education demands new approaches in teaching and new opportunities for learning. *Practical TV & Video Systems Repair* addresses those approaches and opportunities.

ORGANIZATION

Practical TV & Video Systems Repair is a text designed for use in technology curricula by second-year students who emphasize electronics. While providing an overview of basic concepts, the text also establishes an in-depth treatment of each subject. Because of this, at least one course in analog and digital fundamentals is suggested as a prerequisite for any course using *Practical TV & Video Systems Repair*.

Chapters are presented in individual learning segments that begin with elementary concepts and build until each chapter provides complete coverage of the subject matter. Instructors may choose to present lectures in short discussions or divide each chapter into 40- or 50-minute lectures. The layout of the book ensures that students will have a firm grasp of basic concepts before moving onto more complex concepts. by dividing the chapters into five basic sections, the book follows the design of a television from the integration of power supplies to the incorporation of signal-processing circuits and CRTs. The concluding chapter allows the instructor to emphasize the type of skills needed to become a successful technician.

The fourteen chapters of *Practical TV & Video Systems Repair* are divided into five parts: 1) Building a Foundation, 2) Power Supplies and Deflection Circuits, 3) Processing Signals, 4) Reproducing the Image, and 5) Focusing on the Customer. In turn, each chapter contains performance-based Objectives and Key Terms. Each chapter concludes with a chapter-end review that includes objective questions, essay questions, and multiple-choice questions. A comprehensive glossary is located at the end of the text. A number of real-world schematics are placed in the Appendix and in the attached Fold-outs at the end of the book. These numerous drawings are referenced and used within the book.

FEATURES OF THE BOOK

Balance of Theory and Troubleshooting Practice

Combining a clear, easy-to-read style with a technician's intuitive knowledge of the circuitry, *Practical TV & Video Systems Repair* integrates basic electronic theory with descriptions for circuits and components that make up television receivers and computer monitors.

While the text provides a theoretical base, there is an emphasis on practical troubleshooting applications. The text stresses a hands-on, troubleshooting approach to servicing consumer electronics. Each chapter includes lessons on Troubleshooting techniques and description of Service Calls. In addition, a number of the chapters provide lessons on the proper use of test equipment. The many Fundamental Reviews throughout the text maintain the link between the advanced readings and basic knowledge acquired in previous courses. *Practical TV & Video Systems Repair* includes numerous individual circuit diagrams, waveform illustrations, and manufacturer diagrams that combine to make it a great reference and practical textbook.

Building Block Approach

The text uses a consistent building-block approach that introduces the student to basic concepts and then builds to the complex circuit designs. With an emphasis on fundamental knowledge at the beginning of and throughout the text, the student goes on a journey through each stage of technology needed to reproduce an image or a sound, including switched-mode power supplies, microprocessor-controlled tuners, deflection circuits, luminance and chrominance signal-processing circuits, and projection television circuits.

Technology Advancements

The rapidly changing technology curriculum is addressed in the book with outstanding coverage of the latest technology advancements, including: coverage of circuit designs, circuits, and components used in a broad array of consumer electronic products; switched-mode power supplies; digital video electronics, high-definition television (HDTV), projection television, and more.

LEARNING TOOLS

Practical TV & Video Systems Repair has been carefully designed to enhance the study of consumer electronics products such as televisions and computer monitors. For best results, you may want to become acquainted with the following features that are incorporated into the text:

Real-world Schematics

The text is full of drawings within the chapters, appendix and fold-outs. The text guides the student on reading, understanding and analyzing the drawings to solve troubleshooting problems. This will help to improve students' troubleshooting skills in preparation for entry into the job market.

- Fundamental Reviews
- Service Calls
- Test Equipment Primers

Fundamental Reviews

As students learn how components and circuits operate and then how to troubleshoot problems, they will need to apply knowledge about properties, components, and circuits first seen in other electronics classes. To assist their understanding of subjects covered in this text, "Fundamental Review" boxes accompany many of the reading assignments. When linked with the information provided throughout the chapter, the "Fundamental Review" boxes will help students understand how different properties and components affect the operation of the entire circuit.

Service Calls

While reading about technology concepts and circuit designs, students also need to link that knowledge with applications and troubleshooting. Each "Service Call" takes students into a service area and acquaints them with the thought processes and methods used to successfully solve electronic circuit problems. The "Service Calls" appear at the conclusion of eleven chapters.

Test Equipment Primers

Many of the troubleshooting methods illustrated throughout the text rely on the use and application of electronics test equipment. The sections entitled "Test Equipment Primers" provide clear explanations about the use of logic probes and analyzers, multimeters, oscilloscopes, counters, and signal generators.

ACKNOWLEDGMENTS

This text was prepared with the help of many individuals and corporations. Without their help and their commitment to consumer electronics technology education, the text could not have been properly written. A special acknowledgment is due the following instructors who reviewed the chapters in detail:

Stanley Bejma	Cuillier Career Center
John Carpenter	Sandhills Community College
Sam Chafin	DeKalb Tech
John Cmelko	Bryant & Stratton
Albert DeMartin	Los Angeles Pierce College
Wayne Keesling	Vincennes University
Charles McCameron	Bee County College
Clifton Ray Morgan	Northwest Kansas Technical

In addition, I would like to give thanks and acknowledgment to Michelle Ruelos Cannistraci, Greg Clayton, Larry Main, Christopher Chien, and Nicole Reamer, who provided never-ending support for this project. The quality of the text is enhanced by their expertise as well as the assistance of the Interlibrary Loan department of the Fort Hays State University Forsythe Library.

AUTHOR

The author of *Practical TV & Video Systems Repair,* John A. Ross, is involved in technology education and administration at Fort Hays State University in Hays,

Kansas. He received training through the Devry Institute of Technology, and he holds a Bachelor of Arts degree in English and a Master of Science degree in Public Administration. Mr. Ross owned and operated a consumer electronics service business from 1974 through 1996, managed a University Microcomputer Services service area from 1987 until 1994, and has been instrumental in the implementation of new technologies at the University. He has written extensively for the journal *Electronic Servicing and Technology* and is the co-author of *Principles of Electronic Devices and Circuits*. In addition, Mr. Ross has been a featured speaker on topics such as management and the Telecommunications Act of 1996.

DEDICATION

I would like to thank God for giving me the skills, knowledge, patience, and desire needed to write this text. My parents, John C. and Larraine N. Ross, have listened to my ideas and complaints and have given me priceless support and encouragement. I could not have completed the work without their presence and thank them many times over. I would also like to thank students for their reviews and suggestions, my colleagues for their support, and the staff at Delmar Publishers for their assistance.

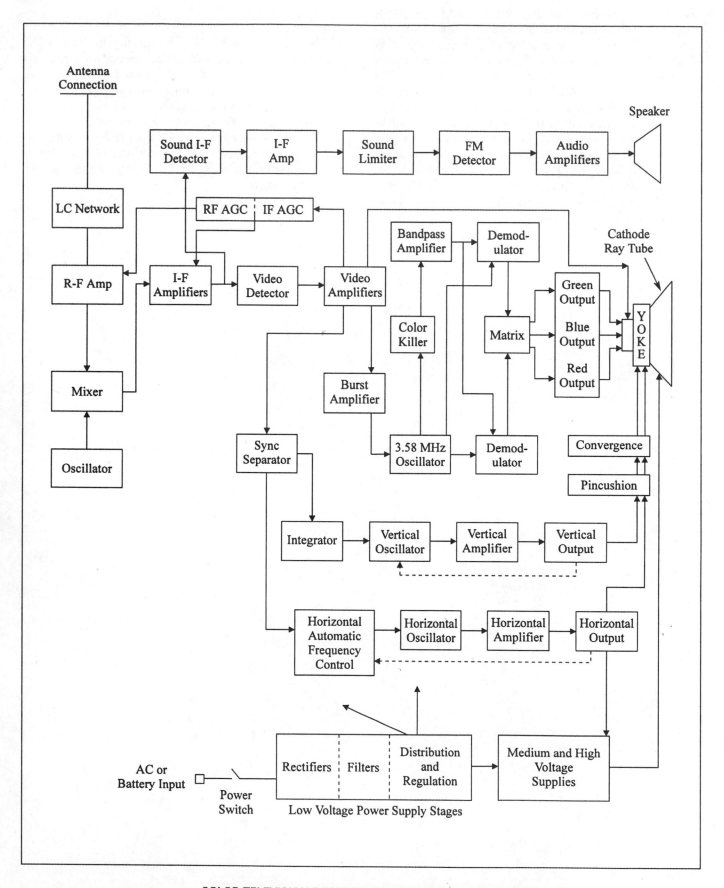

COLOR TELEVISION RECEIVER STAGES AND SIGNAL PATHS

PART 1

Building a Foundation

Chapters one and two emphasize the fundamentals of electronic circuits, analog electronics, and digital electronics. The concepts reviewed in these two chapters provide a foundation for understanding both how modern electronic circuits operate and how the various stages produce functions needed for the proper functioning of the system. With a solid understanding of those fundamentals, you will be able to progress through the remainder of the text with few problems. In addition, you will achieve an understanding of designs, circuits, and processes used in all types of electronic equipment.

Fundamentals

OBJECTIVES   Upon completion of this chapter, you should be able to:

1. Explain electronic systems.
2. Explain source voltages and the difference between low, medium, and high voltages.
3. Discuss energy and signals.
4. Discuss different types of waveforms.
5. Explain the frequency spectrum.
6. Describe harmonics, subharmonics, and noise.
7. Discuss the characteristics of R-F signals.
8. Explain how R-F signals affect circuit components.
9. Define gain, attenuation, bandwidth, coupling, phase relationships, and feedback.
10. Define amplification, oscillation, and switching.
11. Discuss modulation, heterodyning, and demodulation.
12. Understand basic troubleshooting methods and safety practices.

KEY TERMS

amplification	frequency	phase
amplitude modulation	frequency modulation	pulse
attenuation	gain	resonance
bandpass filter	heterodyne	signal
bandwidth	intermediate frequency	source
coupling	modulation	stage
demodulation	noise	superheterodyne receiver
electromagnetic waves	oscillator	switch
electronic system	oscilloscope	wavelength
feedback		

INTRODUCTION

Chapter one defines terms and concepts associated with the operation of electronic systems while establishing the format for the complete text. Beginning with the opening section, the chapter considers electronic systems as a series of input-to-process-to-output blocks.

Using the block diagram of a television shown in Figure 1-1 as a guide, you can see that source voltages and input signals feed into the input of the system. The source voltages become rectified, filtered, regulated, divided, and multiplied before becoming suitable for use in the receiver circuits. Heavy black arrows represent the source voltages in the figure. Energy in the form of high-frequency, information-carrying signals enters the receiver at the tuner and then goes through a series of processes that change the signals into more usable, lower frequency signals. The radio-frequency signals make up part of the frequency spectrum and can be categorized according to frequency bands.

Fine gray arrows represent the signal paths within the receiver. As the signals travel along those paths, a series of processes eliminates portions of the signals, reduces the signals in frequency, and removes information. Those processes become possible through amplification, oscillation, and switching and produce usable information. In turn, amplification, oscillation, and switching depend on basic concepts such as gain, feedback, attenuation, and phase reversal.

This chapter also provides a strategy for troubleshooting electronic problems and introduces basic safety concepts. By applying the concepts introduced in this strategy, your troubleshooting skills will become more efficient and effective. Along with this overall approach to troubleshooting, the chapter also defines the differences between static and dynamic troubleshooting.

As you study chapter one, note that each of the discussed fundamental terms and concepts will have consistent use throughout the text. Your understanding of terms such as gain and concepts such as frequency bands will increase your understanding of the complex circuits seen in the following chapters. All this will make the task of servicing modern electronic products much easier.

1-1 ELECTRONIC SYSTEMS

A group of electronic components interconnected to perform a function or group of related functions makes up an **electronic system**. All electronic systems have the three basic parts shown in Figure 1-2. Those parts

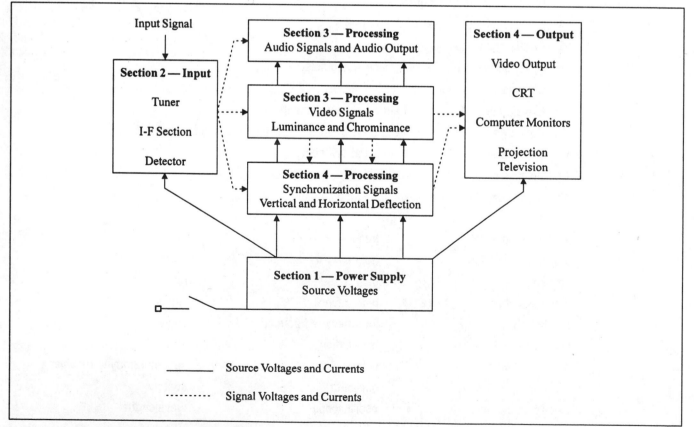

Figure 1-1   Block Diagram of a Television

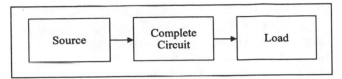

Figure 1-2 Three Basic Parts of an Electronic System

are the source, circuit, and load. The **source** is a device, such as a power supply in a television receiver, that develops a voltage or a combination of voltages.

To take advantage of the force associated with an electric charge and use that force to produce work, we must control the flow of electrons. A load is some type of device that performs a specific function when current flows through the device. Because a load opposes current in an electric circuit, it has resistance. We can define resistance as the property of a device that opposes current in an electric circuit. Resistance is measured in ohms. A resistor is a device that offers a certain amount of opposition to the flow of current.

Electron flow can only occur if a complete path exists from the source to the load and back to the source. Therefore, a complete circuit must exist for electron flow. The insertion of a switch—one of the simplest of control devices—allows the control of the electron flow. A **switch** is a device that opens and closes the path for electrons.

Electronic systems divide into analog and digital circuits. An analog circuit has an output that varies smoothly over a given range and has an infinite number of voltage and values. Each of those values corresponds with some portion of the input. Digital circuits have either an "on" or an "off" output state. The application of an input signal to a digital circuit produces either of the output states but, unlike the analog circuit, no intermediate output conditions.

☑ PROGRESS CHECK

You have completed Objective 1. Explain electronic systems.

Source Voltages

In electricity and electronics, the term "voltage" describes a "difference in potential," or the amount of electric force that exists between two charged bodies. Voltage is defined as either a positive or negative force with reference to another point. A volt is the standard unit of measurement for expressing the difference in potential. Because this force in volts causes the movement of electrons, it is defined as the electromotive force, or emf.

If we have two bodies with opposite charges, or a difference in potential, and then connect those bodies

with a conductor, the charged body with a more positive potential will attract free electrons from the conductor. As the free electrons move from the conductor to the charged body, the conductor assumes a positive charge because of the loss of electrons. Because of this, excess electrons in the other, more negatively charged body begin to flow into the conductor. This flow of electrons or current, continues as long as the difference in potential exists. The basic measure of current is the ampere.

The current in a circuit may be either a direct current (dc) or an alternating current (ac). In a dc circuit, the polarity of the source voltage does not change; electrons flow in only one direction. In ac circuits, the direction of electron flow changes periodically. Because of the changes in direction, the polarity of the source voltage also changes from positive to negative.

Resistors with different values of resistance are connected into circuits to control the amount of current. Because resistance opposes current, the current in a circuit is always inversely proportional to the resistance. If the resistance increases, the amount of current decreases. If the resistance decreases, the amount of current increases.

The ability of a circuit or device to pass current is defined as conductance. Depending on its size and type of material, a conductor carries electricity in varying quantities and over varying distances. A large-diameter wire provides a greater surface area, has less resistance to current, and can carry more current. A material such as copper has better conductance characteristics than a material such as platinum. An inverse relationship exists between conductance and resistance. Conductance is measured in either mhos or siemens.

Voltage Supplies Every stage in an electronic device requires some type of voltage supply because of the signal amplification required to make the system function. **Amplification** is an increase in the voltage, current, or power gain of an output signal. Although systems may utilize an ac power line input, the components within the system rely on dc voltages. We can categorize dc voltages into the low (12–35 Vdc), medium (150–400 Vdc), and high (15–25 kVdc) ranges. Using a television receiver as an example, the low voltage supply provides the necessary voltages for semiconductor operation while the mid-range and high voltages are required for the deflection, focus, and CRT circuits.

Every electronic voltage supply has four distinct parts that involve rectification, regulation, and filtering of the source voltages. Transformers either step up or step down the voltages to the levels required by circuits. Rectification involves the conversion of the required ac voltage value to a pulsating dc voltage. Regulation is defined as the maintenance of a consistent output at the power supply source. With a regulated power supply, changes in the input voltage do not affect the operation of some stages in the system. Filtering smooths the pulsating dc voltage into a usable, constant dc supply voltage.

✔ PROGRESS CHECK

You have completed Objective 2. Explain source voltages and the difference between low, medium, and high voltages.

1-2 ENERGY, SIGNALS, AND FREQUENCIES

Generally, electronic equipment used for communications operates with electrical energy that takes the form of electromagnetic waves. That energy may take the form of radio waves, infrared light, visible light, ultraviolet light, X rays, and other forms. **Electromagnetic waves** are made up of magnetic and electric fields placed at right angles to each other and at right angles to their direction of travel. The wave-like nature of those fields becomes apparent as the magnetic and electric fields vary continually in intensity and periodically in direction at any given point.

Each complete series of variations forms a wave. As one wave travels through space, another wave immediately follows. **Frequency** is the number of waves that passes a point each second and the rate of polarity change. Frequency is measured in hertz (Hz). The distance from any given point or condition in one wave to the corresponding point of the next wave is defined as **wavelength**. We designate wavelength with the Greek letter λ, or lambda, and can calculate a value by dividing the velocity (V) of the waves by the frequency (f) in hertz:

$$\lambda = V/f$$

Because the velocity of radio waves through space equals the speed of light (300,000,000 meters or 984,000,000 feet per second), we can measure wavelength either in meters or feet per second.

Signals

In terms of electronics, we can define a **signal** as a voltage or current that has deliberately induced, time-varying characteristics. A signal voltage or current is different than a source voltage or current for several reasons. Every electrical signal has a distinctive shape described in terms of the time domain, or the amplitude of the signal as a function of time, and frequency domain, or the magnitude and relative phase of the energy. **Phase** is defined as when the repetitions of the signals occur in time, and we further describe signals that have the same frequency and shape in terms of phase. In-phase signals have repetitions occurring at the same time, while out-of-phase signals are displaced along the time axis.

✔ PROGRESS CHECK

You have completed Objective 3. Discuss energy and signals.

Signals can be found at transmission points, may be generated within electronic systems, and may take different forms. We can observe different signal waveforms through the use of an **oscilloscope**, which provides a visual representation of electrical signals. One of the most basic signal waveforms is the sinusoidal wave, or sine wave. Other basic types include the rectangular, ramp, triangular, and sawtooth waveforms, and rectangular pulses.

Sine Waves   The sinusoidal wave, or sine wave, shown in Figure 1-3A represents a mathematical relationship of an alternating voltage or current produced by an alternator, inverter, or oscillator. When a sine wave goes above and below the zero line twice, each combination of maximum positive and negative values equals one cycle. In turn, each cycle subdivides into two alternations, or one-half cycle. We can view an alternation as the rise and fall of voltage or current in one direction.

The value of the voltage or current at any particular point on the sine wave is an instantaneous value.

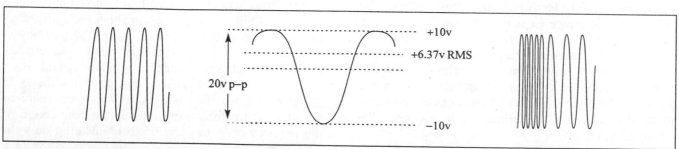

Figure 1-3A   Sine Wave

That is, time passes while the voltage or current goes through its cycle. Each point on the curve occurs only once per cycle and at a particular instant in time. The amount of voltage from zero to either the positive or negative maximum is defined as the peak value and occurs twice during each ac cycle.

When we measure the overall amplitude of a sine wave, we measure the voltage difference between the upper and lower maximum points on the curve. Going back to Figure 1-3A, the upper point on the curve represents +10 volts, while the lower point on the curve represents −10 volts. With the distance between the two points representing 20 volts, the overall amplitude, or peak-to-peak value, equals 20 volts.

If one ampere represents the peak amplitude of alternating current, that value has the same heating effect as .707 ampere of direct current. As a result, we can say that the effective value of an alternating current or voltage is .707 of its peak value. Many times, the effective value is referred to as the root mean square, or rms, value. All values of alternating voltage or current are given as rms values.

When working with alternating voltages or currents, you should also remember that each sine curve alternation has an average value .637 times the peak value. If a curve has a peak value of +10 volts, the average value is less than the rms value and equals:

$$average = .637 \times 10 = 6.37 \text{ volts}$$

Rectangular Waves   A pulse is a fast change from the reference level of a voltage or current to a temporary level, and then an equally fast change back to the original level. Any waveform consisting of high and low dc voltages has a pulse width and a space width and is defined as a rectangular waveform. The pulse width is a measure of the time spent in the high dc voltage state, while the space width is the measure of the time spent in the low dc voltage state. Adding the pulse width to the space width gives the cycle time or duration of the waveform. The values for the pulse width and space width are always taken at the halfway points of the waveform.

As shown in Figure 1-3B, a square wave is a rectangular waveform that has equal pulse width and space width values. While the pulse width and space width for a square wave are equal, the opposite is true for

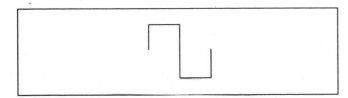

Figure 1-3B   Square Wave

pulses. Shown in Figure 1-3C, pulses usually appear in a series, called a train, and are measured in terms of pulse repetition rate and repetition time period.

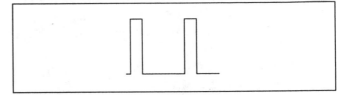

Figure 1-3C   Pulse

Ramp, Triangular, and Sawtooth Waves   Illustrated in Figure 1-3D, a ramp waveform has a slow linear rise and a rapid linear fall. In contrast, the triangular waveform shown in Figure 1-3E rises and falls at a constant rate and has a symmetrical shape. The sawtooth waveform shown in Figure 1-3F appears similar to the triangular wave but has a longer rise time and a shorter fall time.

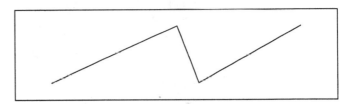

Figure 1-3D   Ramp Waveform

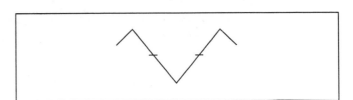

Figure 1-3E   Triangular Waveform

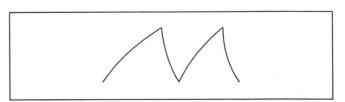

Figure 1-3F   Sawtooth Waveform

☑ PROGRESS CHECK

You have completed Objective 4. Discuss different types of waveforms.

Frequency Spectrum

Electromagnetic wave frequencies cover a wide range that extends from the longest radio waves to the very short waves called cosmic waves. In terms of frequency, this range extends from 10 kHz at the low end to 10^{20} kHz at the high end. Of this range, or spectrum, radio frequencies cover a range from 10 kHz to 3×10^8 kHz.

☑ PROGRESS CHECK

You have completed Objective 5. Explain the frequency spectrum.

Harmonics and Subharmonics

The basic or lowest frequency of a tone is called the fundamental frequency. Harmonics have frequencies that are multiples of the fundamental frequency. As an example, the second harmonic will have a frequency two times higher than the fundamental frequency. If we have a 500-Hz fundamental frequency, the second, third, fourth, and fifth harmonic frequencies equal 1000, 1500, 2000, and 2500 Hz.

As shown in Figure 1-4, each harmonic has a specific amplitude and phase relationship with respect to the fundamental frequency. Point A in the figure shows where the third harmonic has an amplitude of zero, while point B shows where the fundamental and harmonic have equal amplitudes but opposite phases. Given these phase relationships, we can change the shape of any wave by adding or subtracting harmonics or by changing the amplitudes or phases of the harmonics. In addition, we can produce complex waveforms by simultaneously combining several harmonics with a fundamental frequency. Thus, while some waveshapes contain only odd or only even harmonics, other waveshapes contain combinations of odd and even harmonics.

A subharmonic is a precise, fractional division of the fundamental frequency. For example, the master oscillator in a system may have a frequency of 1 MHz. Frequency dividers in the system would subdivide the frequency by ten, with the first divider having a frequency output of 100 kHz, the second an output of 10 kHz, and the third an output of 1000 Hz.

Noise

When we consider the use of signals to communicate, we can define any type of disturbance other than the desired signal as interference, or **noise**. Table 1-1 defines common types of noise. These extraneous signals may take a variety of forms and can affect both the transmission and reception of broadcast signals. The ability of a system to reject noise is defined in terms of signal-to-noise ratio. A system with a high signal-to-noise ratio has a greater ability to reject noise.

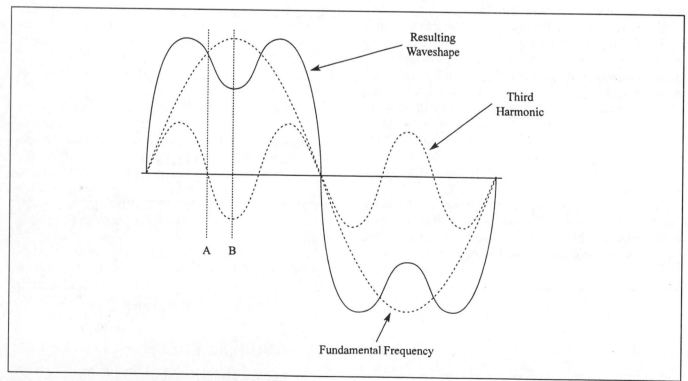

Figure 1-4 Illustration of Harmonics and Subharmonics

Table 1-1   Types of Noise

Atmospheric noise Radio-wave disturbances, such as lightning, that originate in the atmosphere

Common-mode interference Noise caused by the voltage drops across wiring

Conducted interference Interference caused by the direct coupling of noise through wiring or components

Cosmic noise Radio waves generated by extraterrestrial sources

Crosstalk Electrical disturbances in one circuit caused by the coupling of that circuit with another circuit

Electromagnetic interference (EMI) Refers to noise ranging between the subaudio and microwave frequencies

Electrostatic induction Noise signals coupled to a circuit through stray capacitances

Hum Electrical disturbance at the power supply frequency or harmonics of the power supply frequency

Impulse noise Noise generated by a dc motor or generator. Impulse noise takes the form of a discrete, constant energy burst

Magnetic induction Noise caused by magnetic fields

Radiated interference Noise transmitted from one device to another with no connection between the two devices

Radio-frequency interference (RFI) Occurs in the frequency band reserved for communications

Random noise An irregular noise signal that has instantaneous amplitude occurring randomly in time

Static Radio interference detected as noise in the audio frequency (AF) stage of a receiver

Thermal noise Random noise generated through the thermal agitation of electrons in a resistor or semiconductor device

White noise An electrical noise signal that has continuous and uniform power

☑ PROGRESS CHECK

You have completed Objective 6. Describe harmonics, subharmonics, and noise.

1-3 RADIO-FREQUENCY SIGNALS

Radio-frequency, or R-F, signals occupy the frequency range between 10^4 Hz and 10^{11} Hz in the frequency spectrum. As with the low-frequency and dc energy seen with supply voltages, we can transmit radio-frequency energy over wires. However, unlike the low-frequency energy, we can also transmit radio-frequency energy through space in the form of electromagnetic waves.

Certainly, many modern point-to-point transmissions of radio frequencies involve the use of satellite television transmission and reception antennas that send and receive signals through the upper atmosphere and space. Yet we may also consider that conventional R-F transmissions break down into three fundamental categories. While we can define any portion of a radio wave that travels along the surface of the earth as a ground wave, we call the portion of the wave radiated at an angle greater than horizontal the sky wave. Radio waves that travel from one antenna to another without the effect of the ground or upper atmosphere are called direct waves.

The Radio-Frequency Spectrum

Shown in Table 1-2, the R-F spectrum divides into eight categories. When considering the uses of each band, it is important how the frequencies differ in wavelength. As an example, we could transmit frequencies within the EHF band across very long distances because of the extremely long wavelength seen with those frequencies. In practice, the earth would absorb much of the power from the frequency waves. Even with those power losses, EHF transmitters still have a great range.

Frequencies in the LF band are even more prone to losses because of earth absorption. Yet the shorter wavelengths seen with those frequencies allow the use of highly efficient antennas. When considering the MF band, note that the commercial AM broadcast band—533 kHz to 1605 kHz—lies with those frequencies. Because of the wavelength of those frequencies, most AM radio transmissions take the form of ground waves.

The HF band covers frequencies used by foreign broadcast stations and amateur radio stations. Given the frequency band, ground wave transmissions have a limited range of 10 to 20 miles. However, sky wave transmissions between 30 and 60 MHz have a much greater range.

Frequencies found in the VHF band cover the VHF television bands, or 54 MHz to 72 MHz, 76 MHz to 88 MHz, and 174 MHz to 216 MHz. In addition, the VHF band includes frequencies used for commercial

Table 1-2   Radio-Frequency Bands

Designation	Abbreviation	Frequency range (kHz)
Extremely high frequency	EHF	30,000,000–300,000,000
Super high frequency	SHF	3,000,000–30,000,000
Ultra high frequency	UHF	300,000–30,000,000
Very high frequency	VHF	300,000–3,000,000
High frequency	HF	3000–300,000
Medium frequency	MF	300–30,000
Low frequency	LF	30–300
Very low frequency	VLF	10–30

FM broadcast transmissions. Most conventional VHF transmissions involve direct wave transmissions. The UHF band covers UHF television band frequencies, or 470 MHz to 890 MHz. Conventional UHF transmissions also involve direct wave transmissions. The last band of frequencies, the SHF band, has extremely short wavelengths and is recognized as microwave frequencies.

 PROGRESS CHECK

You have completed Objective 7. Discuss the characteristics of R-F signals.

R-F Signals and Component Properties   Circuits used to amplify and process radio frequencies differ from other circuits used for lower frequency voltages and currents because of the properties associated with the components that make up the circuits. Those properties are resistance, inductance, and capacitance.

For example, any R-F current flowing through a conductor creates a changing magnetic field around the conductor. When the field builds and collapses, flux lines cut the conductor. From this, a counter emf is induced that has a directly proportional relationship to frequency. The counter emf, or inductive reactance, opposes the current in the conductor. Therefore, components may have a low inductive reactance at low frequencies and a high inductive reactance at the higher R-F frequencies.

Skin Effect   Another characteristic of R-F signals, called skin effect, originates because of an alternating current flowing through a conductor which is defined as the increasing resistance of a conductor at very high frequencies. The flow of the ac current through the conductor causes magnetic flux lines to move from the center of the conductor and then expand outward. With the lines concentrated more at the center of the conductor than near the surface, the counter emf at the center is also greater. As a result, current concentrates along the surface of the conductor. In turn, the effective cross-sectional area of the conductor decreases. The skin effect results because the resistance of the conductor is inversely proportional to its cross-sectional area.

R-F Signals and Resistors   Radio frequencies can cause resistors to exhibit the properties of resistance, inductance, and capacitance. Inductance occurs because of the connecting leads and the resistance of the device. Capacitance occurs because the R-F signal causes the resistor to have two conductive points separated by an insulator.

R-F Signals and Capacitors   Like resistors, capacitors also exhibit the properties of resistance, inductance, and capacitance. Resistance exists because of the resistance of the capacitor plates, the connecting leads, and the dielectric. Inductance occurs because of the connecting leads and the dielectric plates. The inductance and capacitance form a series resonant circuit.

When the applied signal frequency has a value less than the resonant frequency of the capacitor, the capacitive reactance is higher than the inductive reactance. Thus, at low frequencies, a capacitor presents resistance and capacitance in series. When the applied signal frequency has a value higher than the resonant frequency of the capacitor, the inductive reactance is greater than the capacitive reactance. Therefore, at high frequencies, a capacitor presents inductance and resistance in series.

⟳ FUNDAMENTAL REVIEW

Capacitance

Capacitance occurs with the separation of two or more conducting materials by an insulating material. We measure capacitance with basic units called the Farad (1 Farad), the milliFarad (.001 Farad), the MicroFarad (.0000001), and the PicoFarad (0000000000001 Farad or pF). The capacitance value depends on the amount of total surface area taken by the conducting materials, the amount of spacing between the conducting materials, and the thickness and the type of insulating material.

RF Signals and Inductors   Inductors also offer resistance, capacitance, and inductance. Resistance exists because of the resistance found within the conductor used to wind the inductor. Because inductors have a greater ac resistance than dc resistance, inductors are prone to skin effect. Distributed capacitance in an inductor results from the many small capacitances existing between the turns of an inductor.

The inductance and distributed capacitance of an inductor form a parallel resonant circuit. At frequencies lower than the resonant frequency, an inductor has a higher capacitive reactance and a lower inductive reactance. As a result, the inductor acts like an inductance in series with a resistance. At frequencies higher than the resonant frequency, an inductor has a higher inductive reactance and a lower capacitance reactance. Therefore, the inductor acts like a capacitance in parallel with a resistance.

Resistance always dissipates energy in the form of heat. Reactance returns absorbed energy to the circuit without loss. In an inductor, the loss of energy is proportional to the effective resistance. The ratio of reactance to effective resistance is defined as the quality, or Q, of the inductor and measures the efficiency of the inductor. The mathematical representation of Q is:

$$Q = X_L/R$$

where Q equals quality, X_L equals inductive reactance in ohms, and R equals the effective resistance in ohms.

This method of measuring inductor efficiency is important because Q remains constant over a wide range of frequencies. Both reactance and effective resistance increase with frequency. An inductor or circuit that has a higher value of Q has greater efficiency.

⟳ FUNDAMENTAL REVIEW

Inductance

Inductance is a value associated with coils and is measured in a basic unit called the Henry (1 H), the milliHenry (.001 Henry or mH), and the MicroHenry (.0000001 or uH). The inductance value of a coil depends on size, number of windings, and type of core material. Small coils have lesser values than large coils, while coils with a higher number of windings have larger values. When considering core materials, powered iron cores yield a higher inductance than brass or copper cores.

Inductances also have reactance and impedance values. The impedance value of an inductance varies with the value of the ac frequency and the value of the inductance. Any increase in the frequency or the inductance values also increases the inductive reactance.

Resonance   **Resonance** occurs when a specific frequency causes the inductances and capacitances in either a series or parallel ac circuit to exactly oppose one another. With resonance, a single particular frequency emerges as the resonant frequency, and three basic rules follow:

1. Capacitors and inductors with larger values have lower resonant frequencies.

2. Capacitors and inductors in series have a low impedance at the resonant frequency.

3. Capacitors and inductors in a parallel circuit have a high impedance at the resonant frequency.

✓ PROGRESS CHECK

You have completed Objective 8. Explain how R-F signals affect circuit components.

1-4 CIRCUIT FUNCTIONS

Analog and digital circuits perform the basic tasks of amplification, oscillation, and switching, and contain combinations of passive and active elements. Passive elements—resistors, capacitors, and inductors—route voltages, provide feedback, and perform a variety of functions but do not amplify or oscillate. Active elements, or transistors and integrated circuits, perform amplification and oscillation. For the most part, though, amplification and oscillation cannot occur without interconnections between passive and active elements. In an electronic system, each active element and its associated passive elements make up a **stage**.

Basic Concepts

The three functions of amplification, oscillation, and switching performed by electronic stages and circuits lead to other processes such as modulation, demodulation, heterodyning, frequency multiplication, and frequency synthesis. To varying degrees, all those functions involve basic concepts such as gain, attenuation, bandwidth, coupling, phase relationships, and feedback.

Gain   The term **gain** expresses the ratio of input signal voltage, current, or power to the output signal voltage, current, or power of an amplifier. Mathematically, voltage gain appears as:

$$A_V = E_O/E_{in}$$

where A_V represents the voltage gain, E_O represents the output signal voltage across the load, and E_{in} equals the input signal voltage. To show the current and power gain, we substitute the output and input current or power values into the equation for:

$$A_I = I_{out}/I_{in} \text{ and } A_P = P_{out}/P_{in}$$

The use of more than one stage for amplification produces total amplifier gain, or the product of the voltage gains of the individual stages.

Attenuation   **Attenuation** is the opposite of gain and is shown as loss. With attenuation, the output signal from an electronic circuit has a lower amplitude or signal level than the input signal. Attenuation is specified as loss through either a linear ratio or in decibels.

Bandwidth   Every amplifier has **bandwidth**, or a frequency range over which the amplifier has relatively constant gain. As an example, typical voltage amplifiers have a bandwidth defined as the range of frequencies over which the output voltage is at least 70 percent of maximum when a constant amplitude input signal is applied. In this type of amplifier, the output voltage decreases rapidly at frequencies higher than that range and at frequencies lower than that range.

Coupling   Although a single transistor amplifier can provide a large gain, most electronic devices require more gain than one transistor amplifier can develop. **Coupling** involves the connecting of one or more amplifier stages through various circuit configurations. The output of one stage is applied to the input of a second stage through resistance-capacitance coupling, impedance coupling, or direct coupling.

Phase   The input and output voltages of an amplifier have phase relationships that vary with the configuration of the amplifier circuit. In some amplifier circuits, the output voltage becomes more negative as the input voltage becomes more negative, and then becomes more positive when the input voltage becomes more

positive. With the output and input voltages in step with one another, no phase shift occurs. Figure 1-5A shows an example of in-phase signals.

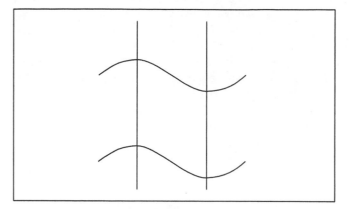

Figure 1-5A   In-Phase Signals

In other amplifier circuits, the output voltage becomes more negative as the input voltage becomes less negative and less negative as the input voltage becomes more negative. With this, the output voltage signals are out-of-phase with the input voltage signals. Figure 1-5B shows an example of two out-of-phase signals. In addition to in-phase and out-of-phase relationships, some applications require that signals follow, or lag, a reference signal. Figure 1-5C shows two signals lagging a reference by 12 and 120 degrees respectively.

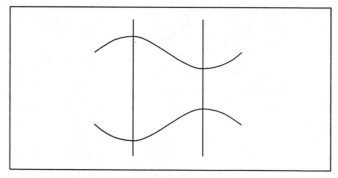

Figure 1-5B   180-Degree Out-of-Phase Signals

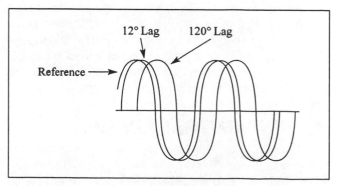

Figure 1-5C   Two Signals Following a Reference

Feedback Almost every type of signal generation technique uses **feedback** to control the frequency, or period, and the amplitude, or level, of an output signal. As shown in Figure 1-6, a feedback signal travels from the output signal to the input of the amplifier. At the input, the feedback voltage modifies the control voltage that determines the size and shape of the output signal. When the feedback voltage has the same polarity or phase as the input signal, we define the feedback signal as positive feedback. Oscillators rely on positive feedback to sustain the oscillations of the amplifier.

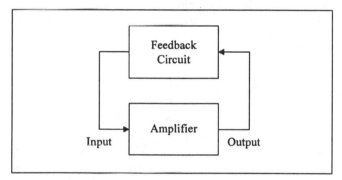

Figure 1-6   Feedback Circuit

The result of positive feedback is the increase of an input signal. A device conducting current and utilizing positive feedback will conduct more current. As the device continues to conduct, the feedback line allows a greater level of feedback signal to travel back to the input until the device saturates. When saturation occurs, the device stops conducting and the positive feedback reinforces the action and shuts the device off.

Other circuits rely on negative feedback, or the feeding back of part of an output signal to reduce the size and shape of the output signal. In this case, the negative feedback signal has an opposite phase or polarity to that of the input signal. Many circuit designs use negative feedback to control the gain of an output signal from an amplifier.

 PROGRESS CHECK

> **You have completed Objective 9. Define gain, attenuation, bandwidth, coupling, phase relationships, and feedback.**

Amplification

Amplifiers increase either the voltage, current, or power gain of an output signal. When a signal passes through two or more stages in sequence, we define the circuits as cascade stages. The efficiency of an amplifier is the ratio of signal power output to the dc power supplied to the stage by the power supply.

We can classify amplifiers by whether the device amplifies voltage or power. The term "amplifier" may describe a single stage, or a transmitter or receiver section consisting of two or more stages of amplification. In a multiple-stage voltage amplifier, all stages contain voltage amplifiers. But, a multiple-stage amplifier designed to provide a power output will include a power amplifier in the last stage. The other stages in the amplifier will consist of voltage amplifiers.

Voltage Amplifiers A voltage amplifier builds a weak input voltage to a higher value but supplies only small values of current. Thus a voltage amplifier always operates with a load that requires a large signal voltage and a small operating current. A cathode-ray tube is an example of this type of load. Every voltage amplifier stage has voltage gain shown as the ratio of the output voltage to the input voltage.

Power Amplifiers Even though voltage amplifiers increase the output voltage, the actual power output may remain at a low value. If the signal must operate some type of current-operated load, such as a speaker, the low power output will not drive the load. Power amplifiers increase the power gain of the circuit by supplying a large signal current. As the following equation shows, power gain is a product of voltage and current:

$$A_P = P_{out}/P_{in} = A_V \times A_I$$

where A_P represents the power gain, P_{out} equals the output power, and P_{in} equals the input power.

Class A, B, AB, and C Amplifiers The classification of a transistor amplifier also describes the length of time that collector current flows during each cycle of the input signal. Both the base bias current and the amplitude of the input signal control the length of time that current exists during the input signal cycle. Given these factors, the classification of transistor amplifiers divides into four classifications: A, B, AB, and C.

If a transistor amplifier operates as a Class A amplifier, the collector current flows continuously throughout the input cycle. During Class A operation, the transistor operates on the linear part of its characteristic curve. Because of this, any distortion of the output signal remains at a minimum. Class A amplifiers have the least distortion of any of the three amplifier classifications. The limitation on the current variations, however, cuts the output signal power of a Class A amplifier to a low level. Because the collector current flows for the entire cycle, a Class A amplifier takes power from the dc supply at a continuous rate. The combined effects of the limited current variations and the constant need for a dc supply drive the efficiency of a Class A amplifier to a range of 25 to 30 percent.

Class A amplifiers have the lowest efficiency of the three amplifier classifications.

Class B transistor amplifiers have collector current during only half the input signal cycle. This occurs because the base bias voltage nearly equals the collector cutoff voltage. As a result, the current has a pulse-like form similar to that seen with half-wave rectification. Thus, the operation of a Class B amplifier is not limited to the linear portion of the characteristic curve. Because the input signal of a Class B amplifier produces larger variations in collector current, Class B amplifiers have a higher efficiency than Class A amplifiers. In addition, a Class B amplifier takes power from the dc voltage supply during only about half of the input cycle. Therefore, Class B amplifiers have a greater signal power output. The combination of greater signal output and the need for less dc power produces the higher efficiency.

Class AB transistor amplifiers require collector current for more than half but less than the full input cycle. During Class AB operation, the transistor is biased between cutoff and the center of the linear portion of its characteristic curve. As a result, Class AB amplifiers permit the use of a larger input signal and provide greater efficiency than do Class A amplifiers, although less than Class B amplifiers. In addition, the Class AB operation produces less distortion than that seen with the Class B operation.

When compared to Class A, B, and AB amplifiers, Class C amplifiers provide the highest efficiency, around 85 percent. During Class C operation, collector current flows for less than half of the input signal cycle. Moreover, the large input signal drives the transistor into saturation and, as a result, produces the highest possible signal output power.

Amplifier Circuit Configuration   Another method for identifying amplifier characteristics involves the circuit configuration, or which element of a transistor amplifier connects to a common point, such as ground. The connection of an amplifier element to a common circuit point can provide either high power gain, impedance matching, or high voltage gain in a single-stage amplifier. The three configuration possibilities are the common-emitter, common-base, and common-collector.

The common-emitter configuration provides the highest power gain, achieved through the use of single transistor amplifiers. In addition, a common-emitter amplifier has a high current gain and a relatively low input impedance. Of the three configurations, only the common-emitter amplifier provides a 180-degree phase reversal.

The common-base amplifier provides a low input impedance of around 30 ohms and an output impedance of nearly 1 megohm. Therefore, common-base amplifiers work well for matching a low-impedance source to a high-impedance load. Common-base amplifiers provide the highest voltage gain of the three amplifier configurations.

Common-collector amplifiers provide a much higher input impedance than that of the other two configurations. In addition, the common-collector amplifier has the lowest output impedance. Given those impedance characteristics, the common-collector amplifier circuit works well for matching a high impedance source to a low impedance load. Despite having a current gain that nearly equals the current gain of a common-emitter amplifier, common-collector amplifiers have a very low power gain.

Amplifier Coupling   As you know, coupling two or more amplifier stages together yields additional gain. Multistage transistor amplifiers are often classified according to the method of coupling. In resistance-capacitance coupling, either resistors or capacitors couple the output of one amplifier to the input of another.

Direct coupling is used in circuits that operate at low frequencies and directly connect one amplifier to another. Another coupling configuration for low-frequency applications is capacitance coupling. This type of coupling allows an ac signal to pass through a capacitor from the output of one amplifier to the input of another. Any dc component is blocked by the capacitor and does not affect any dc voltages at the second amplifier.

Impedance coupling is used with amplifier circuits that must have a wide bandwidth. Rather than use a resistor as the load, an amplifier circuit using impedance coupling uses the inductive reactance of a coil as a load. At high frequencies, the impedance presented by the coil increases and causes the circuit gain to also increase. Transformer coupling is also used in amplifier circuits that operate with high frequencies. It involves the use of the primary and secondary windings of a transformer for the transfer of energy. An ac signal at the output of one amplifier transfers from the primary winding of a transformer to its secondary winding connected to the input of the next amplifier.

Cascaded amplifier stages also feature components that provide neutralization, or the elimination of any effects caused by the internal capacitance of a semiconductor. The capacitance can allow some of the output signal of an amplifier to feed back into the semiconductor as an out-of-phase signal. Given this additional signal, an amplifier will begin to oscillate at a high frequency. The consequences of these parasitic oscillations is the overheating of the amplifier, a loss of gain, and noise distortion.

Amplifier Frequency Ranges The fourth method classifies amplifiers according to the frequency range in which the amplifier operates and the width of the frequency response band. Audio amplifiers amplify audio frequencies ranging from 20 Hz to 20 kHz. In a television, video amplifiers amplify frequencies ranging from a few hertz to 4.2 MHz. An R-F amplifier amplifies radio-frequency signals and may have either a very narrow or a very wide bandwidth.

Oscillation

Electronic systems require one or more alternating voltages or currents to reproduce sound and pictures. In a radio communications transmitter, the system applies a radio-frequency voltage or current to an antenna for the purpose of producing radio waves. When considering the different types of voltages in a receiver, always remember that the alternating current used to produce the radio waves is different than the alternating current present in power lines. Radio waves have a much higher frequency than ac power lines and may require different shapes of waveforms than the sine waves seen with power lines.

As you know, 60 Hz ac line voltages produce sinusoidal waveforms. However, the proper operation of circuits within a receiver may depend on a sawtooth, triangle, or a rectangle waveform. Because of this, the alternating voltages and currents must be produced by circuits within the electronic system called oscillators.

An **oscillator** consists of an amplifier with a feedback circuit connected to the output of the amplifier and converts a dc voltage to an ac signal. With the feedback circuit feeding back a portion of the output to the input of the amplifier, the amplifier begins to produce its own input. For this to occur, the feedback circuit must produce a voltage or current of the *proper phase* and *proper amplitude* to the input.

Referring to Figure 1-7, the amplifier has an output greater than the magnitude of the input. Moreover, the output signal has the opposite polarity of the input signal. The opposite-phase, greater-amplitude output signal travels through the feedback circuit, where a phase reversal and a reduction in amplitude occur. As a result, the signal that feeds back to the amplifier input is identical to the input signal and works as the amplifier input.

The oscillator operation is triggered by small changes in voltage or current at the input when the oscillator turns on. Once the oscillator starts, the whole operation repeats itself to maintain the oscillations. Oscillators vary in stability or the ability to maintain an output that has a constant frequency and amplitude. Different circuit configurations and oscillator types affect stability, with well-designed oscillator circuits

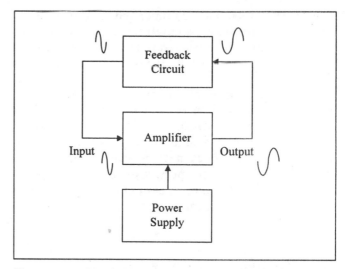

Figure 1-7   Block Diagram of a Basic Oscillator Circuit

maintaining a constant rate of oscillation over a long period of time. The loss of amplitude seen for progressive cycles of oscillation is called damping.

Switching

Switching circuits either generate or respond to nonlinear waveforms such as square waves, and are a fundamental component of many different types of electronic systems. For example, transistors work as switches when driven between saturation and cutoff. In practice, a square wave input causes the transistor to bias off when the input signal has a specific negative level, and to saturate when the input signal has a specific positive level. As a result, the transistor performs the same role as the opening and closing of a mechanical switch.

During operation of a switch, the specific negative value of the input signal equals the emitter supply connection. Regardless of whether the emitter supply equals 0, 5, or 10 volts, the negative value of the input signal for cutoff to occur equals the emitter supply value. The specific positive value of the input signal equals the supply voltage to the collector. Again, depending on the circuit requirements, the positive value of the input signal needed to cause saturation equals the collector supply voltage value.

Along with bipolar transistors, also effective as switches are field-effect transistors (FETs), metal-oxide silicon FETs, and operational amplifiers. Transistor switches are measured in terms of rise time, or the amount of time required to go from cutoff to saturation, delay time, or the time required to come out of cutoff, storage time, or the time required for a transistor to come of saturation, and fall time, or the time required for a transistor to go from saturation to cutoff. FET switches are measured in terms of turn-on time, or the sum of delay time and rise time, and turn-off time, or the sum of storage time and fall time. Operational amplifier switches are measured in terms of slew rate, or the maximum rate at which the device can change its output level.

Basic switching circuit configurations include the inverter and the buffer. An inverter produces a 180-degree voltage phase shift and responds to a low-input voltage signal with a high-output voltage signal. In contrast, a buffer does not introduce a 180-degree phase shift. From the perspective of component configuration, a buffer is either an emitter follower, a source follower, or a voltage follower. Chapter two discusses buffers in more detail.

☑ PROGRESS CHECK

You have completed Objective 10. Define amplification, oscillation, and switching.

1-5 MANIPULATING SIGNALS FOR COMMUNICATION

In the first three sections of the chapter, you learned that every electronic system used for communications requires a voltage source and a signal source found at the input of the device. Signals within the radio-frequency band carry intelligence from the transmission point to the receiver through the process of modulation. **Modulation** involves the encoding of a carrier wave with another signal or signals that represent some type of intelligence. However, because of the high frequencies used for transmission, circuits within the receiver must have the capability to drop the frequencies to a lower level and then remove the intelligence from the signal through **demodulation**.

Modulation

With modulation, an audio frequency signal affects the frequency or amplitude of radio-frequency waves so that the waves represent communicated information. The carrier wave is the sinusoidal component of a modulated wave and has a frequency independent of the modulating wave. As the name suggests, the carrier wave carries the transmitted signal.

Radio waves can carry signal information only when modulated by another signal. While a perfectly unmodulated carrier has zero bandwidth and contains no information, the modulated signal occupies a bandwidth at least comparable to the modulating signal. Modulation combines the waveforms of the combined signals and yields different combinations of those signals.

Several different types of modulation methods exist and may be seen in various types of communications equipment. As an example, AM, or amplitude modulation, and FM, or frequency modulation, are used to transmit the picture and sound information in television systems. Along with the oft-used AM and FM methods, some communication systems use phase modulation and single-sideband modulation.

Amplitude Modulation   With **amplitude modulation**, the amplitude of the R-F carrier shown in Figure 1-8A changes while its frequency remains constant. The changes in the carrier amplitude vary proportionately with the changes in the frequency of the audio-frequency modulating signal shown in Figure 1-8B. Commercial AM radio stations operate with broadcast frequencies in the 550 to 1650 kHz range. When considering television signals, the video signal is amplitude-modulated. In addition, many amateur radio operators use amplitude modulation for their signal transmissions.

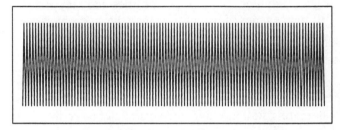

Figure 1-8A   R-F Carrier with No Modulation

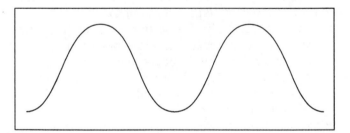

Figure 1-8B   Audio-Frequency Modulating Signal

As mentioned earlier, the constant amplitude carrier wave, or CW, does not contain any modulating information. The thin outline drawn along the peaks of the modulated carrier wave is called the modulation envelope. Exact reproduction of the transmitted signal at the receiver requires that the modulation envelope produced at the transmitter has the same waveform as the modulating signal.

The waveform shown in Figure 1-8C shows an amplitude-modulated waveform resulting from the combination of a 1-MHz local oscillator frequency and an audio signal produced by speaking into a microphone

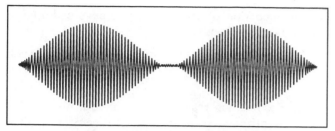

Figure 1-8C Amplitude-Modulated Carrier

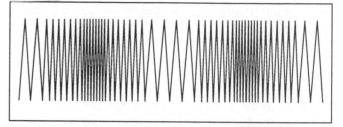

Figure 1-9 Frequency-Modulated Carrier

at an R-F amplifier. A modulator in the system varies the effective voltage obtained from the power supply so that the supply voltage to the R-F amplifier either doubles or drops to zero on audio peaks. The doubling of the power supply voltage causes the corresponding amplitude of the output R-F signal to double. When the power supply drops to zero, the output R-F signal also drops to zero. Consequently, the output waveform found at the amplifier is a modulated radio frequency. Although we can continue to see the local oscillator frequency, the amplitude, or envelope, of the carrier changes at a rate determined by the modulator and the voice signal found at the microphone. Referring again to Figure 1-8B, low-level audio signals produce smaller peaks and valleys in the signal, while larger-level audio signals cause larger changes to appear.

The modulated R-F output signal divides into three individual components, the original oscillator frequency and two sideband frequencies. The upper sideband frequency results from adding the modulating and oscillator frequencies together, while the lower sideband frequency is the difference between the modulating and oscillator frequencies.

Frequency Modulation With **frequency modulation,** the frequency of the carrier changes while its amplitude remains constant. As the modulating signal increases to a maximum positive value, the carrier frequency also changes. When the modulating frequency drops to zero, the carrier frequency decreases to its original value. As the modulating frequency increases to its maximum negative value, the carrier frequency decreases.

Looking at Figure 1-9, the positive peak of the sine wave coincides with an increase in oscillator frequency. As a result, the change in frequency of the frequency-modulated carrier wave corresponds with a change in amplitude of the input signal. The rate of the frequency changes corresponds with the modulating frequency. In contrast to amplitude modulation, the amplitude of the frequency-modulated waveform in the figure remains constant.

As an example of frequency modulation, a microphone controls the capacitance of an oscillator. Thus any voice fluctuations picked up by the microphone

vary the spacing within the capacitor and change the oscillator resonant frequency. With this change, the desired frequency modulation is introduced into the circuit.

Figure 1-9 also shows that the carrier frequency varies above and below a center frequency. We refer to the amount of positive or negative change in the carrier frequency as deviation. While the amplitude of the modulating signal determines the deviation, the frequency of the modulating signal determines the rate that the carrier varies above and below the center frequency.

Phase and Single-Sideband Modulation Both the phase and single-sideband modulation methods suppress the carrier and one sideband while transmitting one sideband from the original sideband signal. Because the upper and lower sidebands also include the modulating information, a typical AM demodulation scheme suppresses the carrier wave and one sideband and transmits the remaining sideband. This type of transmission is referred to as single-sideband transmission, or SSB.

Pulse Modulation Digital communication systems utilize another method called pulse modulation. The system converts the intelligence held within the modulating signal into a pulse. After the conversion occurs, the system pulses the R-F signal for the type of pulse modulation used. The modulating pulses may control the amplitude, frequency, on-time, or phase of the carrier. We refer to the pulse-modulation methods as pulse-amplitude modulation, frequency shift keying, pulse-width modulation, and pulse-phase modulation.

Heterodyning

During the process of **heterodyning,** the output frequencies from two signal sources combine in a mixer circuit. The resulting output signal is an algebraic combination of the two input frequencies. Many different types of receivers use heterodyning, or "beating," to combine the output of a local oscillator with an incoming R-F signal for the production of the intermediate-frequency, or I-F, signal. **Intermediate frequencies** have much lower values that those in the radio-frequency band.

Referring to Figure 1-10, the output signals from an R-F amplifier and from the local oscillator mix at the mixer input and begin the non-linear process of heterodyning. The mixer stage output signal consists of the original R-F signal, the original local oscillator signal, the sum of the original signals, and the difference of the original signals. However, because the operation of the system dictates that only the difference of the two input frequencies is desired, **a bandpass filter** at the output of the circuit allows only the difference frequency to pass. In this case, the radio frequency of 1000 kHz heterodynes with a local oscillator at a frequency of 1456 kHz and produces the desired difference frequency of 456 kHz.

Receiver Operation

Many different types of electromagnetic waves induce voltages and currents at the antenna of a radio receiver. To ensure proper operation, a receiver must have the characteristic of selectivity, or the ability to select one specific frequency while rejecting all others. In addition, a receiver must have sensitivity, or the ability to adequately amplify a weak desired frequency or R-F carrier. When we consider the basic ingredients needed for the reception of an R-F signal, several building blocks fall into place.

1. First, the receiver requires some method, such as a tuned circuit, for selecting the desired frequency.

2. Second, the receiver must have a method for amplifying the desired frequency and then for converting the R-F signal into a lower, easily-handled intermediate frequency (I-F). The conversion of the R-F signal into an I-F signal reduces the need for multiple tuned circuits.

3. Third, the receiver requires a method for removing the intelligence from the modulated signal.

Intermediate Frequencies   As mentioned above, the heterodyning of the R-F signal and local oscillator output signal produces a band of intermediate frequencies. Each frequency in the intermediate-frequency band equals the difference between the oscillator frequency and the frequency of a corresponding wave in the R-F channel. Because the center frequency of the received R-F carrier also heterodynes with the local oscillator output, it produces the center frequency for the intermediate-frequency, or I-F, band. The center frequency of the intermediate-frequency band is called the I-F carrier.

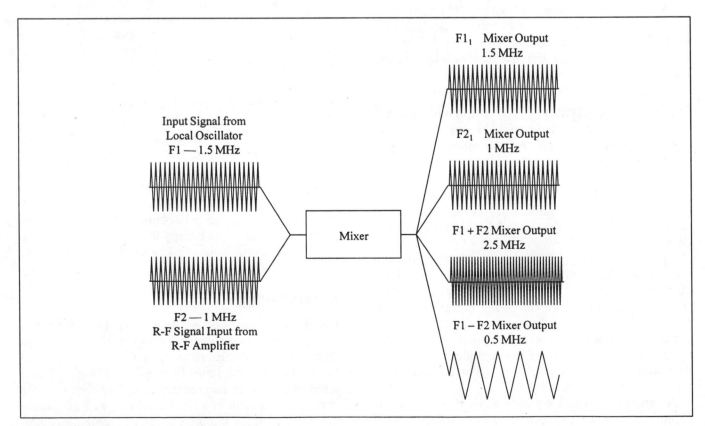

Figure 1-10   Illustration of Heterodyning

AM radios have an I-F of 455 kHz, while FM radios have an I-F of 10.7 MHz. In a television, picture information is carried by an amplitude-modulated carrier, while sound information is carried by a frequency-modulated carrier. Intermediate frequencies in a television receiver range from 20 MHz to approximately 40 MHz.

Superheterodyne Receiver Operation Receivers that operate with a fixed intermediate frequency are called **superheterodyne receivers**. Figure 1-11 uses a block diagram to depict the operation of a superheterodyne receiver. Used in all radio and television receivers, this design provides high selectivity, sensitivity, gain, and reliability because the circuits operate at the relatively low intermediate frequencies. As shown in Figure 1-11, tuned circuits select the desired frequency, while the R-F amplifier amplifies the R-F carrier and the carrier

sidebands. Then, the incoming R-F signal heterodynes with the frequency of the local oscillator contained within the receiver. The mixer stage converts the R-F signal to the I-F signal. After heterodyning occurs, the intermediate frequencies go through several stages of I-F amplification.

Demodulation

Demodulators, also called detectors, decode and recover the intelligence from the carrier signal. When considering the demodulation of amplitude-modulated signals, the demodulator circuits detect the envelope that corresponds with the modulating signal and eliminate the carrier wave. In the simple diode detector circuit depicted in Figure 1-12, the modulated R-F signal travels from output of the I-F stages and encounters the detector. The diode detector allows only half of

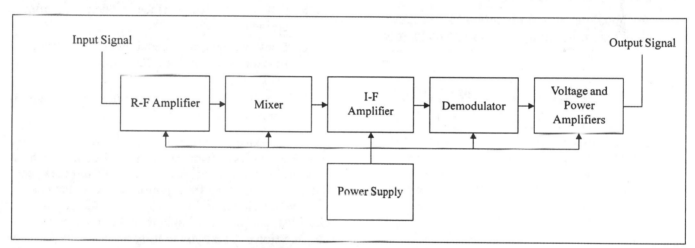

Figure 1-11   Block Diagram of a Superheterodyne Receiver

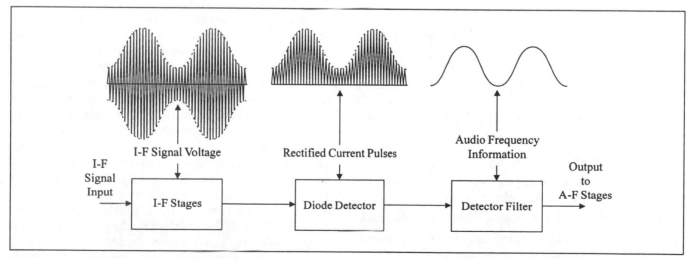

Figure 1-12   A Basic Demodulator Process

the modulated R-F waveform to pass. Because the filter capacitor cannot follow the R-F signal, it passes only the information-carrying envelope of the waveform.

The detection of a frequency-modulated signal for the recovery of audio intelligence is more difficult than the detection of an amplitude-modulated signal. Because of the characteristics of frequency modulation, a circuit called a frequency discriminator combines with a limiting circuit to ensure that the amplitude of the signals remains constant. In those circuits, changes in frequency cause a change in the output voltage from the sound carrier. Other demodulator circuits decode and recover color information contained in sidebands of the subcarrier. Because the color subcarrier is phase-modulated, the demodulator uses the phase relationship of the color difference signals to decode the modulated signals.

✓ PROGRESS CHECK

You have completed Objective 11. Discuss modulation, heterodyning, and demodulation.

1-6 TROUBLESHOOTING METHODS

When troubleshooting video electronics, begin by obtaining a thorough understanding of the circuit operation, symptoms, and possible faults. In most cases, the best servicing equipment cannot be found on the test bench. Rather, it consists of your eyes, ears, nose, touch, and brains.

Safety First!

Whenever you work with an electronic system, a potentially deadly shock hazard exists. Your body establishes a conducting path for electricity that leads through your heart. Even if the electrical shock does not kill, it can cause you to jerk involuntarily and come into contact with sharp edges, fragile picture tubes, or other electrically live parts. As a result, the reaction to the shock could cause damage to the equipment undergoing service.

When working with electronic systems, follow these simple guidelines:

- Wear rubber-bottom shoes.
- Never wear metallic jewelry that could contact circuitry and conduct electricity.
- Never rest both hands on a conductive surface.
- Perform as many tests as possible with the power off and the equipment unplugged.
- Never work on an electronic system when tired.

- Connect and disconnect test equipment leads with all equipment powered down and unplugged. Use clip leads when accessing difficult-to-reach locations.

Logical Troubleshooting

When working with electronic equipment, write a checklist as you begin to service the product. A simple checklist will have a format that asks:

1. What is wrong and what are the symptoms?
2. Is the symptom constant or intermittent?
3. Can you see/smell/sense anything that may indicate where the problem exists?
4. Did something external cause the symptom?
   a. vibration?
   b. overvoltage?
   c. misadjustment?
5. Did the problem surface suddenly or after a period of time?
6. Does the symptom occur when the system is first turned on or does the system need to operate for a time?
7. Are the external and internal electrical conditions in good condition?

As you may suspect, this format translates easily into a flowchart and establishes a specific direction for troubleshooting the system. Each question from the checklist pushes you away from jumping to conclusions.

If we overlay a problem with a television onto the checklist, the questions begin to make sense. For example, question one asks you to fully describe the problem and make a choice about your troubleshooting path. If a problems exists with the audio circuits, you would not want to spend time checking the horizontal deflection circuitry. Question two asks whether certain conditions, such as temperature or vibration, cause the problem or if a component breakdown has occurred. By answering questions three and four, you narrow the choices even further to external or internal causes.

With questions five, six, and seven, you ask specific questions about the symptoms. Bad solder joints, corrosion on control surfaces or connectors, or leaky capacitors may cause problems to occur gradually or after a long period of time. Shorted transistors, open resistors, and faulty rectifier circuits cause sudden failures and may point at other circuit problems. Temperature-related problems appear after the system has functioned for a period of time, while a defective component will cause the problem to show at start-up. Bad connections may cause a problem after movement.

As you move towards specific problems, use techniques that promise efficient use of your skills and equipment. The checklist points toward a process called symptom-function analysis. At this point, you should avoid looking at the entire system and concentrate on the stage pinpointed by the symptom. More advanced methods of troubleshooting require knowledge about the circuit and, in many cases, good documentation.

Test Equipment Primer—Hand Tools

At one time, a television repairman did not need more than a few screwdrivers, a 1/4-inch nut-driver, a pair of long-nosed pliers, a wire cutter, and a soldering gun to complete a service job. Now, the evolution of the electronics industry has affected the way that we purchase and use hand tools. When you begin to plan for those purchases, you must also consider the present and the future applications for the tools. For example, a business that emphasizes video game repair may expand into television, VCR, and personal computer servicing. This expansion may mean that technicians work not only in the shop but also in the customer's house.

Given the wide range of service possibilities, an inventory of hand tools should include:

* Needle-nose pliers
* Precision diagonal cutters
* Wire stripper
* Heat sinks
* Regular pliers
* Scissors
* Hobby knife
* An assortment of picks and scrapers
* Spring-loaded parts retrievers
* Metric-measure nut-drivers
* English-measure nut-drivers
* A set of Allen wrenches
* Different sizes of Phillips screwdrivers
* Different sizes of flat-blade screwdrivers
* Different sizes of precision screwdrivers
* Different sizes of clutch-head drivers
* Plastic alignment tools
* Grounding wriststrap and grounding mat
* Low-wattage soldering iron
* High-wattage soldering gun
* Desoldering device
* Surface-mount device soldering iron
* Protective drop cloth
* Small lamp
* Hand-held mirror

In addition to those tools, the service kit should also include cleaning supplies such as spray tuner cleaner, screen cleaners, and magnetic head cleaners. Most kits also add cotton swabs, screen wipes, soft brushes, electrical tape, solder, and lubricants to the selection. Whether in the shop or on the road, the complement of hand tools should allow a technician to complete a repair without searching for the proper tool.

Of particular note, the list of hand tools includes three different types of soldering tools. The soldering of standard semiconductor components requires the use of a low-wattage iron. Other applications may require the additional heating given by a high-wattage gun. However, installing or removing the third category—surface-mount devices—requires a special type of soldering iron. Because surface-mount devices may have as many as 48, 64, or more pins mounted to a high-density circuit board with extremely thin circuit leads, a soldering or desoldering device must deliver a precise amount of heat to a small location without overheating or causing a static discharge. Surface-mount soldering tools are available as either soldering guns or hot-air tools.

Static and Dynamic Troubleshooting Methods

We can consider troubleshooting from two different perspectives. Static testing involves the checking of circuit conditions with no signal applied to the unit. Voltage current, resistance, and continuity tests are examples of static tests and provide the advantage of isolating the problem to a specific section or stage. As opposed to static testing, dynamic testing involves the observation of how defective circuits in a unit affect the signal. Signal injection and signal tracing are common dynamic testing methods and provide the advantages of isolating the problem to a single stage and conducting the test under actual operating conditions.

Signal Injection The process of signal injection requires a test signal generated by an external source such as a signal generator. A specific test signal is introduced at various points in a circuit or unit and the output is observed. For example, we could inject an audio frequency signal at a test point preceding the AF amplifiers in a radio receiver and then monitor the sound produced by the speaker. Figure 1-13 uses a block diagram of a radio receiver to illustrate the signal injection method.

With the generator probes connected to the test point and ground, we could inject a low-level signal and then listen for a normal audio tone at the speaker. A normal audio tone indicates that the stages following the test point are operating normally. No sound or an abnormal

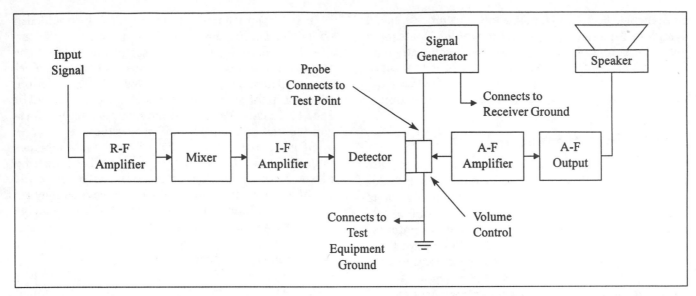

Figure 1-13   Illustration of Signal Injection

sound would indicate that a problem exists in the stages following the test point. From there, we could inject signal between the stages in a suspected section until narrowing the problem to a specific stage or component.

Signal Tracing   Illustrated in Figure 1-14, signal tracing asks whether needed electrical signals are present at key points of a circuit, and involves checks at specific test points for correct signal amplitudes and waveform shapes. While study and training provide answers about the types of signals, documentation provides the checkpoint locations and signal amplitudes. For example, experience may tell you that an R-F signal should

be present at a receiver stage. The specific circuitry, however, may contain impedances that cause levels to vary. Equipment for signal tracing ranges from speakers and indicator lights to multimeters and oscilloscopes.

Other Testing Methods

Along with static and dynamic tests, you can use parts substitution and open connections as testing methods. Parts substitution consists of replacing a suspected defective component with a known good component and observing the results. The proper operation of the circuit after the replacement confirms that the original part is defective.

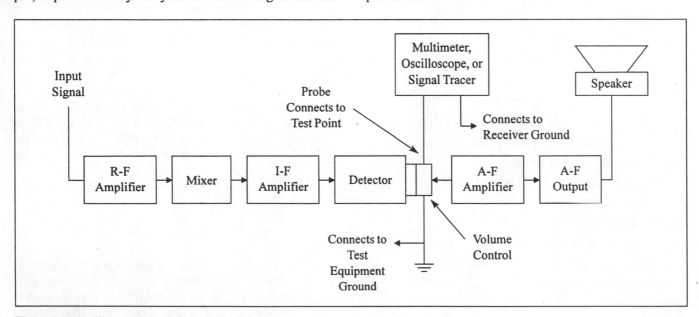

Figure 1-14   Illustration of Signal Tracing

Along with parts substitution, another technique—opening the connection at a point where a component connects to the circuit—can solve a troubleshooting problem. The open connections test involves opening a circuit connection and observing whether the operation of the circuit improves. If it does, the disconnected component may be the source of the problem.

 PROGRESS CHECK

You have completed Objective 12. Understand basic troubleshooting methods and safety practices.

SUMMARY

Chapter one reinforced your knowledge of the fundamentals of the science of electronics. It defined an electronic system, highlighted the need for source voltages throughout the system, and described the ingredients that make up a typical power supply. The chapter then considered electromagnetic waves, signals, and frequencies, and the relationships among them. It then moved on to waveforms, frequency bands, harmonics, and signal noise.

Next, the emphasis shifted to radio-frequency signals and their effect on different types of components. You found that R-F signals affect components such as capacitors, inductors, and resistors in varying ways, in-

cluding skin effect and resonance. Circuit functions were discussed, and you learned that analog and digital circuits perform the basic functions of amplification, oscillation, and switching. Circuit operations were defined, along with modulation, heterodyning, and demodulation.

The chapter concluded by introducing you to general safety procedures and troubleshooting methods. Static and dynamic testing were covered, you learned that troubleshooting circuit problems involves a logical process of asking questions to narrow the search for a problem cause in a circuit.

REVIEW QUESTIONS

1. Resonance occurs when _____ .
2. The three parts of an electronic system are:
   a. the source, load, and complete circuit
   b. the source, load, and display
   c. the source, complete circuit, and display
   d. the antenna, load, and complete circuit
3. Define the term "bandwidth."
4. Why is coupling an important concept for amplifier circuits?
5. An amplifier that has a constant gain from about 20 Hz to just above 20 kHz is classified as a:
   a. video amplifier
   b. R-F amplifier
   c. audio amplifier
   d. I-F amplifier
6. An amplifier develops an output of 20 Vdc with an input of 1 Vdc. What is the voltage gain of the amplifier?
7. An analog circuit uses:
   a. passive components
   b. feedback
   c. switching
   d. bistable elements
8. True or False   A triangular wave is the same as a sawtooth waveform.
9. True or False   A square wave is the same as a pulse.

10. True or False   The average value of a sine wave is the same as the peak value.

11. What is the difference between passive and active elements?

12. Harmonics are:
    a. the lowest frequency of a tone
    b. the same as noise
    c. multiples of the fundamental frequency
    d. fractions of the fundamental frequency

13. List the major characteristics of Class A, Class B, Class AB, and Class C amplifiers.

14. Why would you couple two or more amplifiers together?

15. The R-F spectrum extends from:
    a. 10^4 Hz to 10^{11} Hz
    b. 10^1 Hz to 10^6 Hz
    c. 10^{11} Hz to 10^{14} Hz
    d. 10^4 Hz to 10^6 Hz

16. Name and define three different types of noise.

17. The eight categories of the R-F spectrum differ in:
    a. wavelength
    b. transmission type
    c. resonance
    d. response

18. The VHF and UHF frequency bands are _____ (VHF) and
    _____ (UHF).

19. True or False   Skin effect is a desirable phenomenon.

20. Why is the effect of an R-F signal different for a capacitor than an inductor?

21. The opposite of attenuation is:
    a. resonance
    b. coupling
    c. feedback
    d. gain

22. The ability of a receiver to select one particular frequency and to reject all others is called:
    a. sensitivity
    b. selectivity
    c. gain
    d. attenuation

23. The ability of a receiver to adequate amplify small signals is called:
    a. sensitivity
    b. selectivity
    c. gain
    d. attenuation

24. The two main requirements of the feedback voltage or current to maintain oscillations in an oscillator are:
    a. must produce a voltage or current in phase with the input and of the same amplitude as the input
    b. must produce a voltage or current out-of-phase with the input and of the same amplitude as the input
    c. must produce a voltage or current in phase with the input and higher amplitude than the input
    d. must produce a voltage or current in phase with the input and lower amplitude than the input

25. An oscillator produces:
   a. a direct current or voltage
   b. an alternating current or voltage
   c. no current or voltage

26. In an AM transmitter, what characteristic of the R-F carrier is varied in accordance with the modulating signal?

27. In an FM transmitter, what characteristic of the R-F carrier is varied in accordance with the modulating signal?

28. The process of impressing information onto a carrier wave is called:
   a. heterodyning
   b. demodulation
   c. modulation
   d. amplification

29. Switching circuits rely on:
   a. square waves
   b. sine waves
   c. triangular waves
   d. modulated waves

30. Operational amplifier switches are measured in terms of:
   a. rise time and delay time
   b. rise time, slew rate, and fall time
   c. slew rate and turn-on time
   d. slew rate

31. The circuit used to recover the modulating signal from a modulated R-F carrier is a:
   a. frequency converter
   b. modulator
   c. amplifier
   d. demodulator

32. True or False   Superheterodyne designs are no longer used in modern electronic equipment.

33. The most common intermediate frequency in AM radio is:
   a. 1455 kHz
   b. 455 kHz
   c. 10.7 MHz
   d. 1000 kHz

34. The circuits required for an oscillator are the:
   a. voltage supply and feedback circuit
   b. amplifier and frequency converter
   c. R-F amplifier and I-F amplifier
   d. amplifier and feedback circuit

35. An inverter produces:
   a. no phase shift
   b. 120° phase shift
   c. 180° phase shift
   d. 90° phase shift

36. Name the steps for logical troubleshooting in order.

37. Why should you observe safety precautions when working with electronic systems?

38. Name the six basic safety precautions in order.

39. The I-F signal of a receiver is:
    a. an unmodulated carrier wave
    b. a modulated wave
    c. part of the source voltage
    d. a sine wave

40. Name two examples of static troubleshooting tests.

41. What are two types of dynamic troubleshooting tests?

42. How many signals does heterodyning produce at the mixer output?
    a. one
    b. two
    c. three
    d. four

43. How does signal injection differ from signal tracing?

44. The soldering or desoldering of surface-mount components requires a:
    a. low-wattage soldering iron and a desoldering bulb
    b. high-wattage soldering gun and specialized desoldering wick
    c. specialized soldering paste
    d. specialized soldering/desoldering tool

CHAPTER 2

Digital Fundamentals

OBJECTIVES Upon completion of this chapter, you should be able to:

1. Discuss Boolean algebra.
2. Define digital number systems.
3. Define digital gates.
4. Analyze the operation of flip-flops.
5. Define multiplexers, demultiplexers, and buffers.
6. Define counters, registers, dividers, and prescalers.
7. Discuss ADC and DAC circuits, encoders, and decoders.
8. Discuss transistor-transistor and CMOS logic ICs.
9. Explain frequency synthesis.
10. Define microprocessors, microcontrollers, and embedded processors.
11. Discuss methods for troubleshooting digital circuits.

KEY TERMS

analog-to-digital (ADC circuit)	Byte	flip-flop
AND function	CMOS logic	frequency division
astable multivibrator	combinational logic	frequency multiplier
asynchronous counter	comparator	frequency synthesis
binary-coded decimal	counter	gate
binary number system	decoder	latch
bistable multivibrator	delay element	microprocessor
bit	demultiplexer	monostable multivibrator
Boolean algebra	digital-to-analog (DAC) circuit	multiplexer
buffer	divider encoder	multivibrator

NAND function	register	switching element
NOR function	Schmitt trigger	synchronous counter
NOT function	sequential logic	transistor-transistor logic
OR function	shift register	voltage-controlled oscillator
phase-locked loop	storage register	XOR function
prescaler		

INTRODUCTION

In chapter one, you studied the fundamentals of electronic systems. Although this preliminary investigation touched on digital information in the form of pulses, it emphasized analog systems. In brief, an analog system operates with continuous waveforms over a given range such as the sine wave shown in Figure 2-1A. Analog information is a quantity that may vary over a continuous range of values.

Digital information, in contrast, is data that has only certain, discrete values. The digital information shown in Figure 2-1B offers precise values at specific times and changes step by step. When considering electronic systems, this preciseness translates into immunity against electrical noise distortion and variations in component values. Digital information consists of data held in one of two states: low and high. Noise and component variations do not cause the low and high states of a digital signal to appear as opposite values.

In offering a brief study of digital signal fundamentals, chapter two introduces the concepts of Boolean algebra and binary numbers. In addition, the chapter provides an overview of gates, flip-flops, and microprocessor control, leading to further discussions about frequency multiplication, frequency division, and frequency synthesis. As the following chapters move to actual circuit applications, you will find that many individual circuits and systems—such as tuner control, vertical deflection, and horizontal deflection circuits—rely on the precision offered by digital technologies.

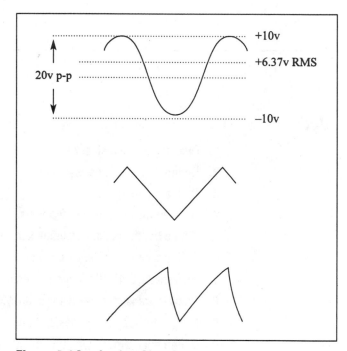

Figure 2-1A   Analog Signals

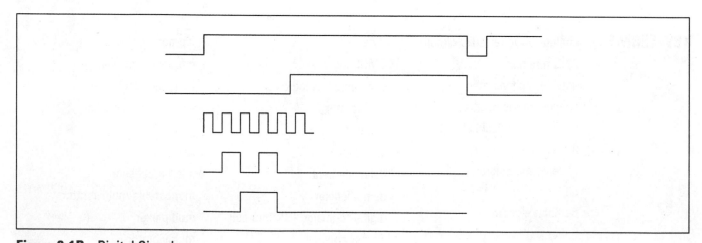

Figure 2-1B   Digital Signals

2-1 BOOLEAN ALGEBRA

All of the operations performed through the use of digital signals originate with a concept called Boolean algebra. Named after mathematician George Boole, **Boolean algebra** is a mathematics system designed to test logical statements and show if the result of those statements is true or false. An example logical statement may appear as: "If it is raining and I have my raincoat, I will stay dry." Boolean algebra will test the validity of the statement by asking if the person is wet or dry because of rain or no rain, and if the person has a raincoat.

Logical statements divide into two possible values, 0 and 1, with 0 representing either false, negation, disable, inhibit, or no, and 1 representing either true, assertion, enable, or yes. Truth tables describe how the output of a logic circuit depends on the logic levels present at the input of the circuit. Each of the truth tables shown in this section lists all the possible combinations of logic levels present at the inputs and the corresponding output conditions.

In electronic equipment, voltage and current levels represent the two possible logic states. As an example, a +5 Vdc level may represent a logic 1 or true statement, while a −5 Vdc level may represent a logic 0 or false statement. In another example, the presence or absence of current could represent the logic 1 and 0 states. Most of the time, though, voltage levels express logic conditions.

Boolean Expressions

If the operation of a circuit can be defined by a Boolean expression, a logic circuit diagram could be constructed from the expression. The fundamental Boolean expressions—AND, OR, NOT, NOR, NAND, and XOR—are pictured in Figure 2-2. Truth tables accompany each expression in the sections that follow, and illustrate the logic characteristics.

The AND Function The **AND function** says that an output is a 1 only if all the inputs are 1. If a 0 appears at any of the inputs to an AND operation, then the output is 0. In English, AND statements would appear as "We will go only if we decide to go" or "The car engine will start only if gasoline and a spark reach the engine." In both cases, a part of the statement is not enough to validate the statement. In terms of electronics, we could illustrate an AND function by wiring two switches in series with a load. Both switches must be on to allow the flow of current.

AND functions may have three or more inputs. Yet, as the number of inputs increase, the AND statement must always remain true. That is, the output is 1 only if

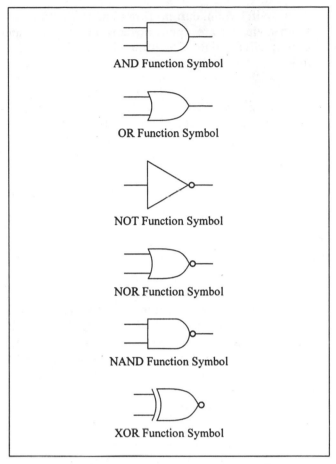

Figure 2-2 Boolean Expression Schematic Symbols

AND Function Symbol

OR Function Symbol

NOT Function Symbol

NOR Function Symbol

NAND Function Symbol

XOR Function Symbol

all the inputs are 1. The truth table for the AND function is:

A	B	Y = A AND B
0	0	0
0	1	0
1	0	0
1	1	1

The OR Function An **OR function** may have two or more inputs and produces a 1 at the output if any of the inputs are 1. The OR function produces a 0 at the output only if all the inputs are 0. In English, an OR statement would appear as "If any one of us goes, then we'll need a car." When we analyze the statement from the perspective of Boolean algebra, it says that if any one or more goes, then a car will be needed. Yet if no one goes, a car is not needed.

If we examine the OR function from an electronics perspective, two switches connected in parallel control the flow of electrical power to a lamp. With either or both switches on, current flows and the lamp lights. If both switches are open, no current can flow and the lamp will not light. The truth table for the OR function is:

A	B	Y = A OR B
0	0	0
0	1	1
1	0	1
1	1	1

The NOT Function   A **NOT function** states that the output is the opposite of the input. That is, the input becomes inverted and 0 is converted to 1 and 1 to 0. NOT functions may have only one input. If shown as an electrical switch, the NOT function is represented by a simple switch placed in parallel with a lamp. Opening the switch allows current to flow to the lamp, while closing the switch short-circuits the lamp and allows current to flow around the device. The truth table for the NOT function appears as:

A	Y = NOT A
0	1
1	0

The NOR Function   The **NOR function** adds a NOT function to the output of an OR function. With this, an output is 0 if any one or more of the inputs is 1. If all the inputs are 0, then the output is 1. The truth table for a NOR function appears as:

A	B	A OR B	NOT (A OR B)
0	0	0	1
0	1	1	0
1	0	1	0
1	1	1	0

The NAND Function   A **NAND function** adds a NOT function to the output of an AND function. With this, the output of a NAND is 0 only if all the inputs are 1. Otherwise, the output is 0. As the following truth table shows, the AND function is completed first, with the NOT function performed at the end of the operation.

A	B	A AND B	NOT (A OR B)
0	0	0	1
0	1	0	1
1	0	0	1
1	1	1	0

The XOR Function   The **XOR function** states that if either input is 1, but not both, the output is 1, and that if both inputs are 1 or both are 0, then the output is 0. From the perspective of electronics, an XOR function can be illustrated by using two switches at different locations to control a light. With both switches closed or open, the light remains off. With either switch closed, the lamp is on. The truth table for the XOR function is:

A	B	A XOR B
0	0	0
0	1	1
1	0	1
1	1	0

☑ **PROGRESS CHECK**

You have completed Objective 1. Discuss Boolean algebra.

2-2 DIGITAL NUMBER SYSTEMS

Computer systems work with the **binary number system**, a system that allows only two values, 0 and 1. The use of binary numbers in those systems breaks information down into elementary levels. In addition, because the binary number system relies only on zeroes and ones, it provides a very basic method for counting and accumulating values.

If you look at the digits of a binary number, each value in the columns equals a value based on the powers of 2. As an example, a binary number represented by 111 has $2^0 = 1$ or 1×1 for the rightmost column, $2^1 = 4$ or 2×1 for the middle column, and $2^2 = 8$ or 4×1 for the leftmost column. In the decimal system, this value would equal $1 \times 1 + 1 \times 2 + 1 \times 4$ or 7. A binary number of 1101 equals $1 \times 1 + 0 \times 2 + 1 \times 4 + 1 \times 8$ or a decimal equivalent of 13.

As mentioned earlier, an electronic system uses high and low signals to represent binary numbers. Each high and low signal is separated by an area of voltage that has no binary meaning. While a high signal has a value of 3 to 5 Vdc, a low signal has a value of 0 to 1 Vdc.

Bytes and Bits   The simplicity of the binary system allows computer systems to move numbers from one part of a system to another and to work with large numbers. Each binary position is called a **bit**, while a group of eight bits is called a **byte**. The sum of bits provides a method for assigning a value. As a result, the number of bits required to complete a task depends on the magnitude of the number. With each bit existing as either a 1 or 0, the value of each successive bit can increase by a maximum value of 2. As a result, the individual bits of a binary number translate to the following decimal values:

32	16	8	4	2	1	$\frac{1}{2}$	$\frac{1}{4}$	$\frac{1}{8}$	$\frac{1}{16}$
2^5	2^4	2^3	2^2	2^1	2^0	2^{-1}	2^{-2}	2^{-3}	2^{-4}

Going back to the addition of binary numbers, the sum of values in a byte equals a decimal value. That is, a byte that appears as 11001101 has an equivalent decimal value of 206.

Binary-Coded Decimal   Another counting system, called **binary-coded decimal**, or BCD, combines the efficiency of the binary system with the familiarity of the decimal system. Keyboards, LED displays, and switches rely on the BCD system. The BCD system uses four bits to represent each digit of a decimal number. As an example, decimal number 759 uses 0111 for the 7; 0101 for the 5; and 1001 for the 9, and appears as 0111 0101 1001.

☑ **PROGRESS CHECK**

> **You have completed Objective 2. Define digital number systems.**

2-3 DIGITAL BUILDING BLOCKS

Four basic logic elements establish most of the functions seen in electronic systems. Those logic elements are gates, amplifiers, switching elements, and delay elements. Each element follows a specific set of rules and has a specific schematic symbol, as shown in Figure 2-3.

Gates

Integrated circuits (ICs) called **gates** perform fundamental Boolean logic functions and combinations of the logic functions. A gate "sees" the logic values at

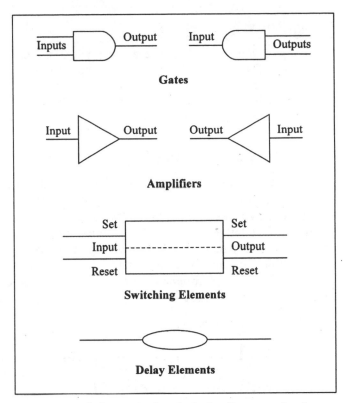

Figure 2-3   Schematic Symbols for Common Logic Elements

its inputs and produces the output that corresponds with the correct logic function. Gates can be connected together so that the output of one becomes the input for another. The connecting of the gates provides more complicated Boolean logic functions and is called **combinational logic**.

An IC with less than 12 gates fits within a category called small-scale integration, while ICs with between 12 and 100 gates are called medium-scale integration. Modern electronic systems utilize both the simplicity of small- and medium-scale integration and the flexibility offered by the millions of gates seen with very large-scale integration, or VLSI.

With a large number of gates connected for combinational logic tasks, propagation delays can occur and, if significantly large, can cause false results in a system. This occurs because signals passed by the gates arrive at slightly different times in different parts of the system. Every digital system includes functions called enable and inhibit to ensure that the propagation delays do not accumulate and the signals remain synchronized.

☑ **PROGRESS CHECK**

> **You have completed Objective 3. Define digital gates.**

Amplifiers

In chapter one, you found that amplifiers increase the value of a signal as it travels from input to output. We expressed the amount of increase as gain. When working with digital circuits, amplifiers drive digital pulses to the next location.

Switching and Delay Elements

A **switching element** causes the logic state found at the output of a device to switch, or change to the opposite condition. With this, an output that originally has a logic 1 will change to a logic 0. In digital electronics, switching elements take the form of flip-flops, astable and monostable multivibrators, and Schmitt triggers. A **delay element** establishes a time delay between the input and output signals.

2-4 MULTIVIBRATORS, SCHMITT TRIGGERS, AND FLIP-FLOPS

In chapter one, you learned that oscillation is one of three basic functions performed by an electronic stage. Three types of oscillators have operating characteristics useful for digital applications: multivibrators, Schmitt triggers, and flip-flops.

Multivibrators

A **multivibrator** is a type of oscillator built around two active devices. A multivibrator may have square waves or abrupt changes between logic states at the output. Three basic types of multivibrators—the astable, monostable, and bistable—have different stability conditions, interstage coupling, biasing, and input/output characteristics.

Astable Multivibrators An **astable multivibrator** (also called free-running) will lock into a logic state when triggered by an external signal. The device does not require a signal at the input. During operation, the astable multivibrator takes advantage of a power supply as it converts dc energy into an output train of square waves.

Free-running multivibrators work as a source for constant-amplitude square wave signals. Circuit designs may use a fundamental frequency of a sine wave or a subharmonic of a sine wave to synchronize the multivibrator. Although astable multivibrators do not provide an accurate frequency at the output, the devices work for applications where an approximate frequency is desired.

Monostable Multivibrators An **monostable multivibrator** (also called one-shot) has one stable condition. When excited by an external signal, the monostable multivibrator will go to its unstable state and then automatically return to its stable state. Monostable multivibrators often function as a delay device and rely on other circuit components to establish the length of the time delay. With this time delay in place, the monostable multivibrator goes back to its stable state after external triggering within a fixed delay time.

Monostable multivibrators lengthen or delay input pulses. In computer systems, a monostable multivibrator will provide a precise timing delay for reference signals. When used with a stable synchronizing signal at the input, monostable multivibrators establish a precise time delay for the operating circuit.

Flip-Flops

During your study of basic gates and logic functions, you found that complex functions required the expansion of the basic operations. The inputs for gates occur at instants in time. In addition, a gate IC does not form any type of memory. To move past those obstacles, we can connect gates into different configurations so that an input pulse will cause the gates to move from one logic state to the other. The next input signal causes the states to reverse.

This configuration forms a **bistable multivibrator**, or **flip-flop**. The output of one gate in a flip-flop is the input for the other gate. Thus, a flip-flop is a cross-connection of two gates that acts like a 1-bit memory element and generally consists of NOT and NAND gates. Because the outputs have opposite states, each complements the other as the flip-flop locks onto the bit. As opposed to the monostable multivibrator, which offers temporary memory, the flip-flop provides a permanent memory of the input signal.

Sequential Logic The use of two NOT gates in a flip-flop yields two possible states. That is, a 0 at the input of the first NOT gate establishes a 1 at the output of that gate, a 1 at the input of the second NOT gate, and a 0 at the output of the second gate. Most flip-flops utilize NAND gates rather than NOT gates and rely on more than two gates for operation. In addition, those flip-flops employ a means for maintaining the synchronization of all input signals at the gates. Without the synchronization, the signals could change states at different times and produce false conditions at the output.

We define the logic used during the synchronization of the input pulses at the gates as **sequential logic**. When gates of a flip-flop look at the inputs simultaneously, the gate circuit is sequential. In a sequential system, a clock oscillator outputs a continuous string of pulses that go to all parts of the system. This type of logic is much more popular than the combinational logic seen with the simple flip-flop model. With the pulses synchronized with a clock/oscillator, the ICs are active at only given times. In between the pulses, the integrated circuits remain inactive and no logic changes occur. Instead, the logic circuits use the inactive period to allow any signals to reach the next point and to reset.

Flip-Flop Types   The previous discussion of flip-flops covered a type called the S-R flip-flop. The designation "S-R" signifies that the flip-flop has a Set condition at one input and a Reset condition at the other input. A 1 at the Set input establishes a 1 at output, while the Reset input clears the condition established by the Set input.

Another type of flip-flop is the J-K flip-flop. J-K flip-flops have additional AND gates at each input and feed the signals found at the output back to the input. The external inputs are designated as J and K. Clock signals to each input set the logic states for the flip-flop inputs and are designated by the subscript n and n+1. With this, the J-K flip-flop has three input signals—the output signals from the AND gates and the clock signals. Given three inputs, the flip-flop has eight possible input possibilities and the potential for the logic states shown in the following truth table:

J_n	K_n	Q_n	Q'_n	S_n	R_n	Q_{n+1}
0	0	0	1	0	0	Q_n
0	0	1	0	0	0	Q_n
1	0	0	1	1	0	1
1	0	1	0	0	0	Q_n
0	1	0	1	0	0	Q_n
0	1	1	0	0	1	0
1	1	0	1	1	0	1
1	1	1	0	0	1	1

The truth table shows that the output at one time depends on the output seen at the previous time. Therefore, at power-up it is impossible to predict what logic state will appear at the outputs. In addition, the Preset and Clear inputs give the circuit design a method for forcing the outputs to a desired state. For example, if a 1 is desired at the Q output, the logic states at the inputs are: Preset = 0, Clear = 1, and Clock = 0. If a 0 is desired at the Q output, the logic states at the inputs are: Preset = 1, Clear = 0, and Clock = 0.

The forcing of the output states occurs because neither the Preset nor the Clear inputs synchronize with the clock. Instead, both inputs tie directly to the flip-flop and bypass the input stage. The direct connection of the inputs is defined as asynchronous because no relationship exists between the inputs and the synchronization provided by the clock signals. Synchronization can occur at any time.

J-K flip-flops have many roles in microprocessor control applications. One common and very important application involves using the flip-flop as an interface between different systems, different circuits within a system, or between computer systems and people. As an example of the last application, a set of J-K flip-flops will inform a microprocessor when a person has pushed a control switch.

D-type flip-flops evolve from J-K flip-flops by attaching a NOT gate between the J and K inputs. With this, the D flip-flop provides a time delay needed by some circuit applications. As an example, the lines carrying logic information to a microprocessor may come from different sources or from various parts of the circuit. As a result, information carried on those lines may arrive at the microprocessor at different times. The insertion of a D-type flip-flop into each line and the connection of all the clock inputs for the flip-flops to the system clock ensure that the digital information will arrive at the processor as a group. This occurs because of the clocking of the input pulses—data is not released until the correct instant—and the delay characteristics given by the flip-flops.

The T-type, or Toggle, flip-flop results from connecting the J and K inputs of the J-K flip-flop together. With this, an output produced after the clock pulse has the opposite state as the output before the clock pulse. As a result, the outputs toggle between 0 and 1 with each clock pulse. An example of a T-type flip-flop application is a system where the pushing of a button alternately turns a light on and off.

Flip-Flop Applications and Accuracy   Flip-flops provide perfect accuracy because the output of a flip-flop reverses logic states with each successive input pulse. The addition of a clock signal to a J-K flip-flop also ensures that the device will remain immune to noise. This immunity occurs through the requirement that the input signal and clock pulse arrive simultaneously before the device will change states.

Because of the accuracy and noise immunity characteristics, flip-flops have widespread use in all digital applications. Along with providing a permanent memory of the input signal, flip-flops are the basis for counting in computer systems. When cascaded, flip-flops work as binary counters.

Schmitt Triggers

A Schmitt trigger is a bistable element built around a multivibrator circuit. During operation, Schmitt triggers work as voltage-level sensors and for the shaping of signals. Because the device switches between two states, it can reshape a signal with a poor rise time—such as a sine wave—into a square wave. When the Schmitt trigger senses an input voltage below a given reference level, the element remains in a set logic state. When the input voltage goes above the reference level, the element switches to the opposite state.

☑ PROGRESS CHECK

You have completed Objective 4. Analyze the operation of flip-flops.

2-5 CONTROL

One major problem with the use of digital circuits is that we live in an analog world. Most physical quantities have a continuous nature. Because of this, electronic circuits require devices that can convert analog inputs into a digital form. Once the information is converted, other circuits process the information. Then a final set of circuits converts the digital outputs back into an analog format.

Multiplexers and Demultiplexers

A **multiplexer** allows a user or a circuit to control the combining of signals from logic gates. A **demultiplexer** takes signals from a single line and allows the direction of the signals to one of several output lines. During operation, a multiplexer receives a large cluster of input lines and allows only the logic state of one of the lines through to the processor. A multiple-bit code in the multiplexer or signals given directly from the multiplexer establish which logic state progresses to the processor. Illustrated in Figure 2-4, the multiplexer consists of an arrangement of AND, OR, and NOT gates.

Buffers

Digital circuits use ICs called **buffers** to establish isolation, fault-protection, the simultaneous distribution

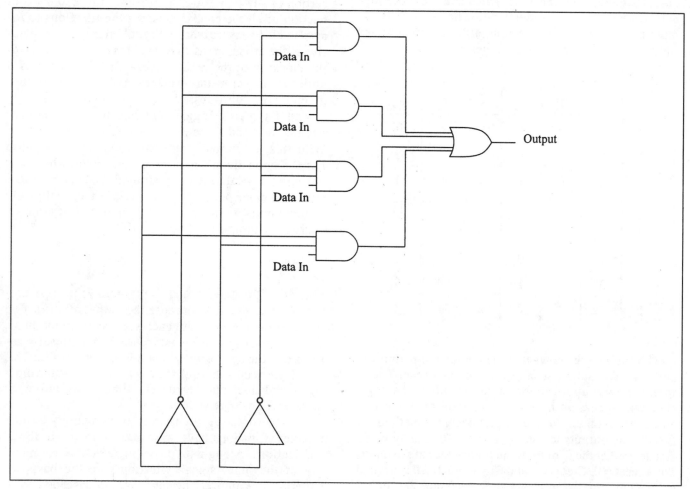

Figure 2-4   Multiplexer Circuit

of gate output signals, and the accommodation of different voltage levels throughout a system. In addition, a buffer reduces the electrical load on the output of an IC while providing a known load. The most basic application of a buffer is the provision of additional electrical drive power on signal lines.

Within the functions of isolation and fault protection, a buffer cleans up signals weakened or corrupted by noise. The changing of signal voltage levels allows the different parts of an electronic system to meet the requirements of the entire system or interconnected systems. While buffers provide additional drive power for signals, the simple ICs also allow a single signal path to accommodate more than one gate. As a result, the path can be used for two-way signal flow.

Pictured in Figure 2-5A, non-inverting buffers do not change the Boolean logic value of a signal passing through the device. The inverting buffer shown in Figure 2-5B always changes the state of the signal. Buffers may be used at the input or output of gates.

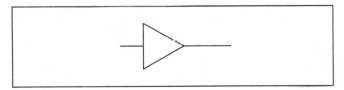

Figure 2-5A   Noninverting Buffer

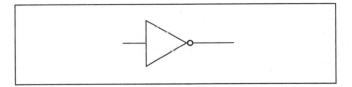

Figure 2-5B   Inverting Buffer

✔ PROGRESS CHECK

You have completed Objective 5. Define multiplexers, demultiplexers, and buffers.

2-6 TIMING AND COMPARISON

Every digital operation relies on precise timing. The basic timing pulses in a digital system are clock pulses used to synchronize the processing of information. Synchronization of the circuits in a digital system occurs through allowing logical operations to proceed only when clock pulses are applied. Because signals cannot change instantaneously, the clock pulses synchronize the circuits so that each circuit performs a logical operation after the signals have changed.

Counters, registers, dividers, and prescalers provide methods for precisely timing digital operations. While

counters count pulses, registers allow the clocking of data in at one rate and out at a different rate. Thus, a register works much like a buffer. Dividers and prescalers divide input signals. Comparators and adders detect the output from the registers and either compare the logic states or the polarity of the signal voltages.

Counters

Systems that rely on digital electronics always require a method for counting input pulses as they arrive and then passing the total to another circuit or a part of the same circuit. In decimal terms, a count up to 500 requires that we begin with 000 and then add 001, 002, 003, and so on until we reach a maximum value of 009 in the first column. Once we reach 009, the next addition carries a 1 into the next column for a 010. Then, the sequence starts over as we move from 010 to 011, 012, 013, and so on until we reach 099. With that, the next addition carries a 1 into the third column, and the numbers continue adding until reaching the value of 500. The first column is the number of ones, the second is the number of tens, and the third column is the number of hundreds.

In the digital world, counting involves binary numbers and different values for the columns. Starting with the least-significant value at the right, the column values are 1, 2, 4, 8, 16, 32, 64, and so on. However, the maximum value with binary numbers is a 1. Therefore, when a column and the columns to the right of it have values of 1, we need to carry a 1 to the next column to the left. Binary counting appears as:

0001	
0010	Carry from 1s column to 2s column
0011	
0100	Carry from 2s column to 4s column
0101	
0110	
0111	
1000	Carry from 4s column to 8s column
1001	
1010	
1011	
1100	
1101	
1110	
1111	

A **counter** is a series of gates and flip-flops that perform the binary counting function. An **asynchronous counter** receives the pulses at any time and operates with applications where the inputs may occur at any time. For example, an asynchronous counter may be attached to a door and have the purpose of counting people as they enter a store. **Synchronous counters** are always accompanied by a system clock and count pulses only if the pulses occur along with the system clock. A synchronous counter uses the input pulse, the clock signal, and clocked flip-flops.

Microprocessor-based systems use synchronous counters to pace ICs at frequencies ranging from 1 to 10 MHz and to set the timing for microprocessor tasks. Generally, the counter input pulses arrive from an accurate and stable oscillator with a known frequency. The addition of an AND gate at the input of the counter either allows the pulses to enter the counter or blocks the pulses.

A clock signal at the other input to the AND gate gates the input pulses. With the clock signal present, the counter counts the number of pulses. The absence of the clock signal tells the counter to stop counting. As a result, the counter measures the number of clock pulses or frequency that occurs during the clock time period. With the input pulses arriving during a set period, the frequency of the input signal can be identified as the number of pulses per second. The use of different gate times and more stages can allow a counter to measure frequencies from 1 to up to several million hertz.

Counter Applications Electronic counters are the basic units for digital measurement and control systems. Applications for counters include the counting of subatomic particles; the calibration of oscillators through the counting of the number of cycles produced per second at the oscillator output; and the tracking of operations in a computer system. In addition, many systems use counters for the tracking of the number of pulses of a known frequency. A very common counter application involves the tracking of the number of pulses applied to the counter input by the system clock over a given time.

Registers

Registers work as a physical interface or buffer between different systems and circuits, and convert serial data to a parallel format or parallel data to a serial format. In addition, a register may temporarily store data, or a binary number during a calculation, or a signal.

Consisting of a series of addressable flip-flops, a **storage register** temporarily stores two to sixteen bits so that the information represented by the bits is used as a group. A **shift register** may perform multiplication and division or may convert binary pulses from a serial to a parallel format. In comparison with a counter, where a user has access to the output of each flip-flop, a user has access to both the input and output of each flip-flop in a register. Registers work as a physical interface or buffer between different systems and circuits and convert either serial data to a parallel format or parallel data to a serial format. In addition, a register may work as a temporary storage for data, for a binary number during a calculation, or for a signal.

Registers often work as scratchpad memory for microprocessors, with data arriving in parallel, or as a group, and then exiting in parallel. When the microprocessor issues an instruction routine, a sequence in that routine may use the data group during a routine operation, or may use the group to **latch**, or set desired conditions. Because the register offers only temporary storage, the system discards the stored data after completing the operation.

Dividers and Prescalers

A **divider** divides input pulses by an output pulse. Inside the divider IC, a circuit of gates and flip-flops measures the number of binary 1s and 0s entering the circuit. Because this internal circuit relies on flip-flops, timing occurs as the output changes state. The addition of NAND gates at the outputs of some of the flip-flops causes the counter to generate an output pulse for every (n) inputs. The output pulse becomes a dividing factor for the input pulses, and the counter becomes a divide-by-n divider. A **prescaler** offers precise division of an input signal by variables preset by a mode selection.

Comparators and Adders

Comparators are found throughout digital circuits and may determine the sign of binary numbers or compare the polarity of dc voltages. Adders may operate as half-adders or full-adders and either perform mathematical functions or make up a comparison circuit. A half-adder adds binary numbers that have two digits, while a full-adder adds binary numbers with three digits. When used in comparison circuits, adders detect errors. For example, the feeding of a logic 1 and 0 into an adder will produce a sum output, while the presence of two 0s or two 1s at the input will produce no sum output. An adder detects equality or inequality when used to compare the contents of registers.

☑ PROGRESS CHECK

You have completed Objective 6. Define counters, registers, dividers, and prescalers.

2-7 CONVERSION

Many system applications require the conversion of analog signals to a digital format or the conversion of digital signals to an analog format. Analog-to-digital and digital-to-analog conversion circuits perform those operations while acknowledging requests from microprocessor controllers and other circuits. In addition to the analog/digital conversion, those circuits also require the conversion between digital formats and the translation of digital data to a display by encoders and decoders. For example, the display of a channel number, the time, and other settings on a VCR requires the use of encoding and decoding.

ADC Circuits

Analog-to-digital-conversion, or **ADC, circuits** take an analog signal and convert it to a digital number that corresponds to the value of the analog signal. Because ADC circuits have different types of operation and interfaces, the operation of an ADC is more complicated than that of a DAC. ADC types include the successive approximation ADC and the integrating ADC.

The conversion time for an ADC is called the sampling rate, which represents the number of conversions per second. ADC circuits connect to a microprocessor through signal lines and wait for requests for conversion or notify the microprocessor that the operation is complete. The digital number provided by the ADC may take the form of a regular binary number, a binary number with a sign bit, or a special format.

During the operation of an integrating ADC, an internal voltage signal, called a ramp, begins with a 0 and then increases to a maximum voltage over a period of time. With the rate of increase constant and precise, the voltage relates to a specific time. Because of this relationship, the circuit also requires a clock signal, a binary counter, and a comparator. Conversion begins with the counter resetting to zeroes while the ramp voltage increases. The counter receives input pulses from the clock and, once the ramp voltage exceeds the analog input voltage, stops. As a result, the value left at the counter has a direct relationship to the analog input.

DAC Circuits

Digital-to-analog conversion, or DAC, circuits produce a dc output voltage that corresponds to a binary code, and convert digital properties to analog voltages. Each set of binary pulses applied to the converter sets the switches within a conversion ladder to either on or off positions. With the switches in the ladder forming AND gates, a reference voltage is applied when the enable pulse and the incoming binary pulse arrive simultaneously.

For example, the conversion of the digital number 1010 means that the first and third pulses to the ladder switches enable the first and third switches. The binary and enable pulses coincide. As a result, reference voltages are applied to the specific ladder resistors, and output voltages are produced. However, the second and fourth pulses do not contain binary pulses. Because a binary pulse cannot coincide with the enable pulse, no voltages are produced at those specific outputs.

Encoders

Encoders either convert BCD data into a multiple-line data set or the multiple lines into a BCD format. An encoder may have any number of inputs and any number of binary bits at the output. At the input, only one input may have a true state at a given point in time. The output will have various combinations. When coupled to a comparison circuit, a comparator will combine a reference voltage with the voltage to be encoded. The output from the comparator represents a binary code found at the input to the encoder.

Decoders

Decoders take a combination of digital inputs and convert those inputs into another combination of digital circuits. Because of this, decoders also operate with other specialized circuits to provide a desired output for a user. Conversion formats for a decoder include:

- decimal to BCD format
- BCD to seven-segment display
- decimal to seven-segment display
- binary to hexadecimal seven-segment display

☑ PROGRESS CHECK

You have completed Objective 7. Discuss ADC and DAC circuits, encoders, and decoders.

2-8 MAJOR INTEGRATED CIRCUIT FAMILIES

Integrated circuits are made from the same type of semiconductor materials used for the construction of transistors. A transistor, however, does not contain additional components such as resistors, diodes, capacitors, and other transistors. An integrated circuit package contains passive and active elements and performs a complete circuit function or a combination of complete circuit functions.

IC packages may house anywhere from a dozen to millions of individual components. Logic systems rely on the use of combinations of the AND and OR functions. Because computers and microprocessor control systems require hundreds of thousands of those functions, cost and size savings are critical. Integrated circuits provide the most efficient method of packaging those functions within a small area. Standard digital ICs are available for every conceivable logic application. Along with those benefits, integrated circuits also offer:

- Reduced power consumption
- Increased reliability
- Higher operating speeds

The monolithic fabrication of integrated circuits offers a method for forming all components found within an IC—transistors, diodes, capacitors, and diodes—inside of a single chip of semiconductor material. Monolithic fabrication takes advantage of the fundamental electrical characteristics of materials and P-N junctions.

In terms of digital logic, integrated circuits separate into **transistor-transistor logic**, or TTL, and **CMOS logic**. The two different groups offer a choice of operating speeds, power consumption, the capability to connect with other devices, and operating voltages. When considering ICs as members of a family, those members have common characteristics that include:

- same power supply voltages
- identical signal levels at inputs and outputs
- internal speed
- functions
- identification numbers

TTL circuits offer versatility and are not sensitive to damage from static electricity. In addition, TTL circuits operate from a single, fixed voltage of +5 Vdc and operate at high speeds. CMOS logic circuits do not consume as much power as TTL circuits but have slower operating speeds. Moreover, CMOS logic circuits can run on voltages ranging from 3 to 18 Vdc. A CMOS logic circuit is much more sensitive to static electricity than a TTL circuit.

☑ PROGRESS CHECK

You have completed Objective 8. Discuss transistor-transistor and CMOS logic ICs.

2-9 USING DIGITAL CIRCUITS TO MANIPULATE FREQUENCIES

Almost all modern communications devices rely on some type of frequency division, multiplication, or synthesis. In television receivers, frequency synthesis appears in deflection circuits as a method for determining the horizontal and vertical oscillator frequencies; tuner circuits as a method for selecting channels; and in luminance/chrominance circuits for setting the proper operating frequencies. Frequency synthesis involves the use of counters, dividers, prescalers, encoders, decoders, and comparators. The type of frequency synthesis used in consumer electronics also introduces voltage-controlled oscillators and phase-locked loops.

Frequency Division and Multiplication

Frequency division involves producing an output signal that has a fractional relationship—such as $\frac{1}{2}$, $\frac{1}{3}$, or $\frac{1}{11}$—to the input signal. **Frequency multiplication** is accomplished by 1) a time-varying circuit that introduces harmonics at the output along with the fundamental frequency; and 2) a resonant circuit tuned to the desired output frequency. The resonant circuit passes only the desired output frequency to the load while rejecting other frequencies, including the fundamental frequency.

As an example, a frequency tripler uses a 1-MHz oscillator to convert dc power from a power supply line into a 1-MHz sine wave. When the sine wave feeds a nonlinear amplifier, the amplifier symmetrically distorts the signal so that the positive and negative peaks flatten. With this flattening, the amplifier introduces odd harmonics into the signal. As a result, the output signal consists of the 1-MHz fundamental frequency along with a sequence of odd harmonic frequencies. Starting with the third harmonic, each harmonic is progressively less than the fundamental. At the output, an LC tank circuit tuned to 3 MHz passes the 3-MHz signal while rejecting the fundamental and all other harmonic frequencies.

Frequency Synthesis

Frequency synthesis is a method of digitally generating a single desired, highly accurate, sinusoidal frequency from the range of a highly stable master reference oscillator. The desired frequency corresponds with a precise function of subharmonic and/or harmonic relationships found in the reference oscillator frequency. With this, a frequency synthesizer translates the performance of the reference oscillator into useful frequencies. When designs employ several reference oscillators, the possible number of output frequencies exceed the number of oscillators.

Most low-cost frequency synthesis circuits use frequency division to produce the desired frequency. The frequency division occurs through the use of a counter because of the ability of the device to function over a wide bandwidth. Frequency synthesis accomplished through this method can only produce an output frequency that has a lower value than the input frequency. Because crystal oscillators have a top frequency of 200 MHz, frequency synthesizers using frequency division are capable of working at frequencies less than 100 MHz.

Other frequency synthesizer circuits rely on frequency multiplication. With this technique, an oscillating signal passes through a diode, transistor, or varactor diode. Then, the non-linear semiconductor elements produce harmonics of the original signal. After that, the harmonically rich signal passes through a sharp, narrow-band filter that attenuates any undesired harmonics.

Another frequency synthesis method involves the mixing of frequencies. A mixer multiplies frequencies and generates a signal that contains both the sum and the difference of the two input frequencies. The circuit uses one frequency as the desired frequency and discards the other through filtering.

Frequency Synthesizer Operation   Figure 2-6 illustrates the operation of a basic frequency synthesizer. In the figure, the master reference oscillator has two outputs. Two divide-by-ten circuits divide the oscillator frequency by ten and then subdivide those dividends by ten. As a result, two accurate subharmonics—a tenth and a hundredth—of the master oscillator frequency exist.

Within this frequency synthesis system, the output of a stable reference oscillator divides into precise subharmonics. Then some type of discrete or integrated

switching arrangement would select the proper harmonics, add the frequencies, and output the signal through a bandpass filter. The filter would allow only the desired frequency to pass and has an extremely precise frequency at the output.

The locking portion of a frequency synthesis circuit depends on the precise measurement, division, and comparison of two input frequencies. While the tuner oscillator supplies an input signal, a reference oscillator provides a known and stable reference frequency. When we begin to look at the frequency synthesis circuit as a whole, dividers and a prescaler measure the input pulses as a series of binary numbers and divide the numbers so that the output is a usable low frequency. Each off and on of the pulse represents a binary number.

Indirect Frequency Synthesis   Indirect frequency synthesis relies on an oscillator controlled by a phase-locked loop (see diagram on next page) to generate its output frequency. Depending on the circuit requirements, the output frequency may consist of extracted clock signals from a data sequence or a multiple of a reference clock signal.

Voltage-Controlled Oscillators   Indirect frequency synthesis circuits usually rely on a **voltage-controlled oscillator**, or VCO. With the VCO, a square-wave output frequency has an inversely proportional relationship to the input voltage. Therefore, a low-input control voltage produces a high-output frequency, and

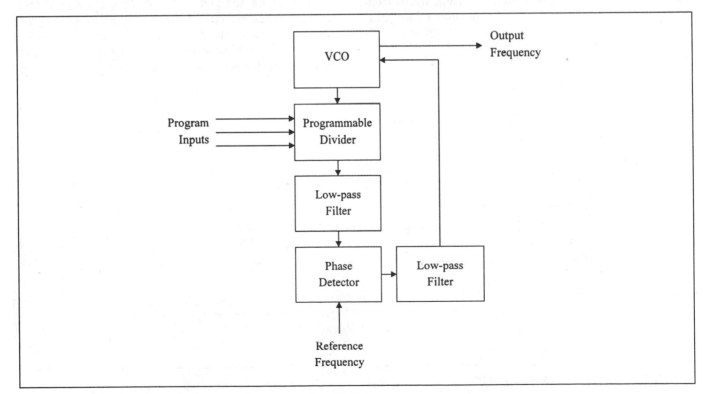

Figure 2-6   Block Diagram of a Basic Frequency Synthesizer

a high-input control voltage produces a low-output frequency. Some VCO configurations may consist of two emitter-follower transistors combined with a differential amplifier; a fixed crystal that resonates at approximately 4 MHz; and a frequency adjust capacitor. However, most VCO designs rely on the precise frequency relationships given by a 555 astable timer.

Phase-Locked Loops

A **phase-locked loop**, or PLL, contains a voltage-controlled oscillator; prescaler and divider circuits, a comparator, and a quartz crystal. It provides a low-cost alternative to frequency synthesis. As Figure 2-7 shows, the PLL receives an input signal and then compares that signal with the feedback of an internal clock signal generated by the VCO. Given the tuning provided by the PLL, the VCO oscillates at a frequency where the two divided signals are equal, and adjusts the feedback signal so that it matches the reference signal applied to the phase detector in both frequency and phase. As a result, the internal and external clock signals synchronize.

Many PLL designs rely on a "charge pump" consisting of inverters, switches, and a passive RC low-pass filter. Looking again at the figure, the input signal from the first frequency divider clock enters a phase detector. In this example, the phase detector consists of a set of buffers and D-type flip-flops. The phase detector compares the clock input with a feedback signal from the VCO. The frequency divider portion of the PLL may utilize a basic digital counter or something as complex as a single-sideband modulator and translates the input and output frequencies to usable levels.

PLL Advantages and Disadvantages   The phase detectors used in PLLs offer the advantage of easy implementation because of the use of digital techniques. In addition, phase-locked loops offer good noise characteristics because of the conversion of noise to phase noise. Moreover, phase-locked loops offer extremely good frequency locking. The key disadvantage of PLLs is the length of time needed to switch from one output frequency to another.

☑ PROGRESS CHECK

You have completed Objective 9. Explain frequency synthesis.

2-10   MICROPROCESSORS

A microprocessor is a complex integrated circuit that provides the functions of a full-scale computer and contains registers, memory, and buffers. The interconnection of the elements within a microprocessor allows the device to perform different preprogrammed operations and to use different sets of instructions that depend on the results of previous operations. Microprocessors execute instructions through a precise method and follow a sequence of instructions.

Within a system, a microprocessor receives inputs from externally connected devices and generates outputs to the same devices. Given the functionality of microprocessors, the applications range from intelligent control of other circuits to the processing of complex routines. In televisions, a microprocessor may work as part of a tuner control unit. In computer systems, the microprocessor is the heart of the processing power and

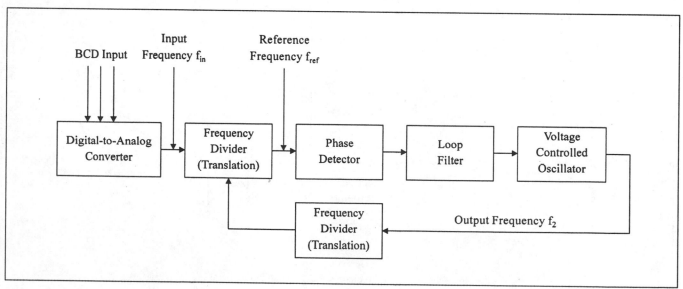

Figure 2-7   Block Diagram of a PLL

interfaces with the computer bus, system memory, and peripheral devices such as video cards, disk drives, keyboards, and printers.

Figure 2-8 shows a block diagram of a basic microprocessor. The microprocessor contains the following:

- a program counter for developing addresses that go onto the system address bus
- an instruction decoder that interprets instructions fetched from memory and then causes the implementation of the instructions
- internal registers that store intermediate data while the microprocessor is implementing complex instructions
- an arithmetic logic unit that performs calculations or Boolean operations
- clocking circuitry that paces each operation at precise rates
- control and sequencing logic for the management of external control signals
- buffers for the address, data, and control buses

Microcontrollers

Modern television tuning control systems include a dedicated microprocessor, or microcontroller, that performs preprogrammed tasks. While microcontrollers perform tasks that do not require a large amount of processing power, the circuits often include standard microprocessors that offer 4-, 8-, 16-, and 32-bit interfaces. In video products, the microcontroller:

- controls other devices
- places and removes information into and out of memory
- handles a wide range of instructions, including comparisons of numbers, arithmetic operations, and changing the meaning of a digital word
- makes decisions based on instruction sets and on data

Each of these tasks becomes apparent when we consider the overall operation of modern video products. The microprocessor responds to customer commands via either a keyboard or remote control device; directs channel selection and bandswitching information to the proper locations; sends instruction sets regarding tuner functions; and controls the channel display. In addition, a microprocessor moves information in and out of memory temporarily during operations such as channel changes. In some applications, such as a programmable satellite receiver where the control system establishes locations, polarities, and audio frequencies, the customer can use the microprocessor to modify existing parameters.

Most of these designs rely on an embedded controller, or a controller embedded within a greater system. Any embedded controller can be defined as a computer on a chip. However, the operation of an embedded controller depends on external components such as memory, buffers, and controllers. Embedded controllers are found in televisions, VCRs, stereos, computers, laser printers, and modems. Typical embedded controller tasks include communications, keyboard handling, and signal processing.

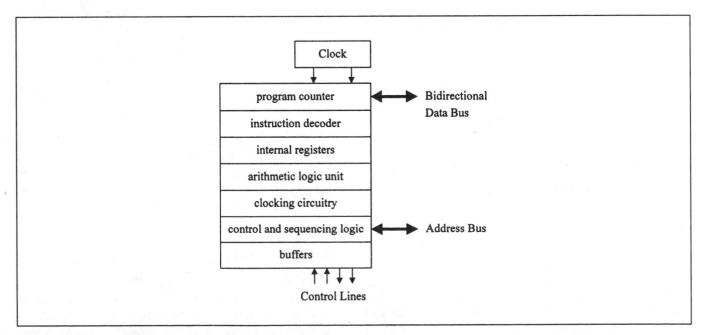

Figure 2-8   Block Diagram of a Basic Mircoprocessor

☑ PROGRESS CHECK

You have completed Objective 10. Define microprocessors, microcontrollers, and embedded processors.

2-11 TROUBLESHOOTING DIGITAL CIRCUITS

Troubleshooting digital circuits involves an understanding of pulses and the monitoring of those pulses with a logic probe, logic clip, or an oscilloscope. Logic probes and analyzers show the presence or absence of pulses and the relationships between those pulses. An oscilloscope can display the amplitude, duration, and frequency of those pulses. In addition to measuring those characteristics, digital troubleshooting also includes the measurement of delays between pulses and the operation of gates, flip-flops, and switching and delay elements.

Pulses have measurable characteristics, including:

- Rise Time (T_R)—the interval in time where the amplitude of an output voltage changes from 10 percent to 90 percent of the rising portion of the pulse.

- Fall Time (T_F)—the interval in time where the amplitude of an output voltage changes from 90 percent to 10 percent of the falling portion of the pulse.

- Time Delay (T_D)—the interval in time between the beginning of the input pulse and the time when the rising portion of the output pulse begins to have an output.

- Storage Time (T_S)—the interval in time between the trailing edge of the input pulse and the beginning of the falling portion of the output pulse.

- Pulse Width (T_W)—the duration in time of the pulse measured between two 50 percent amplitude levels of the rising and falling portions of a waveform.

Test Equipment Primer—Logic Probes and Logic Analyzers

Logic Probes   A logic probe is a small test instrument that can detect and indicate logic states in digital circuits. A logic probe can also detect the presence or absence of a signal polarity of a longer-duration pulse. Most logic probes use an indicator lamp that lights when the tip of the probe is touched to a high logic level. A dark probe indicates that a low logic state exists. When connected to a circuit that has a pulse train, the lamp will produce an alternating display of brilliance. A logic probe operates with an external voltage supply or connects to the supply of the circuit under test.

Logic probes work in situations where short duration pulses and low repetition rates make the use of an oscilloscope for signal tracing more difficult. Two general uses for a logic probe are pulse-train analysis and real-time analysis. Although an oscilloscope is the best tool for pulse-train analysis, a logic probe will show that a pulse train exists at a specific point in the circuit. With real-time analysis, the clock speed of the circuit under test is slowed to a very low rate. The decrease in clock speed is accomplished through the use of an external pulse generator. Then a logic probe is used to monitor both the input and output pulses at the circuit.

Logic Clips   A logic clip is a small test instrument that clips over an integrated circuit and displays the logic states at all the pins of the IC. As with the logic probe, a lighted diode indicates a high logic level while a dark diode indicates a low logic level. A logic clip does not require a power supply and has internal circuitry that determines the location of the ground and power supply connections on the IC.

Logic clips work well for real-time analysis. With the circuit clock again slowed to a very low speed, the clip can display timing relationships as those relationships occur in the circuit. Any problem with gates, flip-flops, or counters becomes apparent through the clip display LEDs.

Testing Gates   Testing an AND gate with a logic probe or clip involves a check for pulse trains and real-time analysis. With the probe in place, check for a pulse train at both the input and output. With real-time analysis, simultaneously inject pulses at both circuit inputs and monitor the output. If comparing the outputs to a truth table, the output should not appear if only input of the appropriate pulse is present.

Testing an OR gate involves simultaneously monitoring all inputs and the output. OR gates produce an output with the presence of any input. Therefore, the checking of OR gate operation should show a pulse train or single pulse at the output whenever a pulse appears at the input. NOR and NAND gates require the same type of testing methods but have inverted signals at the output.

Testing Multivibrators and Flip-Flops

Many times, the monitoring of multivibrator inputs and outputs will show the operating condition of the device. Astable multivibrators require signal monitoring at only one output, while monostable multivibrators and flip-flops require the monitoring of both the input and the output. Usually, the presence of a pulse train at the output of the multivibrator shows that the multivibrator is operating.

In addition to monitoring pulses at the input and output, signal injection also provides a method for testing multivibrator operation. With injection, the technician should remember how many pulses are required to cause the complete multivibrator cycle. For example, a Schmitt trigger requires two input pulses to change the output states and then return the states back to the original condition.

PROGRESS CHECK

You have completed Objective 11. Discuss methods for troubleshooting digital circuits.

SUMMARY

Chapter two provided a detailed overview of terms and concepts used in digital electronics. Beginning with a definition of Boolean algebra, the chapter moved through logic functions and binary numbering systems. From there, the chapter illustrated the basic building blocks of digital circuits through discussions about gates, amplifiers, switching elements, and delay elements.

The chapter described a number of terms and processes that will become evident in each of the following chapters, including buffers, decoders, and comparators. Each of those devices plays a key role in micro-

processor control circuitry. Also examined were storage devices such as registers, and conversion devices such as analog-to-digital and digital-to-analog conversion circuits.

In the discussion of frequency synthesis and microprocessor control you gained important knowledge about circuits and concepts that are applied throughout a modern television system. Chapter two concluded with troubleshooting methods, and covered two simple test instruments used for troubleshooting problems with a digital circuit: a logic probe and a logic clip.

REVIEW QUESTIONS

1. Digital information is read as:
   a. continuous waveforms over a given range
   b. discrete values

2. What are the advantages in using digital information?

3. Boolean algebra is _____.

4. A logic 1 represents:
   a. a high state
   b. a low state
   c. a null state

5. A logic 0 represents:
   a. a high state
   b. a low state
   c. a null state

6. An AND function has a logic 1 at the output if _____.

7. An OR function has a logic 1 at the output if _____.

8. How does an OR function produce a zero at the output?

9. The NOT statement _____ a logic condition.

10. A NAND function will have a 0 at its output only if _____.

11. A binary number system has how many values?
    a. one
    b. two
    c. three
    d. four

12. Each digit of a binary number represents a value based on the _____.

13. A bit is _____ and a byte is _____.

14. Why is the BCD numbering system used in electronic equipment?

15. An integrated circuit that performs fundamental Boolean logic functions is called a:
    a. multivibrator
    b. gate
    c. flip-flop
    d. switch

16. What is the difference between combinational and sequential logic?

17. Explain the differences between astable, monostable, and bistable multivibrators.

18. A flip-flop is a bistable multivibrator that:
    a. consists of NOT and NAND gates
    b. uses combinational logic
    c. operates as a memory element
    d. switches from one logic state to another
    e. all of the above
    f. a, c, and d
    g. a and d

19. Define how a J-K flip-flop works.

20. What is a Schmitt trigger?

21. What are the four things a buffer can do.

22. Synchronization in a digital circuit is accomplished through the use of:
    a. clock pulses
    b. sync pulses
    c. input pulses

23. What is the difference between an asynchronous counter and a synchronous counter?

24. Describe the operation of shift and storage registers.

25. A comparator _____.

26. Compare the operation of an ADC and DAC circuit.

27. True or False   A TTL circuit has a slow operating speed.

28. True or False   A TTL circuit operates from a fixed +5 Vdc supply.

29. True or False   A CMOS logic circuit is not susceptible to static electricity.

30. What is frequency synthesis?

31. Indirect frequency synthesis relies on:
    a. an oscillator controlled by a PLL
    b. a multivibrator
    c. a master oscillator
    d. precise subharmonics

32. Describe the operation of a PLL circuit and then draw a block diagram of the circuit.

33. Describe a microprocessor and then draw a block diagram of the device.

34. Name three tasks completed by a microcontroller.

35. Pulses have measurable characteristics called rise time, fall time, time delay, storage time, and pulse width. Define each one of the characteristics.

36. When would you use a logic probe?

37. A logic clip shows _____ relationships.

38. What is pulse train analysis?

39. When testing a multivibrator, you should _____.

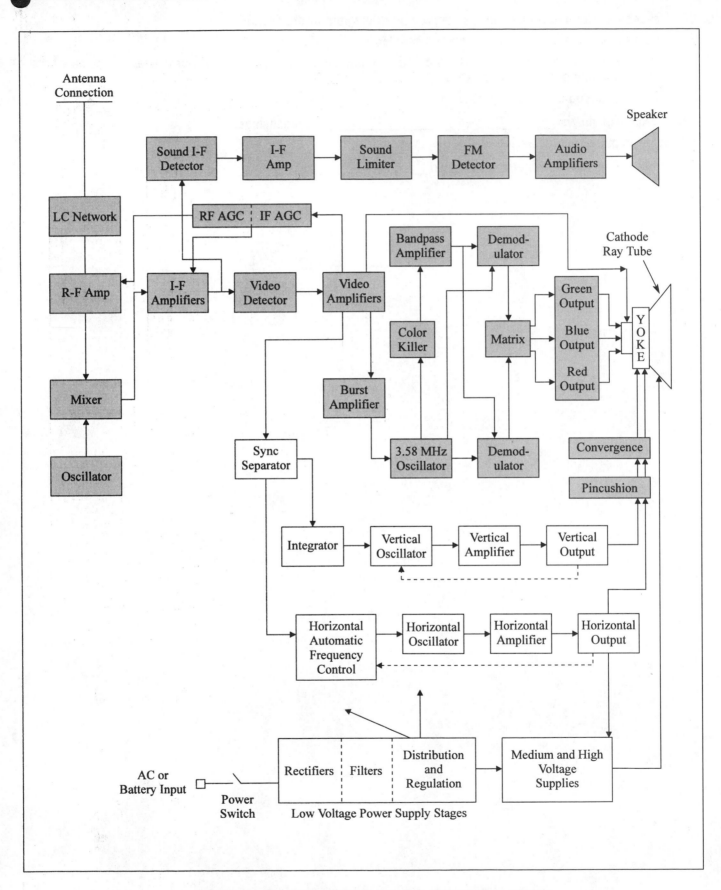

COLOR TELEVISION RECEIVER STAGES AND SIGNAL PATHS

PART 2

Power Supplies and Deflection Circuits

During the first part of chapter one, you found that every electronic system has a power supply. The following four chapters show how the proper voltages are developed throughout a color television. Chapter three defines the characteristics of linear power supplies, while chapter four describes switch-mode power supplies. Both types of power supplies have common uses in all types of modern electronic systems and provide regulated voltages for the entire system.

Chapters five and six shift the emphasis away from common power supplies to the deflection circuits used in color televisions. In part, the proper operation of the scanderived power supplies seen in chapter four depends on the proper operation of the horizontal oscillator circuits shown in chapter five. However, the deflection circuits also provide the drive needed to energize the output sections of the television. The horizontal deflection system provides the drive current needed for the operation of the flyback transformer and the horizontal deflection of the electron beam, while the vertical deflection system provides the properly shaped current needed to energize the deflection yoke and for the vertical deflection of the electron beam.

All this produces the process of sweeping an electron beam vertically and horizontally across the face of a cathode-ray tube. Given a human characteristic called persistence of vision—our visual capability to retain an image for approximately $\frac{1}{16}$ second per image after the image has disappeared or changed—sequential scanning produces a flicker-free representation of the original televised image on the receiver screen. With sequence scanning, the camera breaks the entire picture into elements, and the receiver reassembles the information contained in those elements while reproducing the transmitted image.

CHAPTER 3

Linear Power Supplies

OBJECTIVES Upon completion of this chapter, you should be able to:

1. Discuss the basics of power supplies.
2. Explain power supply transformers.
3. Explain rectifier circuits.
4. Define different types of power supply filters.
5. Define regulation and discuss regulator circuits.
6. Understand voltage dividers and bias voltages.
7. Explain voltage multipliers.
8. Discuss overvoltage and overcurrent protection.
9. Discuss power supply circuits.
10. Discuss troubleshooting methods for linear power supply circuits.

KEY TERMS

bias
direct-coupled doubler
feedback bias
fixed bias
isolation transformer
line regulation
load regulation
phase-control regulator
rectification
regulation

ripple voltage
SCR crowbar circuit
series-feedback regulator
series-pass regulator
series regulator
shunt-feedback regulator
shunt regulator
voltage divider
voltage-divider bias
voltage multiplier

DISCARD

LIBRARY
FORSYTH TECHNICAL COMMUNITY COLLEGE
2100 SILAS CREEK PARKWAY
WINSTON-SALEM, NC 27103-5180

INTRODUCTION

Chapter three introduces you to linear power supplies and establishes background material for your understanding of this chapter and chapter four. As shown in Figure 3-1, linear power supplies consist of a power transformer followed by rectifiers, a filter, and a regulator. In basic terms, linear power supplies have the purpose of delivering correct amounts of ripple-free dc voltages to the transistors and integrated circuits operating within the receiver.

The chapter covers linear supply designs ranging from the simple half-wave rectifier to the full-wave bridge rectifier and the direct-coupled voltage doubler. Although you may have studied rectification and regulation in other classes, chapter three spends some time reviewing those basic principles of power supply designs. The chapter also covers low-voltage transformers, provides a careful analysis of circuit operation, and combines that analysis with a discussion of problem symptoms and troubleshooting methods.

3-1 POWER SUPPLY BASICS

Linear power supply functions conform to the basic block diagram shown in Figure 3-1, and have the basic purpose of converting an ac line voltage to dc voltages. Most linear power supplies feature a power transformer that steps down the 115/230-vac 50/60-Hz line voltage to the lower voltages required by semiconductors, and that isolates the load from the ac line input. Depending on the application, however, a power supply also may operate as a transformerless system.

Transformerless power supplies are found in ac/dc receivers and in nearly all video equipment. The lack of a transformer in the power supply cuts the cost for manufacturing the unit but eliminates the isolation between the ac line input and the load. As a result, transformerless systems require a grounding system that does not rely on a chassis ground.

Regardless of whether the application utilizes a power transformer or not, rectifiers convert the ac line

voltage to pulsating dc voltages while filters smooth the ripples found in the rectified dc voltage. The filter circuits may consist of either capacitors, inductors, resistors, or a combination of passive components. Linear power supplies also feature a regulator stage that is designed to maintain the power supply voltage at constant levels despite any load current variations.

☑ PROGRESS CHECK

You have completed Objective 1. Discuss the basics of power supplies.

Linear Power Supplies That Use Transformers

While working through your basic electronics course, you found that two fixed coils wound on a core make up a transformer, which transfers electric energy from one or more circuits by electromagnetic induction. The inducing of a voltage in the transformer secondary depends on the building and collapsing of a magnetic field. Because of this, the primary windings of a transformer always connect to an intermittent source of current or to a source of alternating current while the secondary connects to the load.

Power Transformer Isolation Linear power supplies using a power transformer provide isolation between the ac line voltage and the load. In this case, a thin layer of insulation separates the primary and secondary windings. Moreover, either side of the secondary windings may connect to the chassis ground while one side of the primary windings connects to earth ground.

Stepping Up or Stepping Down the Line Voltage
Along with providing isolation, the low-voltage transformer either steps up or steps down the ac line voltage and provides windings for auxiliary power supply voltages. Referring to Figure 3-2, the electromagnetic induction produced by the current in the primary windings develops 120 vac across the transformer

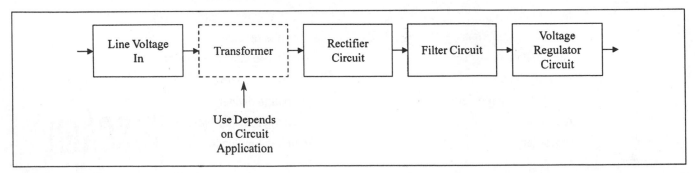

Figure 3-1 Block Diagram of a Linear Power Supply

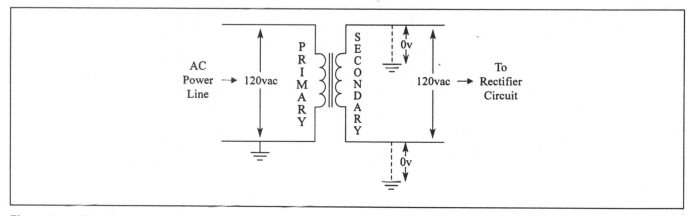

Figure 3-2   Diagram of a Power Transformer

secondary. The current flowing through the primary windings has no relationship with ground. While this may seem confusing, the 120 volts is measured across the windings rather from one winding to ground. A voltage measurement from either winding to chassis ground would show zero volts.

The Cold Chassis Ground   A cold chassis ground has any point on the metal chassis at ground or "earth" potential. Receivers built around the cold chassis rely on either a voltage-regulating transformer or an active device such as a power transformer. With the cold chassis, the ac voltage line connects directly to the power transformer primary. The secondary of the transformer provides ac waveforms for the power supplies located throughout the receiver. In this type of chassis, the space between the primary and secondary windings of the transformer serves as the sole point of power line isolation for the receiver. Figure 3-3 shows a schematic of a typical cold chassis ground system.

✔ PROGRESS CHECK

You have completed Objective 2. Explain power supply transformers.

Transformerless Linear Power Supplies

Most electronic equipment and all electronic devices using ac/dc power supplies do not utilize a power transformer to step up the voltage. While those supplies provide some benefits, such as cost savings, the disadvantage lies within the inability to provide a voltage that has an amplitude higher than the peak of the ac line voltage. A transformerless design has no isolation between the chassis and ground and relies on a grounding system called the hot chassis.

The Hot Chassis Ground   As Figure 3-4 illustrates, a different symbol is used to represent a hot ground than that used for a cold ground in Figure 3-3. The traditional ground symbol seen in Figure 3-3 represents the earth ground, while the inverted open triangle shown in Figure 3-4 represents the hot ground. With the hot chassis, the ground side of the ac line connects to the internal chassis ground and common-signal grounds of the receiver. As a result, one side of the ac power line carries the 120-vac line voltage, while the other side of the ac line connects directly to the chassis and ground. In effect, then, the chassis runs "hot." Because the receiver design does not rely on a transformer, the design offers a low-cost alternative for smaller, more portable receivers.

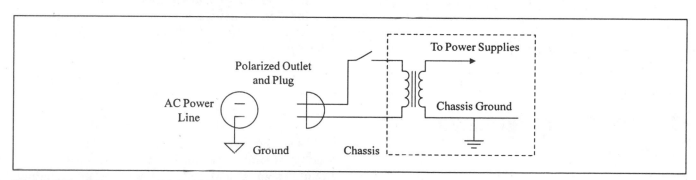

Figure 3-3   Schematic of a Cold Chassis Ground System

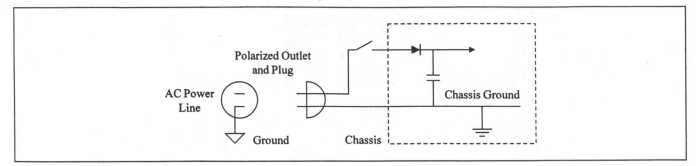

Figure 3-4 Schematic of a Hot Chassis Ground System

From a service perspective, however, the relative lack of protection against plugging in the ac line connection incorrectly can result in the connecting of receiver common-signal points to the hot side of the ac line. With this type of grounding configuration, any connection between the hot ground and any type of cold ground, such as the ground on an antenna system, will set up both a shock hazard and the potential for blowing fuses and ruining components in the receiver. Moreover, a technician may risk ruining equipment or receiving an electrical shock when working with a misconnected hot chassis.

To protect against this danger, all electronic devices have a polarized line plug that prevents the incorrect connection to ground. With one blade wider than the other, the polarized plug can be inserted only one way into a standard ac wall outlet. The wider blade connects to the grounded side of the power line that, in the wall outlet, always corresponds with the white wire. In the receiver, the wider blade always connects to the chassis. Therefore, the use of the polarized plug ensures that the chassis always connects to the grounded side of the ac line.

Rectifying the AC Line Voltage

Rectification involves the conversion of the required ac voltage value to a pulsating dc voltage, which may have either a positive or negative polarity. Depending on the application, linear power supplies may use one of the following four ac-to-dc voltage rectifier circuits to deliver either a half-wave or a full-wave output:

- half-wave rectifier circuit
- full-wave rectifier circuit
- full-wave bridge rectifier circuit
- direct-coupled doubler circuit

♻ FUNDAMENTAL REVIEW

Diodes

A diode is a two-terminal semiconductor device that conducts under specific operating conditions. PN junc-

tion diodes are constructed of an n-type material at the cathode, and of a p-type material at the anode. The n-type material is more negative than the p-type material. An ideal diode acts as an open switch when reverse biased and as a closed switch when forward biased.

Half-Wave Rectifiers Used with hot chassis ground systems in small-screen television receivers, the half-wave rectifier provides the simplest method for rectifying an ac voltage. As shown in Figure 3-5A, the transformer supplies the desired ac voltage to the diode. Connected in series with the T1 secondary and the RL, representing the circuit load, the diode conducts when its anode is positive with respect to the cathode.

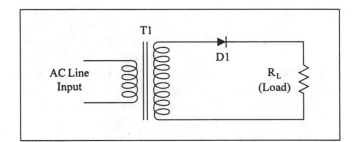

Figure 3-5A Schematic of a Half-Wave Rectifier Power Supply

Figure 3-5B links the conduction of the diode with the alternations of the ac sine wave. During a positive alternation of the ac wave, the top of the T1 secondary is positive with respect to the bottom windings. As a result, the diode conducts, and current flows through the load. During the opposite alternation, the top of the secondary is negative with respect to the bottom windings, and the diode does not conduct. Because the diode only conducts on positive alternations, the current pulses flow through the load in only one direction. Thus, the circuit has a pulsating dc output voltage.

Full-Wave Rectifiers Half-wave rectifiers offer the benefits of simplicity and low cost, but most modern circuit designs cannot rely on the pulsating dc output voltage produced during each positive alternation of

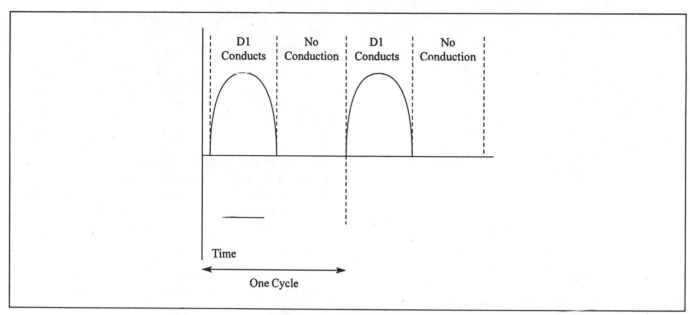

Figure 3-5B   Waveforms for the Half-Wave Rectifier Power Supply

the ac wave. Instead, many designs utilize a full-wave rectifier in the power supply. The full-wave rectifier always features a center-tapped transformer, develops two current pulses, and is used in cold chassis designs.

In Figure 3-5C, the center-tapping of the transformer establishes two half-wave circuits. The diode connected to the top half of the T1 secondary conducts only when the top half is positive with respect to the center-tap, while the diode connected to the bottom half of the secondary conducts only when the bottom half of the secondary is positive with respect to the center-tap. Each alternation of the ac sine wave shown in Figure 3-5D forward biases and then reverse biases the diodes. As a result, the full-wave rectifier circuit produces two current pulses that flow through the load on each positive and negative alternation of the ac sine wave.

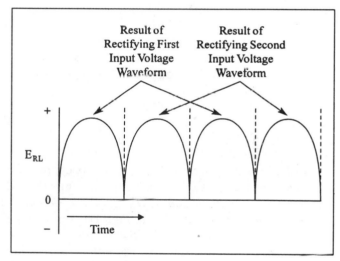

Figure 3-5D   Waveforms for the Full-Wave Rectifier Circuit

Full-Wave Bridge Rectifiers   Pictured in Figure 3-5E, full-wave bridge rectifiers use power transformers that do not have a center-tapped secondary to provide full-wave rectification. Instead, the full-wave bridge features four diodes that rectify the full secondary voltage on each alternation of the ac wave. Compared to the full-wave rectifier diagrammed in Figure 3-5C, the bridge rectifier provides pulsating dc with almost twice the amplitude, while using a secondary with the same amount of turns. The pulsating dc output almost equals the peak voltage found across the transformer secondary. Full-wave bridge rectifiers are used in cold and hot grounding systems and are applied to mostly large-screen televisions.

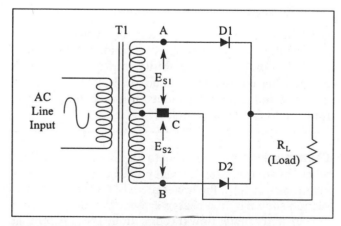

Figure 3-5C   Schematic Diagram of a Full-Wave Rectifier Circuit

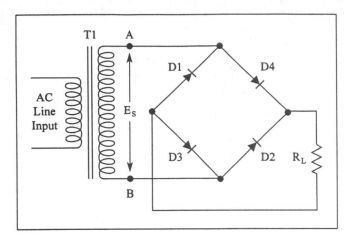

Figure 3-5E   Schematic Diagram for a Full-Wave
Bridge Rectifier Circuit

Referring to Figures 3-5E and 3-5F, we can begin
our analysis of the bridge rectifier operation with the
first positive alternation of the ac wave making the
top of the transformer secondary positive with respect
to the bottom. With this condition, the cathode of D1
and the anode of D4 are positive, and the cathode of
D3 and the anode of D2 are negative. Consequently
diodes D3 and D4 conduct, and electrons flow from
point "B" through D3, through the load, then through
D4 and to point "A."

On the other alternation of the ac wave, the top of the
transformer secondary becomes negative with respect
to the bottom. Because of this change, the cathode of
D1 and the anode of D4 also become negative, while
the cathode of D3 and the anode of D2 become more
positive. Thus, diodes D1 and D2 conduct while D3 and
D4 do not. The conduction of D1 and D2 allows electrons to flow from point "A" through D1, through the
load, then through D2, and finally to point "B." With
D1 and D2 conducting on one alternation and D3 and
D4 conducting on the other, the circuit produces two
current pulses for each cycle of the ac voltage supply,
and provides full-wave rectification.

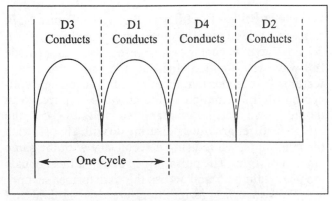

Figure 3-5F   Waveforms for the Full-Wave Bridge
Rectifier Circuit

FUNDAMENTAL REVIEW

Peak Inverse Voltage

The maximum reverse bias of a diode is called the peak
inverse voltage. Semiconductor manufacturers use the
peak inverse voltage as an operational safety limit for
diodes.

Integrated Rectifiers   Advances in the production of
integrated circuits have allowed manufacturers to place
all the active components of a full-wave or bridge rectifier into one semiconductor package. As an example, a
single 1-inch-square package can contain a bridge rectifier capable of handling an average forward current
of 25 mA and surges as high as 400 amps. The performance of an integrated rectifier package is equivalent
or superior to that provided by conventional diode rectifier circuits.

PROGRESS CHECK

**You have completed Objective 3. Explain
rectifier circuits.**

Filtering

Every type of rectifier circuit uses some type of filter
circuit to smooth the pulsating dc output voltage given
through rectification, and to remove as much rectifier
output variation as possible. The most basic type of
filter consists of an electrolytic capacitor connected
across the output of a half-wave rectifier. Other filter
types take advantage of the reactance properties of
inductors or utilize a combination of a power transistor
and a capacitor.

FUNDAMENTAL REVIEW

Capacitance and Inductance

Capacitance is the ability to store an electrical charge.
A capacitor exists whenever two conductors are separated by a dielectric. The ability to produce a voltage
through the cutting of a magnetic field by a current
is called induction. An inductor is a conductor surrounded by a magnetic field.

Capacitor Filters   Referring to Figures 3-6A and 3-6B,
when the first rectifier current pulse begins to increase,
electrons flow into the lower plate of C1, follow the direction shown by the charge arrow, flow out of the upper capacitor plate, flow through diode D1, and through
the secondary of the transformer. As the transformer
secondary voltage decreases, the capacitor discharges.

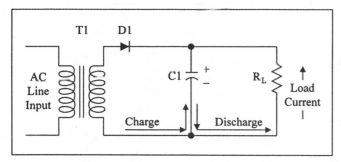

Figure 3-6A   Half-Rectifier Circuit with a Filter Capacitor

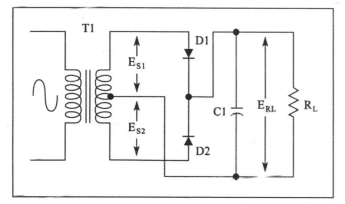

Figure 3-6B   Full-Wave Rectifier Circuit with a Filter Capacitor

However, because D1 allows electrons to flow in only one direction, the capacitor cannot discharge through D1 and the transformer secondary. Instead, the discharge current flows from the lower plate of the capacitor through the load and to the upper capacitor plate. Because the capacitor value does not allow a rapid discharge, the load current and capacitor voltages do not decrease as quickly as the secondary voltage.

Once the secondary voltage decreases to a level less than the voltage across C1, D1 conducts only long enough to charge the capacitor before becoming reverse biased. Thus, C1 can only partially discharge before another ac pulse begins the recharging cycle and causes the capacitor to charge to the peak of the secondary voltage. All this smooths any pulses found across the load. We define the varying voltage across the filter capacitor as the **ripple voltage**.

Inductive Filters   Inductive filters protect the power supply circuit from current surges. Improvement of the filtering action occurs through the addition of a filter choke, an inductor consisting of a winding on an iron core, and another capacitor. The self-induction of the choke opposes any changes in current and— when connected in series with the rectifier and load—carries the entire load current.

LC Filters   Figure 3-7 shows the addition of the choke to the half-wave rectifier circuit. When the applied secondary voltage causes an increase in current, the self-induced voltage within the choke opposes the change and prevents the current from increasing immediately to its peak. As the filter capacitor across the input of the circuit charges to the peak of the applied voltage during the first current pulse and the voltage across the transformer secondary begins to decrease, the input capacitor discharges through the load and the coil. The decreasing load current induces a voltage in the choke in a direction that maintains the current amplitude.

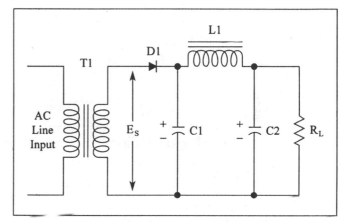

Figure 3-7   Addition of a Filter Choke to a Half-Wave Rectifier Circuit

Figures 3-8A, B, and C show how the waveforms across the capacitors and inductor should appear. While the first figure shows the waveform found across the input capacitor, the second shows the waveform at the inductor. Because the choke has a very low resistance, almost no voltage drops across the coil, and the counter emf of the inductor opposes any ac ripple. Figure 3-8C shows the smooth dc output voltage found across capacitor C2.

RC Filters   Some applications that require only a small current through the load replace the filter choke with the filter resistor shown in Figure 3-9. R1, the filter resistor, connects in series between the rectifier and the load and forms the series arm of a filter consisting of C1, R1, and C2. In the figure, the voltage EC1 is applied to R1 and C2 in series. Because capacitor C2 passes ac frequencies and blocks dc voltages, part of the direct voltage and a larger part of the ac line voltage appear across the filter resistor.

Active Filters   Many power supply circuit designs will supplement a filter capacitor with an active power filter circuit. Usually consisting of either a single power transistor or a combination of a filter driver transistor and a power transistor, the active filter circuit

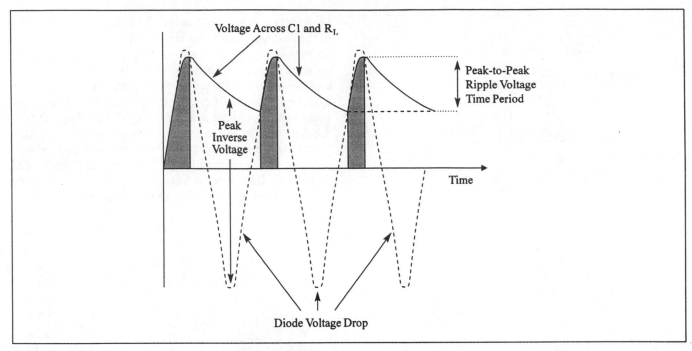

Figure 3-8A   Effect of Capacitor Charging on the Half-Wave Rectifier Circuit

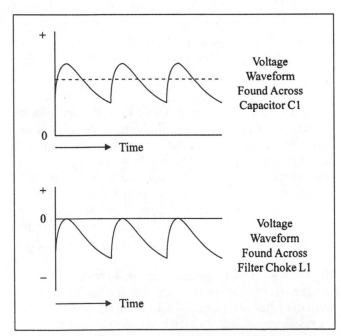

Figure 3-8B   Waveforms Found Across the Filter Choke and Capacitors in a Half-Wave Rectifier Circuit

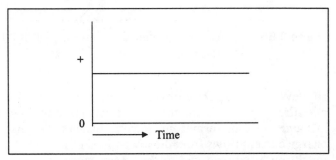

Figure 3-8C   Waveform Found at the Output Capacitor in a Half-Wave Rectifier Circuit

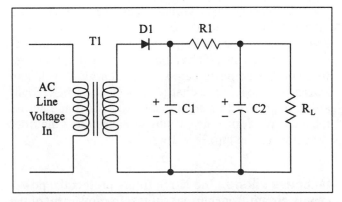

Figure 3-9   Replacement of a Filter Choke with a Filter Resistor

eliminates 60- and 120-Hz ripple voltages and any residual audio frequency or horizontal frequency voltages. With the first type of active filter shown in Figure 3-10, a power transistor connects in series with the rectifier output. Any ripple voltages cause increases in the current flowing through the transistor. In turn, the increased current causes an increase in the reverse bias

of the transistor and a decrease in the amount of conductance. As a result, the voltage at the output of the transistor also decreases.

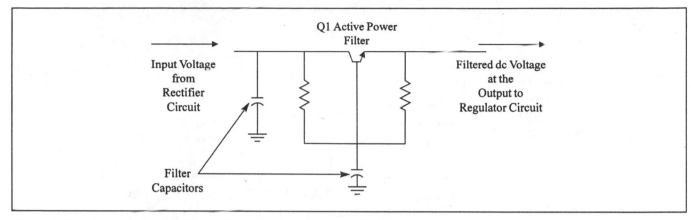

Figure 3-10   Schematic of a Single-Transistor Active Filter Circuit

In the second circuit, shown in Figure 3-11, Q1, a filter driver transistor, amplifies and inverts any ripple voltages. Then the driver transistor applies the inverted voltages to Q2, an active power filter transistor. With the ac voltage applied to the base of the power transistor remaining in phase with the voltage arriving from the rectifier circuit, the two out-of-phase voltages cancel one another. As a result, the ac component is eliminated and only a dc voltage exists at the output.

☑ PROGRESS CHECK

You have completed Objective 4. Define different types of power supply filters.

Regulation

Regulation is the maximum change in a regulator output voltage that can occur when the input voltage and load current vary over rated ranges. We can also define regulation in terms of line regulation and load regulation. While **line regulation** of a voltage regulator shows the amount of change in output voltage that can occur per unit change in input voltage, the **load regulation** rating of a regulator indicates the amount of change in output voltage that can occur per unit change in load current. An ideal voltage regulator will maintain a constant dc output voltage despite any changes that occur in either the input voltage or the load current.

Regulation is required because every device in the circuits that feeds off the low voltage supply has an internal resistance and draws some amount of current. Without some type of voltage regulation, the combination of internal resistance and current flow creates a voltage drop across the resistance and a resulting decrease in the output voltage. Regulator circuits stabilize the rectified and filtered power supply voltages so that the dc level of the voltage does not vary with changes in the line or load.

The two basic types of regulators are the shunt regulator and the series regulator. A **shunt regulator**—shown in Figure 3-12A—is in parallel with the load. Zener diode regulators are a form of shunt regulators.

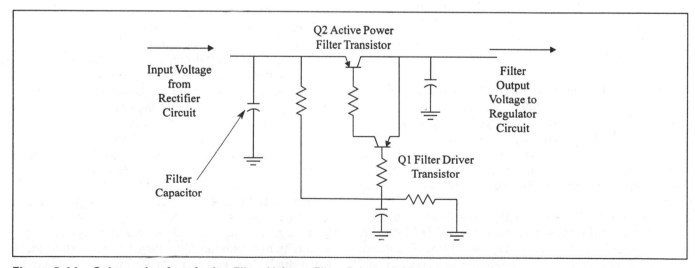

Figure 3-11   Schematic of an Active Filter Using a Filter Driver and Power Transistor

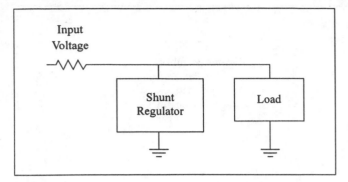

Figure 3-12A   Block Diagram of a Series-Regulator Circuit

As shown in Figure 3-12B, a **series regulator** is in series with the load. Pass-transistor regulators are a form of the series regulator. Both types of particular applications have advantages and disadvantages.

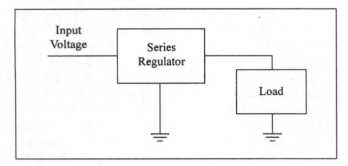

Figure 3-12B   Block Diagram of a Shunt-Regulator Circuit

Regulator Circuits   Regulator circuits for linear power supplies range from a zener diode, transistor circuits working as series-pass regulators, silicon-controlled rectifier (SCR) circuits working as phase-control regulators, and integrated three-terminal regulators. Most modern television receivers use linear regulators for higher voltages such as +115 Vdc and +125 Vdc, while VCRs utilize hybrid regulators for voltages in the +5.1 to +12 Vdc range.

Zener Diode Regulators   The most basic type of voltage regulation occurs through the use of a zener diode and takes advantage of the reverse breakdown characteristic of the diode. That is, when a zener diode operates in the reverse breakdown region, the diode will have a constant voltage across it as long as the zener current remains between the knee current and the maximum current rating.

As shown in Figure 3-13, a zener diode regulator circuit places the load resistance in parallel with the diode. Therefore, the load voltage will remain constant as long as the zener voltage remains constant. If the zener current increases or decreases from the allowable

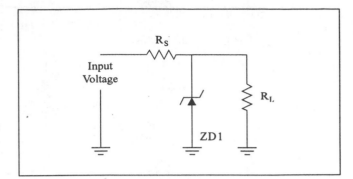

Figure 3-13   Zener Diode Regulator

range, the zener and load voltages change. Consequently, the success of a zener diode regulator in maintaining the load voltage depends on keeping the zener current within its specified range.

⟳ FUNDAMENTAL REVIEW

Zener Diodes

A zener diode is a special type of diode that operates in the breakdown region of the diode characteristic curve. Because of this, a zener diode maintains a relatively constant voltage despite any variations in the diode current. Therefore, zener diodes work well as voltage regulators.

Shunt-Feedback Regulators   Rather than rely on a zener diode for regulation, a **shunt-feedback regulator** uses an error detector to control the conduction of a shunt transistor. Shown in Figure 3-14, output voltages from sample and reference circuits go to the error detector/amplifier. While a voltage divider circuit makes up the sample circuit, the reference circuit consists of a diode and resistor.

In effect, the use of the error detector treats the shunt transistor as a variable resistor. When the transistor conducts, it has a minimum value of resistance. When the transistor does not conduct, it has a maximum value of conduction. Load regulation occurs as any increase or decrease in the load resistance causes a change in the conduction of the error detector. Any decrease at the error detector causes an increase in conduction at the regulator, while an increase at the detector causes a decrease in conduction at the regulator.

Series-Pass Regulators   Referring to Figure 3-15, a **series-pass regulator** circuit features a power transistor to regulate the load voltage and a voltage reference circuit that senses any source or load variations. A zener diode and resistor make up the reference circuit and maintain a constant value of output voltage by

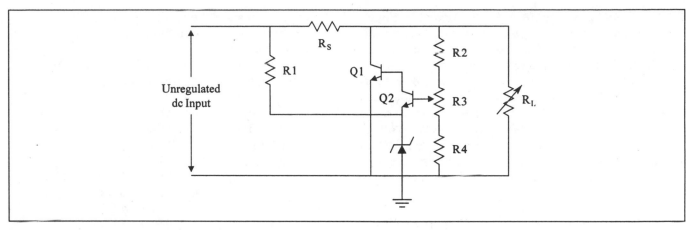

Figure 3-14 Shunt-Feedback Regulator Circuit

adjusting the amount of current flowing into the base of the power transistor. To accomplish this, the control circuit constantly adjusts the emitter-to-collector resistance of the series-pass transistor.

Because series-pass regulators drop voltages to dissipate power, the circuit almost always has a large heat sink and a large power transistor at the output. Although many color television receivers use the series-pass regulator circuit in the main dc voltage supply, the design lacks efficiency because of the wasted power. The inefficiency increases as the voltage drops across the series-pass transistor and the current passing through the transistor increases.

While a power transistor serves as the basic part of the control circuit in Figure 3-15, recent designs use either a Darlington pair or a comparator as the controlling device. The change in design results from the problem of having a zener diode dissipate a high amount of power. Using the Darlington pair prevents the load current from causing very little increase in zener current. The use of a comparator rather than a Darlington pair eliminates problems with temperature.

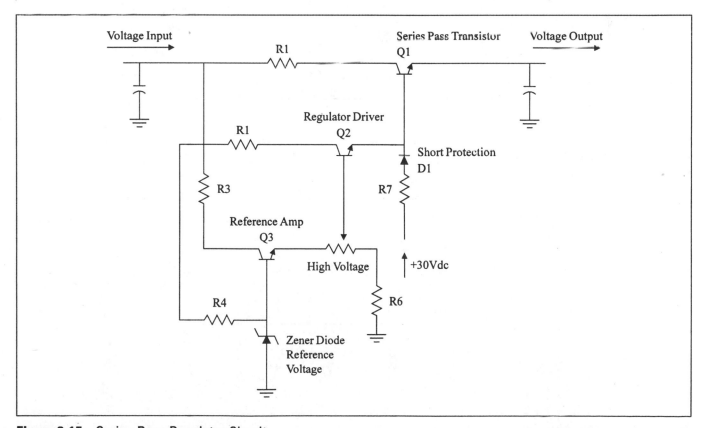

Figure 3-15 Series-Pass Regulator Circuit

Series-Feedback Regulators   Illustrated in Figure 3-16, a **series-feedback regulator** relies on a comparator to work as an error detector, and improves the line and load regulation characteristics seen with traditional series-pass regulators. As the figure shows, a reference voltage input and a sample voltage input feed into the error detector. The error detector compares the two input voltages and establishes an output voltage that is proportional to the difference between the two voltages. Then the error amplifier amplifies the output voltage and uses the voltage to drive the series-pass regulator.

Phase-Control Regulators   Recent regulator designs have replaced the series-pass transistor with a silicon-controlled rectifier. The SCR requires a much simpler triggering circuit because of the need for only a momentary turn-on pulse. The power supply circuit is complicated, however, because the driver subcircuit used to switch the transistor will not switch the SCR off. Rather than use a driver circuit, the newer designs pass the raw B+ voltage taken from the rectifier circuit

through a separate "turn-off" winding on the high-voltage transformer. Then the voltage is applied to the anode of the SCR.

Reducing the anode-to-cathode voltage of the SCR with the negative-going transition of the horizontal output pulse turns the SCR off. From there, the high-voltage transformer turn-off winding inverts and impresses the "hot" pulse onto the raw B+ voltage. In this case, the turning off of the SCR occurs at the horizontal frequency and at the same point in each cycle. Given the fixed point of the turn-off time, the circuit relies on the control of the turn-on time to modulate the pulses. The SCR operates as a phase-control device because it switches on during a specific phase of the ac cycle. Controlling the switch point and the ac current flow in this fashion also controls the average power to the load.

Phase-control regulators vary the conduction angle of a silicon-controlled rectifier; rectification of the ac current controls the level of recovered dc voltage and current delivered to the load. Figure 3-17 shows a sine wave superimposed over the SCR firing angle and illustrates that the phase-control device switches on

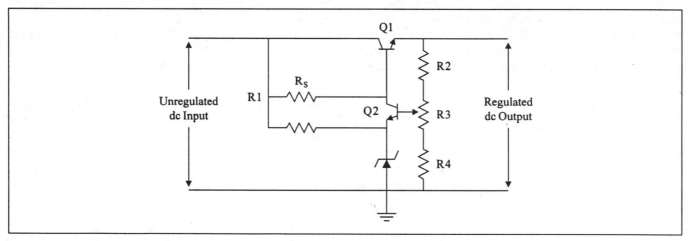

Figure 3-16   Series-Feedback Regulator Circuit

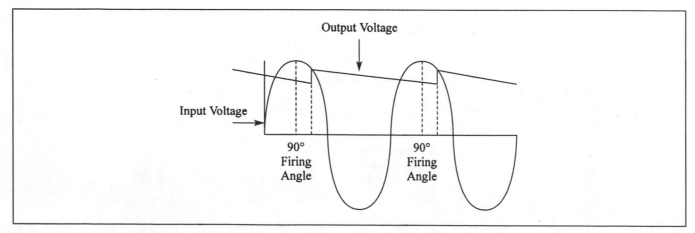

Figure 3-17   Firing Angle of an SCR Operating as a Phase-Control Regulator

during a specific phase of the ac cycle. The SCR operates as a half-wave rectifier and has a turn-on point referenced to the positive peak of the input ac sine wave. Conduction stops on the negative-going zero crossing. Because the SCR is a unilateral device, it will not fire again until the next ac input positive half-cycle and a the trigger pulse occur.

As Figure 3-18A shows, a trigger-pulse generator and a reference voltage control the conduction of the SCR. When the charge voltage of an RC series network reaches the threshold voltage of a unijunction transistor, the UJT fires and creates the trigger pulse found at the SCR gate. Regulation of the power supply occurs by making either the threshold voltage or the RC charging rate proportional to the load voltage.

During our discussion of linear series-pass regulators, we mentioned that a comparator often operated as the control, or drive, circuit. In most cases, an op-amp performs the task of driving the SCR. Looking at the driver circuit shown in Figure 3-18B, the stable reference voltage developed across the zener diode holds the positive input of the comparator to a fixed level. An adjustable voltage divider that includes the B+ voltage

control adjusts the negative input of the comparator through its connection between the output of the voltage regulator, the comparator, and ground. Centering the B+ voltage adjustment causes the negative input voltage to match the positive input voltage that also serves as a reference.

Any drop in the regulated B+ voltage causes the level of the negative input voltage to drop below the positive voltage level and, in turn, causes the comparator output to go high. As a result, the SCR triggers into conduction, and the B+ voltage level increases back to the normal level. The filter capacitor located at the left side of Figure 3-18B and the output of the SCR integrates the pulses into an average B+ voltage, plus or minus half the value of the ripple voltage. The pulse widths generated through this integration control the turn-on time of the SCR.

Switching the SCR off isolates the filter capacitor from the raw B+ voltage. As soon as the horizontal output transistor begins to conduct, it draws current from the capacitor. Consequently, the change in the capacitor voltage allows the negative input voltage at the comparator to drop below the level of the positive input

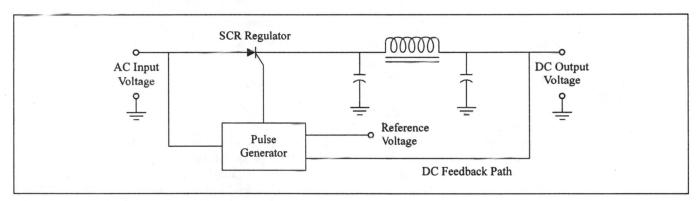

Figure 3-18A   Schematic of a Phase-Control Regulator Circuit

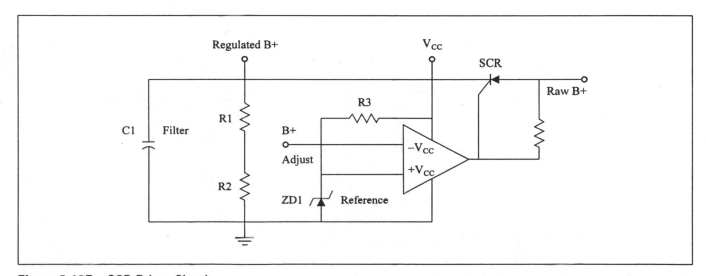

Figure 3-18B   SCR Driver Circuit

voltage. At this point, the SCR again switches on, the B+ voltage level increases, and the comparator changes logic states.

With the SCR latched on, it continues to refresh the charge across the filter capacitor. However, the operation of the horizontal output circuit also drains the charge at the same time. Each of the loads connected to the secondary of the integrated high-voltage transformer combines to make up the total load for the circuit. The current drawn by these loads from the capacitor during each cycle affects the switching point of the SCR.

⮎ FUNDAMENTAL REVIEW
SCRs

The SCR is a three-terminal thyristor that may function as a current-controlled device, a rectifier, or as a latching switch. A thyristor has four semiconductor layers and has either an open state or a closed state. SCRs conduct in only one direction and have an anode terminal, a cathode terminal, and a gate terminal.

Integrated Circuit Three-Terminal Regulators Many modern power supplies rely on an IC voltage regulator because of the characteristics provided by the IC technology. Depicted in Figure 3-19, three-terminal IC voltage regulators hold the output voltage from a dc power supply constant over a wide range of line and load variations. IC three-terminal regulators include the:

- Fixed-positive voltage regulator
- Fixed-negative voltage regulator

- Adjustable voltage regulator
- Dual-tracking regulator.

The first two types of IC regulators provide specific positive or negative output voltages. An adjustable IC regulator allows either a positive or negative output voltage to adjust within specific limits. A dual-tracking IC regulator establishes equal positive and negative output voltages. Depending on the application, different types of dual-tracking regulators may provide fixed-output voltages or output voltages that adjust between specific limits.

While IC voltage regulators are used in many power supply circuits, several limitations exist. When using the fixed and adjustable voltage regulators, the polarity of the input voltage must match the polarity of the output voltage. With the dual-tracking regulator, the IC must have dual input polarities. Also, any IC voltage regulator must include a shunt capacitor for the prevention of oscillations and an output shunt capacitor to improve ripple rejection.

☑ PROGRESS CHECK

You have completed Objective 5. Define regulation and discuss regulator circuits.

Voltage Dividers

A **voltage divider** establishes a method for providing more than one dc output voltage from the same power supply. The most basic voltage divider places a number of resistors across the power supply terminals. For example, the three-resistor voltage divider shown in Figure 3-20 provides three different output voltages. Terminal A supplies +300 Vdc, while

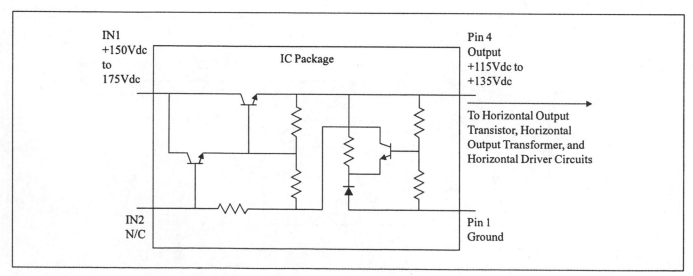

Figure 3-19 Schematic of a Three-Terminal IC Voltage Regulator

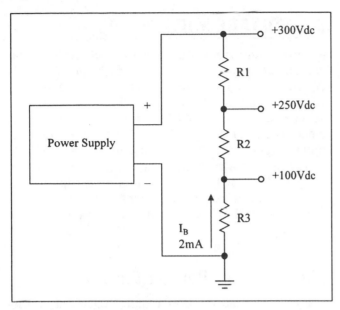

Figure 3-20   Three-Resistor Voltage Divider

terminals B and C supply +250 and +100 Vdc. The values of the resistors are chosen to accommodate the amount of current required by the load connected to the particular terminal.

⟳ FUNDAMENTAL REVIEW

Series Circuit Voltage and Current

In a series circuit, voltage is directly proportional to resistance. Current flow has only one path and is the same in all parts of the circuit.

Bias Voltages

The term **bias** refers to the no-signal dc operating voltage or current between two of the elements of a semiconductor device. For example, the bias of a transistor refers to the bias voltage applied between the base and emitter of a transistor, with the bias points labeled as V_{BE} and V_{EE}. One of the many types of bias used in amplifier circuits is called **fixed bias**. With this type of bias, a dc source external to the amplifier circuit supplies a very stable bias voltage. A regulated fixed-bias supply provides the most stable bias voltage available.

Common fixed-bias circuits include the collector-supply bias and the voltage-divider arrangements shown in Figures 3-21A and 3-21B. In a **voltage-divider bias**

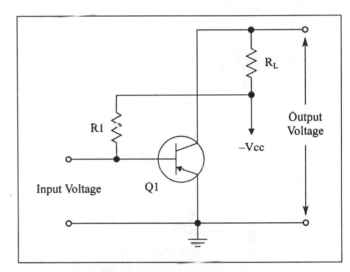

Figure 3-21A   Collector-Supply Bias Circuit

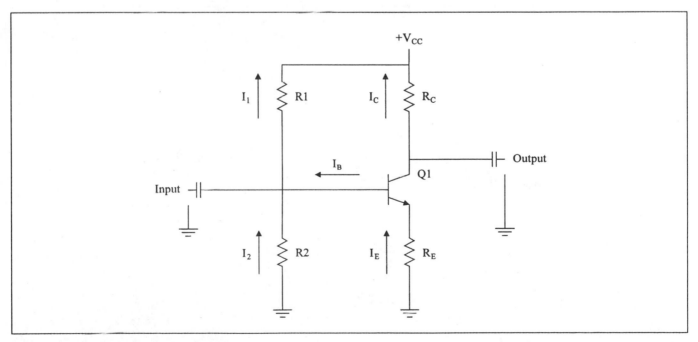

Figure 3-21B   Voltage-Divider Bias Circuit

circuit, resistors R1 and R2 form a voltage divider connected across V_{CC}. Because R2 is in parallel with the emitter-base junction of Q1, two current paths exist. The first extends from the negative terminal of V_{CC} through R1 and R2 and to the positive terminal of the collector bias voltage. The second extends from the negative terminal of V_{CC} through R1, the emitter-base junction of Q1 to the terminal of V_{CC}.

Feedback bias involves the feeding of part of the output signal back to the input as a 180-degree, out-of-phase, or degenerative, signal. Figure 3-21C shows a transistor amplifier circuit employing feedback bias. The use of negative feedback reduces the gain of an amplifier; feeding the output signal back to the input reduces the size of the input signal. Nevertheless, feedback bias improves the stability of the circuit.

↻ FUNDAMENTAL REVIEW

Feedback

A feedback circuit uses signal that travels from the output signal to the input of the amplifier. Degenerative feedback reduces the size of the input signal while regenerative feedback increases the size of the input signal.

☑ PROGRESS CHECK

You have completed Objective 6. Understand voltage dividers and bias voltages.

3-2 VOLTAGE MULTIPLIERS

A **voltage multiplier** provides a dc output that is the multiple of the peak input voltage to the power supply circuit. Although a voltage multiplier will provide an output voltage much greater than the peak input voltage, the operation of the multiplier causes the peak input current to decrease. Commonly used voltage multipliers include doublers, triplers, and quadruplers. As an example of a multiplier function, older television designs often used the combination of a tripler and a separate high-voltage transformer to establish the voltages needed at the CRT focus and high-voltage anodes.

Direct-Coupled Doubler Circuits

A **direct-coupled doubler** circuit offsets that disadvantage by adding the output voltages given by two half-wave rectifier circuits. As a result, the voltage doubler circuit provides a dc output voltage equal to twice the peak of the ac line voltage. Small-screen televisions with hot chassis designs utilize direct-coupled doublers as rectifier circuits.

Looking at Figure 3-22, our analysis of the circuit operation begins with an assumption that the ac alternation has forced point A to become negative with respect to point B. The cathode of D1 is negative with respect to its anode, and the anode of D2 is negative with respect to its cathode. As a result, only D1 conducts, and capacitor C1 charges to the peak of the ac line voltage, or approximately 170 Vdc, through D1.

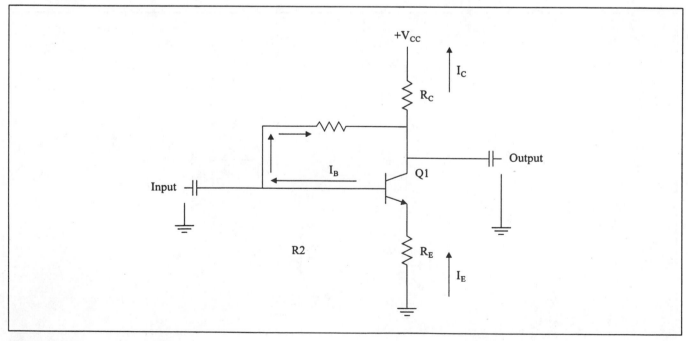

Figure 3-21C   Feedback Bias Circuit

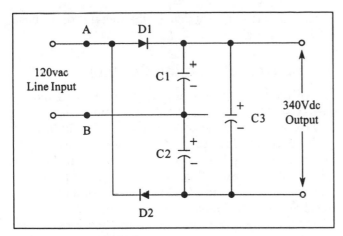

Figure 3-22   Schematic of a Direct-Coupled Doubler Circuit

The other alternation of the ac wave causes point A to become positive with respect to point B. Therefore, the cathode of D1 is positive with respect to its anode, and the anode of D2 is positive with respect with to its cathode. Here, only diode D2 conducts. Because capacitor C1 is in series with D2 and the ac line voltage, the total voltage applied to D2 equals the sum of the ac line voltage and the voltage found across C1. With this alternation, the voltage across C1 series aids the line voltage, and the capacitor charges to the peak of the line voltage. Capacitor C2 then charges to the peak of the line voltage plus the voltage found across C1, and provides a dc output voltage of approximately 340 Vdc.

Voltage Triplers and Quadruplers

The output transformer used in older color television designs has more windings and a larger size because of the need to generate 25 to 30 kilovolts. Efficiency and cost concerns, however, dictate a change away from large transformers and to an alternative method for generating the high voltage. The most common methods involve the use of voltage doublers, triplers, or quadruplers.

Each of the methods relies on the high-voltage transformer as the ac voltage source and uses a combination of diodes and capacitors to build the required voltage level. Within the voltage multiplier, an input capacitor can handle up to three times the applied peak voltage plus the specified dc voltage level. The diodes alternately conduct on the negative and positive half-cycles of the applied ac signal. The conduction of the diodes allows the capacitor to charge to the peak voltage plus the existing charge already on the capacitor.

Figure 3-23 shows a schematic diagram of a voltage tripler. The tripler includes three capacitors and three diodes. Voltage doublers have only two diodes and two capacitors, while voltage quadruplers have four diodes and four capacitors.

☑ PROGRESS CHECK

You have completed Objective 7. Explain voltage multipliers.

3-3 OVERVOLTAGE AND OVERCURRENT PROTECTION

Nearly all types of electronic equipment have some type of fast-acting turn-off protection device. In basic circuits, a fuse or circuit breaker may provide the needed protection. Most circuits, however, feature either an overvoltage or overcurrent protection circuit. Some feature a combination of both types of circuits and a fuse.

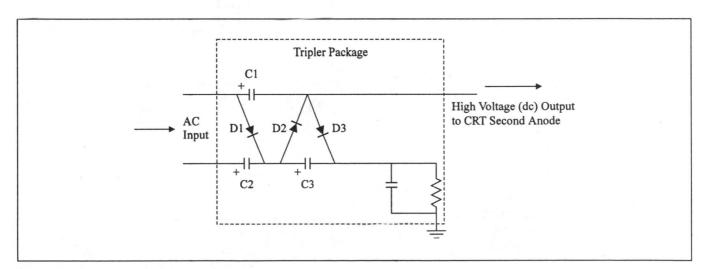

Figure 3-23   Schematic of a Voltage Tripler

Fuses and Circuit Breakers

Any type of fuse or circuit breaker protects the receiver from over-current conditions that can damage components. In short, the fuse or circuit breaker opens when a short or overload condition occurs. Fuses used to protect electronic circuits usually take the form of a small length of wire held within a glass cylinder by two metal caps. While the caps may snap into a fuse holder located on the chassis, some fuses have wire pigtails attached to the caps. The pigtails allow the soldering of the fuse into the correct location.

The fuses used in electronic devices such as televisions, VCRs, and satellite receivers are classified as either fast-acting or slow-blowing fuses. Generally, circuits that do not handle any type of surge or transient currents utilize fast-acting fuses. Slow-blowing fuses couple a small internal heat sink with the fuse wire and offer a minimal amount of time delay before opening. The heat sink allows the fuse to withstand transient overloads. An overload that lasts for a longer period of time will blow the fuse.

Many older televisions rely on small circuit breakers for protection. Within the circuit breaker, a thin metallic strip will remain straight during normal operating conditions but will bend if heated. The straightened-state of the metal strip allows the circuit breaker to remain in series with the power supply circuit. If a short-circuit or overload condition occurs, the heating caused by the overload causes the metal strip to bend and open the circuit.

Overcurrent Protection Circuits

When considering regulator circuits such as the series-pass transistor, you should remember that a short across the load will destroy the pass transistor because of the excessive load current. Most devices that rely on a series-pass regulator also include a current-limiting circuit. As shown in Figure 3-24, the use of an overcurrent protection circuit adds a transistor and a series resistor to the series-pass regulator circuit.

With the resistor connected to the base and emitter terminals, the current-limiting transistor can conduct only if the voltage across the resistor reaches approximately 0.7 Vdc. If the resistor has a value of 1 ohm, this occurs only if the load current or:

$$I_L = 0.7 \text{ Vdc}/1 \text{ ohm} = 700 \text{ mA}$$

Any load current less than 700 mA allows the current-limiting transistor to remain in cut-off. However, any increase of the load current above the 700 mA level causes the transistor to conduct. As a result of the conduction by the current limiter, the voltage at the base of the series-pass regulator transistor decreases. With a decrease in the base voltage, the transistor has a reduced conduction, and the load current begins to decrease to less than 700 mA.

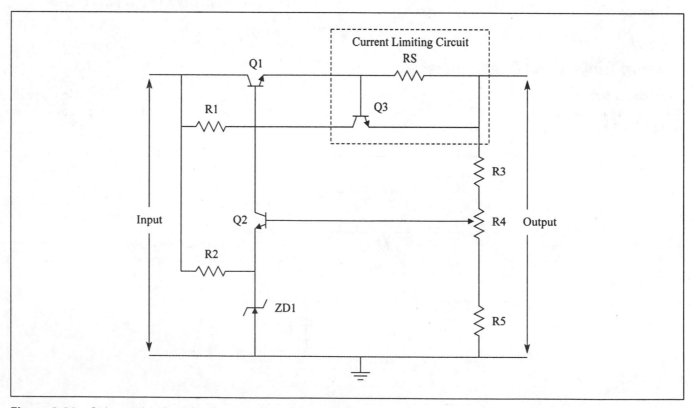

Figure 3-24   Schematic of an Overcurrent Protection Circuit

Overvoltage Protection Circuits

As opposed to series-pass regulators, shunt regulators require protection against overvoltage conditions at the input. If the dc input voltage to the regulator rises above a specified level, the shunt regulator transistor conducts harder so that the output voltage remains constant. With this, the transistor also dissipates more heat.

The **SCR crowbar circuit** shown in Figure 3-25 protects a voltage-sensitive load such as the shunt regulator from excessive increases in the dc power supply voltages. The circuit contains a zener diode, a gate resistor, and an SCR. Under normal circuit conditions, neither the zener nor the SCR conduct and no current flows through the resistor. The SCR acts as an open circuit. But a sudden increase in the dc power supply voltage causes the zener diode to conduct; the current flowing through the zener causes a voltage to develop across the gate resistor. As a result, the SCR begins to conduct, and the dc power supply shorts through the SCR. Although the overvoltage protection destroys the SCR, this action protects the shunt regulator circuit from the overvoltage condition.

☑ PROGRESS CHECK

You have completed Objective 8. Discuss overvoltage and overcurrent protection.

3-4 LINEAR POWER SUPPLY CIRCUITS

Linear power supplies provide excellent regulation, simple circuit designs, a relatively fast transient recovery time, and low output ripple. The drawbacks of linear power supplies include a lower than desired 50 percent efficiency rating because of the amount of power wasted while converting an unregulated dc voltage into a lower, regulated dc voltage. Much of the power loss occurs during high line voltage/high load conditions.

↻ FUNDAMENTAL REVIEW

Transient Recovery Time

Transient recovery time is the time required for a power supply to settle back to a specified output voltage tolerance after a step change occurs in the output load current. With a switched-mode power supply, a slow transient recovery time is desirable. Hold-up time refers to the length of time that the output voltage remains within specified tolerances after the line power fails. When considering power supply circuits, a longer hold-up time is desirable.

A Full-Wave Rectifier Linear Power Supply Circuit

Figure 3-26 shows the schematic for a typical low-voltage power supply circuit that relies on full-wave rectification. In the figure, transistor Q1 operates as a series regulator while transistor Q2 serves as the power output transistor. Rather than rely on a capacitor, an inductor, or a resistor as a filter, the circuit also uses Q1 as an active filter to remove any ac ripple. Transformer T1, diodes D1 and D2, and capacitor C1 work as a full-wave rectifier, while zener diode D3 establishes a reference voltage. Variable resistor R3 is the voltage adjustment control.

The load resistance series connects with the resistance of the regulator across the output of the power supply. Thus, the percentage of the power supply output found across the load depends on the ratio of the load resistance to the total of the load and regulator resistances. Any change in either the load resistance or the series regulator resistance will cause a change in the voltage found across the load.

Without the regulator, an increase in the line voltage would cause a proportional increase in the power supply output voltage. In actual operation, though, the increase in line voltage causes the voltage from R3 to ground to also increase. With the zener reference voltage holding the emitter of Q2 positive with respect to

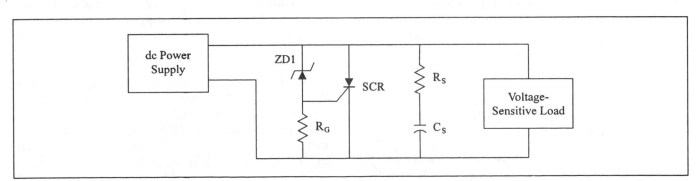

Figure 3-25 SCR Crowbar Circuit

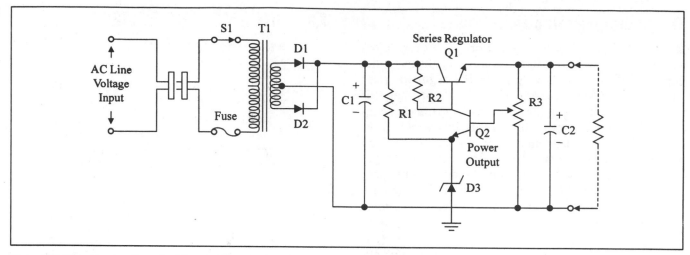

Figure 3-26   Power Supply Circuit Based on a Full-Wave Rectifier

ground, the output transistor conducts harder as the base becomes more positive.

The increased conduction of the output transistor produces a larger voltage drop across the load and makes the base of Q1 less positive with respect to its emitter. Therefore, the regulator transistor offers more opposition to the load current. A larger amount of the voltage found across C1 appears across the regulator and maintains a constant load voltage.

A decrease in line voltage causes the voltage across R3 and the output voltage to decrease and makes the base of Q2 less positive with respect to its emitter. While Q2 conducts less, the voltage drop across the load decreases. As a result, the base of Q1 becomes more positive with respect to its emitter, and the opposition of the regulator to load current decreases.

Changes in load resistance also prompt action within the regulator and power supply output circuits. An increase in load resistance causes a larger drop of power supply voltage across the load. To counter this effect and to maintain a constant load voltage, the regulator resistance must change. With the increased load resistance, less current flows through capacitor C1. Thus, the voltage across R3 increases, the base of Q2 becomes more positive with respect to its emitter, and the Q2 conduction through the load increases. In turn, the forward bias for Q1 decreases and increases the resistance of the regulator transistor. A decrease in load resistance produces the same effect as a decrease in the source voltage.

A Full-Wave Bridge Rectifier Linear Power Supply Circuit

As mentioned in the first section, the full-wave bridge rectifier is commonly used in power supply circuits. Figure 3-27 shows a typical full-wave bridge rectifier

circuit with diodes D1, D2, D3, and D4 making up the bridge. During operation, the portion of the secondary of transformer T1 labeled as L5 provides the 300 vac that, after rectification, produces the dc voltages required by other sections of the receiver. Circuit breaker F3 provides protection for the circuit.

With each of the diodes in the bridge supplying half of the circuit current, the bridge has a maximum current capability of 2 amps. Again referring to the diagram, the dc operating voltages are taken from the output of the bridge, while winding L6 of the transformer provides the 6.3 vac needed for the filaments of the CRT. While capacitors C3, C4, and C5 along with coil L3 provide filtering for the output circuit, coils L1 and L2, capacitors C1 and C2, and resistors R1 and R2 filter any high-frequency R-F noise from the ac signal entering the transformer primary. The impedance of the transformer primary isolates the primary from chassis ground.

A Series-Pass Regulator Circuit

The ac/dc power supplies found in portable receivers operate from either a battery or the combination of a four-diode bridge rectifier and series-pass transistor regulator. The power transformer provides cold chassis isolation from the power line, while either the battery or the ac line connection yields a constant main voltage.

Referring to Figure 3-28, most practical series-pass regulator circuits divide into the series-pass circuit, the regulator driver circuit, and a voltage reference circuit. In the figure, transistor QX703 inhibits any effects of load change on the reference voltage circuit consisting of resistor R704 and diode CR701. While transistor QX702 supplies the current for the base of the QX701, the series-pass transistor, resistor R707 and diode

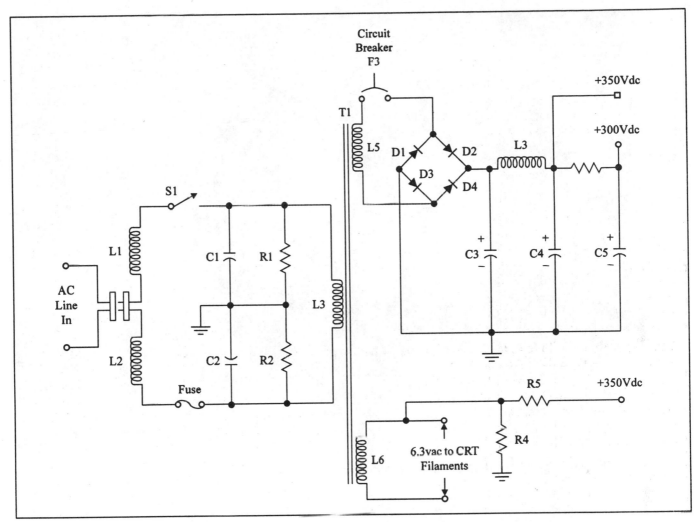

Figure 3-27 Power Supply Based on a Full-Wave Bridge Rectifier

CR702 provide enough current for QX701 to keep the transistor powered in case a short occurs in the supply output. Power dissipation for the circuit occurs across R701.

Figure 3-28 shows the schematic of a series-pass regulator circuit used within a Zenith 5-inch ac/dc television. The receiver works with either six internal flashlight batteries, a rechargeable battery, a car cigarette lighter adapter cord, or an external ac adapter. Moving to the lower left corner of the schematic, external power enters the receiver through a specially designed input connector that closes when no input is connected. The closing of pins 1 and 3 on the power jack connects the internal batteries into the circuit. Switch S701 at the bottom of the schematic closes and connects a battery charger into the circuit when rechargeable batteries are used.

During operation, QX705, the main regulator transistor, operates as the series-pass element. With a positive voltage applied to the circuit, QX705, along with the error amplifiers Q703 and Q704, conduct. The conduction of the transistors also causes zener diode

CR704 to conduct and provide a +2.4 Vdc reference voltage for the regulator circuit. Variable resistor R708 operates as a voltage sensor and allows the adjustment of the +6.5 Vdc B+ voltage for the receiver. Diode CR703 and the error amplifiers protect the circuit against external faults by cutting off the power supply if a short-circuit occurs.

A Voltage-Doubler Power Supply Circuit

The circuit shown in Figure 3-29 operates as a half-wave voltage doubler with diodes D1 and D2 and capacitors C1 and C2 forming the half-wave doubler portion. Closing switch S1 applies the ac line voltage through resistor R1 to C1 and D1. Within this first set of components, resistor R1 has a low resistance value and serves both as a fuse and as a current limiter for the two diodes. Capacitors C3 and C4, coil L1, and resistor R2 make up a filter network. As the schematic shows, various points in the filter network serve as take-off points for the +270, +260, and +150 Vdc voltage supplies.

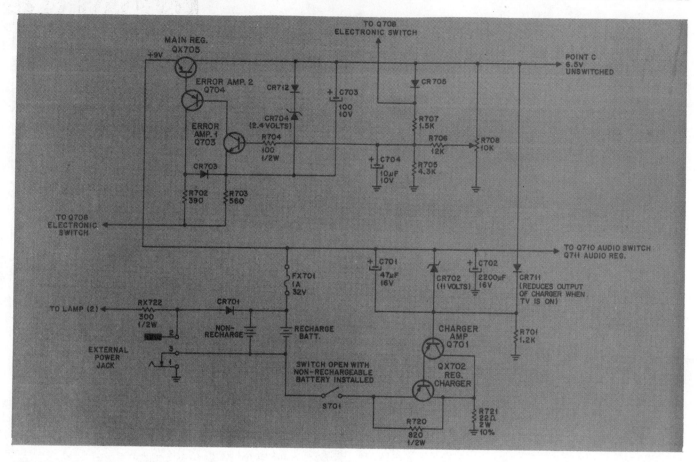

Figure 3-28   Power Supply Based on a Series-Pass Regulator Circuit

During the first ac alternation, the line terminal labeled A becomes negative with respect to terminal B, and electrons flow from point A through the switch and R1, into capacitor C1, and finally to the cathode of diode D1. Then the conduction of the diode allows the electrons to continue to the B terminal. With this flow of current, capacitor C1 charges to the peak of the line voltage.

The next alternation of the ac wave causes terminal B to become negative with respect to terminal A. Therefore, the line voltage and the voltage found across C1 act in series to charge C2 through diode D2. Current flows from terminal B to ground, through C2 and D2 to C1, through R1 and S1, and finally to terminal A. As a result, capacitor C2 charges to almost twice the line voltage. Looking at the schematic in Figure 3-29, the voltage found across C2 forms the output voltage of the entire circuit and is applied to the components that make up the filter circuit.

☑ PROGRESS CHECK

You have completed Objective 9. Discuss power supply circuits.

3-5 TROUBLESHOOTING POWER SUPPLY CIRCUITS

Any problem that occurs within the power supply section of a receiver affects other parts of the receiver. As the troubleshooting examples throughout this chapter have shown, logical troubleshooting always begins with check of the supply voltages. Although the first impulse is to look only for a no-voltage or low-voltage condition, the following sections also illustrate how excessive voltages can affect the performance of the system.

Common power supply problems involve shorted rectifier diodes, blown fuses, open resistors, and open or leaky capacitors. Each of those defects can cause a wide variety of symptoms that may link back to several different stages or circuits. As with many of the circuits that you have already studied, troubleshooting power supplies involves a great deal of logic; the use of multimeters to check voltage, current, and resistance levels; the use of an oscilloscope to observe waveforms; and the strict adherence to safety rules. Your safety procedures should include leakage tests, the use of an isolation transformer, and knowledge about the type of grounding system used in the receiver.

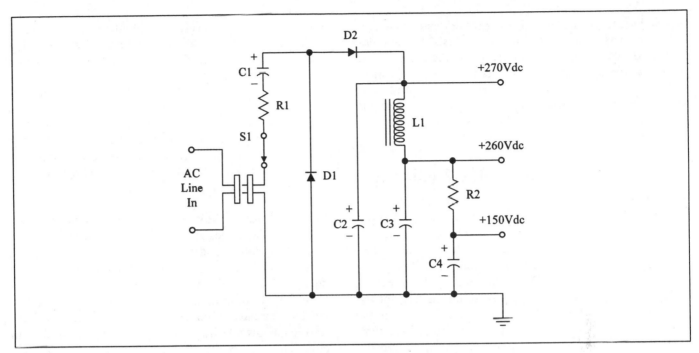

Figure 3-29   Schematic of a Voltage-Doubler Power Supply Circuit

Safety First!

Every video receiver is full of potential sources for serious electric shock. In televisions, the high dc voltage found at the CRT second anode and the high voltage output stages of the receiver can cause a jolting shock. While the high voltage supply has a low current and relatively poor regulation, your reaction to the shock may cause you to either break other components in the receiver or come in contact with other power sources.

Low-voltage power supplies have much lower dc output voltages but much higher current levels. As a result, any contact between the supply and ground can cause a serious electric shock. The same warning holds true for any exposed power line voltage. With most receivers using a hot chassis ground, any contact between the ac line hot ground and the cold chassis ground can cause a life-threatening electrical shock. In addition, contact between the chassis ground of test equipment and the ac hot ground will ruin the equipment.

Isolation Transformer   An **isolation transformer** provides the necessary isolation of the cold chassis ground and the ac power line hot ground through separate, isolated, primary and secondary windings. The use of an isolation transformer protects both you and your equipment from accidentally contacting the ac line hot ground. Every isolation transformer offers a 1:1 turns ratio and provides the standard line voltage at the secondary outlet.

Leakage Current Test   Every manufacturer prescribes leakage tests at the beginning of the service literature. A leakage test ensures that the receiver is properly assembled and that it offers no shock hazard to the user. The test setup involves the use of a resistor bypassed with a capacitor, with one end of the RC combination attached to the metal chassis and the other attached to earth ground. The amount of leakage current from any exposed metal part of the receiver to earth ground must not exceed 0.5 mA RMS. Always follow the manufacturer's guidelines for the leakage test.

Safety-Sensitive Components   Most manufacturers use shading in a schematic diagram to identify components that have special safety characteristics. Those components include RC networks and LC filters used in the ac power line connections; circuits used to provide overvoltage and overcurrent protection, and components in the high-voltage power supply. Never substitute general replacement parts for the safety-sensitive components.

Voltage, Current, and Resistance Measurements

The basic troubleshooting tests described throughout this chapter refer to measurements completed with a multimeter and an oscilloscope. When checking voltage and current readings with either instrument, always verify the type of grounding surface used in the receiver. Along with ruining the equipment, a serious shock hazard can develop if the ground lead of the test instrument is connected to the hot chassis. When measuring dc voltages, always verify that the meter is set to the correct polarity and the correct range. Finally, always check the test leads used during the measurements for any possible breaks. Even the smallest cut in a test lead can touch an electrically hot area of the chassis.

Test Equipment Primer—The Multimeter

The design and capabilities of modern multimeters allow technicians to perform tests in a variety of work environments and to cover a wide range of measurement needs. To this end, modern multimeters provide large LCD numeric displays and some type of bar graph display. While the LCD display offers the preciseness of a digital readout, the bar graph display shows changes inherent with analog signals. Digital multimeters also feature a wide range of functions that supplement the standard voltage, current, and resistance measurements. Some of those functions include a capacitance meter, a frequency counter, true RMS measurements, continuity indicators, and high-voltage warning indicators. When purchasing a multimeter, consider functionality, overload protection, the type of indicators featured on the multimeter, warranty, cost, as well as present and future needs. While an expensive, full-featured digital multimeter may not seem appropriate for your present application, the added features may become valuable in the future. Table 3-1 lists many of the important features to consider when purchasing a modern multimeter.

Multimeters usually arrive with one set of standard test leads but will accept several different types of specialized probes. The test probes allow measurements of high voltages, currents flowing through cables, and temperature. As with oscilloscopes, demodulator probes connected to the multimeter convert a radio-frequency signal into a dc voltage.

Voltage Testing   Voltage testing involves the measurement of voltages at various points in a circuit operating in a static condition. With voltage testing, the term "static" describes a condition where the circuit does not have a signal input but has power applied. Generally, limit voltage tests to sections or stages that you have already pinpointed as having a defect. Voltage testing narrows the search to specific points in the circuit.

Most service manuals depict test points for all circuit layouts. You can find defects if a measured voltage has a value far above or below the value given in the service manual. The service manual provides average values obtained by measuring the voltages at each point in a large number of units. Also, the values shown in schematic drawings are based on specified power line or battery voltages. In almost every case, measuring the

Table 3-1   Purchasing a Digital Multimeter

Functions	Indicators
DC voltage range	Continuity
DC voltage measurement accuracy	High-voltage warning
AC voltage range	Peak hold
AC voltage measurement accuracy	*Additional Features*
Ohms measurement ranges	Logic probe
AC and DC current ranges	Diode junction test
Maximum measurable current	Transistor beta test
Current measurement ranges	Light-emitting diode test
Sensitivity	True RMS measurement
Accuracy	Capacitance measurement and range test
Type of Overload Protection	Frequency counter
Resistance overloads	High-voltage measurement
Voltage overloads	*Service*
Display(s)	Warranty
Oversize LED display	Ruggedness (drop-proof, waterproof)
Bar graph	Battery life
Number of display digits	Physical size

actual voltage will yield a value slightly higher or lower than the average.

Figure 3-30A shows the proper connection of a multimeter for a voltage test. The dc voltage test requires the use of the voltmeter function and requires the placement of test probes according to the polarity of the points under test. In the figure, the positive meter probe connects to the emitter of the transistor (point A) while the negative, or common, probe connects to ground (point B). Thus, the voltage is measured with respect to ground.

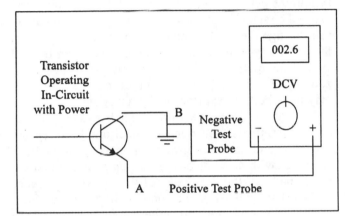

Figure 3-30A   Connection of a Multimeter for a Voltage Test

When using the voltmeter function, always remember that the meter has selectable voltage ranges. Generally, a schematic diagram for the unit under test should provide average voltage values for specific test points. To ensure an accurate reading, the meter should have a setting several times the value shown in the schematic in case a circuit malfunction causes the measured voltage to rise well above the listed value. For example, you could set the meter at the 20 Vdc range when measuring at an indicated value of 1.3 Vdc.

Current Measurement   Often the measurement of total current shows where a problem source exists in a circuit. As shown in Figure 3-30B, the measurement of current is made through the use of the milliammeter function of a multimeter and is accomplished by placing the meter probes in series with the source. The polarity of the probes at either point A or point B does not matter.

More than likely, if the total current has a value larger than normal, a short in some circuit has caused the drawing of too much current. In transistor circuits, only the output stage carries enough collector current to produce a noticeable change in the total current when a defect occurs. Emitter and collector currents of around 1 mA are normal for transistor receiver stages other than the output stage.

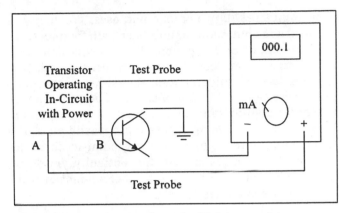

Figure 3-30B   Connection of a Multimeter for a Current Test

Checking for a change in collector current sometimes requires the use of Ohm's Law because of the difficulty involved with inserting probes in series with the transistor. When designs insert a resistor into the collector circuit, you can compute the collector current by dividing the voltage across the resistor by its resistance. You can measure emitter current indirectly because it has almost the same value as collector current. To measure the emitter current of a transistor, measure the voltage across the emitter resistor. Then divide this value by the emitter resistance to obtain a value for the emitter current value.

Resistance Measurement   Troubleshooting often involves disconnecting the power to the system and a check of the resistance of components in the signal path. Although some components will yield questionable in-circuit resistances, the tests provide a rough idea about component operation. Moreover, the resistance testing may allow you to pinpoint faulty solder joints or breaks in circuit traces. Figure 3-30C shows the proper connection of a multimeter for a resistance test. As with the current test, polarity is not a factor in the resistance test.

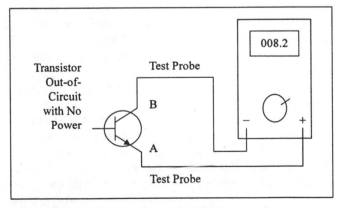

Figure 3-30C   Connection of a Multimeter for a Resistance Test

Continuity Testing For our purposes, continuity is defined as a continuous or unbroken path for direct current. A circuit or component that has continuity will conduct direct current. Any normal resistor, coil, connecting lead, or soldered joint should have continuity.

Generally, use the ohmmeter function of a multimeter for continuity testing. You could, however, connect an indicator lamp across the opposite ends of a component, circuit, or connecting point to check for continuity. When testing only for continuity, you do not need to be concerned with the exact amount of resistance between the test points.

Common Linear Power Supply Problems

In television receivers that utilize a separate, linear, low-voltage supply, a power supply problem can affect the raster, picture, and sound. The most frequent type of problem experienced with low-voltage power supplies involves blown fuses or open circuit breakers caused by a shorted rectifier or regulator transistor. Other problems include leaky or shorted filter capacitors. In many cases, the careful inspection of components in the power supply area can disclose the location of stressed or burnt components. In others, troubleshooting involves carefully checking how the operation of the power supply affects the operation of auxiliary circuits.

When checking a power supply that provides voltages through a voltage divider, disconnect one end of the resistor that supplies power to the load. By disconnecting only one end, you can find if the receiver will remain powered on. While using this type of check, remember that more than one circuit may operate from the same voltage line. As a result, this troubleshooting method may involve disconnecting several circuits in sequence.

Current, voltage, and resistance measurements also provide another method for checking the performance of auxiliary circuits. With the positive end of a diode rectifier disconnected from the circuit, connect one multimeter probe to the disconnected terminal, the other to the circuit connection, and check for the amount of current drawn for the given circuit. An unusually high amount of current—such as 20 or 30 mA— drawn by a circuit indicates that a faulty component is causing an overload.

Lower than normal voltages may show that an overload has occurred in the circuit or that a filter capacitor has opened. In addition, a low voltage may point to a leaky transistor or diode in the regulator circuit. Resistance measurements will verify the open or leaky condition by showing resistance to common ground. Again, disconnect the rectifier diode from the suspected auxiliary circuit and measure the voltage at the

source. If the source voltage remains normal even as the receiver shuts down, then the auxiliary circuit has an open or leaky component. If the source voltage remains at 0 volts, a component in the supply circuit has opened or become leaky. As always, you can also use resistance tests to check a suspected diode, transistor, or capacitor.

Filter Capacitor Problems Many of the problems associated with low voltage power supplies occur because of defective electrolytic filter capacitors. From a symptoms perspective, a defective filter capacitor can cause the raster to lose width, allow black hum bars to roll through the picture, cause black lines to appear in the picture, generate vertical black bars along the left side of the picture, and cause the picture to pull horizontally. Depending on the symptom, the capacitor may have become shorted, leaky, or open.

Defects in filter capacitors occur because of dried, broken terminals, and open connections between the capacitor and common ground. For example, a filter capacitor with dried electrolytics can produce vertical black bars at the left side of the picture because of the effect on boost and cathode voltages. Open filter capacitors in scan-derived power supplies can drop the supply to much below normal and cause the loss of raster, a darker than normal screen, or a badly distorted raster.

Low-Voltage Regulator Problems Every type of video receiver has low-voltage regulator circuits that regulate or stabilize dc voltage supplied to stages and circuits throughout the receiver. A defective low-voltage regulator can cause symptoms such as high-voltage shutdown, blown fuses, horizontal pulling, incorrect picture height and width, poor or no color, hum bars, and poor-quality sound. At face value, each of those symptoms indicates that a defect has occurred within a given section. In many cases, however, the regulator supplying the voltage to the sections has developed a problem.

When you consider regulator circuits, your troubleshooting efforts may involve regulators that use discrete components or regulators using integrated circuits. The first type usually features a large power transistor operating as a variable series-pass resistance between the raw B+ voltage and the regulated B+ voltage. A control circuit varies the bias of the series-pass element to maintain the proper level of regulated B+ voltage at the output. The second type places the regulator circuit within an IC package.

Troubleshooting the discrete regulator circuit involves basic voltage and resistance tests combined with in-circuit and out-of-circuit diode and transistor tests. For example, you can measure the dc voltage at the collector of a horizontal output transistor to determine if the low power supply is operating normally. A very low voltage would indicate that a defect has occurred with the output transistor or the regulator. The same higher than normal voltage found at all three terminals of a regulator transistor points to a leaky condition within the transistor. Rotation of the regulated B+ adjustment should cause the output voltage to vary. If the output voltage does not change, the regulator transistor has become defective.

Although you can follow specific patterns with discrete regulators, the use of IC-based regulators complicates troubleshooting efforts. Referring to Figure 3-31, the output voltage from the four-terminal IC regulator feeds into the horizontal output transistor, the horizontal output transformer, and the horizontal driver circuits. The input voltage at pin IN1 of the IC varies from +150 to +175 Vdc, while the output voltage at pin 4 ranges from +115 to +135 Vdc. Pin IN2 has no connection and pin 1 connects to ground. The complexity of the IC stems from the placement of three regulator transistors within the package.

In most cases, you can only use in-circuit tests to find shorted or leaky components within the IC package. As with all IC packages, you can also check the dc voltages and the resistance from pin to ground of each terminal. The voltages found at the IC pins should match voltage readings found on the schematic drawing. Low resistance readings may occur because of leakage within another component connected to the same terminal. To verify leakage within the IC package, disconnect any capacitors, zener diodes, or resistors that parallel the selected IC pin.

Common Television Power Supply Problems and Symptoms

In addition to the common checkpoints seen in the prior section, always consider all the symptoms given by the unit under test. Because of the complexity of the entire system, television receivers provide a good foundation for symptom analysis. The complexity becomes apparent when one symptom points in several directions. As a result, we always combine our preliminary symptom analysis with solid troubleshooting techniques. In each of the following examples, note how several problems have the same symptom.

Hum Bars Moving Through the Picture   The common symptom of hum bars moving through a picture usually indicates the presence of a leaky series-pass transistor or zener diode. Again considering a television receiver, a misadjusted or dirty B+ voltage control will cause the same symptom or may cause horizontal pulling. In addition, a defective filter capacitor in the regulator circuit may also cause the picture to pull from one side.

Excessive High-Voltage and Current Conditions
Within the horizontal output and high-voltage sections, excessive high voltage and excessive current conditions also indicate problems with regulator circuits. Both the excessive voltage and current will cause the receiver to go into shutdown and can result from a shorted CRT gun, shorted turns in a horizontal output transformer winding, or excessive output from a low-voltage regulator. With the last problem, the

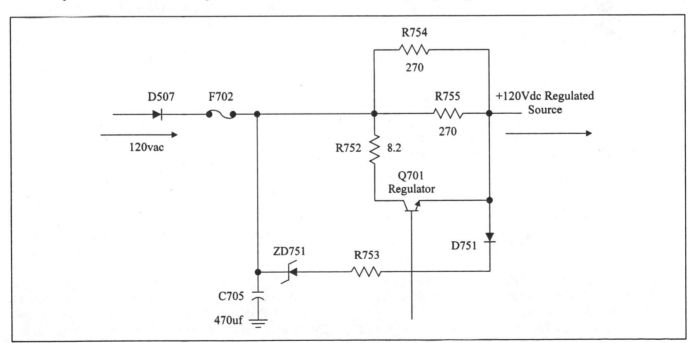

Figure 3-31   Sharp SKC-1310A Regulator Circuit (*Courtesy of Sharp Electronics Corporation*)

low-voltage regulator supplies a higher than normal B+ voltage to the horizontal output transistor collector.

Other symptoms involve abnormal raster displays because of the lack of control over the regulated voltage. As an example, the raster and sound of a television receiver may "pop" in and out at a rapid rate because of a pulsating, abnormally high voltage found at the horizontal output collector. In this case, the regulator circuit has not properly reduced or regulated the B+ regulated voltage. While the intermittent blacking out of a raster could indicate a problem with the CRT grid and screen voltages, CRT socket, or a problem in the video circuit, the problem may trace back to a zener diode that provides the reference voltage for the regulator control circuit.

➲ SERVICE CALL

Magnavox Model 19C301-BA Television with No Sound and No Raster

The Magnavox 19C301-BA (19C3) chassis arrived in the shop with no sound and no raster. Figure 3-32 shows the schematic for the power supply. Checks of the voltages at the power supply showed that the receiver had not gone into shutdown and that the output of the voltage-doubler startup circuit measured correctly at +335 Vdc. In addition, the receiver had a normal horizontal oscillator signal and normal oscillator voltages at the sync processor IC. However, voltages at the Q402 driver transistor showed +330 Vdc at the collector (the normal voltage is +60 Vdc), +1.9 Vdc at the base (the normal voltage is +6.9 Vdc), and +1.1 Vdc at the emitter (+6.8 Vdc is normal).

A check of the schematic showed that a diode regulated the voltage found at the Q402 emitter, and that the transistor had lost forward bias. In-circuit tests of the zener diode and the horizontal driver transistor showed no defects within those components. Out-of-circuit resistance checks of resistors R417 and R418 showed that R417 had opened. Because of the design of the power supply, the technician replaced both resistors. The operation of the television receiver returned to normal.

➲ SERVICE CALL

Sharp Model SKC-1310A Television with a Blown Fuse

After the customer brought the Sharp SKC-1310A portable television into the shop, the technician found that fuse F702 had opened and that the operating voltage for the Q602, the horizontal output transistor, had dropped to +18 Vdc. The technician removed Q602 from the circuit to see if the regulated B+ voltage would

increase. The lack of increase with the removal of the transistor indicated that a defect had occurred within the power supply or the regulator circuits.

In-circuit tests of Q701, the regulator transistor, and D751, a rectifier, showed no problems with those two components. Not completely satisfied, the technician removed one end of D751 to check for possible leaky conditions and shunted C705 with another capacitor to check for problems. Still, no problems surfaced. An in-circuit check of zener diode ZD751 disclosed that the regulator diode had become leaky and had caused the decrease in the horizontal output transistor voltage. Replacement of the +55v zener diode restored normal operation for the receiver.

➲ SERVICE CALL

Sylvania VCR Model #VC8952AT01 Has No Front Panel Display

A customer brought a Sylvania VCR Model #VC8952AT01 in with the complaint that the unit had no front panel display. After checking the power supply voltages, the technician found that the voltages were lower than normal. The replacement of an open resistor in the regulator circuit brought all the supply voltages back to normal levels except for the 4-vac line. Because the 4-vac output supplies the filament of the front panel display, the unit would operate and all front panel controls worked without a display.

Further inspection of the unit showed that the power transformer was making a hissing noise. An oscilloscope check of the base of Q1, a regulator, showed some oscillation at that point. An additional multimeter check of the voltages supplied to the display showed that the voltages had decreased to zero. A further check of the power supply components disclosed that two of the 330uf 6.3 Vdc capacitors that filtered the display voltage output line had opened. Replacing the capacitors restored the normal operation and display of the VCR.

➲ SERVICE CALL

RCA CTC187 Chassis with an Hourglass-like Picture

When the customer brought the RCA television into the repair shop, he showed the technician that the picture bowed in on both sides. Other than the bowing problem, the television had perfect sound and a perfect picture. A check of the power supply with a multimeter disclosed that the dc voltages were almost normal. However, a check of the voltage lines with an oscilloscope showed the presence of an ac voltage in the voltage line feeding into the horizontal output stage of the

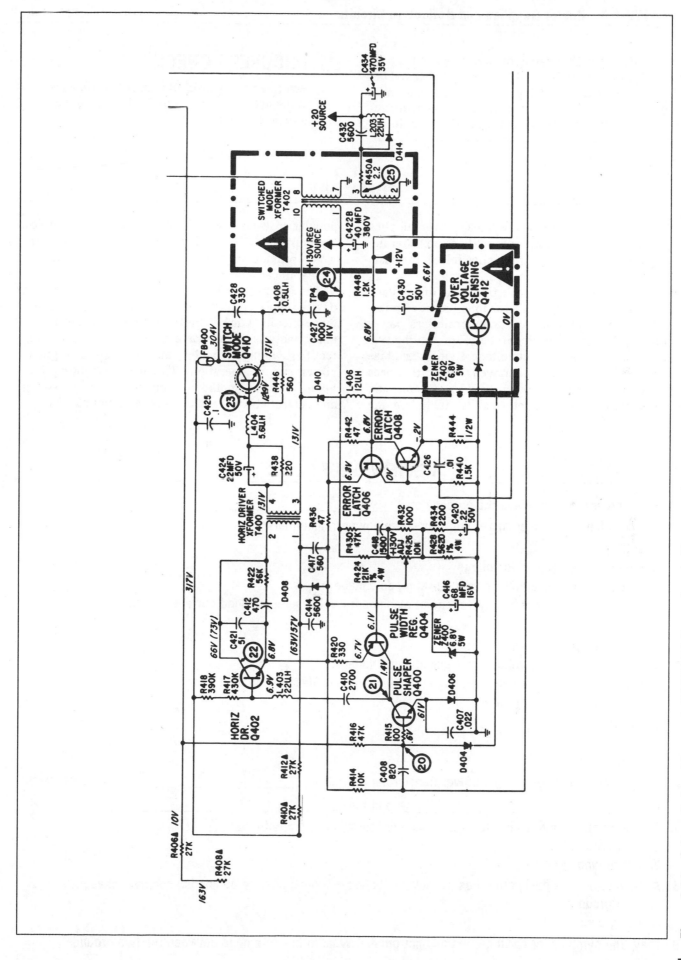

Figure 3-32 Schematic for a Magnavox 19C301-BA Power Supply (*Courtesy of Philips Electronics N.V.*)

receiver. Because filter capacitors are supposed to pass dc and block any ac voltage, the technician checked the main filter capacitor for the power supply and found a leaky condition. The replacement of the capacitor restored the receiver to its original operating condition.

 PROGRESS CHECK

You have completed Objective 10. Discuss troubleshooting methods for linear power supply circuits.

SUMMARY

Chapter three introduced you to linear power supplies and the fundamental concepts of rectification, filtering, and regulation. Linear power supplies follow the basic model of input-rectification-filtering-regulation, and have the basic purpose of converting an ac line voltage to the required dc voltages. The chapter provided information about transformer and transformerless power supplies, and the two different types of grounding systems. You also learned about voltage dividers and bias voltages. The discussion then moved to the differences between cold and hot chassis grounding systems, and overvoltage and overcurrent circuits, including the popular crowbar circuit. You next learned about voltage multipliers—doublers, triplers, and quadruplers—which provide voltages higher than those found in a typical linear power supply.

The last two sections presented a real-time view of power supply circuits and problems encountered with those circuits. You had the opportunity to perform a detailed analysis of full-wave rectifier, full-wave bridge rectifier, voltage doubler, and series-pass regulator power supply operation. The troubleshooting techniques covered included symptom analysis, voltage and current checks, and component-level checks.

REVIEW QUESTIONS

1. Linear power supplies have the purpose of _____.

2. True or False   All linear power supplies utilize power transformers.

3. A rectifier:
   a. smooths ripples in a dc voltage
   b. converts an ac line voltage to a pulsating dc voltage
   c. maintains the power supply voltage at a constant level despite any load current variation

4. A regulator:
   a. smooths ripples in a dc voltage.
   b. converts an ac line voltage to a pulsating dc voltage.
   c. maintains the power supply voltage at a constant level despite any load current variation.

5. A transformer operates through:
   a. resistance
   b. capacitance
   c. inductance

6. Linear power supplies using a power transformer provide _____ between
   the _____ and the _____.

7. Any point on the metal chassis is at ground or "earth" potential with the:
   a. cold ground
   b. hot ground

8. A transformerless design that has no isolation between ground and the ac line is used by a chassis with a:
   a. cold ground
   b. hot ground

9. Draw the symbols for earth ground and hot ground. What is the difference between the two grounds?

10. Draw the output waveforms for a a) half-wave rectifier circuit; b) full-wave rectifier circuit; and c) full-wave bridge rectifier circuit.

11. Describe the operating differences between a filter capacitor, filter choke, and filter resistor.

12. What is ripple voltage?

13. Why do the waveforms shown in Figures 3-8A, B, and C have different shapes?

14. What is the purpose of an active filter?

15. What is the difference between a series regulator and a shunt regulator?

16. To control the conduction of a shunt transistor, a shunt feedback regulator uses:
    a. a zener diode
    b. an error detector
    c. a divider
    d. a linear device

17. A series-pass regulator consists of:
    a. a power transistor and a voltage reference circuit
    b. a zener diode and a voltage reference circuit
    c. a power transistor and an error detector
    d. none of the above

18. A _____ circuit features a power transistor operating in the linear mode and a voltage reference circuit that senses any source or load variations.

19. In Figure 3-18B, the SCR operates as a _____ and has a turn-on point referenced to the _____ of the _____.

20. During each positive alternation of the ac wave, a pulsating dc output voltage is produced by a:
    a. full-wave bridge rectifier circuit
    b. full-wave rectifier circuit
    c. half-wave rectifier circuit

21. True or False   The full-wave bridge features four diodes that rectify the full secondary voltage on each positive alternation of the ac wave.

22. What type of rectifier circuit uses a center-tapped transformer?

23. Describe the operation of a direct-coupled voltage doubler.

24. A fixed-IC regulator provides _____ voltages, while an adjustable regulator provides _____ voltages.

25. What is the purpose of a shunt capacitor in an IC voltage regulator circuit?

26. A voltage divider provides:
    a. only one dc output voltage
    b. negative and positive polarity output voltages
    c. more than one dc output voltage

27. What is a bias voltage?

28. With fixed bias, a _____ supplies a very stable bias voltage.

29. Feedback bias involves:
    a. feeding part of the output signal back to the input as a 180-degree out-of-phase signal.
    b. the use of resistors configured as a voltage divider
    c. feeding part of the output signal back to the input as a 90-degree out-of-phase signal.
    d. the use of an external dc source as a bias voltage

30. Describe the operation of a voltage multiplier.

31. An SCR crowbar circuit consists of
    a. a capacitor, a gate resistor, and an SCR
    b. an inductance, a gate resistor, and an SCR
    c. a zener diode, a gate resistor, and an SCR
    d. a transistor, a gate resistor, and an SCR

32. If an SCR crowbar circuit senses a sudden increase in the dc power supply voltage, the SCR begins to conduct. Describe why this occurs and how the conduction of the SCR affects the operation of the power supply.

33. In Figure 3-26, Q1 operates as a series regulator and as an _____

    _____.

34. Refer to Figure 3-26. True or False   The percentage of the power supply output found across the load depends on the ratio of the load resistance to the total of the load and regulator resistances.

35. Refer to Figure 3-26. A _____ provides the reference voltage for the circuit.

36. Refer to Figure 3-27. Coils L1 and L2, capacitors C1 and C2, and resistors R1 and R2 filter any

    _____ from the ac signal entering the transformer primary.

37. Refer to Figure 3-28. During operation, QX705, the main regulator transistor, operates as the

    _____.

38. Describe the operation of the voltage doubler circuit shown in Figure 3-29.

39. A high voltage tripler contains _____ and provides

    _____.

40. Name four common problems associated with linear low voltage power supplies.

41. A leaky, shorted, or open filter capacitor can cause problem symptoms such as _____

    _____, _____,

    _____, and _____, to appear.

42. Describe methods used to troubleshoot a discrete regulator circuit.

43. Why is the troubleshooting of an IC-based regulator circuit more difficult than the troubleshooting of a discrete circuit?

44. True or False   All regulator circuit problems are associated with lower than normal voltages at the output.

45. Refer to the first Service Call (Magnavox 19C301 Chassis with No Sound and No Raster). What preliminary actions were taken by the technician before beginning the troubleshooting process? What methods were used to find the source of the problem? Why did the defective resistor cause the problem?

46. Refer to the first Service Call. What caused the zener diode to lose forward bias?

47. In the second Service Call (Sharp Model SKC-1310A with a Blown Fuse), a lower than normal voltage at the horizontal output transistor was caused by a leaky:
    a. rectifier diode
    b. regulator transistor
    c. zener diode
    d. filter capacitor

    Why?

48. Describe the following safety practices:
   a. the use of an isolation transformer
   b. leakage checks
   c. proper attachment of test equipment leads
   d. high voltage tests

49 What caused the failure of the front panel display in the third Service Call (Sylvania VCR Model #VC8952AT01 has no front panel display)?

50. In the fourth Service Call (RCA CTC187 Chassis with an Hourglass-like Picture), the symptom was caused by
   a. a shorted regulator that allowed ac to pass into the horizontal output stage of the receiver
   b. an open resistor that allowed ac to pass into the horizontal output stage of the receiver
   c. a leaky capacitor that allowed ac to pass into the horizontal output stage of the receiver
   d. a leaky zener diode that allowed ac to pass into the horizontal output stage of the receiver

51. How did the technician in the third Service Call find that ac was passing into the horizontal output stage of the receiver?

CHAPTER 4

Switched-Mode Power Supplies

OBJECTIVES   Upon completion of this chapter, you should be able to:

1. Explain how switched-mode power supplies work.
2. Discuss switched-mode input, regulation, and transformer operation.
3. Discuss isolation in the SMPS system and the iso-hot grounding system.
4. Describe the operation of scan-derived power supplies.
5. Describe the operation of start-up voltage circuits.
6. Define high-voltage transformers.
7. Explain overvoltage and overcurrent protection in SMPS circuits.
8. Analyze switched-mode power supply circuits.
9. Troubleshoot switched-mode power supply problems.

KEY TERMS   demodulator probe   pulse-rate modulation

flyback   pulse-width modulation

graticule   scan-derived power supply

integrated high-voltage transformer   start-up power supply

iso-hot chassis   switched-mode power supply

kick-start circuit   trickle-start circuit

metal-oxide varistor   variac

optoisolator

INTRODUCTION

Chapter three introduced you to linear power supplies that use a power transformer, a rectifier circuit, filters, and a regulator circuit. Linear power supplies continue to have applications in audio equipment, but do not offer the efficiency needed by many other types of electronic circuits. As electronic devices have become portable, the optimal utilization of power has become increasingly important.

All modern televisions, monitors, personal computers, VCRs, and many other types of electronic equipment rely on a different type of power supply called the switched-mode power supply, or SMPS. Switched-mode power supplies offer advantages such as reduced size, weight, and cost. The high frequency operation of an SMPS allows the use of smaller and lighter components than those seen in linear power supplies. In addition to those benefits, an SMPS offers greater efficiency than a linear power supply. Because an SMPS operates either fully on or fully off, this type of supply loses little power and has an efficiency of approximately 85 percent.

4-1 SMPS BASICS

When you studied linear power supplies, you looked at a block diagram that showed a line input voltage traveling into a power transformer and then through a rectifier circuit, filter, and regulator circuit. With **switched-mode power supplies**, the block diagram changes slightly. Rather than begin with a transformer, the SMPS begins with a full-wave rectifier circuit connected directly to the line and then progresses to a high-frequency transformer, a power transistor, and a pulse generator. Figure 4-1 shows a block diagram for a typical switched-mode power supply. The SMPS supplies 132 Vdc for the sweep circuits, 12 Vdc for a remote control preamplifier, 12 Vdc for the turn-on, and 35 Vdc for the audio stages of a television receiver.

SMPS Components

As with linear power supplies, switched-mode power supplies contain a mix of passive and active components. These include bipolar-junction transistors, rectifiers, silicon-control rectifiers, shunt-regulator ICs,

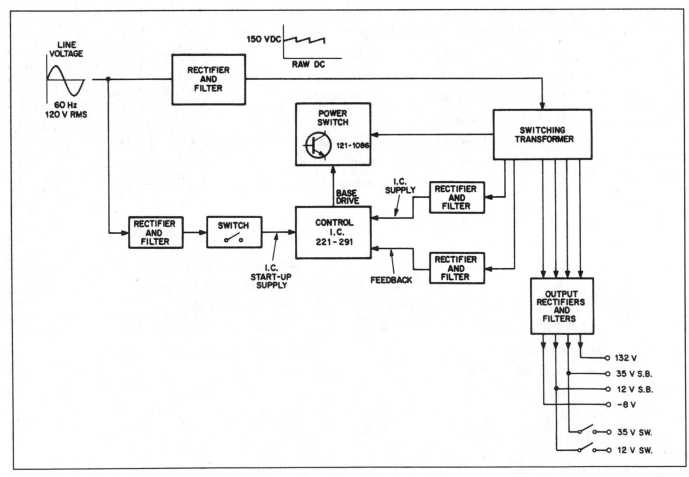

Figure 4-1   Block Diagram of a Switched-Mode Power Supply

optoisolators, filter and bypass capacitors, resistors, metal oxide resistors, and thermistors. Each individual component type affects the performance of the switching power supply and involves tasks such as feedback, control, rectification, overvoltage and overcurrent protection, regulation, isolation, filtering, and voltage division.

When studying the operation of an SMPS, it is easier to consider the components by function. For example, filter capacitors either filter the rectified—and sometimes doubled—ac line input voltage, or filter the output voltages from the SMPS. Other types of capacitors in the circuit provide bypass paths. SMPS power supplies also contain a combination of general type resistors and flameproof resistors, metal-oxide varistors, and thermistors. While the general type resistors are often found in voltage-divider circuits, the flameproof resistors are found in the return circuit for the switching regulator or in the ac line circuit. **Metal-oxide varistors,** or MOVs, and thermistors provide protection against severe surges and appear in the ac line circuits, while optoisolators or optocouplers establish isolation.

Active components such as bipolar transistors metal-oxide semiconductor field-effect transistors, (MOSFETs), and silicon-controlled rectifiers (SCRs) may operate as part of a feedback circuit, as regulators, or in overvoltage and overcurrent protection circuits. Bipolar junction transistors work as either components in a feedback circuit or may function as the SMPS switching device. The type of transistor used in the particular circuit varies with the function. For example, a power transistor capable of handling high voltages will work as a switching device, as will MOSFETs and SCRs. Rectification occurs through the use of either discrete or packaged diodes. Most SMPS units use diodes for ac line rectification or in voltage doubler circuits. The switched power supplies usually rely on some type of three-pin IC regulator for regulation of the output voltages.

⟳ FUNDAMENTAL REVIEW —————

Transistor Switches

A bipolar-junction transistor can operate as a switch by driving the device back and forth between saturation to cut-off. Switching transistors have a delay between each input transition and the time when the output transition begins. The output transition for a switching transistor takes a given amount of time to occur. Propagation delay for a transistor is measured in terms of:

1. Storage time, or the time required for the device to come out of saturation
2. Delay time, or the time required for the device to come out of cut-off

3. Rise time, or the time required for the transistor to go from cut-off to saturation
4. Fall time or the time required for the transistor to go from saturation to cut-off

Because transistor switches have internal resistance and leakage resistance, the device never completely restricts the flow of current or allows all current to pass. Capacitance found within transistors prevents the device from switching instantaneously. Usually, a transistor switch will have a common-emitter configuration because of the current and voltage gain characteristics given by that configuration.

A common-emitter transistor switch conducts because a positive voltage placed at the base of transistor causes the device to enter saturation. The collector-to-emitter voltages decrease to near zero, and current flows through the load. Removing the voltage from the transistor base causes the device to enter cut-off, upon which, the collector-to-emitter voltage rises to the same level as the collector supply voltage and current ceases to flow through the load.

SMPS Operation

All switched-mode power supplies use a high-frequency switching device such as a transistor, MOSFET, insulated gate bipolar transistor (IGBT), SCR, or triac to convert the directly rectified line voltage into a pulsed waveform. An SMPS that has a lower power requirement will feature a conventional transistor or MOSFET as a switcher, while high-power SMPS units will rely on an IGBT, SCR, or triac. Each of the last three components offers latching in the on state and high power capability. This type of capability, however, also requires more complex circuitry to ensure that the semiconductors turn off at the correct time.

The switching on and off of the transistor closes and opens a path for dc current to flow into the transformer. With the flow of current producing a changing magnetic field in the transformer primary, a changing magnetic field also develops in the transformer secondary winding. As a result, voltage is induced in the secondary winding. Rectifiers and filters in the secondary circuit rectify and filter into stable supply voltages.

⟳ FUNDAMENTAL REVIEW —————

MOSFETs, IGBJTs, and Triacs

MOSFETs (metal oxide semiconductor field-effect transistor) are special types of field-effect transistors. A FET has three terminals, labeled as the source, drain, and gate, and consists of a single p-type material and a single n-type material. When compared to bipolar-junction transistors, the source corresponds with the emitter, the drain corresponds with the collector, and the gate corresponds with the base.

A MOSFET has a low resistance when conducting, a very high resistance when not conducting, low leakage currents, low capacitance, and good high-frequency response. Depletion MOSFETs conduct with a forward, reverse, or zero gate bias, while enhancement MOSFETs must have a forward bias. MOSFETs operate as amplifiers, switches, and controlled-current devices.

Like the MOSFET, an insulated gate bipolar transistor (IGBT) is controlled by a gate voltage. In addition, the IGBT has a double diffusion of a p-type region and an n-type region. With the correct voltage applied to the gate, an inversion layer forms under the gate. The use of a positive p-type substrate layer for the drain causes the device to become bipolar. An IGBT turns on when the gate voltage increases to a level greater than the threshold voltage.

A triac is a three-terminal, five-layer device with the same type of forward and reverse operating characteristics as seen with the MOSFET. The primary conducting terminals of a triac are the main terminal 1, or MT1, and the main terminal 2, or MT2. In effect, the construction of a triac provides complementary SCRs connected in parallel.

☑ PROGRESS CHECK

You have completed Objective 1. Explain how switched-mode power supplies work.

SMPS Input After the rectification of the line voltage, the SMPS may have two possible dc inputs. With the first, 150 to 160-Vdc arrives at the SMPS after the direct rectification of 115 to 130-vac line voltage. Some SMPS units, however, require a higher input voltage. In this case, a voltage doubler supplies 300 to 320 Vdc to the SMPS input. Other designs rectifiy a 220 to 240-vac line voltage and also supply the 300 to 320 Vdc to the SMPS input.

While rectification of the line voltage occurs through the use of a full-wave bridge rectifier or a voltage doubler, the input to the SMPS also includes inductors and capacitors for the purpose of filtering line noise and any voltage spikes. Those components also eliminate the transmission of any radio-frequency interference generated by the power supply back into the ac line. As mentioned, most designs feature metal-oxide varistors across the input lines for additional protection against surges.

Switched-Mode Regulators Switched-mode regulator circuits provide the advantage of having a control device that has minimal power dissipation for the entire duty cycle. In particular, switched-mode regulator circuits provide:

- the capability to produce an output voltage higher than the input voltage
- the capability to produce either a positive or negative output voltage from a positive input voltage
- the capability to produce an output voltage from a dc input voltage

A switched-mode regulator circuit uses a control device such as a bipolar transistor, a field-effect transistor, or a silicon-controlled rectifier to switch the supply power in and out of the circuit and to regulate the voltage. Switching occurs because of the ability to send the device into either saturation, the completely on state, or into cut-off, the completely off state.

The duty cycle of the device, or the ratio of "on" time to "off" time, establishes the regulation of the output voltage level. Therefore, regulation in a switched-mode power supply occurs through the pulse-width modulation or the pulse-rate modulation of the dc voltage. **Pulse-width modulation** varies the duty cycle of the dc voltage while **pulse-rate modulation** varies the frequency of the dc pulses.

Figure 4-2 provides an illustration of pulse-width modulation. The on cycles of the pulse train energy double as the time periods for storing energy in a magnetic field. During the off cycles of the pulse train, the stored energy provides output power and compensates for any changes in the line voltage or the load. The pulse-width modulation of the switching transistor changes the conduction time of the device by varying the pulse frequency.

With a low load resistance and the line voltage within tolerance, the switched-mode power supply switches the power into the power supply for only a short period of time. Either a high load resistance, a low line voltage, or a combination of both conditions will cause the switched-mode power supply to transfer more energy over a longer period of time into the power supply. As a result, the switching frequency varies from a higher frequency for lower loads to a lower frequency for higher loads.

↻ FUNDAMENTAL REVIEW

Pulses

A pulse is a fast change from the reference level of a voltage or a current to a temporary level and then an equally fast change back to the original level.

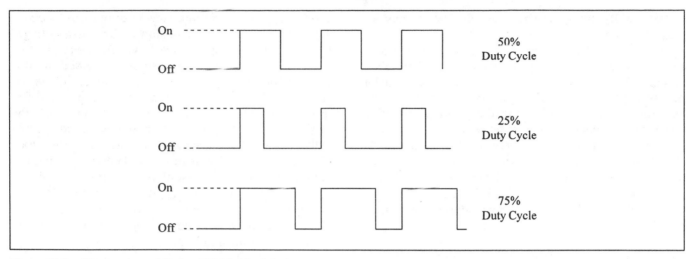

Figure 4-2   Illustration of Pulse-Width Modulation

SMPS Transformer Operation   Switched-mode power supplies do not include any type of conventional power transformer and, as a result, do not have line isolation. At the input of the power supply, a small, high-frequency transformer converts the pulsed waveform taken from the switching device into one or more output voltages. Other components following the high frequency rectify and filter the voltages for use by signal circuits.

☑ PROGRESS CHECK

You have completed Objective 2. Discuss switched-mode input, regulation, and transformer operation.

Isolation in the SMPS System   Although the SMPS does not provide line isolation, the use of the high-frequency transformer establishes an isolation barrier and the type of characteristics needed to operate in the flyback mode. Depending on the circuit configuration, a small pulse transformer or an optoisolator sets up feedback across the isolation barrier. An **optoisolator** is a combination of an LED and a photodiode in one package. The feedback controls the pulse width of the switching device and maintains regulation for the primary output of the SMPS.

Most small switched-mode power supplies such as those used for VCRs use optoisolators for feedback. The optoisolator establishes an isolation barrier between low-voltage secondary outputs and the ac line. Whenever a primary output voltage reaches a specified value, a reference circuit in the output turns on the LED. In turn, the photodiode detects the light from the LED and reduces the pulse width of the switching waveform. This establishes the correct amount of output power and maintains a constant output voltage.

Along with the primary output winding, the transformer has six or more separate windings that provide positive and negative voltages for the electronic system.

The Iso-Hot Grounding System

Illustrated in Figure 4-3, an **iso-hot chassis** combines cold and hot grounding systems through the connection of one portion of the receiver to earth ground and the other portion to the ac line ground. Because of the existence of the three different grounding systems, technicians must take special care when using test equipment and when moving from one system to the other. An iso-hot chassis combines the better points of the cold and hot chassis into a different power supply approach, allowing the isolated connection of video accessories such as video games and VCRs to the standard television receiver. Because of this, the iso-hot design continues to use an isolated transformerless system, but also includes a separate cold chassis ground. Any direct connection between the two types of grounding

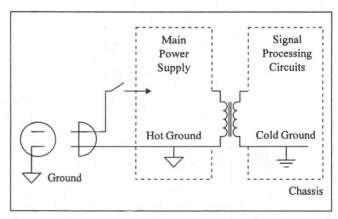

Figure 4-3   Schematic of an Iso-Hot Chassis

systems will damage the receiver as well as any attached test equipment.

Modern televisions provide a good example of how an iso-hot system affects ground connections within a system. Regardless of the manufacturer, the power supply and the horizontal output stage connect directly to ac power-line ground. With the floating ground, a resistor and capacitor connect between the ground and the hot line and allow dc noise and high-frequency ac signal voltages to pass to ground. The tuner, I-F stages, luminance and chrominance processing stages, and vertical stages connect to the cold chassis, which remains isolated from the main power supply and ac line.

PROGRESS CHECK

You have completed Objective 3. Discuss isolation in the SMPS system and the iso-hot grounding system.

4-2 SCAN-DERIVED POWER SUPPLIES

Sections 4-1 and 4-2 describe the fundamental theories used for switched-mode power supplies and the components commonly seen within the SMPS circuits. Those descriptions can be applied to switched-mode power supplies found in a large number of electronic devices. This basic SMPS design supplies low voltages for the device and operates at lower frequencies.

Scan-derived power supplies described in this and the following sections are derivatives of the basic SMPS design. Scan-derived power supplies differ in the application, the operating frequencies, and the method of providing a supply potential. Used specifically in television receivers and shown in the block diagram of Figure 4-4, scan-derived power supplies operate at the horizontal oscillator frequency of 15,750 Hz and supply much higher voltages and currents. The supply potential for the scan-derived power supply is taken from a portion of the horizontal output voltage in the form of voltage pulses.

Deriving Pulses from Scan Circuits

Because the horizontal scan section of a television receiver provides the pulses, the circuit is called a scan-derived power supply. The use of scan voltages depends on an energized horizontal oscillator that provides the drive voltage for the horizontal output circuits. Examining the design and operation of the power supply more closely, we find that it contains the SMPS circuits described earlier. In television systems, the use of a scan-derived power supply increases efficiency and cuts power consumption because of the capability to supply higher current loads.

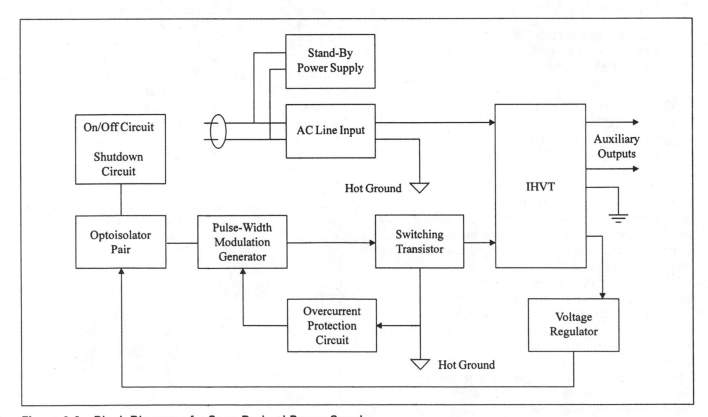

Figure 4-4   Block Diagram of a Scan-Derived Power Supply

A scan-derived power supply has four basic features:

- the rectification of the ac line voltage and the conversion of the rectified line voltage into an unregulated power supply
- the feeding of the unregulated power source into a dc regulator
- the use of a start-up supply circuit to supply voltage for the receiver horizontal oscillator
- the use of a high-voltage transformer, or flyback, to supply the high voltage needed for the receiver CRT and the low voltages for the receiver circuits

While you have become familiar with the first two, the last two features offer several new twists.

 ## PROGRESS CHECK

You have completed Objective 4. Describe the operation of scan-derived power supplies.

Providing a Start-up Voltage

Taking advantage of the excess energy from the horizontal scan section eliminates the need for both a low-voltage power transformer and anything more than a basic low-voltage rectifier circuit. However, the output circuits in a television cannot operate unless properly driven by the horizontal oscillator. This becomes more complicated when we find that the voltages needed for the operation of the oscillator come from the output stage.

To counter this situation, manufacturers include a start-up voltage circuit as part of the scan-derived power supply. The **start-up power supply** supplies a small amount of voltage or current to the horizontal oscillator so that the oscillator can energize and drive the output stage. The start-up voltage or current stops once the horizontal output stage begins to operate. Regardless of whether the start-up circuit relies on a kick-start or a trickle-start, the design has the same purpose.

Kick-Start Circuits A **kick-start circuit** supplies a small amount of voltage to the horizontal oscillator after the receiver is turned on. Referring to Figure 4-5A, the kick-start circuit includes a start transformer, T3274, and a large capacitor. The secondary of the transformer is connected in series with the capacitor. After the receiver is initialized, the charging pulses from the capacitor energize the windings of the transformer. In turn, the transformer delivers +26 Vdc startup voltage to the horizontal oscillator and driver circuit.

Multivibrator Start-up System Figure 4-5B shows another type of kick-start system. This system relies

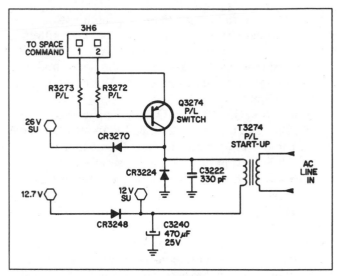

Figure 4-5A Kick-Start Circuit Used in an RCA Television (*Reprinted with permission of Thomson Electronics, Inc.*)

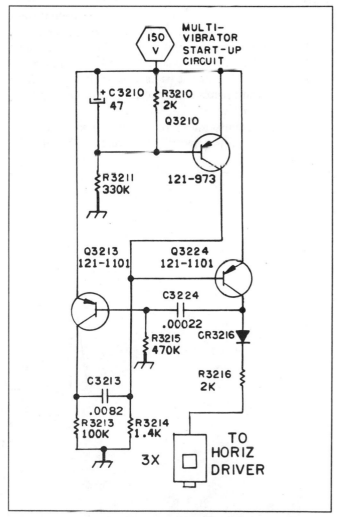

Figure 4-5B Multivibrator-Based Kick-Start Circuit (*Courtesy of Zenith Data Systems, Zenith Electronics, and Rauland*)

on a multivibrator rather than the combination of a start-up transformer, rectifier diodes, and filtering to produce the necessary start-up voltages. The multivibrator start-up system kick-starts the horizontal output transistor and the sweep circuit directly from the 150-Vdc supply.

Transistors Q3213 and Q3224, capacitors C3213 and C3224, and resistors 3213, 3214, and R3215 make up an astable PNP multivibrator circuit that operates off a 150-Vdc supply. The power supply develops at the turn-on of the television receiver and causes the multivibrator to oscillate at that point. During operation, the output voltage at the collector of Q3224 is a square wave with a value oscillating between the B+ voltage value and ground. The frequency of the oscillation varies with the ac line voltage and, with a normal line voltage, has a value between 10 and 20 KHz, while the square wave has a duty cycle of 70 percent on and 30 percent off.

The output current flows through diode CR3216 and resistor R3216 and connects through connector 3X to an added winding on the horizontal driver transformer of the television receiver. As the square wave current oscillates and flows through the winding, it also flows into the horizontal output transistor and causes the transistor to conduct. Once the transistor begins to conduct, the sweep circuit starts and develops the operating voltages for the receiver.

After the development of the operating voltages, the start-up circuit turns off. This occurs through the operation of transistor Q3210, C3210, and resistors R3210 and R3211. As soon as the charge across C3210 reaches a level of 0.7 Vdc, Q3210 begins to conduct because the emitter is more positive than the base. With the conduction of Q3210, the emitter and base of Q3224 short-circuit, Q3224 stops conducting, the oscillation of the multivibrator stops, and current through the start-up winding of the driver transformer ceases to flow.

⟳ FUNDAMENTAL REVIEW
Multivibrators

A multivibrator is a circuit that is designed to have zero, one, or two stable output states.

Trickle-Start Circuits   The **trickle-start circuit** shown in Figure 4-6 uses a slightly different method for starting the oscillator. A large resistor connects in series with either a diode or regulator transistor and in between the unregulated positive voltage obtained at the rectifier circuit and the horizontal oscillator. The value of the resistor is chosen so that it cannot supply enough current to operate the receiver. Instead, only enough current flows so that the oscillator is energized. Once the oscillator energizes and drives the output stage so that the output stage can supply dc power, the current ceases to flow through the resistor.

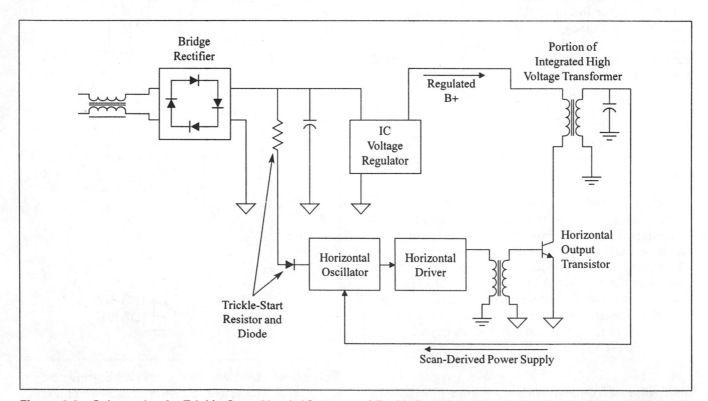

Figure 4-6   Schematic of a Trickle-Start Circuit (*Courtesy of Zenith Data Systems, Zenith Electronics, and Rauland*)

✓ PROGRESS CHECK

You have completed Objective 5. Describe the operation of start-up voltage circuits.

High-Voltage Transformers

In older television receivers, the high-voltage power supply featured a combination of a voltage multiplier—usually a tripler—and a separate high-voltage transformer called the **flyback**. This combination established the voltages needed at the CRT focus and high-voltage anodes, while a separate power supply provides the low voltages needed for the operation of the remainder of the receiver. Modern televisions use an integrated flyback transformer that combines the two older technology concepts into one package while also supplying auxiliary power supply voltages.

Monochrome and color televisions rely on almost the same type of horizontal output stage. A key difference between the two lies with the amount of voltage needed for the second anode of the CRT. While monochrome picture tubes require 10 to 15 kv at the second anode, color televisions require 25 to 30 kv. Because the two types of televisions use almost identical output sections, the flyback in a color television will have more turns and a slightly larger physical size.

In color televisions, the flyback performs three essential tasks:

- Provide the 25 kVdc high voltage required for the second anode of the CRT
- Provide sufficient current for the three electron beams generated within a color CRT
- Regulate the high voltage

While the high voltage generated through the operation of the horizontal output stage and the flyback feeds into the CRT second anode, a portion of the voltage is also applied as focusing potential through a bleeder network. The unregulated focus voltage usually ranges from 10 to 12 kVdc and requires some adjustment through a potentiometer located on the transformer.

While the focus voltage is unregulated, the high voltage at the CRT second anode requires precise regulation; it must remain constant despite any variations in current because the second anode serves three electron guns. If we consider the separate operation of each gun and then the electron guns as a set, the reason for precise regulation becomes more apparent. The red gun, for example, may draw a substantial amount of current during the display of a large, bright, red area on the screen. Allowing the second anode voltage to decrease because of the current usage would cause the blue and green guns to operate at lower potential and would lower the brightness of any color associated with

green or blue. Therefore, the current drain due to one or two of the electron guns must not affect the voltage available for the remaining electron beam or beams.

A Typical High-Voltage Transformer Circuit   Figure 4-7 shows a schematic diagram of a common horizontal scan circuit and the flyback transformer. At the top of the high-voltage winding, a series of pulses delivered during the retrace interval combines with stored energy in the yoke coils to build the high voltage. The yoke voltage pulses multiply by the turns ratio of the high voltage winding to the primary winding and establish the peak voltage.

Again referring to the diagram, the high-voltage supply is constructed around a direct-coupled half-wave rectifier circuit. A single diode rectifies the pulses at the top of the transformer winding. Then the capacitance of the CRT anode region works as a storage device and stores the high-voltage charge.

Because of the large step-up from the primary voltage to the secondary high voltages, the flyback has higher leakage inductance in the high-voltage winding and lessened transformer efficiency. To counteract the decrease in efficiency, the circuit relies on harmonic tuning to increase the efficiency of the total inductance and distributed capacitance of the winding and improve the circuit regulation. The flyback transformer shown in this schematic relies on third harmonic resonance of the flyback pulse frequency to achieve the proper leakage inductance and establish the correct pulse waveform.

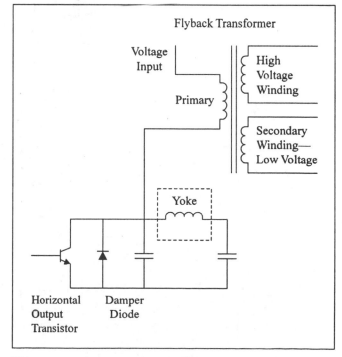

Figure 4-7   Schematic of a Horizontal Scan Circuit Including the Flyback Transformer

Integrated High-Voltage Transformers In a television receiver, the operation of the scan-derived power supply also depends on the use of an **integrated high voltage transformer**, or IHVT. If we take the functions of the flyback transformer and the high-voltage tripler and combine them into one package, we have the essential ingredients for an integrated high-voltage transformer, or as some manufacturers label the device, an integrated flyback transformer. An integrated high-voltage transformer segments the high-voltage windings into several parallel-wound sections that series-connect with one another through diodes. One housing contains both the segments and the diodes. The pulse rectification provided by the IHVT produces all the voltages needed by the receiver chassis.

☑ PROGRESS CHECK

You have completed Objective 6. Define high-voltage transformers.

Auxiliary Power Supplies

Figure 4-8 uses a block diagram of a color television power supply system to illustrate how a system operating at the horizontal sweep frequency can establish the low and high voltages needed for the entire receiver. The integrated flyback transformer is one part of the horizontal-output switched-mode power supply. A 6.3-vac winding on the flyback transformer drives the CRT filament, while a horizontal pulse from another winding on the flyback resets an SCR operating as a control device in another switched-mode power supply. Because the horizontal pulse forces the SCR anode to fall below the level of the cathode during retrace, the SCR resets every horizontal cycle. Turn-on of the SCR occurs during the next horizontal scan period. All this leads to the generation of auxiliary voltage supplies for the:

- tuner and tuner control system
- I-F signal processing systems

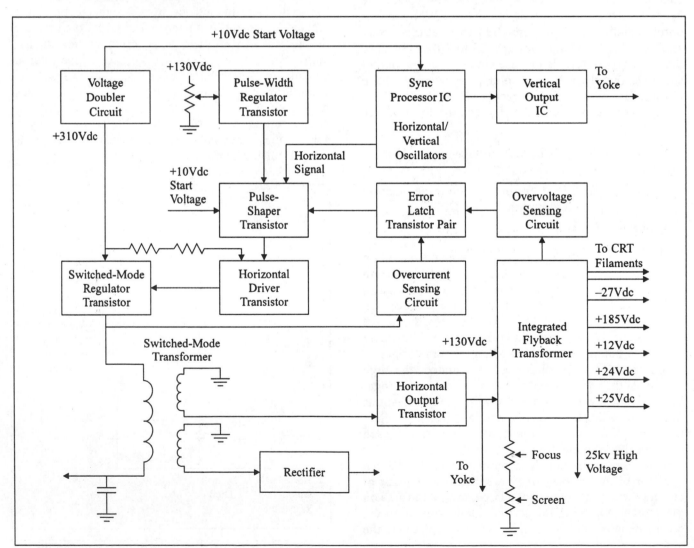

Figure 4-8 Block Diagram of a Modern Color Television Power Supply System

- video and chrominance processing systems
- audio systems
- vertical deflection systems
- CRT anode and focus voltages
- CRT screen, or G2, control

Certainly, each type of system requires a slightly different power supply. To accommodate these differences, scan-derived systems utilize transformer winding and grounding schemes to establish different supply voltages. The voltage supplies utilize the waveform shown in Figure 4-9 to produce a:

- positive-retrace rectified supply
- positive-scan rectified supply
- negative-polarity-retrace rectified supply
- negative-polarity-scan rectified supply

The figure shows which points of the waveform correspond with the four power supply types.

The positive-retrace rectified supply rectifies a large portion of the waveform to yield a high voltage with the low-current loading used to supply dc voltages for the video and chroma output circuits. Reversing the direction of the flyback winding produces a mirror image of the waveform as shown in Figure 4-9, and the lower-voltage, higher-current supply needed for the vertical output circuits, I-F signal processing circuits, and audio circuits. While the negative-polarity-retrace rectified supply continues to utilize a reversed flyback winding, it rectifies negative-going retrace pulses, and produces the negative-polarity high voltage with low-current loading needed by the tuner and tuner control systems. The negative-polarity-scan rectified generates low voltages with a higher current drain.

Looking next at Figure 4-10, the +12, +24, +25, and −27 Vdc supplies are scan-rectified, while the +185 Vdc, the overvoltage-sensing voltage, the focus voltage, and the 25k CRT anode voltage are retrace-rectified voltages. The two types of scan-rectified supplies operate at a duty cycle of approximately 80 percent and produce higher current loads. Because of this, diodes used in the scan-rectified supplies must be able to block reverse voltages that have nine to ten times the amplitude of the output voltage. In addition, to minimize power dissipation during the turn-off interval, the diodes must have fast recovery characteristics.

Overvoltage Protection Circuits

In a television receiver, the fast-moving electron beam strikes the phosphor on the inside of the CRT screen, and expended energy in the form of light photons exits from the picture tube as low-intensity current. With larger picture tubes, the anode voltage has levels of 25 to 35 kv, and the combination of beam current and anode voltage generates X rays. If the X-ray levels reach above 1015 Hz, the emissions can harm human tissue.

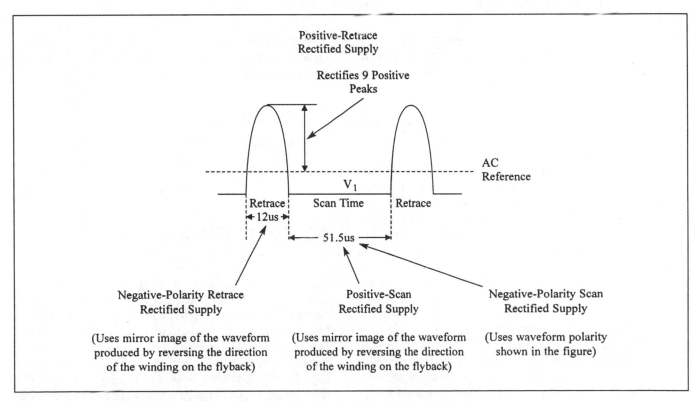

Figure 4-9 Auxiliary Supply Voltage Waveform Taken from the Flyback Transformer

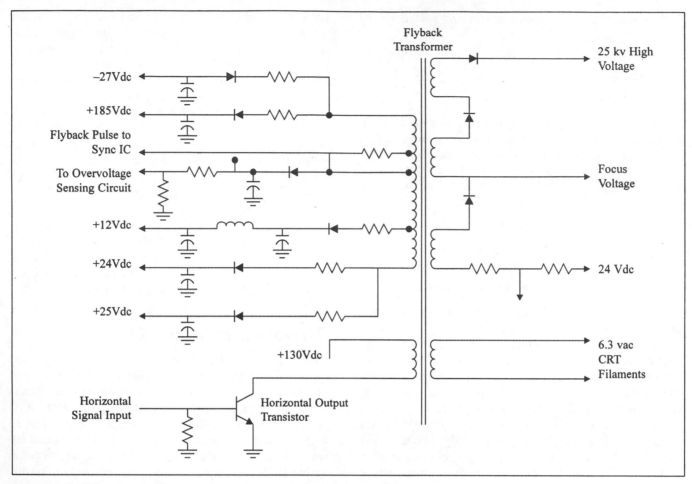

Figure 4-10 Schematic Diagram of a Scan-Rectified Auxiliary Power Supply

As part of the federal government regulation of the video industry, all receivers have strict limits regarding the level of X-ray emission. Those limits involve the regulation and amount of the high voltage found at the CRT anode.

Overvoltage and overcurrent protection circuits shut down the receiver horizontal deflection system if the high voltage goes beyond the specified level for a particular CRT. It does this by linking the high- and low-voltage supplies, as an increase of the high voltage beyond the specified level causes the low voltage increase. At the overvoltage threshold value, the conduction of semiconductor devices in the protection circuit either shuts down the horizontal driver stage, disables the high-voltage stage, or lowers the high voltage to an acceptable level. An overcurrent protection circuit monitors the amount of current flowing through the high-voltage winding and shuts down the horizontal driver if the amount exceeds the specified level.

Other overvoltage and overcurrent protection circuits either shut down the horizontal oscillator or push the 15,750-Hz horizontal frequency off-scale. As the high voltage begins to rise above the acceptable level, the picture breaks into diagonal bars. When the high

voltage exceeds the minimum safe level, the protection circuit shuts the oscillator circuit down and stops the generation of high voltage.

☑ PROGRESS CHECK

You have completed Objective 7. Explain overvoltage and overcurrent protection in SMPS circuits.

4-3 SWITCHED-MODE POWER SUPPLY CIRCUITS

Figures 4-11A, B, and C illustrate the operation a step-down switched-mode regulator, a step-up switched-mode regulator, and an inverting switched-mode regulator through equivalent circuits. In each of the circuits, the inductor stores energy when the switch—representing the control device—closes. In Figure 4-11A, opening the switch allows diode D1 to establish a path for the flow of I_{L1}, or the load current. The circuit transfers energy during the time that the switch is open.

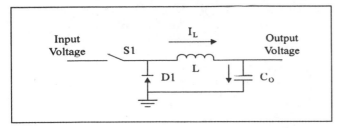

Figure 4-11A Step-Down Switched-Mode Regulator Equivalent Circuit

In Figure 4-11B, the coil represents the primary of a transformer. The dc-isolated secondary of the transformer provides the output voltage and current for the circuit. Depending on the design needs, the output voltage and current may set up either an alternating current or a rectified and filtered direct current. Defined as a parallel switched-mode regulator, the circuit shown in Figure 4-11C often appears as the flyback circuit for a color television receiver. A switched-mode regulator used as a flyback transformer in a color television receiver operates at the horizontal frequency rate of 15,750 Hz.

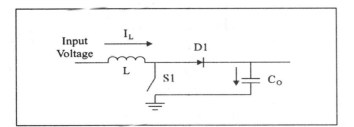

Figure 4-11B Step-Up Switched-Mode Regulator Equivalent Circuit

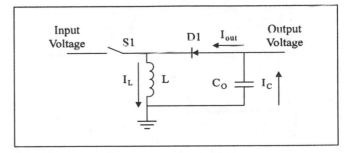

Figure 4-11C Inverting Switched-Mode Regulator Equivalent Circuit

Figures 4-12, 4-13 and 4-14 illustrate circuit used in a large-screen Sylvania television receiver. Looking first at the Figure 4-12, the voltage doubler develops the +310 Vdc main voltage source for the C5 series chassis. Moving to Figures 4-13 and 4-14, we find that the +310 Vdc serves as the source voltage for Q463, the horizontal driver transistor, and Q464, the switched-mode regulator transistor. In addition, the forward bias for the horizontal driver transistor is also taken from the main voltage supply.

At the lower center of Figure 4-13, the ac input voltage applies to T453, the start transformer, which also provides isolation between the chassis and ground. The two ground symbols found at T453 illustrate the isolation occurring between ac ground and chassis. Taking a careful look at all three figures, we find that the switched-mode power supply and the horizontal drive system reference to ac ground. IC340, the sync processor; Q457, the horizontal coupler driver; the input to IC469, the horizontal optic coupler; and the output voltages from T408, the integrated flyback transformer, reference to chassis ground. Figure 4-15

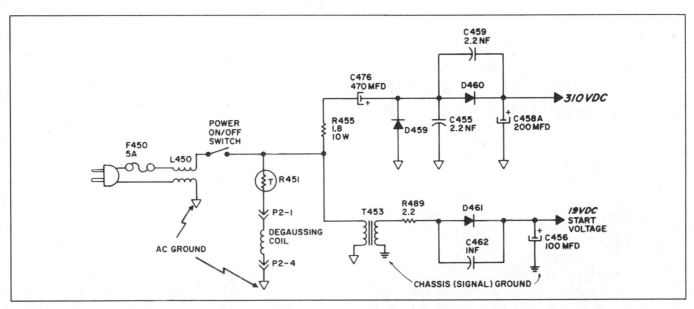

Figure 4-12 Schematic of a Sylvania Projection Television Start-Up Circuit (*Courtesy of Philips Electronics N.V.*)

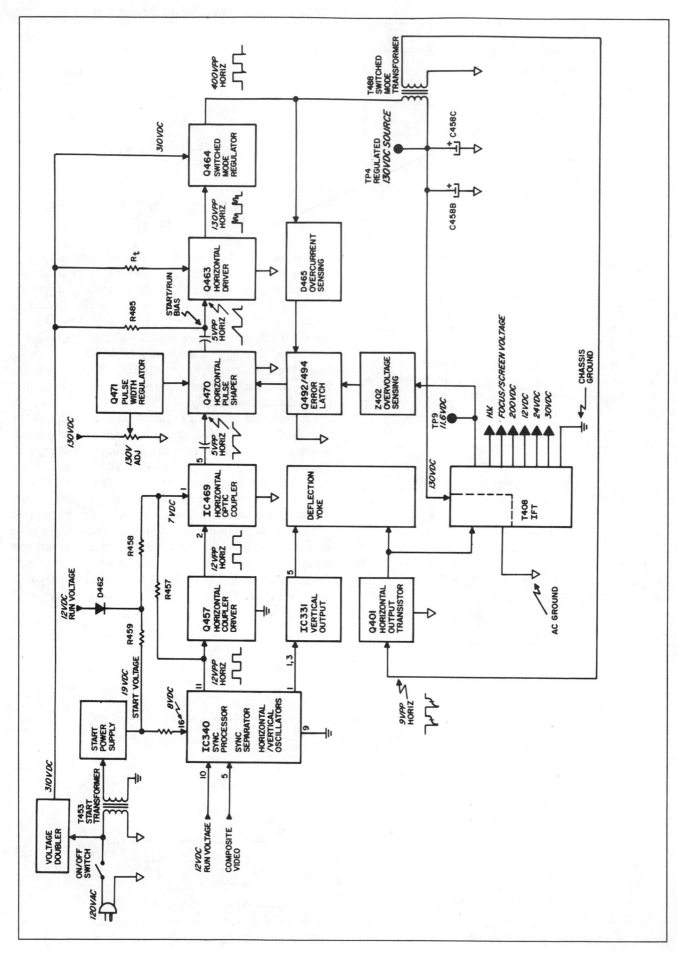

Figure 4-13   Block Diagram of a Sylvania Projection Television Switched-Mode Power Supply *(Courtesy of Philips Electronics N.V.)*

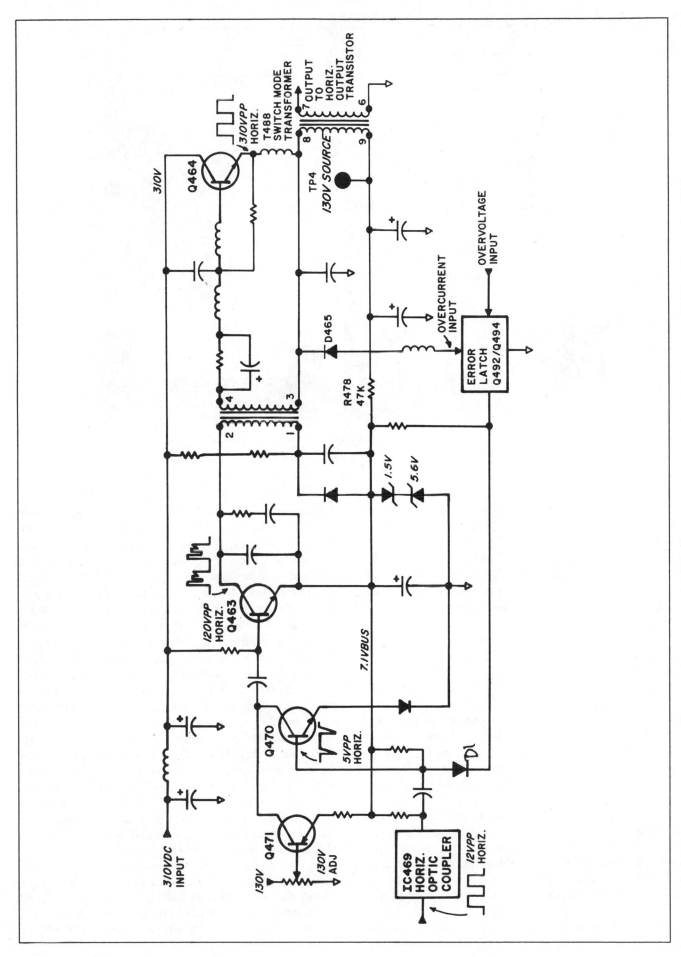

Figure 4-14 Schematic Diagram of a Sylvania Projection Television Switched-Mode Power Supply (*Courtesy of Philips Electronics N.V.*)

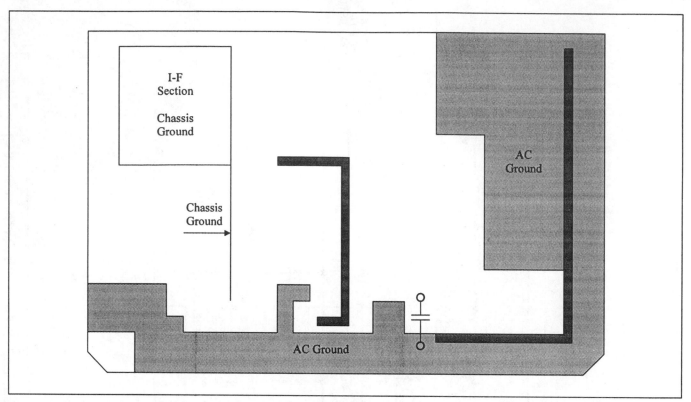

Figure 4-15   Schematic Diagram of a Sylvania Projection Television Ground System (*Courtesy of Philips Electronics N.V.*)

shows a diagram of the ground system, with the shaded areas representing ac ground and the I-F section representing chassis ground.

The combination of D459, D460, C458, and C476 rectifies and filters the ac line voltage applied to T453 and provides a start voltage of +19 Vdc. With the start voltage applied to the sync processor through R334, the horizontal oscillator within the IC begins to output a low-level horizontal square wave signal at pin 11, driving the switched-mode power supply.

The start voltage is also applied to the horizontal optocoupler during this period. The horizontal optocoupler provides further isolation between the two ground systems, and contains a phototransistor. Because the system uses a feedback loop, the optoisolator is necessary to isolate the dc output from the ac line. While this particular system features the optocoupler, other switched-mode power supply designs rely on a small transformer for isolation.

Moving back to the sync processor, the low-level horizontal square wave signal travels from the IC to Q457, the horizontal coupler driver. As the coupler driver conducts, it drives a light-emitting diode within the horizontal optic coupler. The phototransistor contained within the coupler outputs a signal that couples to Q470, the horizontal pulse shaper.

Q471, the pulse-width regulator, determines the on and off time for the horizontal signal by monitoring

the condition of the +130 Vdc source voltage, and controls the operation of Q470. The pulse-width regulator monitors, adjusts, and regulates the +130 Vdc source voltage because the voltage is directly derived from the switched-mode power supply. Once the horizontal oscillator generates the horizontal square signal, the signal drives the switched-mode transformer. While capacitors C458B and C458C filter the square wave and produce an average dc voltage level, the on/off ratio of the square wave regulates the voltage. The on time of the switched-mode regulator transistor determines the amount of voltage found across the filter capacitors, with long on time producing a larger voltage.

In addition to the input from Q471, Q470 also receives another input from the combination of Q492 and Q494, an error latch. The error latch shuts the chassis down if an overvoltage or overcurrent condition occurs. Thus, the horizontal pulse-shaper transistor has a controlled output that depends on the correct source voltage from the switched-mode power supply, the correct waveshape of the horizontal square wave, and the correct voltage and current conditions in the high-voltage section of the receiver. This controlled output couples to the base of the horizontal driver transistor; the driver shapes the signal and drives Q464. From there, the switched-mode regulator drives Q401, the horizontal output transistor, through T488, the switched-mode transformer.

The horizontal output transistor drives the IHVT which, in turn, produces the high and low dc voltages required throughout the chassis. In addition, Q401 drives the yoke horizontal deflection coils. While the scan-derived voltages increase to the correct values, and the circuit applies +12 Vdc source voltage to pin 10 of the sync processor, the IFT switches from a start condition to a run condition. This switching action produces the vertical drive signal found at pin 1 of the sync processor. IC331, the vertical output IC, amplifies the vertical drive signal and applies the signal to the yoke vertical deflection coils.

In this circuit, the horizontal output transistor drives the primary of the IHVT. In Figure 4-16, the processed horizontal signal travels from the 130 Vdc supply to pin 1 of the IHVT, while the horizontal drive feeds into the base of the horizontal output transistor and produces the drive for the IHVT. The transistor also contains a built-in damper diode and connects to the IHVT primary at pin 5 of the transformer. The output from pin 14

of the transformer feeds back to the sync processor and maintains the frequency locking of the automatic frequency control system.

☑ PROGRESS CHECK

You have completed Objective 8. Analyze switched-mode power supply circuits.

4-4 TROUBLESHOOTING SWITCHED-MODE POWER SUPPLY PROBLEMS

Many technicians consider troubleshooting a switched-mode power supply as the most difficult troubleshooting task. Part of this feeling stems from the interdependence of the components found in the supply. Proper operation of the supply requires that the components function as a unit; the failure of an SMPS often claims a number of components in the supply.

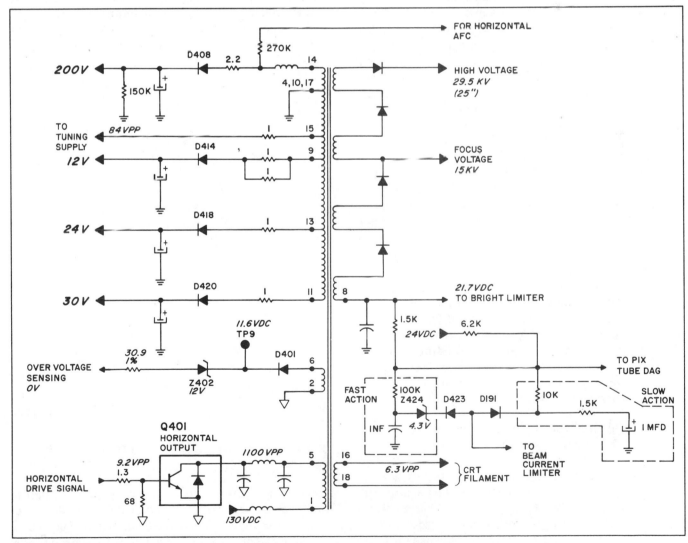

Figure 4-16   Schematic of the Sylvania IHVT Circuit (*Courtesy of Philips Electronics N.V.*)

In addition, many SMPS units do not have any type of overload protection and may suffer a catastrophic failure under heavy load conditions. For example, a heavy load condition may place switching devices such as bipolar transistors under additional stress and cause an early failure. Many times, a power line spike that occurs during turn-on can destroy the switcher.

Safety First!

Every video receiver is full of potential sources for serious electric shock. In televisions, the high dc voltage found at the CRT second anode and the high voltage output stages of the receiver can cause a jolting shock. While the high voltage supply has a low current and relatively poor regulation, your reaction to the shock may cause you to either break other components in the receiver or come in contact with other power sources.

Low-voltage power supplies have much lower dc output voltages but much higher current levels. As a result, any contact between the supply and ground can cause a serious electric shock. The same warning holds true for any exposed power line voltage. With most receivers using a hot chassis ground, any contact between the ac line hot ground and the cold chassis ground can cause a life-threatening electrical shock. In addition, contact between the chassis ground of test equipment and the ac hot ground will ruin the equipment.

Switched-mode power supplies contain connections and components that can provide a lethal shock. When troubleshooting an SMPS, always remember that the input side of the supply connects directly to the ac line. In addition, this type of supply always includes large electrolytic capacitors that may charge to more than +300 Vdc.

When attaching test equipment to an SMPS, always ensure that the test probe is securely fastened to the component under test. General test procedures call for connecting or disconnecting any test leads with the unit under test unpowered and unplugged. If you must connect test probes under live conditions, cover all but the tip of the probe with electrical tape. In addition, clip the reference, or ground, of the multimeter or oscilloscope to an appropriate ground point so that only one hand is required when testing the circuit.

Discharging the Main Filter Capacitors An SMPS should discharge capacitors quickly when powered off. Good test procedures, however, always require the discharge of the filter capacitors. While the capacitors connect to ground through bleeder resistors and should drain quickly, the resistors can fail. To discharge a filter capacitor, connect a high-wattage resistor with a value that matches the working voltage of the capacitor from the positive terminal to ground. As an example of the

appropriate resistor value to use, a 2-kilohm 10-watt resistor will discharge a 400-uf 200-Vdc capacitor.

Rather than connecting the resistor to the capacitor by hand, use either a capacitor discharge tool or a capacitor discharge indicator circuit. You can construct a capacitor discharge tool by soldering one end of an appropriately sized resistor to a well-insulated, 2- to 3-foot long clip lead. Solder the other end of the resistor to a well-insulated contact point. In most cases, a length of #14 bare copper wire running through an insulator will work as the contact point. An insulator such as a small PVC pipe will also work as a handle for the discharge tool.

You can supplement the discharge tool with a discharge indicator circuit. By building two sets of four 1N4007 diodes and two LEDs into the circuit shown in Figure 4-17, you can have a probe that gives a visual indication of charge and the polarity from a maximum level down to a few volts. The diodes will provide a near-constant voltage drop of about 2.8 to 3 Vdc across the positive LED as long as the input to the circuit remains around +20 Vdc. The brightness of the LEDs will not decrease until the input voltage drops below +20 Vdc, and will extinguish at an input voltage of approximately +3 Vdc.

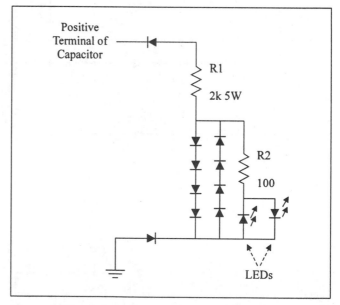

Figure 4-17   Schematic of a Capacitor Discharge Tool

Troubleshooting SMPS Systems with Test Equipment

Even though you can rely on your sight and smell to find damaged components in a switched-mode power supply, the application of test equipment becomes a valuable resource for locating a defective part or parts. The efficient troubleshooting of an SMPS requires the use of a variable ac transformer, a multimeter, an oscilloscope, a semiconductor tester, and a capacitor tester.

Variable AC Transformers   An isolation transformer and a variable ac transformer, or **variac**, are two different pieces of test equipment. While the variac allows the varying of the line voltage, it does not provide any isolation. When troubleshooting SMPS systems, a variac provides an easy method for testing a circuit or unit without applying the full line voltage, and for testing the ability of system to regulate. After repairing an SMPS, use the variac to run the input voltage for the SMPS up to 1.2 times the normal ac line voltage and then down to nearly half the line voltage.

Dummy Loads   Any test of an SMPS should be conducted without the connection of the original load. The use of a dummy load protects not only the circuits that follow the power supply but also the power supply in case a fault exists in the original load. Many technicians use a series light bulb as a dummy load during the testing of switching power supplies.

When working with dummy loads, connect a load to each of the supply lines found in the SMPS; some SMPS designs will not initialize without a load on each line. A test of the power supply under operating conditions usually requires a load that is about 20 percent of the original, full load. Then using the variac, slowly increase the input voltage to the supply. The primary

capacitors should charge and disclose any possible shorted conditions or open capacitors.

Test Equipment Primer— The Oscilloscope

An oscilloscope works as a useful, versatile addition to the test bench for any technician attempting to trace or monitor electrical signals. With an oscilloscope, a technician can check the display of a waveform and look for variations in amplitude and time; determine phase relationships between voltages and currents; or look at the frequency response of a circuit. When using an oscilloscope, notice that waveforms display as a plot or graph of the variations. The horizontal movement of the oscilloscope trace is proportional to time, while the vertical movement of the trace is proportional to voltage.

Figure 4-18A shows the front panel of a basic oscilloscope, which features a display and a set of controls. Interestingly, the oscilloscope has many of the features commonly associated with a television receiver. For example, a signal feeding into the oscilloscope causes the device to display waveforms. Controls allow the user to set the intensity, focus, and horizontal and vertical centering of the displayed waveform. Also, the oscilloscope requires some type of synchronization for the proper stabilization of the display.

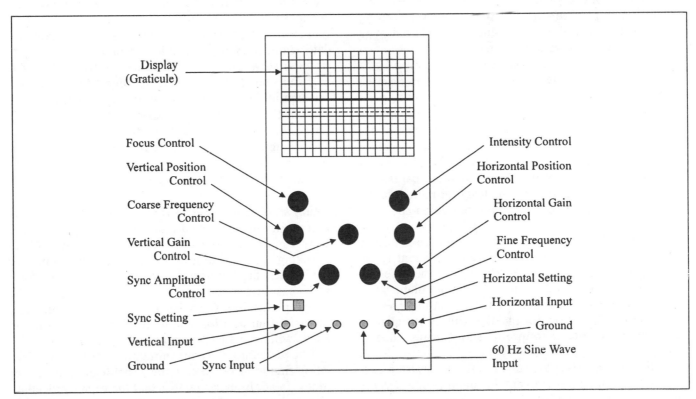

Figure 4-18A   Front Panel of a Basic Oscilloscope

Aside from those similarities with a television receiver, the oscilloscope offers some unique functions. An internal sweep oscillator produces a sawtooth wave and the horizontal trace of the CRT beam. The action of the sawtooth wave produces linear deflection and sets up a time base for the displayed signal. If the time base has one-half the frequency of a measured sinewave, the oscilloscope will show two cycles of the sinewave. Vertical deflection is controlled by the amplitude of the test signal. Vertical amplifiers allow the oscilloscope to have the capability to measure the bandwidth of a signal. Typical oscilloscopes have a vertical response ranging from 5 to 200 Mhz.

The frequency adjustments allow you to set a frequency range and phase for the oscillator. Depending on the application, you can set the horizontal selector switch so that the horizontal trace is generated internally or externally. For accurate display of a waveform, adjust the frequency of the sweep oscillator so that it corresponds to the test signal. The display signal will appear to become stationary if the sweep oscillator runs at either the same or a submultiple of the test signal frequency. Most oscilloscopes will offer the capability to simultaneously display and compare two or more traces.

A sync selector switch allows you to select different sync sources for the control of the oscillator. An internal sync setting causes the oscillator to self-synchronize with a sample of the test signal. A line— or 60-Hz setting—synchronizes the oscillator with the 60-Hz cycle of the power line. An external sync setting provides the flexibility of connecting external sync sources directly to the oscillator.

The display of the oscilloscope is divided into a grid called a **graticule**. Because the oscilloscope offers a linear display, the device can display the peak-to-peak value of any test signal. Those values are measured as volts per centimeter of vertical deflection. In addition, video waveform monitors and oscilloscopes feature a dotted line across the display face that represents the same luminance setup level. In most cases, the setup level, referred to as the pedestal or picture black level, signifies the difference between video and sync signals.

Because of the different types of signal and circuit conditions, oscilloscopes also feature different types of probes. A **demodulator probe** allows the observation of modulated R-F signals by demodulating the signal before its application to the vertical input terminals. Rather than allow for the checking of the quality of the signal, demodulator probes show the presence of a signal.

Other probes offer a built-in isolation resistor or a low-value capacitor to prevent the loading of the circuit under test. High- and low-impedance probes that contain various values of capacitors allow the testing of high- and low-frequency circuits. While attenuator probes include a range switch for the proper attentuation of an input signal, direct probes allow a straight-through connection to the test circuit.

Each of these probes allows the technician to test circuit operation through signal tracing and the logical process of deduction. With this technique, a technician can follow the stage-by-stage progression of a signal and find the point where the signal weakens or disappears. Signal tracing begins with a measurement at the input of a stage and then a measurement at the output of the same stage. A normal response at the output shows that the stage has no problems. A weak or absent signal at the output indicates that a problem exists within the stage.

Schematic diagrams often show key test points for a circuit as well as the normal waveform and amplitude for the signal at that point. In addition, knowledge about the operation of particular stages also becomes increasingly important. For example, a technician should know about the amount of gain expected from a stage. High-gain R-F and video stages will have a higher amount of amplification than power output amplifiers.

Many times, the use of signal injection increases the chance for the accurate measurement of gain. While the changes in amplitude seen with normal operating signals may make the checking of a waveform difficult, an injected signal provides a steady signal source. With the signal injected at the input of the stage, a measurement can be taken at the stage output.

In the example shown in Figure 4-18B, the demodulator probe is used to check for the presence of a signal at the first two test points. Because of the circuit loading and detuning caused by the demodulator probe, an amplitude measurement is nearly impossible. At the third test point, a resistance probe provides low-pass filtering and sharpens the display. The use of the low-capacitance probe at the video amplifier stage output minimizes circuit loading while allowing the technician to check for the proper amount of gain.

Using an Oscilloscope to Test the SMPS Given the complexity of a switched-mode power supply, a wideband oscilloscope becomes especially useful. When checking the operating performance of the switching device, use the oscilloscope to evaluate the waveform at the base and collector of a switching transistor or the drain of a MOSFET. For scan-derived power supplies, the use of an oscilloscope to confirm that oscillation exists will save a large amount of repair time.

The procedure for confirming the presence of oscillation first involves setting the variable ac transformer to 0 v. In addition, remember that the repair procedure also involves working on the "hot" side of the iso-hot chassis. Always verify that the connections of the

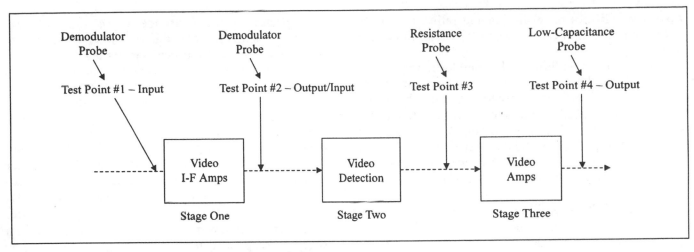

Figure 4-18B   Example of Signal-Tracing Test Points

test equipment attach to the proper ground. If testing a switching transistor, attach the oscilloscope test probe to the collector of the transistor. If testing a switching MOSFET, attach the oscilloscope test probe to the drain.

Slowly increase the variable ac transformer setting during the measurement at the switcher and monitor the shape of the waveform. The measured waveform should resemble the waveform in Figure 4-19. While increasing the variac setting, also listen for squealing noises. Any type of unusual noise indicates that other problems exist in the scan-derived power supply.

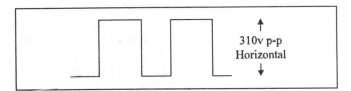

Figure 4-19   Waveform Found at the SMPS Switcher (*Courtesy of Philips Electronics N.V.*)

Typical SMPS Problems and Symptoms

Troubleshooting an SMPS problem requires a consistent problem-solving procedure. That procedure should include:

1. Preliminary check of the B+ voltage
2. Verification of the presence of start-up voltages in a scan-derived power supply
3. Verification of the presence of oscillation in a scan-derived power supply
4. Check of the SMPS output voltage
5. Check for regulation

By checking for the presence of B+, you can narrow the search for the problem source from the entire SMPS

to the switching device, the bridge rectifier, or the transformer. The additional check for a start-up voltage in a scan-derived power supply discloses whether or not the power supply has the proper voltage-current source. Check also for the presence of oscillation in the scan-derived power supply and the appearance of the waveform. After verifying the operation of the SMPS, check for proper voltage levels throughout the power supply and for the proper regulation of the output voltage.

Common Problems   Typical problems causing the failure of a switched-mode power supply include blown supply fuses, open fusible resistors, high amounts of ripple in one or more output lines, an audible whine with a lower than normal voltage at one output, and intermittent power cycling. In many cases, bad solder connections within the SMPS can cause symptoms to appear that mimic component-caused failures.

The blown supply fuse may occur because of a shorted switched-mode power transistor or other semiconductors found in the supply. While a fault in the start-up circuit for the supply may cause fusible resistors to open and shut down the supply, the main power supply fuse will not open. When considering ripple in the output lines, check for ripple at the line frequency of 60 Hz or at the switching frequency of 10-kHz or more. A dried filter capacitor connected in the main supply will cause an output line to have a 60 Hz ripple, while a dried filter capacitor connected in a specific output line will cause the higher frequency ripple.

The last two symptoms—audible whine with a lower than normal voltage and periodic power cycling—involve shorted semiconductors, a fault in the regulator circuitry, a fault in the overvoltage sensing circuitry, or a bad controller. Usually, the failure of a switching transistor is accompanied by the failure of other semiconductors in the circuit. At times, though, a switching transistor will not have the voltage rating needed to withstand the strain caused by the constant on and off switching.

As you have read through the troubleshooting materials, it should become obvious that a large number of problems can cause the receiver to go into shutdown. Troubleshooting the SMPS involves signals that begin at the ac voltage input, the horizontal oscillator, and voltage doubler circuits. As the switched-mode power supply operates, an integrated flyback transformer produces dc voltages for the remainder of the chassis.

Locating SMPS Switching Problems

If the SMPS utilizes a power transistor as a switching device and the power supply fails, always test the transistor for shorted and open junctions. The partial failure of a switching transistor often results from leakage or a change within the operating parameters of the semiconductor. Most new SMPS units rely on either an SCR or a MOSFET as a switching device. Testing either an SCR or MOSFET requires a multimeter for basic tests such as a shorted condition, and additional test equipment for any other tests. When replacing a switching device, always use an exact replacement as recommended by the manufacturer of the SMPS.

SMPS Capacitor Problems

Any switched-mode power supply design allows a large amount of current to flow through electrolytic capacitors. In some cases, the repeated operation of the SMPS system will cause the capacitors to short internally or develop an intermediate open condition. Under high load conditions, a capacitor may open and then "heal" at line rates. Many times, discoloration or a slightly bulged appearance will show that the capacitor has begun to overload.

Figure 4-20 shows a pi filter normally found on the auxiliary power supply lines of an SMPS. When the capacitor on the input side of the filter fails, the inductor absorbs most of the switching voltage from the transformer and the rectifier diodes. As a result, the regulator transistor works harder to generate the +5 Vdc desired at the output. With this additional load on the power supply, other capacitors in the supply begin to open.

SMPS Power-Cycling Problems

Many SMPS problems involve a dead supply and a sound that either resembles a *tweet-tweet-tweet* or a *flub-flub-flub*. In addition, a fault of this nature may cause display LEDs to flash or, with televisions, may allow a partial raster to appear. Most power-cycling problems result from a shorted component in the auxiliary power supply. Those components include diodes, capacitors, and SCRs in the overvoltage crowbar circuit.

A failure in the overvoltage sensing circuit will also cause a power-cycling problem. If suspecting a failure of this type, check the SCR in the crowbar circuit. Low-power SCRs often operate as control devices in the crowbar circuits used for overvoltage and overcurrent protection. Generally, out-of-circuit tests on the SCRs will disclose any faults. If the power supply fails to energize because of a failure in the protection circuits, the problem may trace back to a shorted SCR.

After checking the component, check for any short-circuit conditions on the output lines connected to the SCR. Then remove the SCR and use a variac to slowly increase the input voltage while monitoring the voltage at the output line. Checking the voltage at this point will show whether the voltage is going past the overvoltage level, remaining clamped at a low level, or staying at the correct level under normal load conditions. A momentary overvoltage spike at receiver turn-on will cause the overvoltage circuit to react.

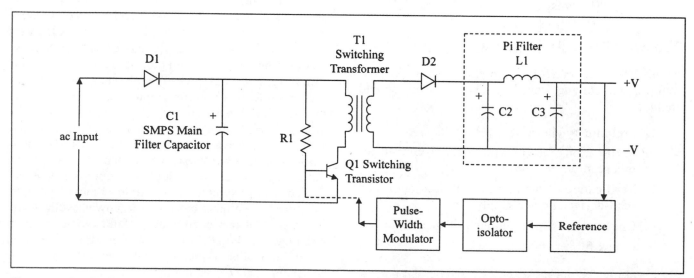

Figure 4-20 Pi Filter Found in the Auxiliary Power Supply of an SMPS (*Courtesy of Philips Electronics N.V.*)

Finally, aside from checking the components in an SMPS, always use your senses to find a possible fault. In addition, after checking the fuse and unplugging the unit, take a moment to thoroughly dust the power supply with a soft cloth and Q-tips. Many times, dust will cause the failure of a power supply. Survey the power supply for any possible open circuit conditions, any possible paths for a short circuit, and burned components. After completing those checks, disconnect the secondary loads from the supply. Then use an external voltage supply to individually power up each auxiliary circuit.

Testing the SMPS After the Repair

After repairing an SMPS, replace the load normally connected to the supply with a dummy load. The primary load requires the most consideration, with smaller loads applied to the other outputs. Depending on the application, the load should vary from 2 to 3 ohms at 15 watts, to 25 to 50 ohms at 2 to 5 watts. Personal computers usually have a smaller load on a +5 Vdc output, while VCRs have a larger load on a +12 Vdc output. Using a variac, slowly increase the input voltage while observing the main output voltage. As the input voltages reaches 50 percent of its normal level, the output should reach or surpass its normal operating value.

Troubleshooting Auxiliary Power Supply Problems

Often the shutdown symptom occurs because of a defect in the auxiliary power supply section. With linear power supplies, separate windings of the power transformer secondary may supply different B+ voltages. A switched-mode power supply usually has low-voltage power supply circuits operating from voltages taken from the flyback transformer. The parts of the receiver tied to the flyback secondary range from the tuner and audio sections to the horizontal driver and output sections. When you consider either type of auxiliary power supply and the associated circuits, note the number of diodes and electrolytic filter capacitors used in those circuits. A defect in any one of the diodes or capacitors can cause either a shutdown or a defect traceable back to several different stages.

Despite the differences between linear and switched-mode power supplies, certain patterns remain. Each auxiliary power supply extending from a separate winding on an IHVT will have rectification, filtering, and regulation. In most cases, you will find that a low-voltage power supply consists of a bridge rectifier, a high-voltage filter capacitor, and a combination of transistor and zener diode regulation. Knowing that

each low-voltage circuit probably has these basic parts makes troubleshooting the auxiliary power circuit easier.

In addition to knowing about the basic parts of the subsystem, also consider the various functions of the stages attached to the auxiliary supply. For example, if the receiver has a symptom of no sound but normal picture and normal raster, concentrate on the auxiliary line that supplies the audio output circuit rather than the line tied to the vertical circuits. An incorrect voltage in the auxiliary line indicates either that the power supply has a defect or that a defect in the supplied circuit has caused an overload.

All this sounds rather simple until you consider that a short or lower than normal resistance in an auxiliary circuit may cause abnormal loading on the power supply, additional damage to the rectifier diode, and the shutting down of the entire system. In late-model televisions, shutdown occurs because the use of voltages derived from the secondary winding of an IHVT. With the coils wound on the same transformer core as the high-voltage windings, the chassis may start up and then quickly shut down. At times, the shutdown may be preceded by a symptom, such as a white horizontal line stretching across the center of screen, which shows the location of the defect.

If the receiver shuts down, shutdown circuits either monitor the amplitude of the flyback pulse or compare the flyback pulse amplitude to the amplitude of the current. All this traces back to the need for regulating the high voltage at the CRT second anode and reducing the chances for X-ray emission at the picture tube.

Troubleshooting Overvoltage/ Overcurrent Shutdown Problems

The troubleshooting of a shutdown problem is complicated by designs that tie the start-up voltages for a television receiver to the horizontal oscillator and driver stages. With the exception of the horizontal output transistor, all stages operate from dc voltages rectified from the horizontal sweep. Any interruption in those voltages allows filter capacitors to discharge and results in a total shutdown of all the circuits. From a troubleshooting perspective, the comparative lack of symptoms—no raster, no sound, no picture—makes problem-solving more difficult.

For that reason, some basic procedures are needed. When encountering a receiver locked into shutdown, power up the receiver and check for any symptoms. A shutdown preceded by normal picture decreasing to a thin horizontal line indicates that a problem in the vertical circuits has overloaded the low-voltage power supply tied to the IHVT. A raster that grows brighter or is limited to only one color just before shutdown occurs should lead to an investigation of the CRT and a possible overload in the video circuits.

Always use an isolation transformer, a variac, and a good high-voltage probe when troubleshooting the shutdown condition. The utilization of a variac allows the testing of the receiver under conditions where the ac line voltage begins at 65 vac. With such a low level of line voltage, the receiver will develop some high voltage. Increasing the line voltage while measuring the high voltage should show whether excessive high voltage has caused the shutdown. An excessive high-voltage condition becomes apparent if the high voltage reaches 26 kv or higher at the 80-vac level. With this type of defect, shutdown will occur as the line voltage reaches 80 to 100-vac.

In addition to monitoring the voltage levels, use an oscilloscope to check waveforms around the horizontal output transformer. Again utilizing the variac and the isolation transformer, increase the line voltage to the point just below where shutdown occurs. Then observe the waveforms at the base and collector of the horizontal output transistor. Distortion of the base waveform should direct your attention to possible overloads in the horizontal oscillator and driver circuits.

Distortion of the collector waveform—such as reduced retrace pulses or distorted pulses between the retrace pulses indicates that an overload exists in the stages following the output transistor. In some cases, shorted turns in the yoke or the horizontal output transformer can overload the power supply. In others, a shorted rectifier diode or shorted filter capacitor in an auxiliary supply line can overload the secondary of the IHVT. As shown earlier, a defective regulator can cause the voltage at the horizontal output transistor collector to increase to the point where shutdown occurs.

Troubleshooting the Sylvania C5 SMPS Circuit

Most problems with the Sylvania C5 switched-mode power supply begin within the SMPS circuitry and usually involve a totally shut down receiver with no raster, no sound, and no picture. Because the receiver utilizes an iso-hot chassis, always take special precautions when attaching test equipment to ground surfaces, and always use an isolation transformer. Service guidelines for the C5 chassis recommend against using any heat sinks in the chassis as a ground connection for test equipment. Instead, the guidelines specify the correct ground points for the chassis ground and the ac ground.

Troubleshooting that circuitry involves the observation of waveforms found at the horizontal pulse shaper, the horizontal driver, and the switched-mode regulator transistors. Figure 4-21 provides examples of how each of the waveforms should appear in a normally operating receiver. The monitoring of the waveforms ensures

that the horizontal signal is traveling from its source point at the horizontal oscillator to the switched-mode power supply.

Along with the waveform observations, other checks involve the presence of correct voltages at key locations. Using a multimeter, check for the +310-Vdc main supply voltage at the cathode of D460 or at the collector of the horizontal output transistor. The absence of the main supply voltage should point the troubleshooting efforts toward either the ac input circuit or the voltage doubler circuit. In addition, check for the +19-Vdc start-up voltage. Without the start-up voltage, the receiver cannot develop the horizontal square wave signal.

In this particular chassis, the switched-mode power supply controls the shutting down from an overvoltage or overcurrent condition. When an overvoltage condition occurs, the overvoltage sensing circuit sends a high signal to the base of Q494, the error-latch transistor, and causes the transistor to conduct. With Q494 conducting, the base of the horizontal pulse shaper is grounded and signal flow stops. Furthermore, Q492, the other error-latch transistor, also begins to conduct and, as it grounds the +7.1-Vdc voltage bus, cuts off the horizontal pulse shaper from its source voltage.

An overcurrent condition generates a negative signal at pin 8 of T488, the switched-mode transformer, and

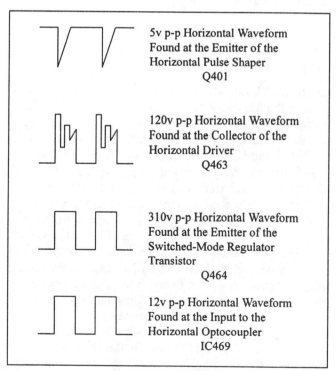

5v p-p Horizontal Waveform Found at the Emitter of the Horizontal Pulse Shaper Q401

120v p-p Horizontal Waveform Found at the Collector of the Horizontal Driver Q463

310v p-p Horizontal Waveform Found at the Emitter of the Switched-Mode Regulator Transistor Q464

12v p-p Horizontal Waveform Found at the Input to the Horizontal Optocoupler IC469

Figure 4-21   Waveforms Found at the Horizontal Pulse Shaper, Horizontal Driver, and Switched-Mode Regulator of a Sylvania 19C5 Chassis (*Courtesy of Philips Electronics N.V.*)

causes the forward biasing of D465, the overcurrent-sensing diode. The negative signal also travels to the emitter of Q494 and causes the transistor to conduct. Again, the conduction of the error latch causes the chassis to shut down. With the chassis in its shutdown mode, the removal of the error-causing condition will cause the chassis to attempt a restart and will result in the chassis making a "ticking" sound.

➲ SERVICE CALL

Sylvania Model RAJ147 Television with No Sound and No Raster

The Sylvania color television came into the repair depot, with no raster and no sound. Before checking any part of the receiver, the technician took time to study the type of power supply used in the receiver. Referring to the schematic diagram for the chassis, the technician found that transistor Q401 switches the primary of chopper transformer and should have a +163-Vdc at the collector. In addition, the technician found that diodes D416, D417, D418, and D419 supplied voltages to the secondary sources. Figure 4-22 shows a section of the scan-derived power supply.

Component checks showed that Q401 had shorted and that resistor R401 and fuse F400 had opened. After replacing the defective parts, the technician used a variable isolation transformer to lower the ac line voltage to 45 vac. Voltage checks showed +62-Vdc at the Q401 collector (normal under the ac line conditions) and +67-Vdc at the cathode of D416. However, a check of the schematic showed that the D416 cathode should have only +11.2-Vdc. While the voltages at the other diodes were closer to normal, all measured high.

Increasing the line voltage to 68 vac caused the voltage at the diodes to decrease back to normal. At this point, a check of the higher flyback secondary voltages also showed normal readings. However, increasing the line voltage to 90 vac again caused Q401 to short, and resistor 401 and fuse F400 to open. To check for a possible intermittent short in the secondary voltage supplies, the technician disconnected one end of R505, D420, R418, R425, R417, and D480 to remove the loads from the chopper transformer. In addition, the technician also disconnected one end of R415 to isolate the error-latch transistors from the chopper transformer. Still, no problems surfaced.

Further checks took the technician to the Q400, the pulse-width regulator. While in-circuit tests disclosed no problems, an out-of-circuit check of the transistor with a transistor tester showed that a leaky condition existed. Replacement of the pulse-width regulator transistor, the chopper-transformer driver transistor, the open R401, and the blown fuse returned the television to normal operating conditions.

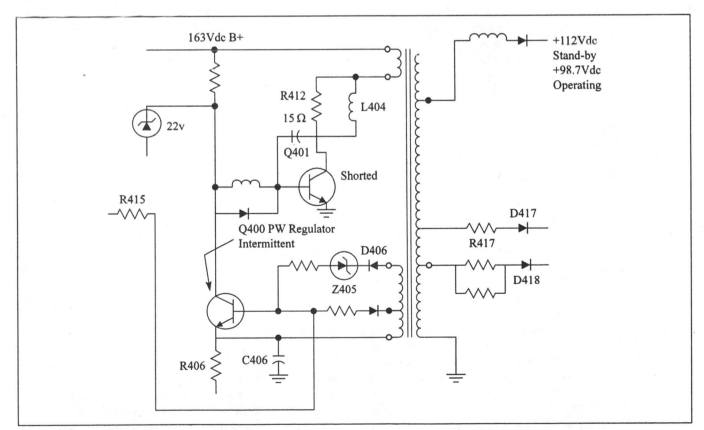

Figure 4-22 Schematic of the SMPS of a Sylvania Model RAJ147 Television (*Courtesy of Philips Electronics N.V.*)

⊙ SERVICE CALL

Samsung Model TXB1940 Television in Shutdown Mode

When the customer returned the Samsung television to the service center, the receiver would not power up. By paying close attention, though, the technician could hear the *tic-tic-tic* sound that indicated a problem with the switched-mode power supply. When checking voltages, the technician found a very high B+ voltage of +157.2-Vdc at the horizontal output transistor collector, but zero volts at the transistor base and no voltage at the horizontal output transistor side that connected to the transistor base.

At first, the technician concluded that the receiver had gone into shutdown because of the excessive voltage at the horizontal output transistor collector. However, a further check at the horizontal oscillator IC showed that incorrect voltages existed at pin 33 of the IC. The voltage at this particular pin is supplied by a regulator transistor. Instead of having the necessary +8-Vdc so that the oscillator could maintain its correct frequency, the voltage check at pin 33 showed +2.8-Vdc.

Rather than jump to conclusions about a defective regulator or oscillator, the technician began checking a voltage-divider circuit connected to the regulator. By performing several out-of-circuit resistance checks on resistors in the voltage divider, the technician found that the value of a 39 K resistor had increased to 130 K and had reduced the line voltage input into the regulator. The replacement of the resistor restored the receiver to its normal operating condition.

⊙ SERVICE CALL

RCA Model CTC136 Chassis Has Only a Tic-Tic-Tic Sound at Turn-on

When the technician applied power to the RCA chassis, a distinct *tic-tic-tic* sound could be heard. The technician also noticed that the tuner indicator LEDs would not illuminate. Additional checks showed that several voltages such as the tuning voltage were missing. Bypassing the start-up SCR in the SMPS allowed the receiver to have a normal raster, but the television would display only snow instead of a normal picture.

The technician consulted the service literature and began static tests on the SMPS, and found that C422 had become leaky. After replacing the receiver load with a dummy load and replacing C422, the technician tested the SMPS by using a variac to slowly bring the line voltage to 70 percent and then 100 percent levels. The SMPS maintained regulation. Replacement of the capacitor restored the receiver to its normal operation.

Post-repair checks of the power supply voltages and the tuner voltages showed that all voltages were within normal tolerances.

⊙ SERVICE CALL

RCA Model CTC169 Television Experiences Intermittent Power-On Conditions

When the technician began checking the RCA chassis, the set would power up normally at times and not turn on at other times. Once the intermittent condition began, the SMPS would cycle on and off three times and then shut down, or would power up for different periods of time and then shut down.

During this intermittent condition, the voltage on the +15-Vdc line measured only 10 to 11-Vdc. In addition, the technician found that the regulated output voltage from the SMPS measured +142.5-Vdc with the receiver operating normally. During troubleshooting of the receiver, the technician replaced transistors Q4105, Q4106, and Q4107 along with the regulator IC. Unfortunately, none of those replacements solved the problem.

After continuing to look for the source of the problem, the technician saw that C4118, a 47-uf filter capacitor, was slightly discolored. As Figure 4-23 shows, the capacitor connects to pin 16 of U4101. In addition, the technician checked C4118, the +15-Vdc rectifier diode, and found that the diode tested good and then broke down with an applied load. Replacing both the capacitor and the diode solved the intermittent power supply problem.

⊙ SERVICE CALL

TTX-3700 17-inch Computer Monitor with Shorted Rectifier

At first glance, the 17-inch SVGA computer monitor seemed to work fine. However, the technician found that a rectifier in the 185-Vdc line would short if the monitor was turned on less than 30 seconds after being turned off. A period of more than 30 seconds between turn-off and turn-on allowed the monitor to operate normally.

The technician suspected that an excessive peak current at the filter capacitor in the SMPS was responsible for the shorted rectifier. To verify this suspicion, the technician checked to see if the 185-Vdc was directly derived from the ac line input and if the monitor contained an IC for the control of the pulse-width modulation. Both checks proved the suspicions correct.

The technician had initially decided to replace the PWM controller, but decided to take another look at the current-limiting circuit and the type of diode used in

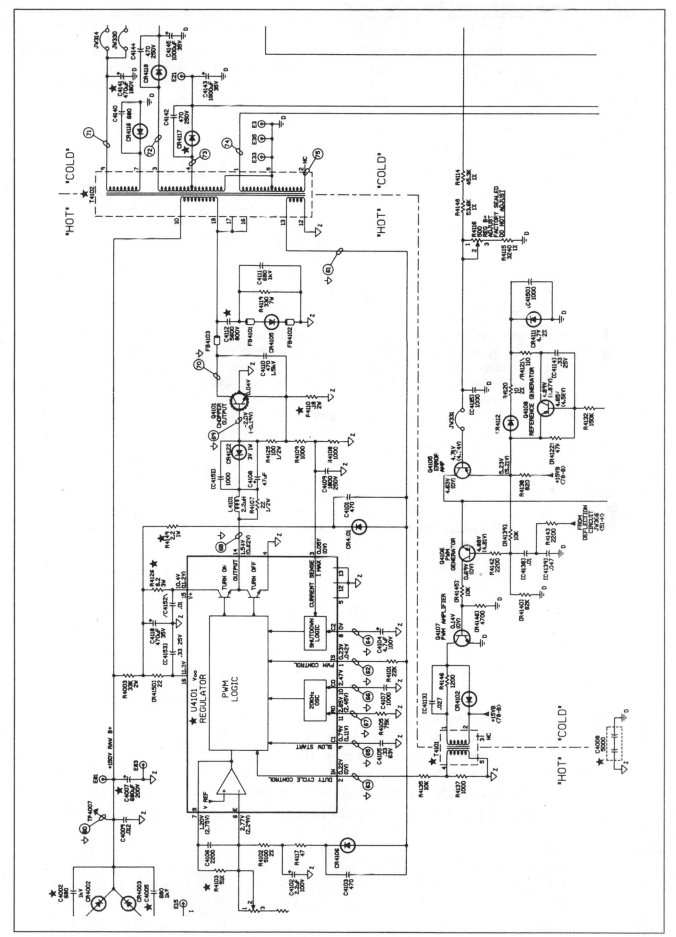

Figure 4-23   Schematic of the SMPS of an RCA CTC-169 Television Chassis (*Reprinted with permission of Thomson Customer Electonics, Inc.*)

the rectifier circuit. This particular circuit placed the degaussing coil in series with the ac line input and used the coil as a current-limiting device. The coil contained a MOV for surge protection. When replacing the diode, the technician also verified that the new diode rated for switch-mode operation rather than a common rectifier.

As a result of the checks, the technician replaced the diode with the correct type of rectifier and replaced the MOV in the degaussing circuit. The degradation of the MOV in the degaussing coil had altered the time constant of the current-limiting circuit. The original installation of an incorrect diode type had placed the diode under stress when the time constant changed and eventually caused the shorted condition. Replacement of the two components restored the monitor to its normal operating condition.

 PROGRESS CHECK

You have completed Objective 9. Troubleshoot switched-mode power supply problems.

SUMMARY

Chapter four continued the discussion of power supplies with thorough coverage of the switched-mode power supply technology currently used in almost all electronic devices. An SMPS uses a rectification-filtering-switching-regulation model and relies on a switching device to pulse voltages obtained from a capacitor. Switched-mode voltage may be regulated in several ways: through pulse-width modulation, a flyback regulator, a combination of the two, or a forward-switching regulator.

The chapter also covered scan-derived power supplies on which all modern televisions rely. With this type of supply, the SMPS model remains intact except for the use of a high-frequency transformer, a drive voltage obtained from the horizontal deflection system, and a start-up voltage. The scan-derived power supply was demonstrated with a Sylvania television circuit. From this, you learned about the voltage doubler start-up circuit, the switching devices, and the iso-hot grounding system. In addition, you saw the shape of waveforms produced at the supply.

SMPS units can present a serious safety hazard if handled incorrectly. An isolation transformer and variac are valuable tools for SMPS troubleshooting. Despite the complexity of an SMPS, the chapter presented specific methods for troubleshooting, which were illustrated in each of the Service Call examples.

REVIEW QUESTIONS

1. How is a switched-mode power supply different than a linear power supply?

2. Capacitors found in an SMPS may provide _____ and

   _____.

3. Active components in an SMPS may operate as:
   a. part of a feedback circuit, as rectifiers, or in filter circuits
   b. part of a feedback circuit, as regulators, or in filter circuits
   c. part of a feedback circuit, as rectifiers, or in overvoltage and overcurrent protection circuits
   d. part of a feedback circuit, as regulators, or in overvoltage and overcurrent protection circuits

4. All switched-mode power supplies use a high-frequency switching device to convert directly rectified line voltage into a:
   a. pulsed waveform
   b. smooth dc voltage
   c. trapezoidal waveform
   d. modulated waveform
   e. smooth ac voltage

5. An SMPS may have:
   a. two dc inputs
   b. three dc inputs
   c. four dc inputs
   d. zero dc inputs

6. Name three benefits of switched-mode regulation.

7. True or False   Pulse-width modulation varies the frequency of dc pulses while pulse-rate modulation varies the duty cycle of the dc voltage.

8. True or False   An SMPS uses a conventional transformer.

9. What schematic symbols are used for the cold ground and the hot ground?

10. In a color television receiver, an auxiliary power supply provides voltages for the

    _____ and _____ stages.

11. A switched-mode regulator circuit uses a _____, such as a bipolar transistor,

    a field-effect transistor, or a silicon-controlled rectifier, as a _____.

12. What advantages are derived from the use of a switch-mode regulator?

13. Which type of grounding system features a hybrid design where part of the chassis ties to a cold ground while the other portion has a floating ground with an above-ground potential?
    a. iso-hot
    b. cold
    c. hot

14. True or False   While the focus voltage is precisely regulated, the high voltage at the CRT second anode is not regulated.

15. Why does the flyback transformer shown in Figure 4-7 require harmonic tuning?

16. The high voltage found at the second anode of a CRT ranges from _____ to

    _____, while the focus voltage ranges from _____ to

    _____.

17. Name the three essential tasks performed by a flyback transformer.

18. Describe an integrated flyback transformer.

19. True or False   An SMPS contains a mix of passive and active components.

20. An SMPS begins with a:
    a. full-wave rectifier circuit connected to the ac line
    b. high-frequency transformer
    c. power transistor
    d. pulse generator

21. Where would you find a MOSFET or SCR in an SMPS?

22. An SMPS operates on the concept of converting _____ to a pulsed waveform.

23. Isolation in an SMPS occurs through:
    a. a conventional power transformer
    b. a small, high-frequency transformer
    c. an optoisolator
    d. inductance coupling
    e. resistance coupling

24. The iso-hot chassis utilizes a combination of the _____ and
    _____ grounding systems.

25. A scan-derived power supply has a:
    a. low operating frequency
    b. high operating frequency

26. Scan-derived power supplies operate because of a drive voltage obtained from the:
    a. start-up circuit
    b. horizontal deflection circuit
    c. vertical deflection circuit
    d. linear power supply

27. What is the purpose of a start-up supply?

28. What is the difference between a kick-start and a trickle-start circuit?

29. Auxiliary power supplies often have a positive-retrace rectified supply and a negative-retrace rectified supply. Why?

30. Describe how you would attach a test equipment probe to an iso-hot grounding system.

31. The use of an isolation transformer protects you from _____.

32. What is a variac?

33. Why would you use a variac for testing an SMPS?

34. True or False   When testing an SMPS after a repair, always connect the supply to the original circuit.

35. An oscilloscope is a useful tool for _____.

36. What is the purpose of a demodulator probe?

37. When testing an SMPS, you could use an oscilloscope to _____.

38. List the five points of the SMPS test procedure.

39. A shorted semiconductor in an SMPS will produce either:
    a. an audible whine or a ticking sound
    b. a ticking sound or normal sound from the speaker
    c. an audible whine or a hissing sound
    d. nothing

40. When troubleshooting the Sylvania RAJ147 television in the first Service Call, the technician disconnected one end of R505 and D420. Why?

41. When troubleshooting the Samsung TXB1940 television in the second Service Call, the technician concluded that shutdown was caused by excessive voltage at the horizontal output collector. Why? Did a defective oscillator cause the problem?

42. When working on the RCA CTC-136 chassis in the third Service Call, the technician found that the tuning indicator LEDs would not light. Refer to your reading materials and the Service Call. What did this symptom indicate?

43. What steps did the technician take before discovering the problem source in the fourth Service Call?

44. When troubleshooting problems with the 19C5 chassis switched-mode power supply, you should observe voltage readings at _____, _____, and
    _____.

45. A ticking sound in a switched-mode power supply is associated with:
    a. normal operation
    b. shutdown
    c. the start-up voltage circuit
    d. the control timing of the regulator circuit

46. Name four methods for solving problems in auxiliary power supply circuits.

47. Define the basic procedure for solving a shutdown problem. Why is a variable ac power supply a valuable tool?

48. Troubleshooting of the Sylvania 19C3 switched-mode chassis involves the observation of waveforms at

_____, _____, and the

_____. The monitoring of the waveforms ensures that the horizontal signal

is traveling from _____ to the _____.

49. A horizontal pulse-shaper transistor has a controlled output that depends on the

_____, _____, and the

_____ section of the receiver.

50. What is an optoisolator and what function does it perform?

51. What type of grounding system is used in the Sylvania C5 chassis?

52. The voltage doubler in the Sylvania chassis develops the +310-Vdc main voltage for the
_____ transistor and the _____
transistor.

53. True or False   Overvoltage and overcurrent protection circuits shutdown the horizontal deflection system if the high voltage goes beyond the specified level for a particular CRT by linking the high and low voltage supplies.

CHAPTER
5

Sync Signals and Horizontal Deflection

OBJECTIVES Upon completion of this chapter, you should be able to:

1. Discuss different types of sync-separation circuits and the reason for sync separation.
2. Describe the effects of noise on sync-separator circuits and the methods used to prevent noise interference.
3. Understand the basic principles of horizontal deflection circuits.
4. Describe horizontal AFC circuits.
5. Describe horizontal oscillator circuits.
6. Describe horizontal driver and output circuits.
7. Explain IC-based horizontal scan systems.
8. Troubleshoot horizontal deflection systems.

KEY TERMS

anti-hunt circuit
damper diode
Darlington stage
differentiator
digital storage oscilloscope
horizontal AFC stage
horizontal driver

horizontal foldover
horizontal hold control
horizontal output transistor
sawtooth voltage
sine wave sweep oscillator
sync separator stage
sync signals

INTRODUCTION

During the discussion of scan-derived power supplies, you found that the operation of the power supply depended on a constant voltage/current source drawn from the horizontal scan section of the television receiver. Moreover, you learned that the generation of the supply source also depended on the proper operation of the horizontal oscillator. Those links become especially evident in this chapter.

We will first examine the separation of sync signals from the composite video signal and the use of those signals during the operation of the horizontal and vertical oscillators. Regardless of whether the circuit handles vertical or horizontal deflection, the sweep oscillators use the separated sync signals to provide output voltages. The voltages have waveforms that produce sawtooth currents in the coils of the deflection yoke. Every sweep stage within the deflection system is followed by an amplifier that produces the relatively large current required by the deflection coils.

The chapter then moves to the horizontal deflection system and explains why and how the system operates. While considering the system as a whole, you will also study the individual stages that make up the system. Those are:

- The horizontal automatic frequency control stage
- The horizontal oscillator stage
- The driver stage
- The horizontal output stage

Horizontal deflection systems have three basic functions. First, the system provides and applies a modified sawtooth-shaped current to the horizontal deflection coils. This action causes the CRT electron beam to deflect back and forth across the face of the CRT. In addition, the system establishes the drive voltage needed by the high-voltage transformer, which provides the voltage needed for the CRT anode. Finally, the horizontal deflection system operates as a source point for many of the low-voltage supplies. Those supplies are often derived from the horizontal output transformer.

The gradual building and collapsing of magnetic flux in the coils causes the CRT beam to deflect from side to side and up and down. While the deflection of the beam causes a spot to trace on the CRT screen, luminance and chrominance signal-processing circuitry produces the picture details. Magnetic flux changes occur because of the sawtooth shape of the waveforms passing through the deflection coils.

Finally, the chapter looks at problems that could occur in the horizontal AFC, oscillator, driver, and output stages, and provides common solutions. It also shows that defects in each one of those sections can affect the performance of the receiver as a whole.

5-1 SYNC SEPARATION

The deflection circuits found in a television receiver rely on separate synchronizing pulses for timing before a raster is produced. The lack of proper timing can cause the reproduced picture to roll vertically or tear horizontally. In many ways, we can consider the **sync signals** as the structural portion of the picture-making signals. The sync signals include horizontal sync pulses, vertical sync pulses, and equalizing pulses.

Sync separation occurs through the demodulation of the video I-F signal, the recovery of the composite video signal, and the amplitude separation of the sync pulses. A **sync-separator stage** eliminates the video and blanking signals while amplifying only the horizontal sync, vertical sync, and equalizing pulses. While these circuits separate sync pulses from the picture and blanking signals, sync separation also involves the removal of 60-Hz vertical sync pulses from 15,750-Hz horizontal sync pulses.

Figure 5-1 shows the complete composite video signal and identifies the sync signals. After demodulation occurs, three types of signals—video, blanking, and synchronizing signals—are emitted at the output of the video detector. Each of the three signals has a proportional amplitude and exactly the same frequency as the video, blanking, and synchronizing signals found at the transmitter. Applied to deflection circuits in the receiver, the sync signals control the oscillation frequencies of the sawtooth current generators.

The generated vertical and horizontal currents ensure that the electron beam inside the receiver CRT remains in step with the electron beam within the transmitting camera. Without this structure, the reproduced picture would consist of nothing more than meaningless colored blobs. The horizontal sync pulse rides at the top of each horizontal blanking pulse as shown in Figure 5-1, and has a duration of 5 microseconds. Spaced 58.5 microseconds apart, the horizontal sync pulses control the side-to-side movement of the electron beam.

↻ FUNDAMENTAL REVIEW

Demodulation

The process of demodulation decodes and recovers the intelligence from a carrier signal.

Diode Sync-Separator Circuits

Figure 5-2 shows the simplest method for separating sync signals with a diode sync-separator circuit. The composite video signal travels into the anode of the diode. During positive alternations of the input signal,

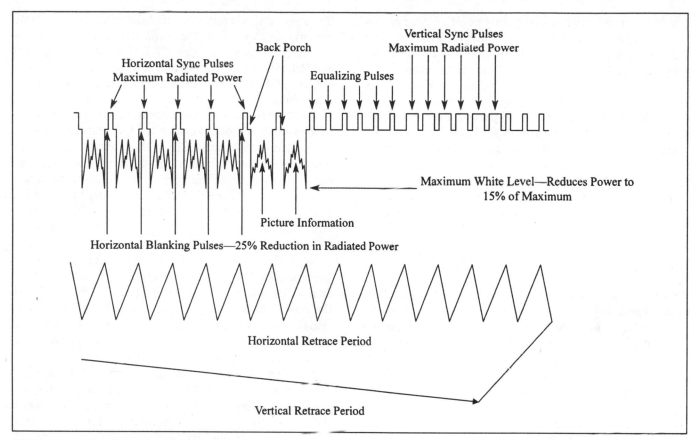

Figure 5-1   Composite Video Signal

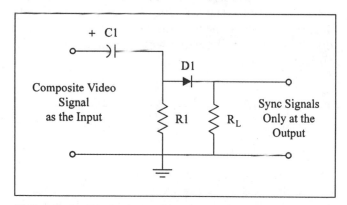

Figure 5-2   Diode Sync-Separator Circuit

electrons flow from the cathode of the diode to its anode and charge capacitor C1. Between each peak of the signal, part of the capacitor charge leaks through resistor R1 and produces a voltage across the resistor. The anode end of the resistor is negative with respect to ground.

Passing only the sync signal portion of the composite video signal requires a circuit that provides the clipping level shown in Figure 5-3. Adjusting the clip level depends on the level of the bias voltage found across the resistor. Because the bias voltage level depends on the amplitude of the input signal and the time constant

produced by R1 and C1, any change of the input signal amplitude can cause the biasing of the diode anode either above or below the clipping level.

The voltage across R1 provides a negative bias for the diode. Because of this, the diode conducts and

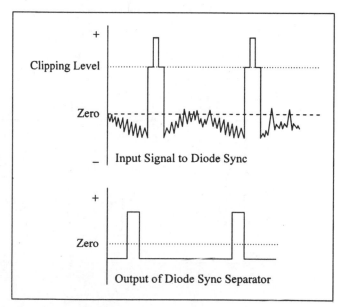

Figure 5-3   Illustration of Desired Sync-Separator Clipping Level

produces current only when highly positive sync pulses occur. During each of these intervals, the electron flow through D1 and R1 recharges the capacitor to an amount equal to the voltage that leaked through R1 between the pulses. As the output signal develops, it has the same shape as the sync pulses.

With a diode sync separator, part of the input signal is lost across the diode. As a result, the sync pulses produced at the output of the circuit have a lower amplitude than the input signal. Yet, given a sufficiently high amplitude at the input, the diode sync separator produces a distortion-free output signal. Because of the reduced noise, the diode sync separator is widely used in many receivers.

Transistor Sync-Separator Circuits

Figure 5-4 shows a schematic diagram of a transistor sync-separator circuit. The video signal has negative sync pulses and travels from the junction of resistors R1 and R2. From there, the signal couples into a combined, dual time-constant network consisting of capacitor C1, resistor R4, capacitor C2, and resistor R3, and then to the base of Q2, the sync separator. Because the Q2 circuit does not rely on a fixed bias and no bias exists at the Q2 base-emitter junction, the transistor remains cut off.

The arrival of a negative sync pulse pushes the sync separator transistor into saturation. As a result, the Q2 base current charges capacitor C1. Because the transistor has a low input resistance, the capacitor has a short charging time constant and charges completely during the time that a horizontal sync pulse occurs. When

the sync pulse ends, the transistor goes back into cut-off. At this point, the transistor has a high base-to-emitter impedance, and C1 discharges through C3 and R4. The values of C3 and R4 allow the automatic control of the Q2 base bias voltage over a wide range of input signals.

Even with a sudden decrease in the amplitude of the input signal, the bias voltage for Q2 remains close to the voltage seen across C3. If the input signal amplitude decreases to a point where the next incoming sync pulse does not drive the transistor into saturation, less current flows through the capacitor during the charging cycle. In this situation, C3 charges to a smaller positive value and allows the Q2 base voltage to move closer to zero.

With C1 discharging, the voltage at the base of Q2 becomes less positive and the cut-off/saturation begins again. Because of the amount of control given through the base bias circuit, the sync separator circuit maintains a constant output signal. In the output circuit of Q2, a coupling capacitor transfers the sync pulse output to the base of Q3, a sync-pulse inverter.

Positive sync pulses developed across R8, the emitter load resistor, travel through a differentiating network consisting of C5 and R10 to the horizontal deflection circuits. As mentioned earlier, the resistor and capacitor in the differentiating network have values that yield a short time constant. If the vertical circuits require positive sync pulses and the horizontal circuits require negative sync pulses, the integrating network would connect to the Q3 emitter circuit, and the differentiating network would connect to the Q3 collector circuit.

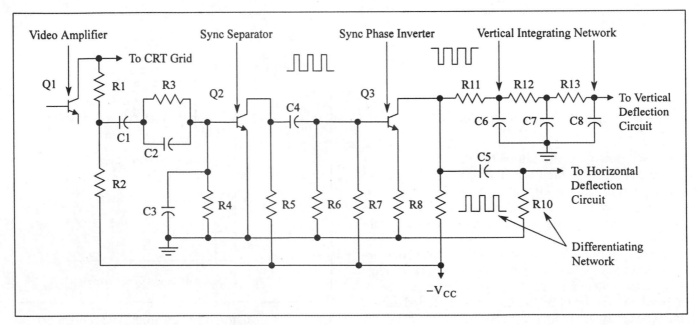

Figure 5-4   Transistor Sync-Separator Circuit

Integrated Circuit Sync Separation

Figure 5-5 represents one part of a larger integrated circuit that contains the sync separation, vertical drive, vertical oscillator, horizontal AFC, and horizontal oscillator circuits. As the diagram shows, the composite video signal feeds into pin 5 of the IC . The sync-separator section of the IC sends the horizontal sync signal to the horizontal AFC circuit and the vertical sync signal to the vertical integrator.

The sync separator circuit relies on a **Darlington stage**, an operational amplifier, and a feedback loop to separate the sync signals from the composite video signal. While the Darlington stage places a smaller load on the video amplifier stage, the op-amp and feedback loop establish a constant slicing level between the sync tip and pedestal level. During operation, capacitor C1 charges to the sync tip level, and transistor Q1 conducts on the pulse tips. By conducting during the tip period, the transistor cancels the pulse as it outputs a signal to capacitor C2. While C2 charges to the pedestal level, a voltage divider consisting of R1 and R2 controls the slice level.

🔁 FUNDAMENTAL REVIEW

Darlington Pairs

A Darlington Pair amplifier is a special type of emitter follower that uses two transistors to increase the overall current gain and amplifier input impedance.

🔁 FUNDAMENTAL REVIEW

Operational Amplifier

An operational amplifier (op-amp) is a high-gain, dc amplifier with high input impedance and low output impedance, which is housed in an integrated circuit package. Op-amps have linear operating characteristics.

☑ PROGRESS CHECK

You have completed Objective 1. Discuss different types of sync-separation circuits and the reason for sync separation.

Noise-Reduction Circuits

Noise spikes within the video signal can cause the bias voltages found in sync-separator circuits to build to a large level and eventually disable the sync separator for a short period of time. If this interruption occurs, the circuit will lose several sync pulses, and the picture will tear both horizontally and vertically. The transistor noise switch, or noise gate, circuit shown in Figure 5-6 reduces the effects of any input noise by monitoring the video signal and canceling any noise spikes.

During operation, a positive video signal with negative sync pulses travels from the junction of R5 and

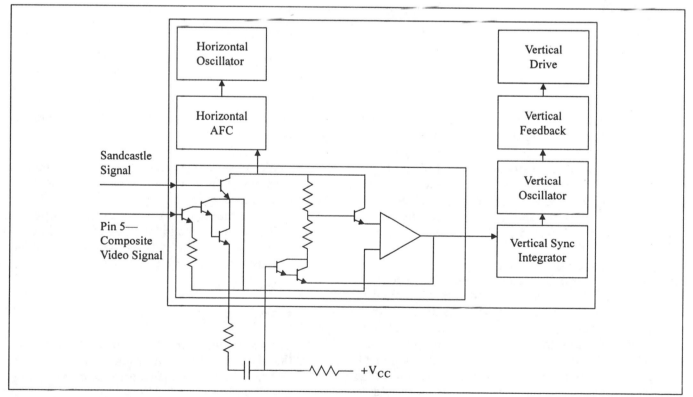

Figure 5-5 Schematic Diagram of a Sync Processor IC

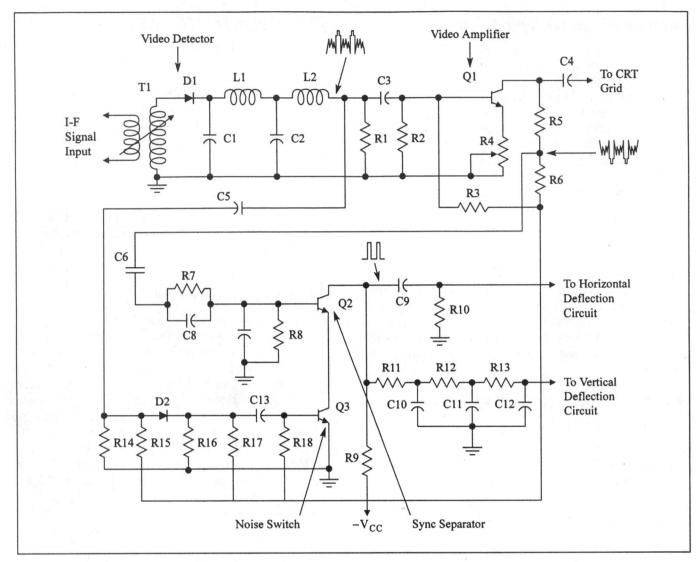

Figure 5-6   Transistor Noise Switch Circuit

R6 in the video amplifier circuit through a dual time-constant network to the base of Q2, the sync separator. With no noise spikes occurring in the video signal, Q3, the noise switch, operates in saturation because of the bias supplied through R18. The noise switch conducts only when a sync pulse exists at the sync separator. When Q3 saturates, the small dc potential between the Q3 collector and emitter causes the Q2 emitter to have a ground.

A second video signal travels from the output of the video detector to the anode of diode D2 and remains 180 degrees out-of-phase with the signal applied to the sync-separator base. Bias voltages obtained from the divider circuits, consisting of R14 and R15, and R16 and R17, cause the cathode of D2 to operate more positively than the anode. If a noise spike occurs with a higher amplitude than the reverse bias found at the diode, D2 conducts and the noise pulse passes to the base of Q3 through capacitor C13.

As a result, the positive-going noise spike opposes the forward bias at the noise switch and causes the transistor to cut off. With the noise switch cut off and in series with the sync-separator emitter circuit, no sync pulses can output from the sync separator. Once the interruption of sync signals occurs, the horizontal oscillator maintains the stability of the horizontal deflection signal. The vertical sync is lost, however, and returns only after the noise spike ends through the application of sync signals to the vertical oscillator.

☑ PROGRESS CHECK

You have completed Objective 2. Describe the effects of noise on sync-separator circuits and the methods used to prevent noise interference.

5-2 HORIZONTAL DEFLECTION BASICS

Figure 5-7 uses a block diagram to illustrate the basic operation of the horizontal deflection system. The system consists of a horizontal deflection oscillator stage, a horizontal driver stage, and a horizontal output stage. In addition, the system relies on the automatic frequency control of the horizontal frequency through a feedback loop connected from the horizontal output stage to the oscillator stage. The system directly couples to the yoke horizontal deflection coils, relies on a damper diode to prevent unwanted oscillations, and provides a high-voltage pulse through the operation of the output stage and the horizontal output transformer.

⟳ FUNDAMENTAL REVIEW

Oscillators

An oscillator consists of an amplifier with a feedback circuit connected to the output of the amplifier. With this configuration, an oscillator converts a dc voltage to an ac signal.

Building the Sweep Voltage

The building of a sweep voltage may occur through the operation of a conventional oscillator or through the use of a countdown circuit. Other designs rely on switching circuits built around silicon-controlled rectifiers. Horizontal oscillator circuits require higher stability and include a blocking oscillator, multivibrator circuits, and sine wave oscillators. SCR-based circuits use a combination of two silicon-controlled rectifiers and two reference diodes to develop the horizontal sweep voltages and the high voltage. While one SCR

generates the voltage needed for trace, the other generates the voltage for retrace. Modern television receiver designs have moved away from the oscillator and SCR-based circuits and use the countdown system. Countdown circuits combine the vertical and horizontal oscillators into one unit and enclose the operation within one semiconductor package.

Because of the amount of power needed to drive the horizontal yoke coils and deflect the electron beam, horizontal deflection systems also rely on a driver and an output amplifier. From there, the systems connect to the yoke coils and the flyback transformer. Magnetic flux produced through the flow of current through the yoke causes the electron beam to deflect across the CRT screen.

The Differentiator Circuit

The horizontal deflection system is preceded by a series RC network called a **differentiator** that prevents vertical sync pulses from entering the horizontal circuit and triggering the oscillator. As shown in Figure 5-8, a differentiator is a high-pass filter. The differentiator has a short time constant and uses the rapid charging and discharging of a capacitor to trigger the oscillator. As a result, the high-frequency horizontal pulses pass through the circuit.

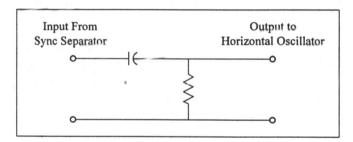

Figure 5-8   Schematic of a Differentiator Circuit

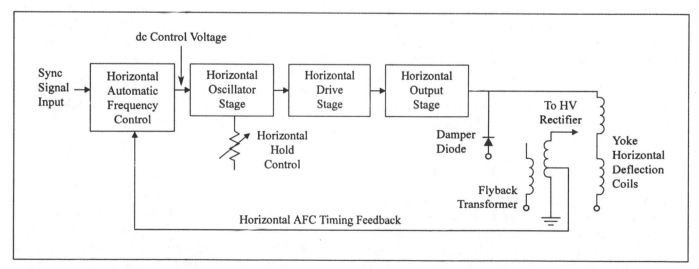

Figure 5-7   Block Diagram of a Horizontal Deflection System

The capacitor charges through the resistor to the peak of the applied voltage. At the instant of the trailing edge of the rectangular input pulse, the capacitor discharges through the resistor. The constant flow of charging and discharging current through the resistance establishes a series of voltages at the output of the differentiator. The spikes have twice the peak-to-peak amplitude of the applied input signal.

The Horizontal AFC Stage

The **horizontal AFC stage** compares the frequency and phase of feedback pulses taken from the horizontal output stage with horizontal sync pulses arriving from the differentiator. Through the comparison of those signals, the AFC circuits maintain the correct 15,750-Hz frequency of the horizontal oscillator. If the oscillator frequency deviates from the correct setting, a mismatch occurs between the signals arriving at input of the AFC circuit. Either an increase or a decrease in frequency will cause the AFC circuit to generate a positive or negative dc control voltage. The voltage change causes the oscillator to lock back into sync.

The Horizontal Oscillator

Horizontal sweep generators must have the stability to resist triggering by noise pulses. In comparison to oscillator circuits that you will see in the vertical deflection system covered in chapter six, the extremely stable operation of horizontal oscillator circuits during phase changes does not detract from the operation of the horizontal deflection circuit. The reason for this lies within the time needed for completion of one horizontal cycle. Compared to the slow vertical cycles, the horizontal oscillator completes 15,750/60 or 262.5 cycles during the time that the vertical oscillator requires to complete one cycle. Therefore, the horizontal oscillator can take the time of almost 200 cycles to change phase with no visible changes occurring the picture. The phase change may occur during a channel change or during a network program change. In many cases, the horizontal deflection circuit will feature either a blocking oscillator or multivibrator. Instead of producing a sawtooth output, though, the horizontal oscillator produces either a rectangular wave or a pulse output.

Blocking oscillator and multivibrator circuits are, by design, relatively unstable. For that reason, the design of the horizontal deflection system supplements that type of circuit with another type of network called the sine wave stabilizing network in an effort to provide additional stability. In other cases, a very stable sine wave oscillator will operate as the horizontal oscillator. The output from horizontal oscillator stage takes the form of a rectangular pulse.

The Horizontal Hold Control The **horizontal hold control** adjusts the free-running frequency of an oscillator. Although most modern television receivers do not rely on a manual horizontal hold control because of advances in solid-state technologies, other televisions retain the hold control as a potentiometer located in the oscillator circuitry. With the horizontal hold control misadjusted, the oscillator frequency will deviate from the 15,750-Hz scan rate and the picture will lose its horizontal sync.

Older television receivers also rely on a horizontal frequency control. While the horizontal hold control provides a fine adjustment of the oscillator frequency, the horizontal frequency control establishes a coarse adjustment. Electrically, the frequency control operates as part of a tank circuit, which sets synchronizing pulses to trigger the oscillator.

Horizontal Waveshaping and Drive Circuitry

The **horizontal driver** reshapes the rectangular pulse taken from the oscillator output to provide the waveshape needed to produce a horizontal output signal. To do this, the driver stage uses a semiconductor device as a switch. Conduction time determines the length of time that the dc supply voltage connects to the horizontal deflection coils of the yoke for each horizontal scan. Thus, the pulse width is an important factor when checking the performance of the driver stage.

↻ FUNDAMENTAL REVIEW

Rectangular Waveforms

A waveform resulting from alternating high and low dc voltages is defined as a rectangular waveform. A square wave is a special-wave rectangular waveform that has equal pulse-width values, which represent the time spent in the dc high-voltage state, and space-width values, or the time spent in the low-voltage state. The pulse-width, space-width, and cycle time of a rectangular waveform are measured at the halfway points on rectangular waveforms. The use of an oscilloscope allows the measurement of four different switching times found with rectangular waveforms. Those times are:

1. Delay time, or the time required for a semiconductor device to come out of cut-off
2. Storage time, or the time required for a semiconductor device to come out of saturation
3. Rise time, or the time required for the device to make the transition from cut-off to saturation
4. Fall time, or the time required for the device to make the transition from saturation to cut-off

The Horizontal Output Stage

The raster will not have any brightness without the functions provided by the horizontal output stage. While the horizontal output stage operates as a class C power supply that produces output pulses, each horizontal line displayed on the raster corresponds with the switching on of the horizontal output.

Because the horizontal output pulse is applied to the high-voltage rectifier that produces the high voltage for the CRT, the horizontal output stage also requires a **damper diode** to minimize the possibility of oscillations in the horizontal scanning current. The damper diode also produces the boosted B+ voltage for the horizontal output amplifier as it rectifies the horizontal deflection output. A boosted B+ voltage results from the series-adding of the rectified voltage with the dc voltage from the power supply. Higher supply voltages at the horizontal amplifier produce higher ac power output.

In addition, the damper diode provides a damping current that establishes the part of the horizontal scanning that occurs on the left side of the CRT and just after flyback. Also referred to as reaction scanning, the use of the damping current for scanning produces almost one-third of the horizontal trace. In keeping the class C operation, reaction scanning allows the horizontal output amplifier to cut off during part of the retrace time. When the amplifier cuts off, reaction scanning produces the trace from the left edge of the screen toward the right.

When the horizontal output amplifier conducts, it completes the trace to the right edge of the screen. In short, the combination of reaction scanning and boosted B+ voltage increases the power efficiency of the horizontal output stage. The horizontal output transistor supplies the horizontal scanning current directly to the yoke coils.

⟳ FUNDAMENTAL REVIEW

Power Amplifiers

The output amplifiers seen in both the vertical and horizontal deflection circuits are power amplifiers. In those circuits, the output amplifier drives low-resistance loads and has an output power rating of 1 watt or higher. The efficiency of a power amplifier is measured in terms of the amount of power drawn from the supply that is actually delivered to the load. Although power amplifiers deliver a large amount of power to the load, some power is dissipated throughout the amplifier circuit. As an example, the current flowing through the base, emitter, collector resistors, and the amplifier transistor in a transistor-based circuit results in power dissipation.

Power amplifiers are divided into four classes. The Class A amplifier conducts during the entire 360 degrees of the input cycle and—depending on the type of coupling used in the circuit—has a maximum efficiency ranging from 25 to 50 percent. Generally, Class A amplifier circuits will feature either RC or transformer coupling.

Class B, Class AB, and Class C amplifiers conduct for less than 360 degrees of the cycle. Any class B amplifier will contain two transistors, with one transistor conducting for the negative portion of the input cycle and the other for the positive portion of the cycle. Class AB amplifiers are a variation of the class configuration. Both the Class B and the Class AB amplifiers have maximum efficiency ratings of 78.5 percent.

In the Class C amplifier circuit, the transistor conducts for less than 180 degrees of the input cycle; reactive components in the circuit produce the remainder of the output. Class C amplifiers offer a maximum efficiency of nearly 99 percent.

☑ PROGRESS CHECK

You have completed Objective 3. Understand the basic principles of horizontal deflection circuits.

5-3 HORIZONTAL AFC CIRCUITS

As you know, vertical and horizontal sweep oscillators have different stability requirements because of different operating frequencies. Both types must synchronize with received sync pulses and must resist the influence of noise pulses. In addition, both must have the capability to change phase whenever the customer changes channels or a network has a station break. The phase change must occur within a reasonable length of time. The difference in stability requirements and operating frequencies, however, means that the horizontal oscillator changes phase gradually while the vertical oscillator must change phase at a rapid pace.

The 15,750-Hz operating frequency of a horizontal oscillator closely resembles noise. When considering differentiator circuits, note that the circuit has RC values designed to pass the horizontal sync output pulses. Unfortunately, those RC values also allow high-frequency noise pulses to pass; the integrating circuit found in vertical processing systems blocks the noise pulses. Because the differentiator allows noise to pass and because the horizontal oscillator requires stability, horizontal oscillator circuits require a control circuit that monitors both frequency and phase. A similar circuit is not found in the vertical oscillator circuit.

Like other automatic frequency control circuits that you have studied, the horizontal AFC circuit uses a dc control voltage developed by either a phase or timing comparator to control an oscillator frequency. With older designs, the comparator circuit consists of two diodes supplied with push-pull horizontal sync pulses from either a phase-splitter or sync-splitter stage. The push-pull sync pulses from one input to the comparator, while a sample of the oscillator frequency—a **sawtooth voltage**—makes up the other input. Sawtooth voltages result from the relatively slow charging of a capacitor through a large resistance and then rapidly discharging the capacitor through a small resistance. Feedback from the horizontal output stage provides the sawtooth voltage.

If the oscillator frequency drifts either too low or too high, the comparator generates a dc correction voltage that locks the horizontal oscillator back to the correct frequency. As you may suspect, newer designs incorporate horizontal AFC circuits into an integrated circuit package and utilize a phase-locked loop to generate the control voltage. Regardless of the design type, the horizontal AFC circuit has the sole task of maintaining the horizontal frequency of the reproduced picture. Without this type of horizontal synchronization, the picture would tear apart.

FUNDAMENTAL REVIEW
Phase

Phase refers to the repetitions of the signals occurring in time.

FUNDAMENTAL REVIEW
Comparators

Comparators are found throughout digital circuits and may determine the sign of binary numbers or compare the polarity of dc voltages.

As you begin to explore horizontal AFC circuits, you should remember that several basic rules govern the operation of the circuit:

1. The magnitude and polarity of the control voltage depend on the long series of incoming horizontal sync pulses.
2. Oscillator frequency depends on the repetition rate of the horizontal sync pulses and is not influenced by high-frequency noise pulses.
3. The horizontal oscillator must operate at the correct frequency, and the oscillator output must have the correct phase relationship with the sync pulses.

Figure 5-9 shows a block diagram of a typical horizontal AFC system.

When the horizontal AFC comparator begins operation, it compares the received sync pulses from the sync processor with either a sine wave or sawtooth voltage taken from the horizontal amplifier. The phase relationship between the two signals establishes the characteristics of the control voltage. Any rapid fall-off of a sawtooth voltage must match with the arrival of the sync pulses so that the flyback motion of the electron beam spot has the correct relationship with the received picture signal. Without the proper phase relationship in place, the modulation of the picture begins either before or after the start of the left-to-right trace. With that situation, the picture would shift toward one side of the screen or the other.

Sawtooth Wave Horizontal AFC Circuits

Figure 5-10 illustrates the operation of a simple type of horizontal AFC system that compares the sync-pulse phase with the phase of the sawtooth voltage obtained from the horizontal deflection circuits. In the first stage of the AFC circuit, a sync-pulse phase splitter produces

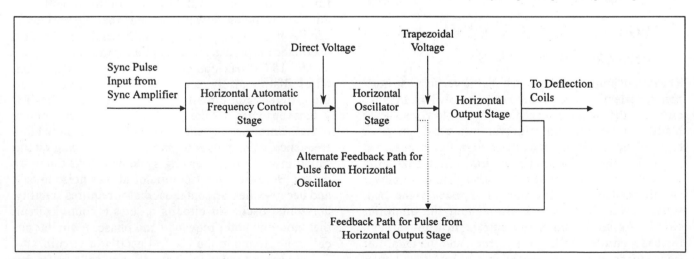

Figure 5-9 Block Diagram of a Horizontal AFC System *(Courtesy of Philips Electronics N.V.)*

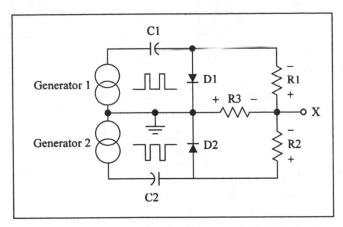

Figure 5-10   Schematic of a Horizontal AFC Equivalent Circuit

positive sync pulses at one output and negative sync pulses at the other output. The combination of these sync pulses controls the conduction of a pair of diodes, which connect as a phase detector.

In actual operation, the sync pulses exist as reference pulses, while the retrace of the electron beam synchronizes with the reference. As the figure shows, the circuit compares the retrace portion of the sawtooth voltage with the sync pulses. The polarity of the dc voltage found at the output of the phase detector depends on the phase of the sawtooth voltage. Used as a control voltage, the dc voltage controls the frequency of the horizontal oscillator.

We can use Figure 5-9 to further analyze the sawtooth voltage AFC circuit by splitting the positive and negative sync pulses into separate generators. Generation of the positive sync pulses causes diode D1 to conduct and, in turn, the charging of capacitor C1 to the indicated polarity. At the same time, generation of the negative sync pulses causes D2 to conduct and the charging of capacitor C2 to its indicated polarity. Given that the capacitors charge only the very low forward resistance of the diodes, both capacitors rapidly charge to the maximum amplitude of the sync pulses. The circuit

has the purpose of developing a voltage at point X and determining the polarity of that voltage through the conduction of D1 and D2.

When there is an interval between the sync pulses, the voltage found across the capacitors reverse biases each diode and allows the capacitors to discharge through resistors R1 and R2. If each set of sync pulses has the same amplitude and has charged the capacitors to the same level, the discharge causes voltage drops to appear across each resistor. The polarity of each voltage drop is indicated in the figure. With this, the circuit operates as a balanced bridge, point X serves as the ground, and no current flows through R3.

Increasing the amplitude of the positive sync pulses to a level greater than the negative sync pulses causes D1 to conduct more than D2, and allows C1 to have a greater charge than C2. During the sync pulse interval, the discharge current flowing through C1 is greater than the amount of discharge current given by C2. As a result, the additional current flows through R3 and develops a voltage across the resistor, and point X becomes negative with respect to ground.

If we increase the amplitude of the negative sync pulses to a level greater than the positive sync pulses, D2 conducts more than D1, and C2 charges to a greater level than C1. Here, C2 has a greater discharge than C1, and the additional current again flows through R3. In this case, however, the current develops a voltage with the opposite polarity of that seen when D1 conducts at a higher rate. Thus, point X becomes positive with respect to ground.

Under actual circumstances, circuitry in a television receiver holds the sync pulses constant and varies the conduction of the two diodes by injecting a sawtooth voltage signal into the double-diode circuit. Figure 5-11 uses another generator to represent the insertion of the sawtooth voltage, while Figure 5-12 shows the relative amplitude and desired phase of the two sync signals and the sawtooth signal. In Figure 5-12, the positive-retrace portion of the sawtooth

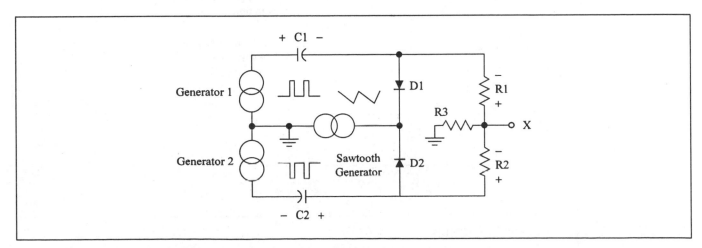

Figure 5-11   Sawtooth Voltage Horizontal AFC Equivalent Circuit

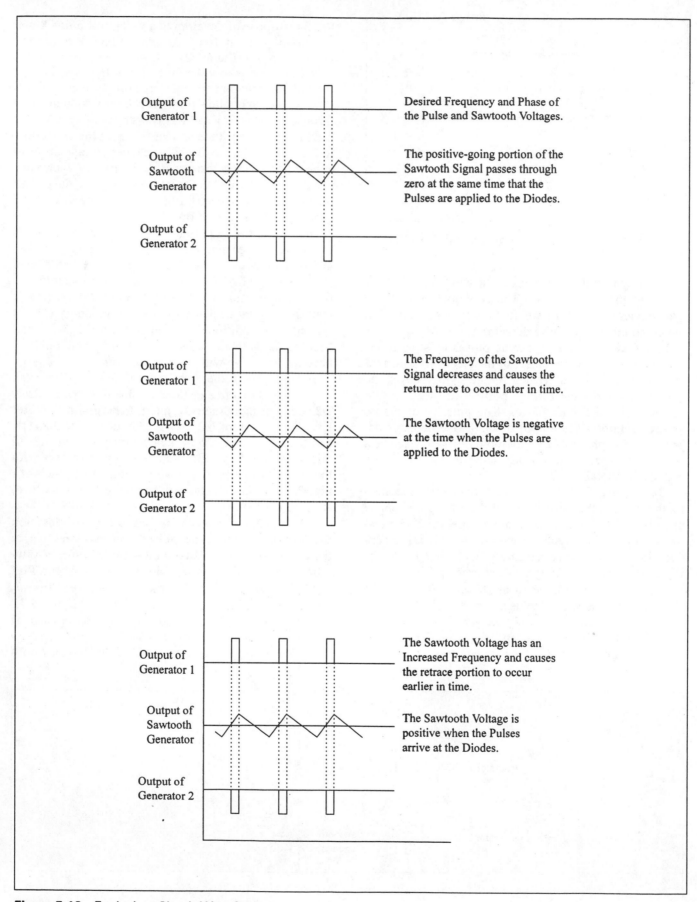

Desired Frequency and Phase of
the Pulse and Sawtooth Voltages.

The positive-going portion of the
Sawtooth Signal passes through
zero at the same time that the
Pulses are applied to the Diodes.

The Frequency of the Sawtooth
Signal decreases and causes the
return trace to occur later in time.

The Sawtooth Voltage is negative
at the time when the Pulses are
applied to the Diodes.

The Sawtooth Voltage has an
Increased Frequency and causes
the retrace portion to occur
earlier in time.

The Sawtooth Voltage is
positive when the Pulses
arrive at the Diodes.

Figure 5-12 Equivalent Circuit Waveforms

passes through zero at the same instant the circuit applies the sync pulses to the diodes. When this occurs, the diodes have the same level of conduction and the capacitors charge equally. As before, this results in a zero voltage reading at point X.

Note that the frequency and phase of the injected sawtooth voltage governs whether D1 or D2 has a higher conduction rate. If the frequency of the sawtooth voltage decreases, the return trace occurs slightly later than normal. In turn, the sawtooth voltage is negative during the time that the sync pulses arrive, and a phase difference occurs between the two signals. Thus, the negative portion of the sawtooth voltage applies to the cathode of D1 and the anode of D2. With this, D1 conducts more than D2 and the charge across C1 increases while the charge across C2 decreases. The unequal charging of the capacitors results in a negative voltage at point X.

An increase in the sawtooth signal causes the retrace portion to start earlier than normal, and an opposite phase difference occurs. As a result, the positive portion of the sawtooth voltage appears during the arrival of the sync pulses and applies to the anode of D1 and the cathode of D2. At this point, D2 conducts more than D1, and the charge across C2 increases while the charge across C1 decreases. Therefore, a positive voltage appears at point X. Whether negative or positive, the voltage found at point X is the dc control voltage for the AFC circuit.

We can take our analysis of the AFC operation to the next level by applying the dc control voltage to the horizontal oscillator. Figure 5-13 shows a schematic diagram of a complete horizontal AFC/horizontal oscillator circuit. Capacitor C3 develops a negative-going sawtooth wave from pulses obtained from the horizontal deflection circuitry. Resistors R3 and R4

and capacitors C4 and C5 filter the control voltage found at the output of the phase detector. The filtered control voltage across C5 controls the bias of the horizontal oscillator and, as a result, varies the oscillator frequency. Because a negative sawtooth waveform feeds into the AFC circuit, the control voltage polarity varies inversely with the changes in the horizontal frequency.

Whenever the horizontal frequency begins to decrease, the lagging phase of the sawtooth voltage causes diode D2 to conduct more than D1. Consequently, the AFC circuit produces a positive-polarity control voltage that decreases the reverse bias on the NPN horizontal oscillator. Given those conditions, the oscillator frequency increases back to 15,750-Hz.

An increase of horizontal frequency past the correct setting causes the sawtooth voltage waveform to lead rather than lag the horizontal sync pulses. Thus, the conduction of D1 produces a negative-polarity control voltage that appears across both C3 and C7. The negative dc control voltage at C7 shifts the horizontal oscillator frequency and phase back to normal by increasing the reverse bias voltage at the base.

Anti-Hunt Horizontal AFC Systems

The original idea of causing an oscillator to shift back to the correct frequency with the application of a dc correction voltage influenced many AFC designs. However, one serious flaw in the original concept caused the development of a supplementary circuit. With any AFC system, the application of the control voltage could occur at the same instant that the oscillator goes out-of-phase. As a result of that timing, the control voltage overcorrects the oscillator phase so that it shifts further than necessary. Thus, the oscillator

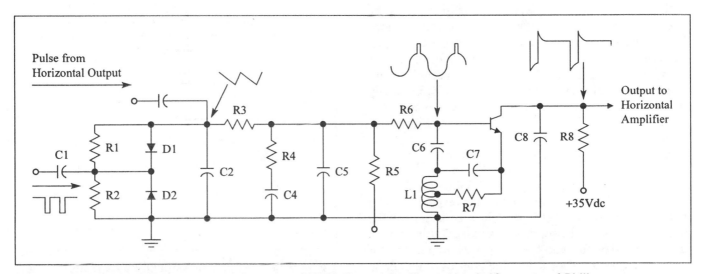

Figure 5-13 Schematic Diagram of a Complete AFC/Horizontal Oscillator Circuit (*Courtesy of Philips Electronics N.V.*)

frequency shifts past the correct frequency and to a point in the other direction.

When this occurs, the control circuit senses the incorrect phase and causes another shift to occur. With each phase shift, the oscillator phase alternately lags and leads the correct phase before eventually locking to the correct phase and frequency. In very basic terms, the oscillator hunts until it finds the correct setting.

Anti-hunt circuits slow the build-up of the correct voltage and allow the dc control voltage to shift the oscillator frequency without overcorrecting. Figure 5-14 shows a schematic diagram of a basic anti-hunt circuit used to delay the rise of the control voltage. Moving back to our study of the horizontal AFC circuit, a series of positive or negative pulses is found at the output of a phase detector when the oscillator goes off-frequency. The pulses cause the oscillator to shift back and lock to the 15,750-Hz horizontal frequency. The direction of the frequency pulses depends on the direction of the frequency error.

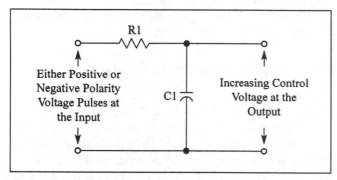

Figure 5-14 Schematic Diagram of an Anti-Hunt Circuit

As an example of the circuit operation, a series of dc control pulses forms the input to the anti-hunt circuit. Each pulse causes the capacitor to charge through the resistor during the pulse interval. The capacitor retains the charge during the periods between the pulses. Moving to Figure 5-15, the curve labeled E_2 represents the voltage across C2. As the charge across C2 increases in small increments, the voltage E_2 gradually increases and functions as the control voltage. Applying the slowly increasing control voltage to the oscillator shifts the oscillator toward the correct phase. This causes a decrease in the amplitude of the control-voltage pulses. At the optimum point, the amplitude of E_2 equals the maximum amplitude of the pulses.

Although the circuit and the voltage response shown in Figures 5-14 and 5-15 provide a solution for the anti-hunt problem, better answers exist. Taking a look at Figure 5-16, the addition of a speed-up capacitor across the resistor allows positive control of the phase and frequency through the application of a much larger control voltage. It also improves switching time. When

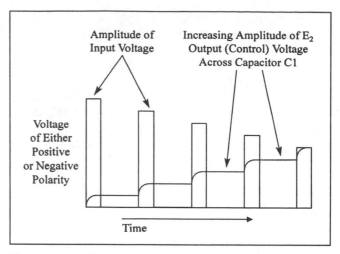

Figure 5-15 Comparison of AFC Output Voltage and the Anti-Hunt Voltage

an input pulse arrives at the circuit, both capacitors charge immediately. The reactance of C1 effectively shorts the resistance presented by R1 while the two capacitors charge in series. As a result, the sum of the voltages across the capacitors equals the amplitude of the applied input voltage.

If we consider the circuit from another perspective, the capacitors—while charging in series—form a voltage divider. While the process of charging in series deposits the same charge on each capacitor during a pulse, the capacitors do not have identical values; the capacitance of C1 is less than that of C2. Thus, because of the voltage divider action, the voltage found across C1 is greater than the voltage found across C2.

With R1 in parallel with C1, a discharge path exists for capacitor C1. As soon as the capacitor initially charges, it also begins to discharge through the resistor. The capacitor continues to discharge until the next pulse arrives. Figure 5-17 shows how the relationship between voltages across the capacitors affects the application of the control voltage. With the arrival of the first pulse, C2 immediately charges to its initial level.

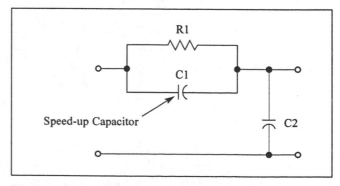

Figure 5-16 Addition of a Speed-up Capacitor to the Anti-Hunt Circuit (*Courtesy of Philips Electronics N.V.*)

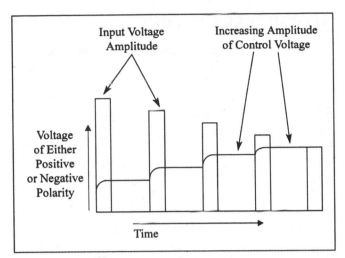

Figure 5-17 Relationship Between Anti-Hunt Input Voltage and Capacitor Charge

The remainder of the input pulse voltage stays across C1. In the interval between the pulses, the charge across C2 remains constant while the charge across C1 decreases to zero. With the next pulse, the partial charging of C2 establishes a divided voltage that equals only the difference between the maximum pulse voltage and the voltage across C2. As the figure shows, the control voltage gradually builds until it equals the maximum amplitude of the input voltage pulses.

☑ PROGRESS CHECK

You have completed Objective 4. Describe horizontal AFC circuits.

5-4 HORIZONTAL OSCILLATOR CIRCUITS

We can begin to classify horizontal deflection circuits in a number of different ways. One of the most common

methods for classification involves the type of active components used in the system. While older receiver designs rely almost totally on transistors, new designs combine nearly all the horizontal functions within one or two integrated circuits. Another method for classification considers the type of oscillator used in the horizontal deflection system. Again looking at the older receivers first, many utilize blocking oscillators, or sine wave oscillators. Newer receivers combine the operation of the vertical and horizontal oscillators into one integrated circuit package.

Sine Wave Horizontal Oscillators

A **sine wave sweep oscillator** uses pulse-generating circuits that produce oscillations; it usually takes the form of a Hartley oscillator. As shown in Figure 5-18, the upper part of the coil that makes up one part of the LC circuit connects to the emitter and base of transistor Q1. The lower part of the coil carries the oscillator collector current. With this in mind, we can trace the collector subcircuit from the emitter through Q1 to the collector, through R3 to the negative-polarity collector supply voltage, through the supply voltage, to ground, and from ground through the lower section of L1 to the Q1 emitter. A voltage-divider circuit consisting of R1, R2, and R5 establishes the base bias voltage.

Any change in the collector current flowing through the lower part of coil L1 induces a voltage in the upper part of the coil. Capacitor C1 couples the induced voltage to the base of Q1. When the transistor conducts, the increasing current through L1 allows the application of a negative voltage to the base. With this negative voltage in place, the base current increases and causes another increase in the collector current. Once the collector current reaches the saturation point and C1 has charged, the collector current has no further increase and the base voltage decreases to zero.

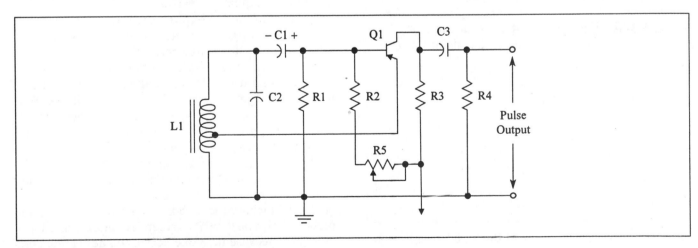

Figure 5-18 Sine-Wave Sweep Oscillator Circuit

With this, C1 begins to discharge through R1 and causes both the base and collector currents in Q1 to cut off. But, the previous large pulse of collector and emitter current shock excites the L1C2 resonant circuit and causes the LC circuit to resonate at 15,750-Hz. Each negative portion of the sine wave couples through C1 to the base of the transistor and overcomes any reverse bias. The resulting pulses of base and collector current recharge C1 and rebuild the energy in the tank circuit. Therefore, the inductance and distributed capacitance of coil L1 and the capacitance of C2 determine the oscillation frequency. While L1 may have a variable inductance to better control the oscillator frequency, the RC components in the circuit have a short time constant and prevent the oscillator from blocking. As a result, each positive and negative alternation of the base voltage has an equal duration.

During normal operation, the variations of the base sine wave voltage have sufficient amplitude to drive the transistor into both saturation and cut-off. With the base voltage swinging back and forth past the positive point of cut-off and the negative point of saturation, the collector current cannot match the variations that occur during base voltage peaks. Given this factor, Q1 operates as an overdriven amplifier and establishes a clipping action that produces a large voltage across R3. The voltage waveform is shown in Figure 5-19A. When applied to the short, time-constant, high-pass circuit consisting of C3 and R4, the voltage sets up the narrowed pulses seen in Figure 5-19B.

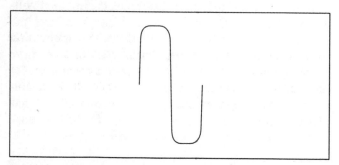

Figure 5-19A   Voltage Waveform Found at the Sine-Wave Sweep Oscillator Circuit

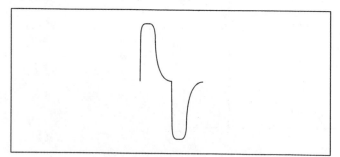

Figure 5-19B   Pulses Found at the Sine-Wave Sweep Oscillator Circuit

☑ PROGRESS CHECK

You have completed Objective 5. Describe horizontal oscillator circuits.

5-5 HORIZONTAL DRIVER AND OUTPUT CIRCUITS

Most transistor-based horizontal deflection systems place a driver transistor between the oscillator and amplifier to provide the needed drive for the output amplifier. Horizontal output amplifiers require considerably more drive than vertical output amplifiers because of the inverse relationship between the power needed to deflect the CRT beam and retrace time. While the vertical deflection retrace time is 560 to 600 microseconds, the horizontal deflection retrace time is only 8 microseconds. Given this short retrace time, the horizontal output stage must have the capability to supply a large amount of peak power.

Horizontal Driver Circuits

Figure 5-20 highlights the portion of the horizontal deflection circuit that contains the driver and the output transistor. In this circuit, rectangular pulses from the horizontal oscillator couple to the base of Q1, the horizontal driver. During the scanning period, the transistor has a strong forward bias and operates in the saturated condition. During retrace, the incoming signal pulses go sharply positive and cut off the horizontal driver.

With the emitter and collector of the transistor connected to transformer T1, sharp cut off of the driver also causes a pulse to appear across the primary winding of the transformer. After the transformer steps the pulse down and reduces the circuit impedance, the pulse cuts off the output stage during the retrace interval. Thus, the driver transistor provides the base current required by the output transistor.

Because of the ac coupling seen between the driver and the output transistor, the signal at the output transistor base rises above and below the zero axis. The use of a transformer to couple the two stages rounds the pulse waveform but does not affect the operation of the output stage. Positive pulses drive Q2 into cut off during the retrace interval. The transistor goes into saturation during the trace interval because the base is negative enough with respect to the emitter.

Blocking capacitor C1 couples the rectangular voltage waveform to the horizontal deflection coils and causes a sawtooth current to flow through the coils. With C1 blocking the dc component of the signal, the picture shown on the CRT screen remains centered horizontally. Because the yoke horizontal deflection coils usually have an inductance value that ranges below

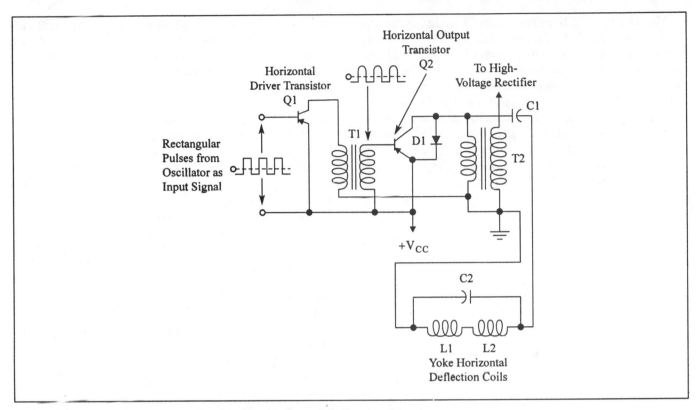

Figure 5-20   Horizontal Driver Portion of a Horizontal Deflection Circuit

1 millihenry, an impedance match exists between the inductance of the coils and the collector circuit of the horizontal output transistor.

Diode D1 functions as the damper diode and reduces the collector-emitter resistance of Q2 during the trace interval. In some circuit designs, the damper diode is located within the horizontal output transistor. The damper diode prevents voltage induced across the deflection coils from causing a reverse current to flow through the output transistor during the retrace period. If the reverse current builds, the trace will become non-linear. The voltage from the horizontal output transistor-collector series feeds through the T2 primary and produces a sharp, high-voltage pulse across the transformer secondary. After rectification, the high-voltage pulse is applied to the anode of the CRT.

Horizontal Output Transistors

When you consider the operation of the **horizontal output transistor**, remember that the transistor has two key purposes:

- develop a sawtooth current through the deflection yoke
- produce a high-voltage pulse, which is rectified and used as the accelerating dc voltage for the CRT

Both of these factors build additional requirements into the type of device used as an output transistor. Every horizontal output transistor must have characteristics that allow the handling of high peak-currents and voltages with a fast turn-off time.

Figure 5-21 uses an equivalent circuit to describe the operation of a transistor-based horizontal output system and the methods used to produce the sawtooth current waveform. In the series-parallel circuit, an open switch interrupts the current flow and causes an induced voltage to appear across coil L1. The induced voltage has the opposite polarity of the source voltage, and as a result, opposes the source voltage. Consequently, the produced current oscillates between L1 and capacitor C1.

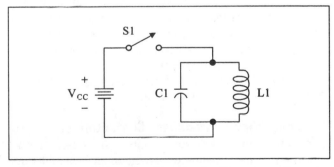

Figure 5-21   Equivalent Circuit of a Transistor-Based Horizontal Output System

The circuit oscillates for one-half cycle, or the period of time needed to move from point two to point three on the waveform chart shown in Figure 5-22. At the conclusion of this time period, the switch closes for an instant and allows the source voltage to act as a load across the coil. When this occurs, the positive voltage found across the coil has a higher value than the source voltage. Therefore, current flows in the opposite direction through the battery. The flow of current decreases until it reaches zero at time 4. From time 4 to time 5, the source voltage is greater than the induced coil voltage, and current flows from the battery and through the coil.

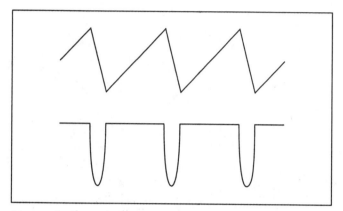

Figure 5-22 Waveforms Found at the Horizontal Output Equivalent Circuit

Each change of current direction occurring between time 3 and time 5 in the equivalent circuit represents the interval needed for the electron beam to steadily scan from left to right. Between times 2 and 5, the beam blanks out and swings quickly back to the left of the screen. Opening the switch causes a sharp pulse of voltage to develop across the coil. After application to a step-up transformer and rectification, this pulse supplies the anode voltage for the CRT.

When we compare the equivalent circuit and an actual horizontal output circuit, the switch represents a switching transistor. Applying a square or rectangular waveform to the base of the switching transistor drives the device from cut-off to saturation automatically. This takes advantage of the small difference seen between the emitter-base junction and the collector-base junction of the switching transistor. With the two junctions nearly identical, the transistor conducts in either direction. The direction of conduction depends on the polarity of the applied voltages.

Horizontal Output Transistor Characteristics Transistors used in the horizontal output stage have a much lower collector resistance than the reactance of the horizontal yoke coils. Because of this, the coils have the characteristics of an inductance and require either a

rectangular or square wave to produce a sawtooth current. This occurs because of the type of waveform needed to produce the sawtooth current. Several reasons exist for these characteristics.

During operation, the horizontal output transistor goes into saturation and presents a low collector resistance. In addition, the copper wire that makes up the horizontal deflection coils also has a very low resistance. If we used the circuit shown in Figure 5-23 to represent the horizontal deflection amplifier circuit, the series RL portion of the circuit is almost a pure inductance. Any signal applied to the base of the horizontal output transistor must have a rectangular waveform to produce a sawtooth current in the deflection coils.

We can use Figures 5-23 and 5-24 to further examine the switching characteristics of the horizontal deflection circuit. In Figure 5-23, the basic transistor circuit produces a sawtooth current in the deflection coils. With the first waveform applied to the base of the transistor from time 1 to time 2, the transistor has forward bias, conducts, and causes current to flow through the deflection coils. As a result, the electron beam moves from the center to the right side of the CRT display. Given the amplitude of the input signal, the transistor goes into saturation, and the internal emitter-to-collector resistance reduces almost to zero.

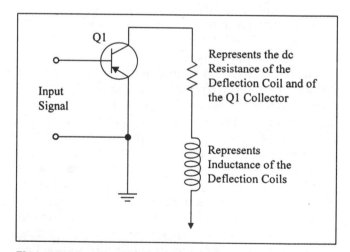

Figure 5-23 Sample Horizontal Output Transistor Circuit

Looking at the first waveform in Figure 5-24, the signal applied to the base of Q1 goes sharply positive at time 2. The base junction of the PNP transistor becomes reverse biased and causes the transistor to cut off. Because the transistor goes from saturation to cut-off in a relatively short time, the collector current does not immediately cut off because of the accumulation of carriers in the base region. The second waveform of Figure 5-24 shows the collector current transition.

From time 2 until time 3, the transistor remains cut off and several key events occur. During the first part of

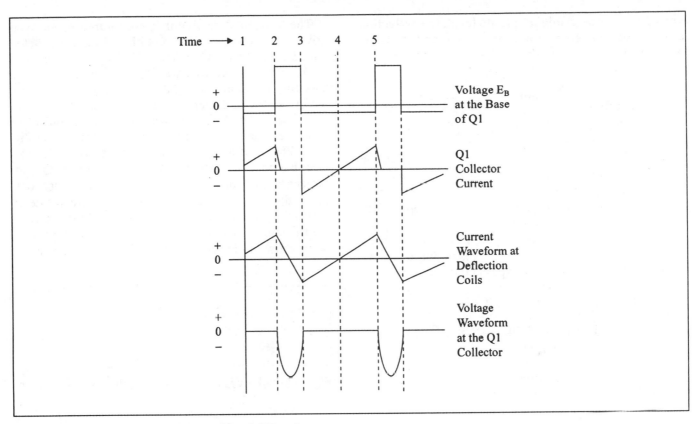

Figure 5-24 Horizontal Deflection Circuit Waveforms

the time period, the current supplied to the deflection coils decreases to zero, and an induced voltage appears across the coils as the magnetic field collapses. In turn, a half-cycle of oscillation occurs between the deflection coils and capacitor C1. The oscillation causes the current flowing through the deflection coils to change direction during the second part of the time period. This action causes the electron beam to move to the left side of the CRT display.

At time 3, the transistor loses the reverse bias and begins to saturate. Here, the current passes through the transistor in the reverse direction because of the polarity of the induced voltage across the deflection coils. The third waveform of Figure 5-24 shows the waveform for the current flowing through the deflection coils. From time 3 to time 4, the electron beam moves back to the center of the CRT display as the collector current shows a linear decrease. As shown in the fourth waveform of Figure 5-24, the induced voltage across the coils decreases to zero at time 4. At this point, the collector current begins to flow.

Overvoltage Protection Circuits

In a television receiver, the fast-moving electron beam strikes the phosphor on the inside of the CRT screen, and expended energy in the form of light photons exits from the picture tube as low-intensity current. With larger picture tubes, the anode voltage has levels of 25 to 35 kv, and the combination of beam current and anode voltage generates X rays. If the X-ray levels reach above 1015-Hz, the emissions can harm human tissue. As part of the federal government regulation of the video industry, all receivers have strict limits regarding the level of X-ray emission. Those limits involve the regulation and amount of the high voltage found at the CRT anode.

Overvoltage and overcurrent protection circuits shut down the receiver horizontal deflection system if the high voltage goes beyond the specified level for a particular CRT. It does this by linking the high- and low-voltage supplies, as an increase of the high voltage beyond the specified level causes the low voltage increase. At the overvoltage threshold value, the conduction of semiconductor devices in the protection circuit either shuts down the horizontal driver stage, disables the high-voltage stage, or lowers the high voltage to an acceptable level. An overcurrent protection circuit monitors the amount of current flowing through the high-voltage winding and shuts down the horizontal driver if the amount exceeds the specified level.

Other overvoltage and overcurrent protection circuits either shut down the horizontal oscillator or push the 15,750-Hz horizontal frequency off-scale. As the high voltage begins to rise above the acceptable level, the picture breaks into diagonal bars. When the high voltage exceeds the minimum safe level, the protection circuit shuts the oscillator circuit down and stops the generation of high voltage.

5-6 COMPLETE HORIZONTAL DEFLECTION SYSTEMS

Each of the prior sections has either emphasized the theory behind the production of horizontal scan voltages or has shown how individual parts of the system operate. In this section, you will have the opportunity to view horizontal deflection as a system. Each function of the horizontal AFC and oscillator stages affects the operation of the drive and waveshaping stages as well as the production of voltages at the horizontal output stage.

A Horizontal Deflection Circuit Based on a Sine Wave Oscillator

Figure 5-25 shows the schematic diagram of an almost complete horizontal deflection system that includes a sine wave oscillator. At the upper left of the drawing, negative sync pulses from the sync separator and a negative sawtooth waveform obtained from the primary of the flyback transformer travel into the horizontal AFC circuit. Resistors R3 and R4 along with capacitors C4 and C5 filter the output of the AFC circuit as it moves to the base of Q1, the horizontal oscillator. The AFC dc control voltage maintains the frequency and phase of the horizontal oscillator.

The horizontal oscillator is configured as an overdriven sine wave oscillator with a frequency determined by coil L17, C27, and C25. Figures 5-26A and 5-26B show the desired waveforms for the oscillator base and collector. The negative-polarity pulse found at the Q1 collector direct couples to the base of Q2, the PNP horizontal driver. In turn, the conduction of the driver produces the positive-polarity pulse seen in Figure 5-26C and the Q2 collector. Transformer T2 couples that positive-polarity pulse to the base of Q3, the horizontal output transistor. With transformer coupling, the phasing of the transformer secondary determines the polarity of the coupled pulse. In this circuit, the transformer couples a 105-v peak-to-peak positive pulse shown in Figure 5-26D to the base of the output

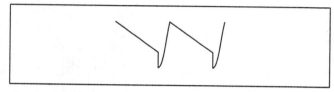

Figure 5-26A   Waveform Found at the Oscillator Base

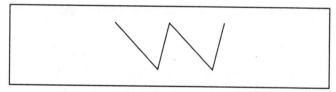

Figure 5-26B   Waveform Found at the Oscillator Collector

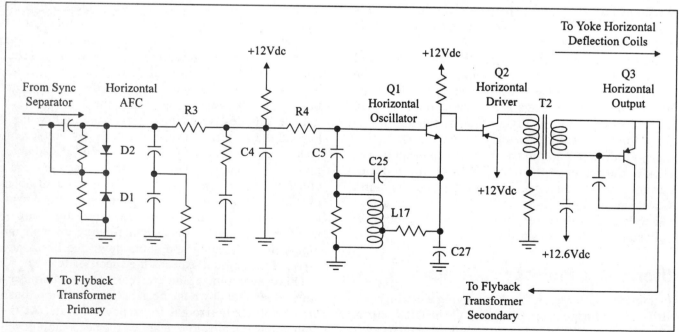

Figure 5-25   Schematic Diagram of a Sine-Wave Oscillator Horizontal Deflection System

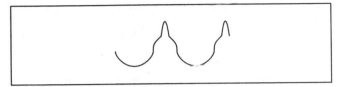

Figure 5-26C Positive-Polarity Pulse Found at the Horizontal Driver Collector

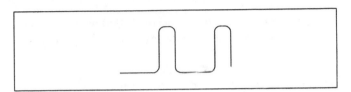

Figure 5-26E Positive Pulse Found at the Emitter of the Horizontal Output Transistor

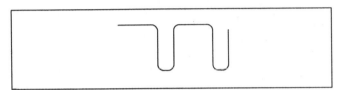

Figure 5-26D Positive Pulse Found at the Base of the Horizontal Output Transistor

Figure 5-26F Positive Pulse Found at the Yoke Coils

transistor. When analyzing the output transistor circuit, two key points emerge:

- The horizontal output transistor is connected as an emitter follower with all signals measured with respect to ground. Therefore, the base signal is approximately the same as the emitter signal.
- With no bias applied between the base and emitter of the PNP output transistor, the wide negative portions of the base pulse cause the transistor to conduct, while the narrow positive portions cause the transistor to turn off.

Any sudden turn-off of the horizontal output transistor develops the large positive pulse at the emitter shown at Figure 5-26E. The pulse is then applied to the horizontal yoke coils and capacitor C109. At this point in the circuit, and as shown in Figure 5-26F, the capacitor prevents dc signals from reaching the coils and affects the linearity of the sawtooth current flowing through the coil windings. Although not shown in the schematic, the Q3 returns through the flyback transformer to the +12.6-Vdc source. A damper diode prevents any oscillations from occurring at the start of the trace.

☑ PROGRESS CHECK

You have completed Objective 6. Describe horizontal driver and output circuits.

5-7 IC-BASED HORIZONTAL SCAN SYSTEMS

The prior sections have emphasized horizontal deflection designs that relied on transistors for the horizontal AFC, horizontal oscillator, and driver functions. Most modern television systems, however, place the sync separator, horizontal phase control, and horizontal oscillator functions within the same integrated circuit package that houses the vertical signal processing functions. This system takes advantage of digital countdown circuits that divide the phase-locked signal of the horizontal oscillator while establishing both the horizontal and vertical scan frequencies. While some designs enclose the driver stage in the same package, all rely on a power transistor to develop the horizontal output signal.

Figure 5-27 shows a block diagram of the horizontal APC system usually found enclosed in the IC package. The circuit synchronizes the horizontal scan by using an automatic phase-control loop and input signals from the sync separator and the horizontal output transformer. Integration of the two input signals produces a sawtooth waveform. At the beginning of the APC loop, a phase detector compares the phase of the two input waveforms; if a phase error exists, the phase detector sends a dc-coupled error signal through the

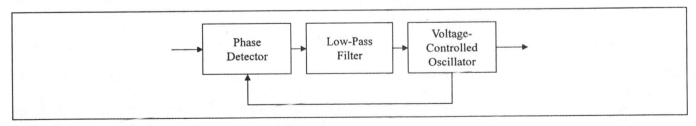

Figure 5-27 Block Diagram of the Horizontal APC System

low-pass filter to the voltage-controlled oscillator. The VCO then shifts input frequency in the direction of the phase error.

Figure 5-28 shows a gated differential amplifier operating as the phase detector in the APC loop. The sync-pulse input drives the common feed from the current source, while the sawtooth waveform derived from the flyback pulse is applied to one side of the differential pair. Rather than use the LC-tuned Hartley oscillator seen in the previous section, the IC-based design uses a relaxation oscillator, illustrated in Figure 5-29, as the VCO. The relaxation oscillator charges a timing capacitor from a current-source timing resistor until the voltage across the capacitor reaches the IC trigger level. Then internal circuitry switches modes and discharges the timing capacitor at a given rate until the capacitor reaches the low threshold level of the IC. Once the IC resets, the charge/discharge routine begins again.

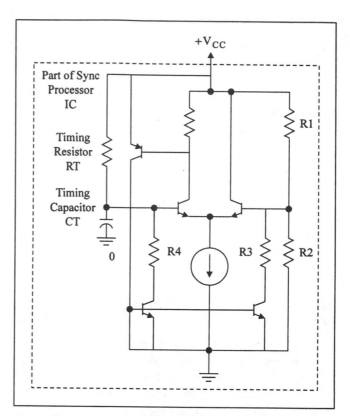

Figure 5-29   Relaxation Oscillator

horizontal sync pulses to the horizontal APC circuit, a horizontal pulse from the integrated flyback transformer also arrives at the APC circuit and maintains the control of the horizontal scan frequency. R9 provides a manual adjustment of the horizontal frequency at the horizontal oscillator circuit.

As you saw earlier, the horizontal APC circuit compares the differentiated sync signal with the sawtooth signal integrated from the flyback pulse. Looking at the diagram, the flyback signal enters the IC at pin 19. If the two signals match, an equal amount of charging and discharging current flows in and out of the APC filter at any given horizontal sync period. Thus, a constant control voltage exists at the output of the APC block. If the two signals do not match, a phase error occurs. For example, a lagging sawtooth wave will increase the amount of discharging current and cause a decrease in the amount of control voltage. In turn, the decrease causes the VCO to change in frequency.

The sync differentiator found within the IC improves the performance of the PLL system by eliminating the broad pulse seen during the vertical sync period. Without the differentiation, the PLL would incorrectly force the sawtooth wave to center with the broad vertical pulses. As a result, the AGC gating and burst gating of the vertical countdown system would have numerous errors. These errors would become especially apparent during channel changes when a wide pulse generated within the AGC circuit would cause the VCO

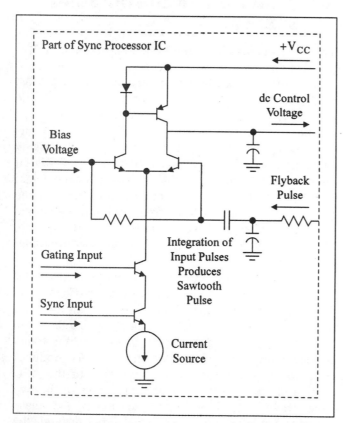

Figure 5-28   Gates Differential Amplifier Operating in the APC Loop

A Horizontal Scan Circuit

Figure 5-30 uses a block diagram of a sync processor IC to illustrate the operation of the horizontal scan circuit. The composite video signal feeds into pin 5 of the sync processor and to the sync-separator circuit. While the sync separator and the sync differentiator pass the

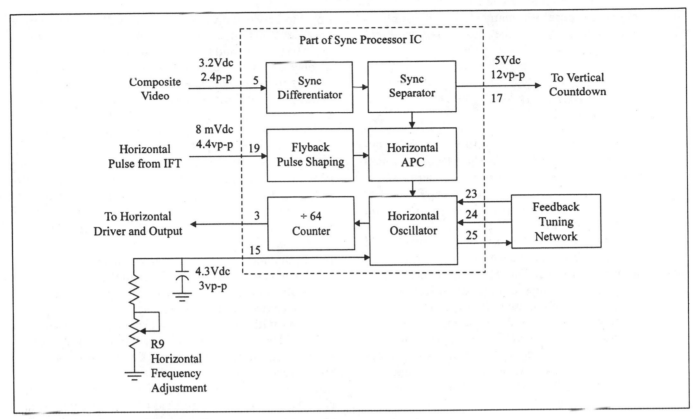

Figure 5-30   Block Diagram of the Sync Processor IC Horizontal Section

frequency to drift. From a customer perspective, the receiver would require a longer time to recover from one signal to the next. The elimination of the broad vertical pulses allows the VCO frequency to stay closer to the free-running frequency and within capture range.

A free-running frequency of 1.007 MHz is found at the output of the VCO. A divide-by-64 counter divides the frequency down to the horizontal frequency, while a 31.5-kHz frequency is tapped from the counter for the vertical countdown circuit. After buffering, the horizontal signal travels as an output from pin 3 of the IC to the horizontal output stage. External to the IC package, a feedback-tuning network connects to the VCO through pins 23 and 25. While a −22.5-degree phase shift occurs at pin 23, a +22.5-degree phase shift occurs at pin 24. The phase shifts establish the proper relationships for the horizontal oscillator output signal.

↻ FUNDAMENTAL REVIEW

Gates and Flip-Flops

Integrated circuits called gates perform the digital logic functions and combinations of the logic functions. A gate "sees" the logic values at its inputs and produces the output that corresponds with the correct logic function. Gates can be connected together so that the output of one becomes the input for another.

When the output of one gate is the input for another the configuration is called a flip-flop. A flip-flop is a cross connection of two gates that acts like a 1-bit memory element and generally consists of NOT and NAND gates. When gates of a flip-flop look at the inputs simultaneously, the gate circuit is sequential. In a sequential system, a clock oscillator outputs a continuous string of pulses that go to all parts of the system.

☑ PROGRESS CHECK

You have completed Objective 7. Explain IC-based horizontal scan systems.

5-8 TROUBLESHOOTING SYNC AND HORIZONTAL CIRCUITS

Troubleshooting sync and horizontal circuits presents a problem for many technicians because of the integration of the circuit functions. For example, a loss of sync may trace back to the sync-separator stages or to a fault within the horizontal or horizontal oscillator stages. In addition, the relationship between the horizontal oscillator stage and scan-derived power supplies may also cause problems for technicians. As you know, a scan-derived power supply cannot operate without horizontal drive. Therefore, many shutdown problems are caused by a defect in the horizontal oscillator circuit.

Test Equipment Primer— The Oscilloscope

Matching the correct oscilloscope to your service application is one of the more important test equipment decisions. Although all of us have concerns about costs, we should also recognize that the capabilities of equipment like an oscilloscope allow us to remain aligned with technological changes.

Certainly, modern oscilloscopes offer a variety of features such as built-in measurements, digital displays, and digital storage. The decision to purchase an oscilloscope should hinge on the capability of the equipment to perform fundamental tasks, the most fundamental being acquiring and displaying waveforms. With that fundamental concept in mind, you can begin matching the waveform acquisition and display capabilities with your bandwidth, rise time, triggering, and capture needs. From there, you can consider specifications such as:

- Input impedance
- Vertical-axis sensitivity
- Horizontal sweep rate
- Signal delay time
- The ability to delay the sweep rate
- The ability to display multiple traces
- Multiple functions
- Warranty

Each of these features adds to the cost of the oscilloscope and to the value it offers a service operation.

As soon as we mention digital circuitry and digital clock rates, the bandwidth and rise time needs of an oscilloscope begin to increase. A typical rule of thumb says that the oscilloscope bandwidth should exceed the highest expected frequency signal. Moreover, the rise time of the oscilloscope should allow the precise measurement of the rise time of a waveform. Rise time is the speed required for a waveform to increase from 10 to 90 percent of its total amplitude. For 5 percent accuracy, the oscilloscope should offer a rise time not more than three times faster than the rise time of often-measured waveforms. A need for 2 percent accuracy changes the oscilloscope rise time requirement to not more than five times the waveform rise time.

As an example, any attempt to measure a digital-pulse waveform shows that the waveform consists of an infinite number of sine wave components that have individual amplitudes and phases. Here the rela-

tionship between oscilloscope bandwidth and rise time becomes most apparent. Using an oscilloscope with a large bandwidth and a precise rise time allows the clean display of the waveform and discloses any problems such as ringing or overshoot. An oscilloscope with low bandwidth and limited rise time will show only a clean pulse.

Rise time and the input impedance of an oscilloscope are also related. Any high-input impedance will lower the rise time. Sensitivity refers to the capability of the oscilloscope to display the peak-to-peak amplitude of a waveform, and is expressed as volts per division, or v/div. To fully understand how the volts per division value works, you should remember that the voltage of any input signal can be measured by multiplying the number of vertical divisions covered by the displayed signal times the v/div signal. If you display a 2-v peak-to-peak signal on a 1-v/div screen, the signal will cover two vertical divisions.

Along with vertical axis sensitivity, you also need to consider the horizontal sweep rate of the oscilloscope. Measured in seconds per division, the horizontal sweep rate is the amount of time required to move the oscilloscope CRT beam across the distance of one division. You can supplement the ability of the oscilloscope to acquire a waveform by adding the capabilities for magnifying and delaying the sweep. With sweep magnification, the oscilloscope magnifies the waveform by switching the horizontal amplifier gain. Delayed sweep, or triggering, allows the oscilloscope to capture and expand on the full rise time of a pulse. With this, a technician has the ability to magnify any portion of the waveform.

Another consideration when purchasing an oscilloscope is the number of acquisition channels needed for a particular task. Oscilloscope manufacturers offer a variety of display choices that provide maximum flexibility for video servicing. Depending on the need, a technician can purchase a single-, dual-, or three-channel oscilloscope for a reasonable cost. Some manufacturers also offer four- and eight-channel oscilloscopes. For most service applications, a dual-trace oscilloscope provides a method for viewing and comparing the input and output signals at the same time. But because most digital circuits have multiple inputs and outputs, four-channel oscilloscopes have become more popular.

The need to observe multiple signals may also lead to the purchase of a higher-cost **digital storage oscilloscope**, or DSO. With the DSO, a technician can store signals in digital memory and recall them later. This capability allows simultaneous viewing and comparing of stored and real-time signals. A feature once found only digital storage oscilloscopes—peak detection— allows almost all analog, high-bandwidth oscilloscopes to capture and display fast-moving noise spikes.

Other features such as automatic setup, store/recall, and built-in measurement devices may make a higher-end oscilloscope a more attractive purchase. The automatic setup feature provides a method for the oscilloscope to automatically sense the characteristics of a waveform and adjust the display for that particular waveform. An integral part of the automatic setup feature, called store/recall, further automates the front panel setup by storing and recalling a large number of front-panel setup adjustments for specific needs.

Measurement functions complement the waveform display by placing voltage and time measurements along the waveform. The voltage settings allow measurements of peak-to-peak values, negative peak values, and positive peak values. During operation, the voltmeter function automatically tracks the waveform and displays the needed readings. Along with the voltmeter, some oscilloscopes also offer a built-in counter/timer that shows measurements of frequency, period, width, rise time, and fall time. Like the voltage measurements, the counter/timer readings can match to any portion of the waveform.

Troubleshooting the Sync Separator

If a sync-separator circuit fails completely, the displayed picture will roll vertically and break into slanting bands simultaneously. Intermittent locking of the picture discloses that the vertical and horizontal oscillators continue to work at the 60-Hz and 15,750-Hz frequencies but that the sync timing is incorrect. Partial failure of the sync-separator circuit may allow some of the black video information to become part of the sync signal. When this problem occurs, the picture will exhibit a bending or weaving effect resembling a symptom caused by overloading in the AGC circuit. Along with weaving, the picture may lose all synchronization and reproduce negative images.

As with the video amplifiers, you use an oscilloscope to find the problem source in a sync-separator circuit. In this case, a triggered-sweep oscilloscope with delayed sweep capability works best for observing horizontal and vertical sync waveforms. When observing horizontal sync waveforms, set the oscilloscope time base to 10 microseconds per division and connect the external trigger input to the source of the horizontal sync waveform. The delayed-sweep capability of the oscilloscope becomes necessary for expanding the waveform into specific segments such as the blanking interval or the sync pulses. When observing vertical sync waveforms, set the time base to 2 microseconds per division and connect the external trigger input to the source of the vertical sync signal. Adjusting the delayed sweep control to the X10 setting will again allow you to see specific portions of the waveform. To observe the blanking interval of either the horizontal or vertical sync waveforms, use the trigger control to unlock the display and then reset the control so that the display locks. Setting and resetting the trigger control allows the display to lock onto a given field.

Troubleshooting Horizontal Deflection Systems

Problems with the horizontal deflection system fall under the following general headings:

1. horizontal frequency problems
2. horizontal output problems
3. scanning voltage-supply problems

The first type of problem causes the picture to break into the mass of slanting black bars seen in Figure 5-31A. Each bar represents one portion of the horizontal blanking signal. Even with the diagonal bars tipping everyone off to the existence of a frequency-related problem, variations exist.

If the problem is with the sync signal, adjustments of the horizontal hold control will lock the picture momentarily, after which the picture will move rapidly from one side to another. The horizontal oscillator can run at the correct 15,750-Hz horizontal scan frequency but is not locked by the horizontal sync pulses. Unfortunately, the same type of symptom will appear if a component failure prevents the horizontal feedback pulse from reaching the AFC system. Both problems cause the same type of symptom because the phase comparator portion of the AFC or APC loop compares the sync and feedback pulses.

Figure 5-31A Horizontal Frequency Problems

The large oscillator frequency error may also affect the high voltage and the width of the raster. Reducing the horizontal frequency error to almost the correct scan frequency results in fewer bars as shown in Figure 5-31B. Each bar also becomes wider as the margin of error narrows. With both symptoms, the direction of the diagonal bar slant indicates whether the oscillator frequency has increased or decreased. Bars that slant left show that the oscillator frequency has shifted too low, while bars that slant right show that an increase in horizontal frequency has occurred.

Troubleshooting horizontal frequency problems calls for waveform analysis and peak-to-peak voltage measurements. Going back to the equivalent circuits used during the explanations of AFC and oscillator circuit operation, the AFC circuit relies on the phase and frequency comparison of sawtooth and rectangular waveforms. When the AFC circuit operates normally, the waveforms should have the same phase and equal peak-to-peak voltages. Defective capacitors in the frequency stages can cause the formation of incorrect waveshapes and/or peak-to-peak voltages. In addition, a breakdown in the phase comparator will cause the AFC/APC circuit to become unbalanced. As a result, the dc control-voltage range will vary from normal, and the oscillator may lose the sync lock.

One symptom of leaky capacitors in the AFC/APC circuit is called "pie-crust" distortion. With this type of distortion, the oscillator remains locked to the correct frequency, and the picture does not break into diagonal lines. The images in the picture, however, lose the crisp edges and appear to consist of a series of wavy lines.

Because of the switching characteristics and large-current handling requirements seen in the horizontal output transistor circuit, the output transistor often breaks down to a shorted condition. The loss of horizontal output results in the "no raster, no sound, no picture" condition depicted in Figure 5-31C rather than a thin vertical line stretching down the center of the CRT screen. In addition, a shorted horizontal output transistor may cause a power supply fuse to blow.

As you learned in chapter four, the horizontal output section—including the output transistor and integrated high-voltage transformer—is the source of the supply voltages for most of the circuits in the modern television. The receiver design eliminates the need for a low-voltage power transformer. Because of this, a problem in the horizontal output stage can affect the operation of the entire receiver.

Specific problems in the horizontal output stage include the failure of safety capacitors; shorted, leaky, or open horizontal output transistors; or a shorted damper diode. Each of those faults can cause the "no raster condition" to appear and can also hamper the generation of both high and low voltages. For now, though, we will concentrate on methods for finding the defective components.

Many times, the appearance of a safety capacitor will disclose that the device has become leaky. Out-of-circuit checks will confirm that the problem exists. Voltage checks at the collector of the horizontal output transistor can also lead to conclusions about a shorted, leaky, or open transistor. Again relying on out-of-circuit checks, a test of the output transistor with a transistor tester will quickly show if a problem exists. Keep in mind that most television manufacturers recommend against measuring the voltage at the collector of a horizontal output transistor. Because the high signal voltages found at that point can quickly ruin a digital multimeter, use an oscilloscope to monitor the horizontal waveform at the output transistor collector.

A breakdown within the horizontal output transistor can cause the receiver to go into a condition called high-voltage shutdown. To find if high-voltage

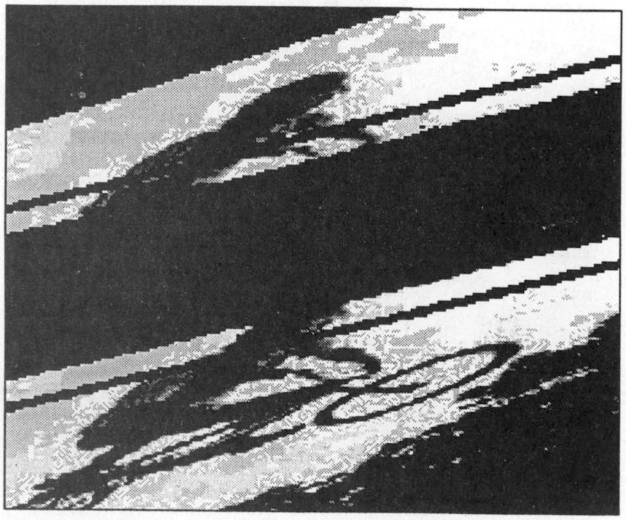

Figure 5-31B   Horizontal Frequency Problems

Figure 5-31C   No Horizontal Deflection Condition

shutdown has occurred, power the receiver through a variable isolation ac transformer. You can start the ac voltage at a low reading of 65 vac and gradually increase the voltage to a point where the receiver operates. A high-voltage shutdown condition will cause the receiver to turn off as the ac voltage nears 120 vac. As you will find in the next chapter, this type of failure involves the regulation of the generated high voltage so that a high level of X-ray emission does not occur.

With the receiver in shutdown, inject a horizontal sweep signal from a waveform generator to the base of the horizontal output transistor. If horizontal sweep results from the injection of the signal, move your voltage and waveform tests back to the circuits preceding the output stage. A defect in the oscillator, APC, or driver stages can prevent the horizontal drive from reaching the necessary levels. If the injection of the signal does not produce the horizontal sweep, your tests should include the horizontal output transistor, flyback or integrated high-voltage transformer, horizontal yoke coils, and associated voltages.

The loss of horizontal drive usually causes a narrow picture or the loss of the raster; a slight decrease in the drive amplitude will reduce the output to the yoke deflection coils. When observing a narrow picture like that seen in Figure 5-31D, also note the decrease in brightness and the distortion in the picture. More than likely, the narrow picture results from a border-line defect such as internal leakage in the horizontal output transistor or a leaky damper diode. Sometimes other symptoms, such as distortion on only the left or right side of the picture, can provide a clue about the source of the problem. With the left side distorted, check the damper diode. Right-side-only distortion usually points to the output transistor.

Two other symptoms that sometimes accompany drive problems are the horizontal foldover seen in Figure 5-31E, and a non-linear horizontal display associated with a narrow picture. **Horizontal foldover** describes a symptom where one portion of the picture folds back over one edge of the display. After checking the driver transistor output waveform and the horizontal output transistor collector waveform, measure the amplitude of the waveform at the yoke horizontal deflection coils. A defective capacitor located across the yoke coils can cause the coils to resonate at the incorrect frequency and distort the waveform. The non-linear display often points towards a short in one of the horizontal coils, and may appear as the keystone raster shown in Figure 5-31F.

Figure 5-31E   Horizontal Foldover

Figure 5-31D   Narrow Picture Condition

Figure 5-31F   Keystone Raster

Troubleshooting Horizontal AFC Systems

When troubleshooting a suspected horizontal AFC problem, first determine the type of oscillator system used in the receiver. As mentioned earlier, older video receivers may have a blocking oscillator, multivibrator, or sine wave oscillator. Generally, a study of the schematic diagram will disclose the type of oscillator used. While a sawtooth wave oscillator usually includes a blocking oscillator and a double-diode phase detector, most sine wave oscillators rely on a Hartley oscillator and a phase detector. Multivibrators will always have some type of feedback path from either the horizontal output or the horizontal driver back to the input to the oscillator.

Horizontal AFC circuit problems often result in the picture tearing into diagonal lines or breaking up, or the production of multiple, distorted images. While these symptoms may seem to narrow the search, keep in mind that problems may exist in the sync separator, the horizontal hold circuitry, or another part of the horizontal deflection circuitry. Therefore, locating a possible horizontal AFC problem involves signal injection and tracing.

↻ FUNDAMENTAL REVIEW

Multivibrators

When a circuit is designed to have zero, one, or two stable output states, the circuit is a multivibrator. An astable multivibrator has no stable output state and is a square wave oscillator. Monostable multivibrators have one stable state and generate a single output pulse with the application of an input trigger signal. Bistable multivibrators have two stable states and switch from one state to the other with the application of an input trigger signal.

Troubleshooting Horizontal Sine Wave Oscillator Circuits

Figure 5-32 repeats the schematic drawing of the sine wave oscillator horizontal deflection system originally seen in Figure 5-25. As we begin our troubleshooting analysis of the circuit, we move through the following possible problems that could affect a horizontal deflection system:

- No raster, no sound, no picture with the power supply fuse blown
- No raster, no sound, no picture but the power supply fuse does not blow
- Raster has normal width and brightness; picture not synchronized
- Distorted raster and picture
- Normal raster and distorted picture

The highlighted component labels in the figure indicate possible problem causes.

When considering a horizontal deflection problem that may cause the loss of raster, sound, and picture along with a blown power supply fuse, many potential

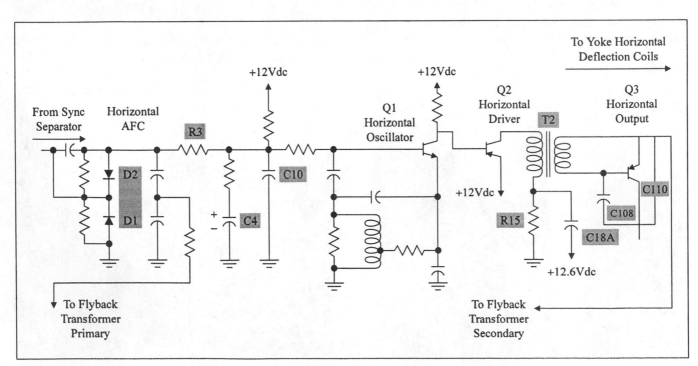

Figure 5-32 Schematic Drawing of a Sine-Wave Oscillator Horizontal Deflection System

trouble spots exist in this particular circuit. For example, a shorted capacitor C18A places the 27-ohm R15 directly across the power supply and causes the power supply fuse to blow. In addition, a shorted capacitor in the damper-diode circuit also places a portion of the flyback transformer across the power supply and causes the fuse to blow. Other components that, when shorted, could cause the power supply fuse to blow are the damper diode, C110, C108, the horizontal output transistor, and the horizontal driver transistor. Moreover, a short between the primary and secondary of T2 or between the primary and secondary of the flyback would blow the power supply fuse. Because of the large number of problems that can cause identical symptoms to occur, and because no clear-cut troubleshooting method exists for quickly finding the problem source, troubleshooting this type of condition often involves the careful observation and testing of individual components.

A normal raster with a picture that lacks synchronization tells you that either the horizontal AFC or horizontal oscillator circuitry has stopped functioning normally. If possible, adjust the horizontal hold control until you eliminate the diagonal bars. The picture should float in the horizontal direction. When this happens, you know that the problem exists in the horizontal AFC circuit rather than in other portions of the horizontal deflection system. Then observe the waveforms found at the cathode and anode of diode D2 and compare your observations with the manufacturer's specifications.

If adjusting the horizontal hold control does not produce a floating picture, isolate the AFC circuit from the oscillator circuitry by disconnecting one end of resistor R3. Then try adjusting the horizontal hold control again for a floating picture. A problem exists in the AFC circuit if the picture floats at this point.

As an example, an open capacitor C4 or C10 affects the filtering of the dc control voltage found at the output of diodes D1 and D2. Resistors R3 and R4 along with the two capacitors remove any sawtooth components from the control voltage before it travels to the base of the horizontal oscillator. In addition, those components prevent rapid changes from occurring in the control voltage and overcorrecting the oscillator. If one of the two capacitors open, the picture will not center horizontally and will have scalloped edges.

Otherwise, a problem exists in the oscillator circuitry. An abnormal waveform at the Q1 collector points to possible problems either in the AFC circuitry or the oscillator circuitry. For example, a shorted capacitor C10 would remove the forward bias from the base of the oscillator and would not allow oscillation to occur.

Carefully observe the waveforms at each element of the horizontal oscillator, horizontal driver, and hori-

zontal output transistors if the raster has disappeared while not causing the fuse to blow. Confirmation that the Q1 collector waveform is good with a zero amplitude Q2 base waveform narrows the search to transformer T2, resistor R15, or the horizontal driver transistor. If all waveforms up to and including the Q3 emitter waveform are correct and the receiver has no raster, a fault lies within the high-voltage supply.

Sometimes, the sawtooth capacitor connected across the yoke horizontal deflection coils will decrease and cause the raster and picture to distort. The capacitance value of the capacitor affects the linearity of the horizontal sawtooth current. Generally, this distortion takes the form of reduced width or a very slight compression at the sides of the raster. An open sawtooth capacitor will result in the display of a single vertical line at the center of the raster.

Troubleshooting IC-Based Horizontal Deflection Systems

The troubleshooting methods used for IC-based horizontal deflection systems compare nicely with the methods used for solving problems with the discrete sine wave oscillator. As before, you can use waveform and voltage observations to narrow the range of your troubleshooting search. At the start of the troubleshooting process, you can monitor the waveforms and voltages found at the countdown and VCO circuits. Along with the IC terminals, you can also check voltages at the base and collector of the driver transistor and at the base of the horizontal output transistor.

Incorrect waveshapes and voltages at any of those points can cause the receiver to lose horizontal sync or cause it to go into shutdown. If you start at the base of the horizontal driver transistor and work back to the deflection IC, you trace the progress of the drive signal from the output of the IC to the transistor. An internal fault within the IC such as a leaky condition can reduce the amplitude of the drive signal or eliminate the signal entirely. As you know, the lack of a drive pulse will cause the horizontal scan to shut down.

One certain clue about the drive signal lies within the supply voltages found at the integrated circuit. A decrease in the supply voltage at the IC almost always points toward a defect within the IC. In addition to checking the IC, logic also suggests monitoring the source of the supply voltages. While a leaky horizontal deflection IC will have low supply voltages, an open condition within the IC will cause the supply voltages to increase.

Along with voltage checks, you can use the oscilloscope to observe the shape and amplitude of waveforms at key points. One method for checking the operation of the horizontal deflection IC involves disconnecting internal power to the receiver, injecting the correct dc supply voltage at the IC, and observing the oscillator waveform. Under normal operating conditions, the oscillator should have a 31.468-kHz sawtooth waveform. After verifying the presence of the oscillator waveform, check for the presence of the horizontal drive output waveform. The lack of a drive waveform at the IC test point indicates that the IC has become defective, while the presence of a waveform shows that the driver stage has a defect.

Although the troubleshooting search seems to focus on the integrated circuit, do not ignore the components associated with the IC. If the picture breaks into diagonal lines, for example, check for electrolytic capacitors located in the oscillator circuit. Many times, an electrolytic capacitor will become leaky and, in turn, will change the voltage found at the oscillator. In addition, always inspect the components within the horizontal deflection circuit for signs of overheating, and always check the circuit solder connections. Coupling resistors often bear the brunt of circuit operation and, after overheating, change resistance. Moreover, diodes located in the voltage supply circuit may develop a leaky condition and reduce a supply voltage to near zero. Solder connections at important points such as deflection IC and the driver transformer always merit checking.

In receivers that use an IC-based horizontal deflection system, any failure within the APC loop, the VCO, or the countdown circuit will cause the receiver to completely shut down.

➲ SERVICE CALL ───────────────

Quasar Model TT6267BW-1 Television—Picture Breaking Up

The Quasar receiver arrived with the complaint that the picture broke up vertically and horizontally at times. After confirming the presence of the correct dc supply voltages in the sync-separator circuit, the next check involved verifying the presence and appearance of the vertical and horizontal sync waveforms with an oscilloscope. After setting the oscilloscope sweep setting to 30 Hz, the technician verified that the vertical sync signal was in place at the input of the circuit. Changing the oscilloscope sweep setting to 7875 Hz allowed the same check of the horizontal sync signals. As with the vertical sync signal, the horizontal sync signal was present and correct at the input to the sync-separator circuit.

The next checks involved verifying the presence of the proper sync waveforms at the integrator and differentiator networks. Again, the waveforms had the correct appearance. As it should, the checks showed only a vertical waveform at the integrator output and only a horizontal waveform at the differentiator output.

Because the customer complaint indicated an intermittent fault, the technician used a cooling spray and hot air on the sync-separator transistor and surrounding components. Still, no indication of a component breakdown was apparent. The search then moved to the transistor-noise switch circuit adjacent to the sync-separator circuit. While the transistor checked out, a check of a coupling capacitor in the base circuit of the switch disclosed that the capacitor had opened. The open capacitor prevented the application of noise pulses to the noise switch and allowed the noise pulses to trigger the deflection circuits.

➲ SERVICE CALL ───────────────

Emerson Model TT0194 Television with No Sound, Raster, or Picture

The Emerson television arrived with no sound, no raster, and no picture. A quick check of the horizontal output and the power supply stages showed that the horizontal output transistor had shorted and that a 3.9-ohm resistor in the power supply had opened. Figure 5-33 shows a schematic of the problem area.

After replacing the transistor and the resistor with original replacement parts, and operating the receiver at a reduced line voltage using a variable ac power supply, the receiver seemed normal. Because the transistor package included the damper diode, the manufacturer's schematic specified the replacement of the horizontal output transistor with an original replacement rather than with a universal replacement.

However, the high-voltage transformer, the horizontal output transistor heat sink, and the horizontal output transistor remained extremely hot. Two other noticeable symptoms were a black border along the right side of the raster and a faint white drive line at the center of the raster. The black border seemed to ripple back and forth.

The last two symptoms seemed to point toward a fault in the horizontal driver stage. A careful inspection of the components in that stage showed that a decoupling capacitor connected to the primary of the horizontal drive transformer had opened. Replacing the capacitor allowed the receiver to operate normally with no extreme heating in the horizontal output stage.

➲ SERVICE CALL ───────────────

Quasar TV Model TT6267BW-1 (chassis# ALDC175) Has a High B+ Voltage and an Unstable Horizontal Hold

The customer brought the Quasar television into the shop with the complaint that the horizontal hold would

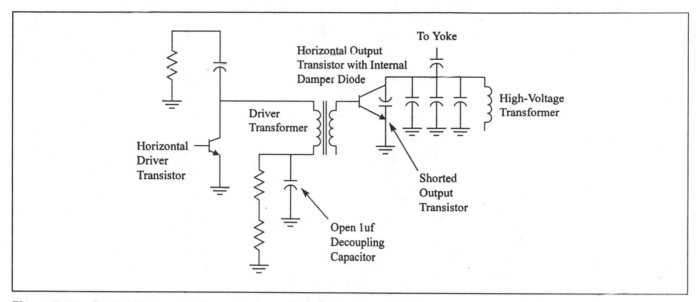

Figure 5-33   Partial Schematic for an Emerson TT0194

not lock. A preliminary check of the horizontal deflection circuits showed that the B+ voltage supplying the circuits was too high.

A check of the service literature for the receiver showed that an overvoltage condition would cause the horizontal oscillator to go off frequency. The service literature pointed out that the receiver included an overvoltage protection circuit and that the circuit would affect the oscillator frequency in an effort to lower the high voltage found at the CRT. If the high voltage goes too high, a viewer could be subjected to dangerous X rays.

Because the horizontal oscillator and its associated components operated correctly, the technician disconnected the circuit and replaced it with a dummy load. This procedure showed that the B+ continued to range high even with the connection of a different load. A component level check in the power supply regulator circuit disclosed a defect in the 3-terminal IC regulator. In addition, the analysis shows that the resistance of a 10 kilohm feedback resistor in the same circuit had also increased. Replacing the regulator IC and the resistor allowed the horizontal oscillator to lock onto the correct frequency and restored the normal operation of the receiver.

➲ SERVICE CALL

Quasar Model TS-977 Chassis—Picture Has No Horizontal Hold

When the customer brought the Quasar TS-977 chassis into the repair center, the picture had broken into many diagonal lines. A quick check of the schematic showed

that the receiver relied on an all-in-one IC package. That is, the IC contained the vertical and horizontal countdown circuitry, AGC circuitry, and video circuitry.

Before drawing any conclusions about the source of the problem, the technician measured the B+ voltage at the output of the regulator. On the chassis, test point TPD91 provided a convenient measuring point. The voltage measured as the correct +123-Vdc at the line voltage of 120 vac. With the correct voltage at the regulator, the technician concluded that the primary power supply regulator was working correctly and that the B+ adjustment remained at the correct setting.

Along with those checks, the technician also took a moment to study the overvoltage protection circuit used in the TS-977 chassis. As the schematic drawing Figure 5-34 shows, diode D513 rectifies the horizontal retrace supplied by the flyback winding at pin 1. When any variations of high voltage occur, the positive dc voltage found at the cathode of D513 varies proportionately with the changes.

Under normal operating conditions, both Q553 and Q554 remain cut off. However, the dc voltage found at the D513 cathode is applied to the base of Q553, the overvoltage protection amplifier, through resistors R531 and R532. Because zener diode D512 clamps the emitter voltage of Q553 and establishes an operating level, any changes in the base voltage above a given level affect the conduction of Q553. If the amplitude of the horizontal retrace pulse increases, the amplitude of the dc voltage at the D513 cathode also increases, causing Q553 to conduct and, as a result, causing the Q553 collector voltage to decrease.

The decrease in the Q553 collector voltage also causes the Q554 base voltage to decrease. Therefore,

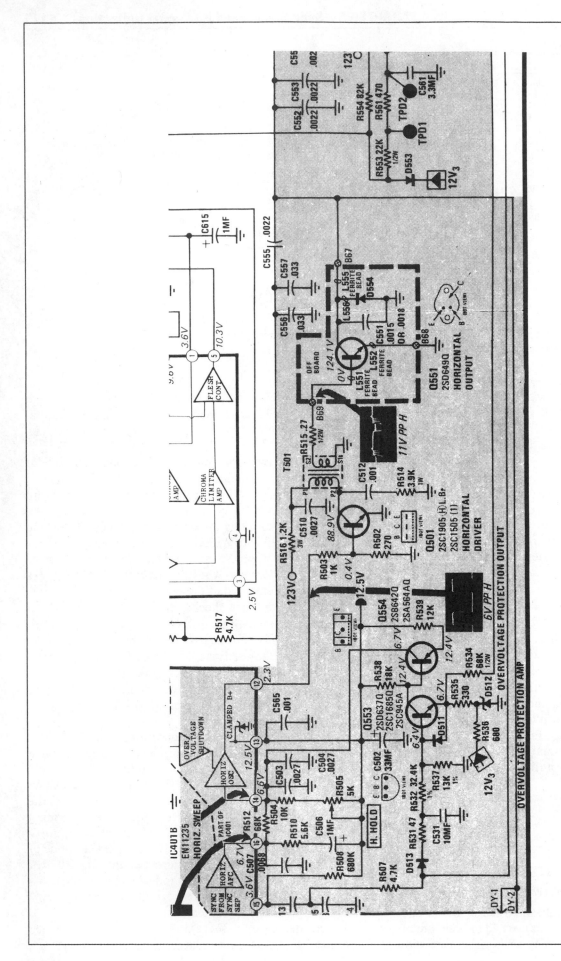

Figure 5-34 Schematic of the Quasar TS-977 Chassis Horizontal Deflection Circuit (*Courtesy of Matsushita Electric Industrial Company*)

Q554 conducts and presents a smaller resistance in parallel with resistor R504 and horizontal hold control R505. With that lowered resistance in place, the oscillator frequency increases and the high voltage decreases.

Referring to the manufacturer's schematic shown in Figure 5-34 and to the service guidelines, the technician found that shorting the base and emitter of Q553 provided a method for confirming whether the problem originated in the overvoltage protection circuitry or the horizontal AFC/oscillator circuitry. If shorting the elements together allowed the horizontal sync to lock, the overvoltage protection circuit was the problem source. If not, a problem had occurred in the horizontal AFC, oscillator, or sync circuit.

In this case, the horizontal oscillator did not lock in sync. Consequently, the technician began checking the waveforms and voltages at the deflection IC. In addition, the checks involved inspections of the solder connections and the components associated with the integrated circuit. Through careful observation of the waveforms at the AFC and oscillator terminals, the technician found that the sync pulses and comparison pulses applied to the AFC circuit were incorrect. Further checks of the components surrounding the IC disclosed that C506, a 1-uf electrolytic coupling capacitor, had become leaky and that the resistance of R504, a 10-K resistor, had increased. Both components connect to pin 16 of IC401. The replacement of both components restored the normal operation of the receiver.

☑ PROGRESS CHECK

You have completed Objective 8. Troubleshoot horizontal deflection systems.

SUMMARY

Chapter five defined each of the stages that make up a horizontal deflection system in a television receiver. Before moving into those stages, though, the chapter discussed the separation of horizontal sync signals from the composite video signal. The chapter showed that several different types of sync separator circuits exist and that the sync signals must be protected against noise.

You learned how a conventional oscillator or a countdown circuit could provide the drive signals for the remainder of the section. The chapter described the differentiator that filters undesirable signals out of the horizontal sync signals, and moved to definitions of the horizontal AFC and oscillator stages. You found that the horizontal AFC stage prevents the horizontal oscillator from shifting off-frequency.

From there, the chapter covered horizontal waveshaping and drive circuitry, where you learned that the driver stage must produce a specific pulse width. The discussion of the functions provided by the horizontal output stage showed that the circuit requires a damper diode and a special type of transistor for the output stage.

The theory behind the horizontal stages was illustrated with descriptions of the ways different horizontal AFC, oscillator, and driver circuits function. The chapter pulled all the stages of a horizontal system together into a circuit, and then moved to newer circuits that enclose all but the horizontal output function within one semiconductor device.

The troubleshooting section added to your knowledge of oscilloscope functions, and illustrated horizontal circuit problems such as horizontal frequency problems, horizontal output problems, and scanning voltage supply problems.

REVIEW QUESTIONS

1. Name the basic stages of a horizontal deflection system.

2. The horizontal deflection frequency is:
   a. 60 Hz
   b. 31.5 kHz
   c. 15,750 Hz
   d. 15,750 kHz
   e. 60 kHz

3. Another name for the horizontal output transformer is the _____.

4. True or False   Horizontal deflection systems rely on an automatic frequency control circuit.

5. Would you use an unstable or a stable oscillator in the horizontal oscillator stage? Why?

6. The output from horizontal oscillator stage is a _____.

7. The horizontal hold control is used to _____.

8. In a horizontal driver circuit, the following determines the length of time that the dc supply voltage connects to the horizontal deflection coils of the yoke for each horizontal scan:
   a. waveshape
   b. conduction time
   c. amplitude

9. Any signal applied to the base of the horizontal output transistor must have a _____

   waveform to produce a _____ current in the deflection coils.

10. True or False   A differentiator has a short time constant and prevents unwanted horizontal pulses from entering the horizontal oscillator circuit.

11. If you are working with an older television, the horizontal oscillator stage would probably not have a:
    a. blocking oscillator
    b. multivibrator
    c. sine wave oscillator
    d. countdown circuit

12. If you are working with a newer television, the horizontal oscillator stage would be built around a:
    a. blocking oscillator
    b. multivibrator
    c. sine wave oscillator
    d. countdown circuit

13. Why is an automatic frequency control circuit needed in the horizontal deflection system?

14. A horizontal AFC circuit uses a _____ developed by either a phase or

    timing comparator to control an _____ frequency.

15. True or False   The diodes found in a sawtooth horizontal AFC system are rectifiers.

16. True or False   In a sawtooth horizontal AFC system, the frequency and phase of the injected sawtooth voltage governs whether D1 or D2 has a higher conduction rate.

17. In a sawtooth horizontal AFC, the sawtooth frequency has decreased. How does the decrease in frequency affect the trace and retrace? Does the circuit produce a negative or positive control voltage?

18. In a sawtooth horizontal AFC, the sawtooth frequency has increased. How does the increase in frequency affect the trace and retrace? Does the circuit produce a negative or positive control voltage?

19. Why is an anti-hunt circuit needed as a supplement for the horizontal AFC system?

20. Draw the anti-hunt circuit that uses a speed-up capacitor and show the location of the speed-up capacitor.

21. True or False   The use of a speed-up capacitor in an anti-hunt circuit allows the control voltage to build up gradually.

22. A horizontal driver transistor provides _____.

23. The damper diode prevents voltage induced across the deflection coils from causing a reverse current to flow through the output transistor during the retrace period. Without the damper diode, the circuit would

    experience _____.

24. Name the purposes of the horizontal output transistor.

25. The equivalent circuit shown in Figure 5-21 illustrates that the horizontal output transistor operates as a

    _____.

26. True or False   Transistors used in the horizontal output stage have a lower collector resistance than the reactance of the horizontal yoke coils.

27. Yoke horizontal deflection coils have the characteristics of an inductance and require

    _____ to produce a sawtooth current.

28. Refer to Figures 5-18 and 5-19. As the collector current increases and then decreases, the electron beam

    moves from _____ and then from _____.

29. Refer to Figure 5-25. The state of the horizontal output transistor varies between

    _____ and _____.

30. Refer to Figure 5-25. The oscillator is configured as _____. The

    _____, _____,

    _____, and _____ control the frequency

    of the oscillator. Draw the waveforms found at the base and collector of the oscillator.

31. Refer to Figure 5-25. The _____ of transformer T2 affects the

    _____ of the signal applied to the base of the horizontal output transistor.

32. Is an integrated circuit a discrete device?

33. True or False   The horizontal output stage is usually enclosed within the same IC package as the horizontal AFC and oscillator circuits in an IC-based system.

34. The input signals for the IC-based horizontal circuit are taken from the _____

    and the _____.

35. The phase detector circuit that makes up the horizontal APC consists of:
    a. a gated differential amplifier
    b. a binary counter
    c. two diodes located external to the IC
    d. an operational amplifier connected to a VCO

36. Refer to Figure 5-27. True or False   The horizontal APC circuit compares the differentiated sync signal with a sawtooth signal.

37. Refer to Figure 5-28. True or False   The sync differentiator found within the IC improves the performance of the PLL system by enhancing the broad pulse seen during the vertical sync period.

38. Refer to Figure 5-28. True or False   A divide-by-64 counter establishes the proper horizontal scan frequency.

39. Refer to Figure 5-28. The free-running frequency for the VCO is _____.

40. Name the three basic categories of horizontal deflection problems.

41. When the picture breaks into diagonal bars, the cause is usually found in _____

    or the _____ stages. If the bars slant toward the left, the horizontal frequency has:
    a. increased
    b. decreased

    Bars that slant towards the right indicate that the horizontal frequency has:
    a. increased
    b. decreased

42. Even a slight decrease in the amplitude of the drive signal will cause _____.

43. In Figure 5-32, an abnormal waveform at the Q1 collector points to possible problems

_____.

44. If the capacitance of the sawtooth capacitor connected across the horizontal yoke coils decreases, the

symptoms include _____ .

45. Refer to Figure 5-32. What is the effect of a shorted C18A?

46. When troubleshooting a problem associated with a horizontal deflection IC, you should
   a. check all waveforms and voltages
   b. check the solder connections
   c. check the status of associated components such as capacitors, diodes, and resistors
   d. all of the above

47. If a sync-separator circuit fails completely, the displayed picture will

_____.

48. Intermittent locking of the picture discloses that the vertical and horizontal oscillators continue to work at the 60-Hz and 15,750-Hz frequencies, and that the sync timing is:
   a. correct
   b. incorrect
   c. unimportant

49. True or False   Partial failure of the sync-separator circuit may allow some of the black video information to become part of the sync signal.

50. Refer to the first Service Call. What were the symptoms that led to the discovery of the open decoupling capacitor? What methods were used to find the defective component?

51. Refer to the second Call. Outline the procedures that the technician followed to solve the problem.

52. Refer to the second Call. If shorting the base and emitter elements of Q553 would have locked the horizontal

sync, the problem would have been in the _____.

53. A narrow picture is usually caused by _____.

54. Horizontal foldover is caused by _____.

55. True or False   When checking a horizontal deflection IC for possible defects, one key test point is the oscillator frequency.

56. True or False   In modern television receivers, the horizontal deflection system is the source for scan-derived voltages needed throughout the receiver.

57. True or False   When a receiver loses horizontal output, the resulting symptom is always a raster reduced to thin white line stretching from top to bottom in the center of the screen.

CHAPTER 6

Sync Signals and Vertical Deflection

OBJECTIVES   Upon completion of this chapter, you should be able to:

1. Understand the basics of vertical deflection systems.
2. Describe vertical oscillator and multivibrator circuits.
3. Discuss vertical driver and output circuits.
4. Discuss vertical scanning and countdown circuits.
5. Troubleshoot vertical signal-processing circuits.

KEY TERMS
equalizing pulse

integrator circuit

triggered sweep

vertical blanking pulse

vertical foldover

vertical linearity

vertical sweep oscillator

vertical sync pulse

INTRODUCTION

When we began our first serious analysis of the composite video signal at the receiver, we noted that the signal contained vertical and horizontal synchronization signals. Those signals separated from the remainder of the composite video signal at the first video amplifier stage because of the operation of the sync-separator stage.

Regardless of whether the circuit handles vertical or horizontal deflection, the sweep oscillators use the separated sync signals to provide output voltages.

The voltages have waveforms that produce sawtooth currents in the coils of the deflection yoke. Every sweep stage within the deflection system is followed by an amplifier that produces the relatively large current required by the deflection coils.

Considering only the vertical scanning circuit, circuit operation—involving an oscillator, driver, output amplifier, and the yoke deflection coils—causes the electron beam to move from the top to the bottom of the raster at the vertical frequency. While the electron beam deflects horizontally, the vertical sawtooth deflection causes the beam to trace downward at a

151

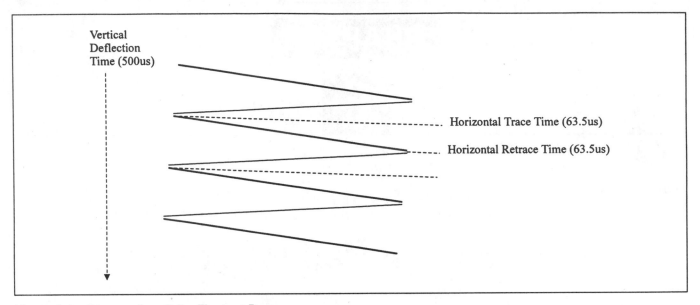

Vertical
Deflection
Time (500us)

Horizontal Trace Time (63.5us)

Horizontal Retrace Time (63.5us)

Figure 6-1   Progression of the Electron Beam

constant speed. A rapid vertical retrace returns the beam to the top of the raster. Figure 6-1 shows the progression of the electron beam.

Technological advances have produced several circuit designs for the processing of the vertical sync signals. Early designs include blocking and multivibrator circuits that use coupling transformers. Chapter six gives a detailed overview of those designs, and also considers newer vertical deflection circuits that utilize an amplifier, current feedback, and a precise current-sensing resistor. The chapter concludes with step-by-step instructions for troubleshooting each of those designs, as well as a discussion about the use of an oscilloscope for finding circuit problems.

6-1 VERTICAL SYNC SIGNALS

Figure 6-2 shows a composite video signal waveform before sync separation occurs. The **vertical blanking pulse**—a long-duration rectangular wave—carries sync and equalizing pulses on top of the waveform. At the end of each vertical scan, a blanking pulse initiates conditions that reduce the electron beam intensity to zero as the beam enters the retrace interval. Once the electron beam scans the last line of picture information, the vertical sync and equalizing pulses occur during the retrace interval.

The **vertical sync pulse** consists of six serrated pulses that occur at the end of each field scan. Within this group of six pulses, each individual pulse has a duration of 27.2 microseconds, with 4.5 microseconds separating each pulse. **Equalizing pulses** are another type of sync pulse that occur directly before and after the vertical pulses. Also grouped as six pulses, the

equalizing pulses have a duration of 2.5 microseconds and a 29.2 microsecond separation. Equalizing pulses determine the exact location of the scanning lines of a field in relation to the lines found in the preceding field.

While horizontal sync signals move through a differentiator circuit, vertical sync signals travel through an integrator before reaching the deflection system. The integrator uses the sync signals to produce a waveform that corresponds with the vertical sync block. Given a reasonably fast time constant, the integrator increases the stability of both the trigger level and the vertical locking. The problem of vertical locking gains importance when we consider that devices such as VCRs may send distorted vertical sync waveforms to the receiver because of nonstandard playback modes or anti-piracy sync waveforms.

Figure 6-3 shows a basic **integrator circuit**—an RC filter that places the capacitor in the shunt path. Integration occurs because the low-pass filter adds the capacitive charge to the voltage already found across the capacitor. The time constant of the filter establishes the separation of the output waveform.

As an example, a network that features a 10-ohm resistor and a 0.005-microfarad capacitor will have a time constant of 50 microseconds. When we compare this to the 4.7-microsecond horizontal pulse width, we can see that the capacitor will not have a significant charge during the horizontal pulses. Moreover, the capacitor will discharge slightly during the intervals between the pulses.

With the wider vertical pulses, the capacitor has more than adequate time to charge to and stay at its maximum level. The serrations found on the vertical sync pulse help the capacitor maintain its charge and

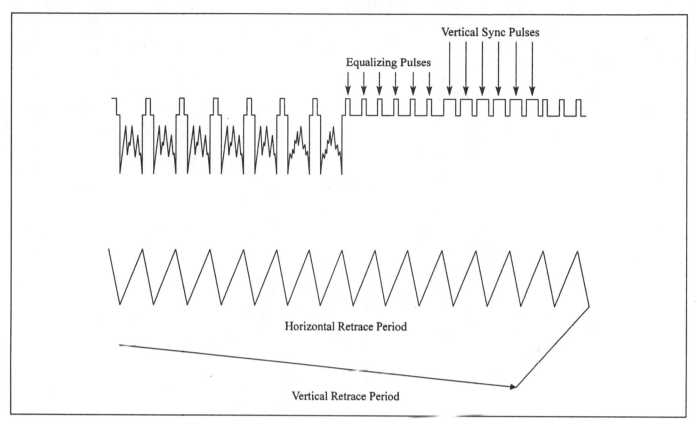

Figure 6-2 Composite Video Signal Waveform

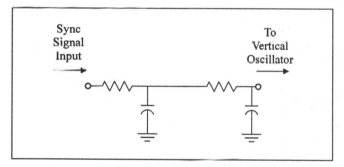

Figure 6-3 Schematic of an Integrator Circuit

allow the horizontal oscillator to maintain synchronization during the longer intervals given by the vertical sync pulses. Because of the longer period found at the end of the vertical sync pulse, the capacitor discharges completely and stands ready for the integration of the next vertical sync pulse.

Vertical Sync Pulses and an Integrating Network

Figure 6-4 again shows the transistor sync-separator circuit first seen in chapter five. Now we will take a closer look at the vertical sync pulses and the integrating network. Negative sync pulses developed across

R9, the Q3 collector load resistor, travel through an integrating network consisting of R11, C8, R12, C7, R13, and C8 to the vertical deflection circuits.

Unlike the differentiator circuit, the integrator uses an RC network that provides a long time constant. When the long-duration vertical sync pulses pass through the integrator, the output voltage of the circuit increases to the peak of the applied voltage and then decreases slightly during the retrace interval. However, as each of the short-duration equalizing pulses passes through the integrator, the output of the circuit increases slightly and then decreases to zero. In contrast to the charging and discharging of the RC network allowed by the low-frequency vertical pulse, the higher-frequency equalizing pulses allow only a slight charge and discharge of the capacitors.

At this point, the purpose of the equalizing pulses riding on the vertical sync pulse becomes apparent. During the relatively lengthy charging and discharging period during the vertical trace and retrace interval, the receiver could lose horizontal synchronization. Because of the long RC time constant, vertical trace and retrace could begin at any time during the application of a horizontal sync pulse, and could occur at different times from one field to another. The short-duration equalizing pulses prevent this from occurring by maintaining the charge and discharge timing of the RC network.

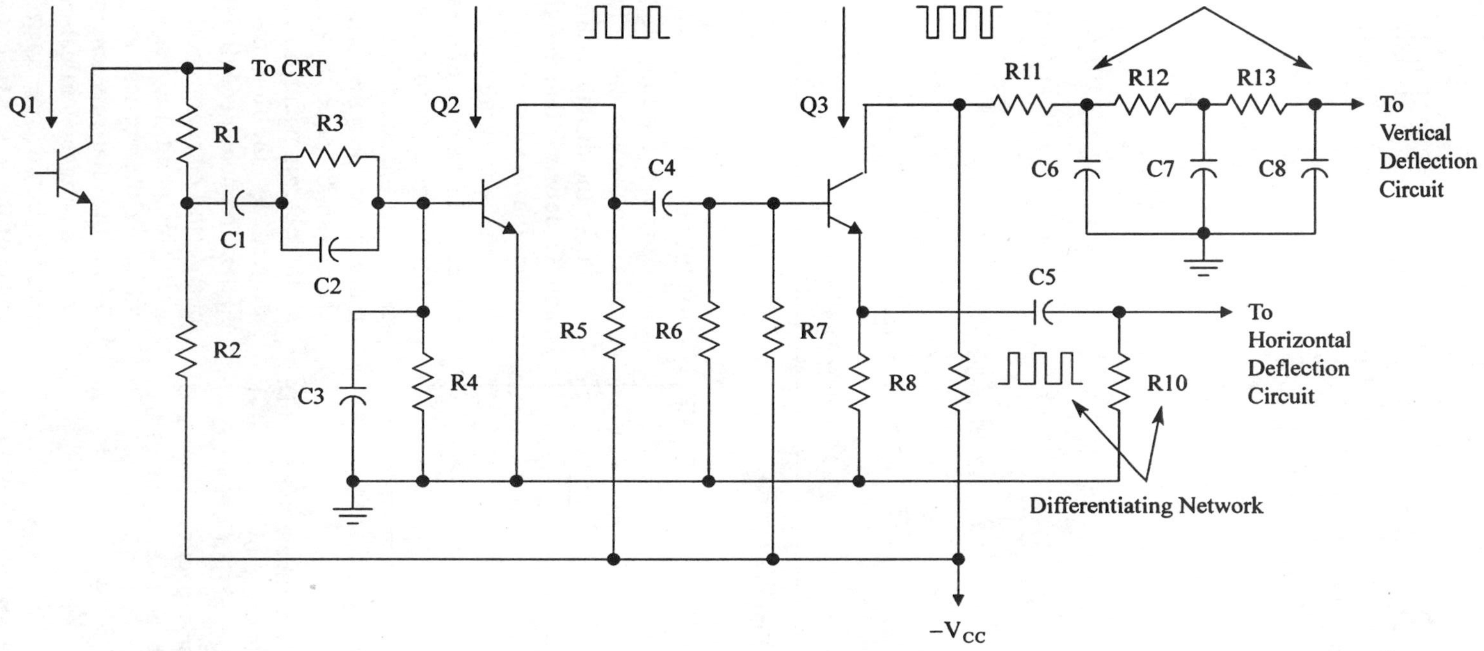

Figure 6-4   Transistor Sync-Separator Circuit

6-2 VERTICAL DEFLECTION SYSTEM BASICS

Any time that a current passes through a purely resistive component, the current will have the same waveshape as the voltage applied to the component. From this, we know that a sawtooth voltage applied across a purely resistive load produces a sawtooth current in the load. Producing a sawtooth current in an pure inductance, however, requires the application of a rectangular wave to the inductor. Deflection circuits contain both purely resistive and purely inductive components. As a result, the deflection system must produce a voltage waveform that combines a sawtooth component with a rectangular or pulse component.

Voltage builds up at a slow rate and falls at a fast rate. If we lengthen the charging time, the capacitor voltage waveform will have the slight curvature seen in Figure 6-5. Most video circuits, however, require a minimum curvature of the sawtooth wave and use capacitance and resistance values that establish a time constant much longer than the capacitor charging time.

Figures 6-6A and 6-6B use two equivalent circuits to illustrate how a sawtooth voltage forms. In the circuit, the alternate opening and closing of a switch connected across the capacitor sets up the conditions shown in the two figures. Opening the switch allows electrons to flow only from ground to the lower plate of the capacitor, and from the upper plate through resistor R1 to the positive terminal of the voltage source. The electron

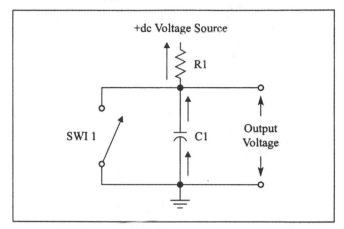

Figure 6-6A   Open-Switch Equivalent Circuit of a Sawtooth-Voltage-Forming Circuit

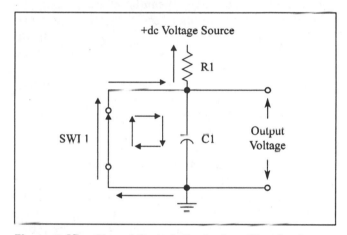

Figure 6-6B   Closed-Switch Equivalent Circuit of a Sawtooth-Voltage-Forming Circuit

flow causes the capacitor to charge to the indicated polarity and, as a result of the capacitive charge, causes a voltage to form across the capacitor. This voltage forms the output of the circuit and the ramp of the sawtooth waveform.

Momentarily closing the switch shorts the terminals of the capacitor and allows electrons to flow from the negative plate of C1, through the switch, and then to the positive plate. All this causes the capacitor to discharge. Because the switch offers much less resistance to current than the resistor, the capacitor discharges in much less time than needed for charging. As a result, the discharge output voltage forms the vertical portion of the waveform. Opening and closing the switch causes a series of sawtooth waveforms to occur.

The Vertical Oscillator

In real-time systems, the construction of the combination sawtooth/pulse waveform begins with the integrating circuit and carries through the oscillator and amplifier circuits. The oscillator operates as the switch

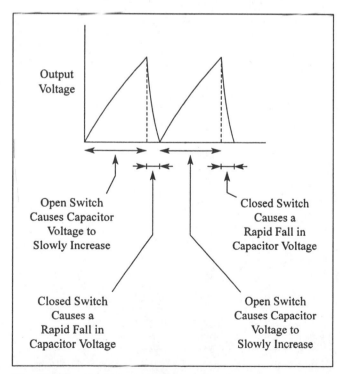

Figure 6-5   Sawtooth Voltage Waveform Found Across Capacitor

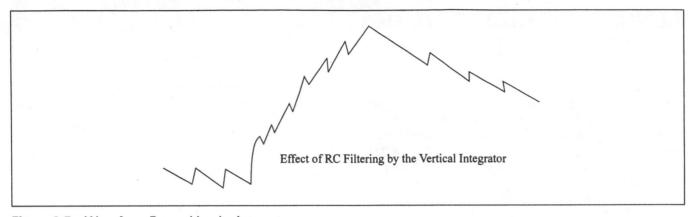

Effect of RC Filtering by the Vertical Integrator

Figure 6-7 Waveform Formed by the Integrator

seen in the equivalent circuits. Depicted in Figure 6-7, the waveform produced by the integrator moves into an oscillator that has a timing interval slightly longer than the vertical frame period. As the oscillator begins to process the vertical ramp waveform, the upper portion of the waveform triggers the oscillator into conduction and, consequently, provides vertical synchronization. At the completion of the waveform, the oscillator resets and waits for the threshold level of the next ramp.

Figure 6-8 isolates the RC sawtooth-forming circuit as it indicates how the transistor circuit forms the sawtooth voltage. During operation, the transistor remains cut off for a long period of time and conducts for short intervals. A controlling voltage at the input to the circuit has a pulse waveform and is applied to the base of the circuit. With this pulse waveform at the input, negative portions of the pulse hold the transistor at cut-off while short-duration positive portions cause the transistor to briefly conduct collector current. When the transistor is cut off, the operation of the circuit matches the operation of the open-switch circuit shown in Figure 6-6A. When the transistor conducts, the operation of the circuit matches the operation of the closed-switch circuit shown in Figure 6-6B.

Turning to Figure 6-9, we can see how the input pulse voltage, or E_{IN} determines the timing of the produced sawtooth voltage, or E_{OUT}. A negative input pulse cuts off the transistor, allows the capacitor to charge, and causes the output voltage to rise in a positive direction. A positive input pulse causes the transistor to conduct, which, in turn, discharges the capacitor and causes the output voltage to rapidly decrease to zero. In short, the negative-pulse periods determine the output voltage waveform rise time, while the positive-pulse periods determine the output voltage waveform fall time. Therefore, each cycle of the pulse voltage produces one complete sawtooth voltage cycle.

The pulse waveform depicted in Figure 6-9 closely resembles the sync pulse signal found in the composite

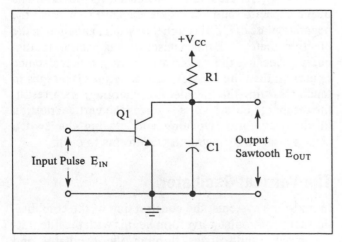

Figure 6-8 RC Sawtooth-Forming Circuit

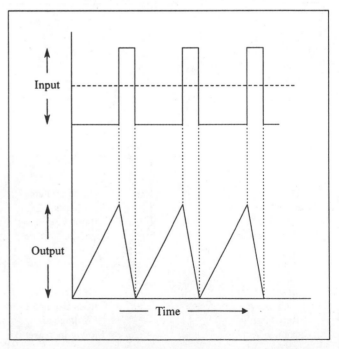

Figure 6-9 Relationship Between Input Pulse Voltage Waveform and Timing of the Sawtooth Voltage

video signal. But sync pulses alone cannot drive a sawtooth-forming circuit, for two important reasons:

1. Sync pulses have a relatively short duration, which does not provide the time needed by the CRT beam to complete vertical and horizontal retrace motions.

2. Sync pulses exist only through a signal generated at the transmission point. Tuning a receiver to a channel with no station would disable the deflection stages and cause the CRT beam spot to remain stationary.

With both reasons in mind, deflection circuits rely on pulse generators that are a part of the sweep oscillator stage. A complete **vertical sweep oscillator** includes a pulse generator, a discharge transistor, and a sawtooth-forming RC circuit. The design of the pulse generator depends on the type of oscillator circuit used in the deflection circuit.

Configured as an astable oscillator, the oscillator switches between two unstable states of conduction and cut-off at the free-running frequency of less than 60 Hz. For this reason, the oscillator has a cut-off period slightly longer than the interval between sync pulses. The astable configuration of the oscillator allows easy synchronization; an application of the vertical sync pulse locks the oscillator so that it holds at 60 Hz. When a feedback loop supplies a control voltage from the vertical output circuit back to the vertical oscillator circuit, the entire system functions as a multivibrator.

The Vertical Hold Control As the block diagram of Figure 6-10 shows, the vertical hold control can adjust the free-running frequency of the oscillator. When used, the customer-accessible vertical hold control consists of a potentiometer located in either the base or emitter circuits of the oscillator transistor. In either case, the potentiometer varies the base-emitter bias of the oscillator.

Adjustment of the vertical hold to one extreme will cause the picture to break away from the vertical sync frequency and roll rapidly. Adjusting the hold control to the other extreme will cause a slow roll in the opposite direction. To properly adjust the vertical hold control under normal operating conditions, adjust the control in one direction until the picture begins to slowly roll downward. Then turn the control in the opposite direction until the picture begins to roll in the opposite direction. Adjustment of the vertical hold control to a setting between those two limits will cause the oscillator to remain synchronized despite any changes in the free-running oscillator frequency.

Vertical Deflection Waveshaping Circuitry

The middle portion of the diagram in Figure 6-10 represents the waveshaping circuitry found within the deflection system. Rather than use the sync-separated signal as the deflection waveform, the system relies on the waveshaping network to provide a ramp waveform that has a gradual, linear rise time. Given this need, the waveshaping section includes frequency-determining components consisting of either an RC or an LC network. Because of the type of waveform produced, the components in the network have values that force the capacitor to charge slowly and discharge rapidly.

The slow charge and rapid discharge produces the linear rise time seen in the ramp, or sawtooth, waveform. Referring back to Figure 6-9, the sawtooth waveform establishes the linear scanning needed during the trace. However, the waveshaping circuit guarantees that the vertical amplifier will go into cut-off during the

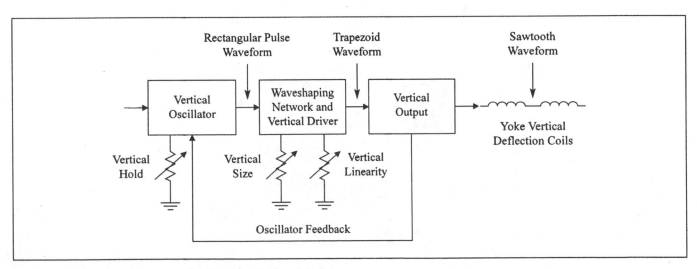

Figure 6-10   Block Diagram of a Vertical Deflection System

retrace interval by adding a pulse to the sawtooth wave-form. As a result, the sawtooth waveform evolves into the **trapezoidal waveform** seen in Figure 6-11.

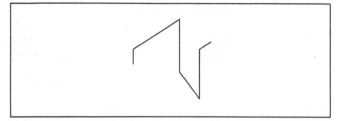

Figure 6-11 Trapezoidal Waveform

Vertical Size Control   Along with the RC circuit, the waveshaping network also includes several amplifiers. Again looking at the block diagram pictured in Figure 6-10, the vertical size, or height, and linearity controls interface with the waveshaping network. The vertical size control adjusts the amount of output given by a vertical amplifier and, as a result, the amplitude of the sawtooth current in the vertical coils of the deflection yoke. Vertical size controls are always set to fill the screen from top to bottom.

The location of the vertical size control varies with the design of the receiver. If located in the vertical oscillator stage, the size control varies the charging time of the sawtooth-forming capacitor and changes the peak-to-peak amplitude of the sawtooth voltage. If located in the driver stage, the size control adjusts the bias applied to the base circuit of the driver transistor.

Misadjustment of the vertical size control will cause the reproduced picture to have too much height, with all objects too tall, or to have insufficient height. Adjustment of a vertical size control usually coincides with adjustments of the vertical linearity control. A picture has proper height when the picture area overlaps slightly more than $\frac{1}{4}$ inch of the CRT face.

Vertical Linearity Controls   **Vertical linearity** adjusts the linearity of the sawtooth ramp waveform, and is set to eliminate the overcrowding or overspreading of the scanning lines at either the top or the bottom of the picture. A picture that lacks linearity will appear out of proportion, with either the top or bottom squeezed or stretched. As you will see in a later section, many vertical deflection systems do not utilize a manually operated size or linearity control.

When used, vertical linearity controls vary either the shape of the sawtooth voltage applied to the base of the vertical driver or the forward bias applied to the base of the vertical output amplifier. Producing a linear, proportional picture requires that the sawtooth of current applied to the vertical deflection coils has a constant rate of change. The rate of change applies to the point of retrace or the maximum peak of the sawtooth waveform.

Because the size and linearity controls interact, adjustment should follow a procedure where the size control is adjusted first for the correct picture height. Then adjustment of the linearity control should yield perfectly round circles and proportional scenes. Many color-bar generators include cross-hatch generators

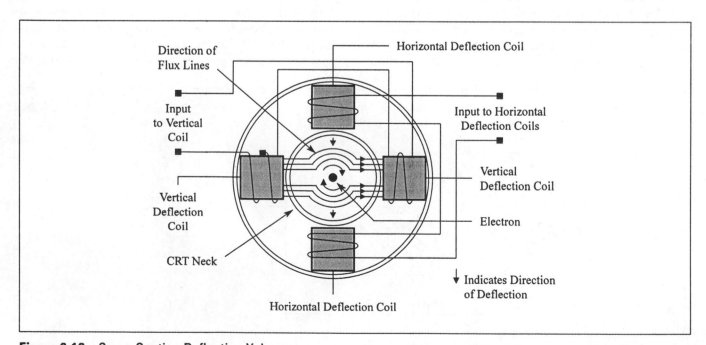

Figure 6-12 Cross-Section Deflection Yoke

that provide a visual aid for the size and linearity adjustments.

Vertical Output Amplifiers

Vertical deflection systems always include some type of large-signal power amplifier in the output stage. During the active scanning period, the amplifier current swings between cut-off and saturation. Thus, the output stage provides the large amount of current needed to drive the vertical scanning coils in the deflection yoke. Depending on the receiver design, the vertical output transistor may operate as a Class A amplifier and conduct during the scanning period, or may be part of a Class B push-pull circuit.

The waveforms seen at the output of the vertical deflection system—as well as the horizontal deflection system—travel to the deflection yoke. Illustrated in Figure 6-12, the deflection yoke contains coils that, when energized, deflect the CRT electron during the trace and retrace intervals. If the magnetic deflection coil presented only resistance, the application of a sawtooth voltage would produce sawtooth current. However, the inductance presented by the deflection coils causes the application of a sawtooth voltage and does not produce a sawtooth current. If the magnetic deflection presented only inductance, the application of a rectangular pulse voltage would produce a sawtooth current. Real-time magnetic deflection coils present both resistance and inductance. Because of this characteristic, the applied voltage must have a trapezoidal waveform that contains both square and sawtooth components.

One method for producing the trapezoidal waveform shown in Figure 6-11 is shown in Figure 6-13A. Sawtooth-forming capacitor C1 charges through the large resistance represented by R1. A second resistor, R2, connects in series with C1 and carries both the charge and discharge currents of the capacitor. When

transistor Q1 enters cut-off, C1 charges, and electrons flow from ground through R2 to the negative plate of C1, and from the positive plate through R1 to the positive-supply voltage terminal. When Q1 conducts, electrons flow from the negative plate of C1, through R2 to ground, and from ground through Q1 to the positive plate of the capacitor. The flow of electrons caused by the conduction of the transistor develops the sawtooth voltage across the capacitor shown in Figure 6-13B.

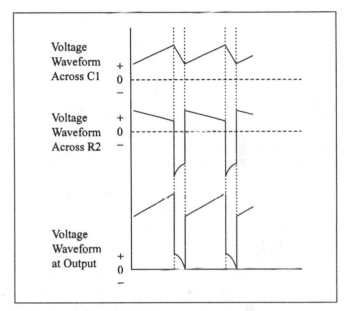

Figure 6-13B   Waveforms from Trapezoidal Waveform-Forming Circuit

As we saw, both the charge and discharge currents of C1 flow through R2. This develops the second voltage waveform shown in Figure 6-13B, which has some sawtooth characteristics and some pulse characteristics. When C1 charges, the direction of electron flow makes the upper end of R2 positive with respect to ground. When C1 discharges, the upper end of R2 is negative with respect to ground.

We can begin our final analysis of the circuit by considering the beginning of the cycle. Here the level of charge current has reached its maximum point, and the voltage found across R2 has its maximum positive value. As C1 begins to charge, the charge current decreases, and the voltage across R2 falls towards zero. At the beginning of the C1 discharge interval, the discharge current has a high value. Thus, a large negative voltage is produced across R2. As the discharge current decreases, the value of the negative voltage decreases. Once Q1 stops conducting, C1 begins to charge, and the polarity of the voltage across R2 switches to positive.

The output voltage for the circuit develops across the series combination C1 and R2. At every instant, the output voltage equals the sum of the voltage across the capacitor and the voltage across the resistor. Adding

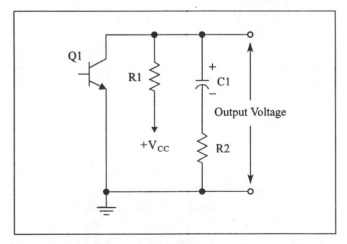

Figure 6-13A   Trapezoidal Waveform-Forming Circuit

the two voltages together produces the third waveform in Figure 6-13B, the trapezoidal waveshape of the output voltage.

PROGRESS CHECK

You have completed Objective 1. Understand the basics of vertical deflection systems.

6-3 VERTICAL OSCILLATOR CIRCUITS

In the discussion of vertical deflection basics, we noted that the system features an oscillator section, and that the oscillator has a free-running frequency. In addition, the oscillator must have characteristics that allow easy synchronization by the sync pulses. During a channel change or a station break in a network broadcast, newly arrived sync pulses generally do not correspond with the starting times of the oscillator cycles. Consequently, the oscillator must quickly change its starting time and match the new sync signals.

Obviously, these requirements rule out an extremely stable oscillator. Instead, the vertical deflection circuit relies on an unstable oscillator circuit that can follow the rapid, large changes of phase of the sync pulses. Blocking oscillators and multivibrators have the characteristics needed for these rapid changes. With each type, a cycle interruption can occur and a new cycle can begin when a large enough pulse arrives at the sync input point. Whether the system utilizes a blocking oscillator or a multivibrator configuration depends on the design of the entire vertical deflection system.

Even though the oscillator circuits are unstable and easily controlled by pulses applied to the sync input points, the vertical deflection circuit remains relatively immune to triggering by noise pulses. As mentioned in the first section of this chapter, the integrating circuit eliminates most noise pulses that pass through the sync circuits. Removing the noise pulses at this point reduces the chances that noise could falsely trigger the vertical oscillator.

A Vertical Blocking Oscillator Circuit

Within the three-stage circuit shown in Figure 6-14, transistor Q1 functions as a vertical blocking oscillator, Q2 is the vertical driver, and Q3 is the vertical output transistor. The integrator at the left of the drawing consisting of R1, R2, C1, and C2 filters the incoming positive polarity sync pulses and produces pulses that have the correct timing intervals for the deflection circuitry. Diode D1 couples the pulses to the secondary of the blocking transformer, T1. With magnetic coupling

through the transformer inverting the polarity of the signals, a negative-polarity sync signal appears at the base of the vertical oscillator.

When we examine the vertical oscillator circuitry of Figure 6-14, capacitor C3 and resistors R3 and R6 determine the free-running frequency for the oscillator. R6, the vertical hold control, adjusts the amount of dc voltage applied to the Q1 base and provides additional locking of the 60-Hz scan rate. The waveform found at the base of Q1 looks like that shown in Figure 6-15A. Moving from the base to the collector of the oscillator, C8 charges through R9 and forms the sawtooth waveform shown in Figure 6-15B.

The positive portion of the waveform seen in Figure 6-15A can shock-excite the windings of the center-tapped transformer found in the input of the oscillator circuit, and the transformer might produce a damped wave. Connected across the inductor L3, diode D2 prevents the generation of the damped wave by conducting and dissipating the excess energy. Diode D2 uses the characteristics of a diode to allow positive sync pulses to travel to the oscillator circuit while blocking any portion of the Q1 emitter waveform from coupling back to the sync circuit.

As we discussed in the last section, R12, the vertical size control, and R11, the vertical linearity control, adjust the amplitude and shape of the sawtooth waveform found at the input of the vertical driver, Q14. Optimum performance of this circuit requires the adjustment of the two controls so that a waveform resembles the one in Figure 6-15B. When we move to the collector, or output, of the vertical driver, we find that the driver has inverted the signal and has produced the positive-going sawtooth waveform seen in Figure 6-15C.

While looking at the figures, you may have noticed that the voltage at the input to the driver has a greater value than the signal found at the driver output. Although this may seem relevant, it is important to remember that transistors amplify current. When we analyze the performance of this vertical deflection system, we find that the vertical driver produces the substantial amount of current needed to drive the base-emitter circuit of the vertical output transistor. With Figure 6-15D, you can see that the vertical output transistor has a collector voltage of +65 Vdc and that the shape of the waveform has changed from a sawtooth to a pulse.

Because of the characteristics of the coils found within the yoke, the pulse at the vertical output collector produces a sawtooth of current through the vertical windings of the deflection yoke. While the bulk of the sawtooth current flows through the yoke windings, the bulk of the direct current flows through the windings of the vertical choke, T3. Applying the direct current to the yoke windings would cause the raster to become off-center.

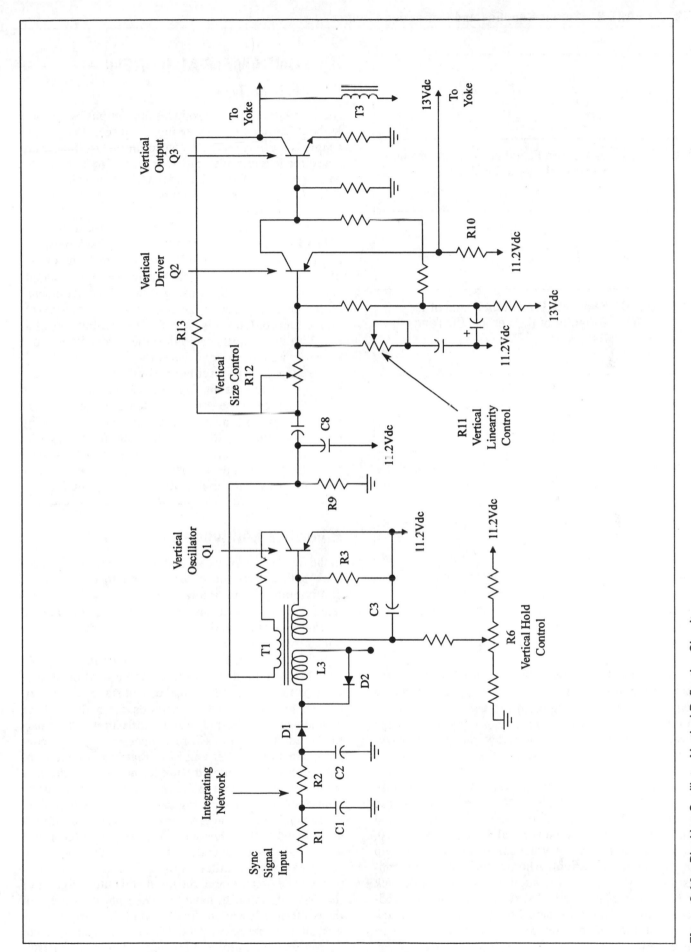

Figure 6-14 Blocking Oscillator Vertical Deflection Circuit

161

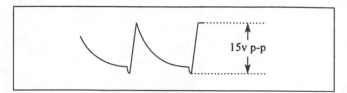

Figure 6-15A   Waveform Found at the Base of the Vertical Blocking Oscillator

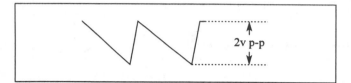

Figure 6-15B   Sawtooth Waveform Found at the Collector of the Vertical Blocking Oscillator

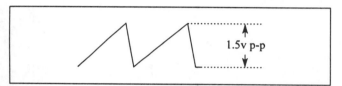

Figure 6-15C   Waveform Found at the Collector of the Vertical Driver

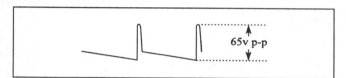

Figure 6-15D   Waveform Found at the Collector of the Vertical Output Transistor

The output has a sawtooth waveform because transistor deflection coils have a much higher resistance than reactance. The applied voltage must have a sawtooth waveform because the current develops across the resistance. As you know, the current waveform through a resistor matches the waveform of the applied voltage.

The linearity of reproduced raster improves because of a feedback circuit that allows an out-of-phase sawtooth voltage to build and oppose the input signal found at the emitter of the vertical driver. Resistor R10, the Q2 emitter, and the yoke form the feedback loop for the sawtooth current. Another feedback loop, consisting of R13 and R12, supplies in-phase feedback to the base of the vertical driver. The in-phase feedback voltage slightly offsets the out-of-phase feedback found at the Q2 emitter and stabilizes the gain of the circuit.

FUNDAMENTAL REVIEW
Oscillator Types

An oscillator circuit produces an output waveform without having an external signal source. A dc power supply is the only input for an oscillator. Oscillators are measured in terms of stability, or the ability to maintain an output that remains constant in frequency and amplitude. Common oscillator types are the Armstrong, Hartley, Colpitts, and Wien-Bridge.

The Armstrong oscillator uses a transformer to provide a 180 degree phase shift in the feedback network. Because of the presence of the transformer, the output of the transformer inverts from primary to the secondary. A discrete amplifier using a pair of capacitors and an inductor to produce the regenerative feedback required for oscillations is classified as a Colpitts oscillator. With the Colpitts oscillator, the feedback network produces a 180 degree phase shift.

Like the Colpitts oscillator, Hartley oscillators rely on tapped inductors and a single capacitor to achieve the 180 degree phase shift in the feedback loop. Wien-Bridge oscillators achieve regenerative feedback by producing no phase shift at the resonant frequency. The Wien-Bridge oscillator has a positive feedback path that supplies the non-inverting input, and a negative feedback path that supplies the inverting input.

A Vertical Multivibrator Circuit

In the first section you saw that the utilization of a feedback path allows the entire system to operate as a multivibrator. A multivibrator pulse generator does not require inductive or transformer coupling. Instead, the multivibrator system depends on positive feedback obtained by RC circuits.

In Figure 6-16, the schematic shows that Q1, Q2, and Q3 continue to function as the vertical oscillator, vertical amplifier, and vertical output stages. However, the design combines two methods for making those components work together as a multivibrator. This particular multivibrator is a resistance-capacitance coupled, common-emitter amplifier that has the output of the second stage coupled back to the input of the first stage. Since the signal in the collector circuit of a common-emitter amplifier is reversed in phase with respect to the input of that stage, a portion of the output of each stage is fed to the other stage in-phase, with the signal at the base as regenerative feedback. The amplified feedback causes oscillation to occur.

With the first method, the input and output signals of the vertical amplifier have the same phase because of the emitter-follower configuration of the transistor. As a result, the phase reversal caused by the conduction of the vertical oscillator and the vertical output transistor

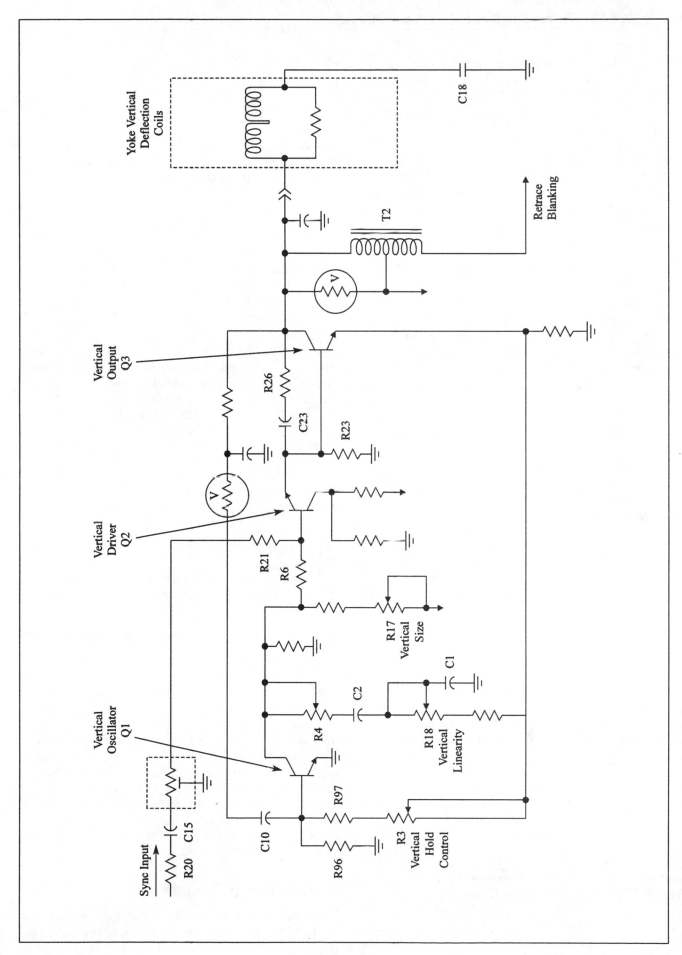

Figure 6-16 Schematic of a Vertical Deflection Multivibrator Circuit

also causes the output of Q3 to be in phase with the input to the oscillator. The second method establishes a feedback path through resistor R5, capacitor C4, resistor R6, and capacitor C5 and—using the proper amount of feedback—causes oscillation to occur.

In the diagram, C10, R7, and R8 work with R3, the vertical hold control, to determine the correct operating frequency for the circuit. The addition of the components in the feedback path also allows those resistance and capacitance values to affect the operating frequency. Therefore, no convenient dividing line exists between the vertical oscillator and vertical amplifier stages. Consequently, a problem in any part of the vertical deflection system can stop oscillations and reduce all vertical waveforms to zero.

As we analyze the circuit, the charging and discharging of capacitor C1 combines with a positive feedback pulse to produce the waveform shown in Figure 6-17A. Looking again at the schematic, the waveform appears at the base of Q1. Capacitors C1 and C2 in the Q1 collector circuit form a series combination circuit and produce the sawtooth waveform. R4 adds variable peaking to the circuit. The operation of this circuit supplies the trapezoidal waveform found at the collector of Q2 and seen in Figure 6-17B.

From there, R6 couples the trapezoidal waveform from the Q2 collector to the Q3 base. As the diagram shows, Q2 is configured as an emitter follower. Because of this, the same trapezoidal waveform appears at the Q2 emitter and the Q3 base. Figure 6-17C shows the waveform found at the collector of Q3. This waveform is applied to the vertical windings of the deflection yoke and produces the sawtooth of current needed to deflect the CRT beam. C18 prevents any direct current

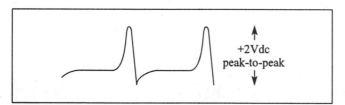

Figure 6-17A   Waveform at the Base of the Vertical Oscillator

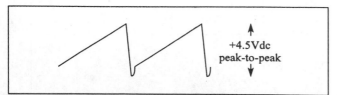

Figure 6-17B   Waveform at the Collector of the Oscillator, Base, and Emitter of the Vertical Driver, and Base of the Vertical Output Transistor

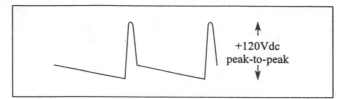

Figure 6-17C   Waveform at the Collector of the Vertical Output Transistor

from flowing through the yoke windings. Transformer T2 inverts the Q3 collector pulse and applies it as the vertical retrace blanking pulse.

Configuring the circuit as a multivibrator adds several interesting side notes to our analysis of the circuit. Synchronization of the oscillator occurs through the coupling of negative-polarity sync pulses through R20, C15, and R21 to the vertical amplifier base. At the vertical output transistor emitter, a 1-v peak-to-peak trapezoidal voltage feeds back to the base and collector circuits of the vertical oscillator through an integrating circuit consisting of R18, or the vertical linearity control, and C1. Because of the integration found prior to the vertical oscillator collector circuit, the sawtooth portion of the trapezoid waveform developed by Q1 maintains linearity. Finally, a dc voltage found at the Q17 emitter establishes the forward bias for the vertical oscillator and stabilizes the dc voltages found around the complete circuit.

☑ PROGRESS CHECK

You have completed Objective 2. Describe vertical oscillator and multivibrator circuits.

6-4   VERTICAL DRIVER AND OUTPUT AMPLIFIER CIRCUITS

Deflection output amplifiers increase the level of the signal generated by the oscillator and supply a sawtooth current to the deflection coils. To accomplish this task, the deflection amplifier must have an input waveform that has the proper shape to cause a sawtooth current waveform in the deflection coil circuits. Because the vertical deflection coils have a very low reactance when compared to the collector resistance of a vertical deflection amplifier, the input signal to the amplifier must contain a large sawtooth component to produce a sawtooth current.

Vertical Driver Circuits

Transistor deflection circuits require the addition of an intermediate amplifier, called a driver, to produce the large signal needed by the deflection coils. The driver stage fits between the oscillator stage and the output stage. With the driver in place, the transistor-based circuit can directly supply the necessary current to the deflection coils without the use of a transformer. The low load impedance of a transistor matches nicely with the impedance of the deflection coil.

Most vertical driver circuits use a common-emitter configuration because of the high-stage gain and the impedance matching given by the configuration. On the latter point, the collector resistance of a common-emitter-configured transistor is much higher than the reactance of the deflection coils. As a result, the circuit operates in a resistive mode, and a sawtooth-driving voltage produces a sawtooth current in the coils.

Moving to the circuit shown in Figure 6-18, the negative portion of the sawtooth voltage applied to the base of Q1 causes the amplifier-collector current to increase and produces a positive collector voltage. The voltage remains positive because of the voltage drop seen across the deflection yoke coils. When the sawtooth base voltage changes direction during retrace, the current flowing through the coils decreases and induces a voltage across the deflection coils. The induced voltage adds to the Q1 collector voltage.

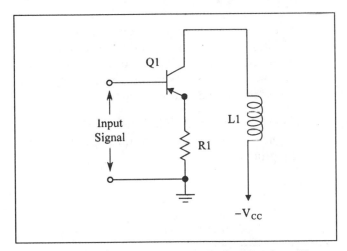

Figure 6-18   Simplified Vertical Deflection Amplifier Circuit

Referring to the waveforms shown in Figure 6-19A, the addition of the two voltages causes the output waveform depicted in Figure 6-19B to go negative with respect to the source voltage found at the collector. While the waveform shown in Figure 6-19A represents the sawtooth-driving voltage applied to the Q1 base, the output waveform appears between the Q1 collector and ground.

When the sawtooth voltage applied to the base of the amplifier goes negative, the amplifier-collector current increases. Because of this increase and the voltage drop across the deflection coils, the collector voltage begins to change in a positive direction. However, when the

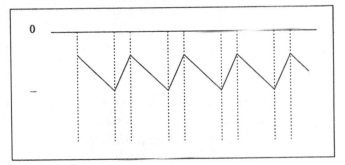

Figure 6-19A   Sawtooth Driving Voltage Waveform (*Courtesy of Philips Electronics N.V.*)

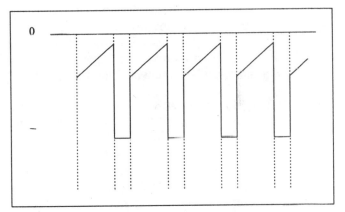

Figure 6-19B   Driver Output Waveform (*Courtesy of Philips Electronics N.V.*)

sawtooth base voltage suddenly changes direction during the retrace interval, the current flowing through the deflection coils decreases at the same rate. From your fundamental knowledge of electronics, you know that this action induces a voltage across the deflection coils. When the induced voltage adds to the source voltage at the Q1 collector, the output voltage of the transistor goes negative with respect to the source voltage. The output voltage is found at the Q1 collector.

During the trace interval, the input voltage waveform slants downward. The polarity of the high-induced voltage across the deflection coils reverse biases the collector junction during the retrace interval. While this may seem complicated, the action has the sole purpose of reducing the vertical retrace time. From the perspective of component performance, the vertical output transistor must withstand the peak voltage that occurs during retrace. Shortening the retrace time eliminates some of the stress on the output transistor. Figure 6-19C shows the sawtooth current waveform found at the deflection coils. As the sawtooth current passes through the coils, it produces the linear vertical trace on the CRT screen.

Figure 6-20 completes our analysis of vertical output amplifier circuits with a schematic drawing of a deflection amplifier and driver stage. Capacitor C1

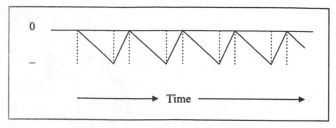

Figure 6-19C Sawtooth Current Waveform Found at the Deflection Coils (*Courtesy of Philips Electronics N.V.*)

couples the vertical oscillator output signal to the base of the Q1 driver stage. Resistors R1 and R2 form a voltage divider and establish the proper base bias for the driver. Capacitor C2 couples the output of the driver to the amplifier. Resistors R5 and R6 operate as a voltage divider and establish the base bias voltage for the amplifier.

Going back to the input of the circuit, a series RC circuit consisting of R4 and C3 connects between the Q1 collector and ground and shapes the Q1 output waveform. In this diagram, potentiometer R4 is the linearity control. Looking at the amplifier circuit, potentiometer R7 controls the amplifier gain by controlling the base-emitter bias. Potentiometer R7 is the height control. Q2 amplifies the driver output and applies the signal to the deflection coils.

Vertical Centering   In some circuit designs, compensation for beam decentering occurs through the shunting of the collector output through a coil and the coupling of the amplifier output through a capacitor tied to the deflection coils. As a result, no direct current

flows through the coils to push the beam off-center. Other designs use permanent magnets to oppose the beam-pulling of the deflection coils. The magnets attach to the rear portion of the deflection yoke. Still other receivers utilize centering potentiometers that vary the amount of direct current flowing through the deflection coils. Regardless of the design approach, centering the picture should not disturb either the picture linearity or the picture size. Incorrect centering of the raster will obscure either the top or bottom of the raster. Proper centering of the picture also involves a quick check of the deflection yoke assembly to ensure that the yoke rests as far forward as possible on the CRT neck.

☑ PROGRESS CHECK

You have completed Objective 3. Discuss vertical driver and output circuits.

6-5 IC-BASED VERTICAL SCANNING CIRCUITS

Each of the early vertical oscillator and multivibrator circuits described in the last section used transformer coupling between the circuit and the yoke. As technologies evolved, vertical scanning circuits abandoned transformer coupling because of the following factors:

- Added expense of using transformers
- Higher power dissipation through transformer coupling

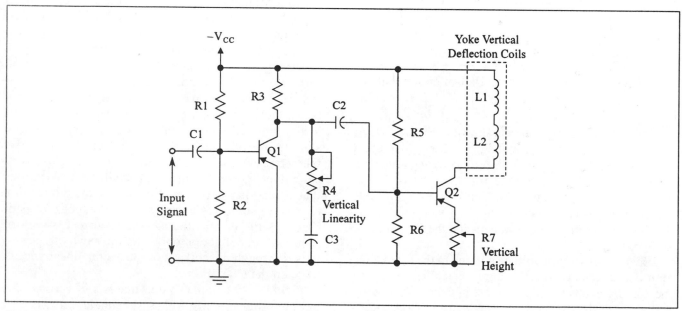

Figure 6-20   Schematic of the Vertical Driver and Amplifier Circuit (*Courtesy of Philips Electronics N.V.*)

- The additional linearity control required by transformer coupling
- The need for a thermistor to compensate for any heat-related resistance changes in the yoke

Newer vertical deflection circuits utilize an amplifier, current feedback, and a precise current-sensing resistor. With this, the circuit no longer requires an adjustable vertical linearity control or the addition of a thermistor in the yoke circuit. As Figure 6-21 shows, the amplifier contains a complementary stage consisting of a single NPN transistor and a single PNP transistor. Like many audio amplifier circuits, the vertical circuit may also rely on quasi-complementary or Darlington configurations.

Regardless of the amplifier configuration, the proper bias for the amplifier is the focal point of the circuit. With too little quiescent current, the circuit will experience cross-over distortion. As a result, the reproduced picture will also display a fine horizontal white line through the center of the picture. Too much quiescent current will cause higher power dissipation within the output transistors and the eventual failure of the transistors.

Returning to Figure 6-21, direct coupling allows the transferral of the trapezoidal waveform from the circuit to the yoke. In theory, the use of direct coupling should require the application of two power supplies. The supplies would allow the adjustment of the vertical centering of the picture and ensure the stability of the amplifier dc-bias point. Yet, although dc coupling provides efficiency, the addition of a coupling capacitor prevents excessive current from building in the output circuit, traveling through the yoke, and damaging the CRT. In some cases, the excessive current has caused CRT necks to break.

Moreover, the addition of the capacitor eliminates the requirement for two power supplies. With the dc current no longer direct coupled to the yoke, the dc-bias point is no longer critical, and the centering of the raster no longer depends on a voltage source. The current source in the circuit causes a ramp of voltage to appear across the capacitor and the current-sensing resistor. Any dc voltage found across the capacitor is controlled by the retrace interval. The coupling capacitor and a voltage-feedback resistor also serve to correct the sawtooth waveform through waveshaping.

Vertical Countdown Circuits

The vertical countdown circuit depicted in both Figures 6-22A and 6-22B uses digital logic to take advantage of the frequency relationship between the horizontal and vertical scan rates. With this, one oscillator running at 31.5 kHz improves the noise immunity of the vertical circuit and the overall picture stability by

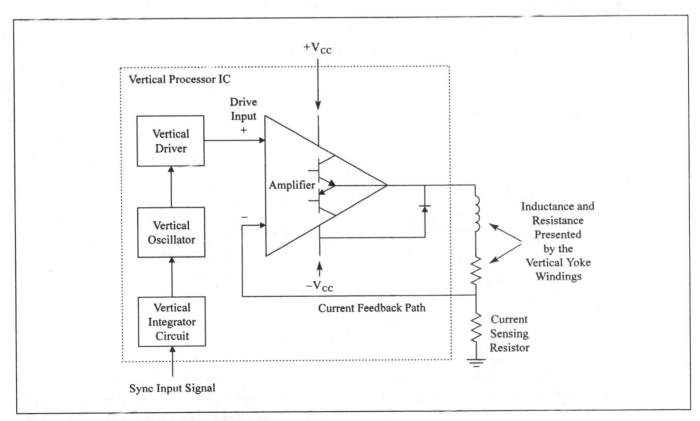

Figure 6-21   Schematic of a Class B Vertical Amplifier Circuit

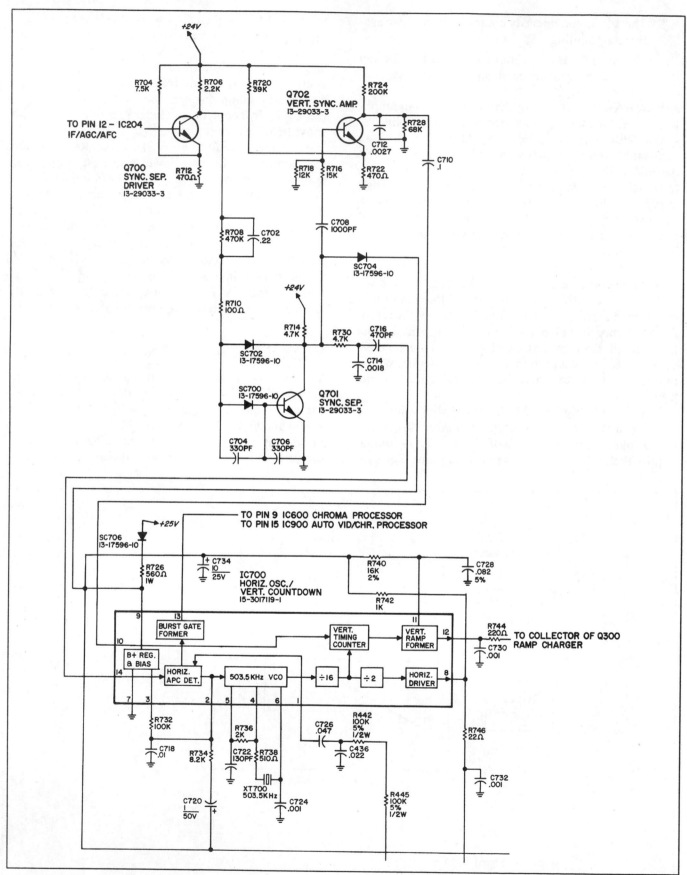

Figure 6-22A   Schematic of a Vertical Deflection Circuit using Vertical Countdown (*Courtesy of Philips Electronics N.V.*)

phase-locking the vertical and horizontal scan rates. In short, the vertical sync is taken from a pulse train having a frequency twice the horizontal scan rate. As a result, the design offers exact timing for both the odd and even fields and provides precise interlace scanning with no need for a vertical hold control.

In Figure 6-22A, the combination horizontal/vertical oscillator IC, IC700, contains a divide-by-two flip-flop that converts the 31.5-kHz clock input into 15,750-Hz horizontal drive pulses. In addition, a vertical timing stage within the same IC reduces the output from the flip-flop to 60 Hz by dividing the 31.5-kHz signal by 525. From this, the vertical ramp appears at pin 12 of the IC. Because the horizontal and vertical circuits rely on a common oscillator and on the counting down of the scan frequencies, the receiver does not need either vertical or horizontal hold controls.

Pausing with our circuit analysis for a moment, a 525-count counter shown in Figure 6-22B establishes the 525 counts needed for interlaced signals. The same counter could provide the 541 counts needed for non-interlaced signals. Depending on the circuit, the number of counts establishes the vertical output pulse. Although televisions utilize an interlaced standard, many VCR cameras, video games, and most computer monitors rely on a non-interlaced format.

In Figure 6-23, Q300, the ramp-charger transistor, charges C303, the linear sawtooth capacitor. Static adjustment through the height control, or R306, and dynamic adjustment through the Q300 base-voltage amplitude control the amount of charge across C303. The Q300 base voltage contains a mixture of the vertical drive waveform obtained from C303 and a dc voltage obtained through R315. Negative-going pulses at pin 12 of IC200 discharge the linear sawtooth capacitor through diodes SC303, SC306, and R744.

Q301, the vertical inverter transistor, inverts and amplifies the sawtooth signal, while Q302, the vertical buffer transistor, drives both the linearity circuit and the voltage divider supplying the input signal for IC304, the vertical output IC. The signal found at pin 4 of the integrated circuit drives the vertical deflection coils directly.

C325 also couples the vertical signal to the coils, while R354 establishes part of a negative feedback path back to pin 2 of IC304. The feedback signal reduces distortion and non-linearity within the reproduced picture. As the yoke coil inductance collapses, diode SC308 clips off most of the positive-going pulse. Reducing the pulse amplitude at this point eliminates the possibility that excessive amplitude could short the vertical output IC.

A Completely IC-Based Vertical Scan System

Figure 6-24 allows us to complete our discussion of IC-Based vertical scan systems through the illustration of a complete system housed within two IC packages. Beginning at the left side of the drawing, the sync processor IC contains the vertical oscillator which, in turn, develops the vertical drive signal. Pin 1 of the IC is the connecting point for the oscillator signal. From there, the signal couples to the vertical output integrated circuit.

This particular circuit features a guard circuit—a circuit that blanks the video signal if a problem occurs within the vertical output stage. Pin 7 of the vertical output IC serves as the guard circuit output and normally has a 1-Vdc reading. If the vertical deflection collapses, however, the voltage at pin 7 rises to 3-Vdc. DC coupling carries the voltage through the luminance

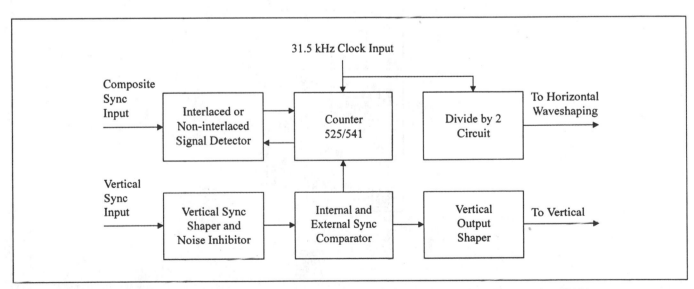

Figure 6-22B   Block Diagram of a Vertical Countdown Circuit (*Courtesy of Philips Electronics N.V.*)

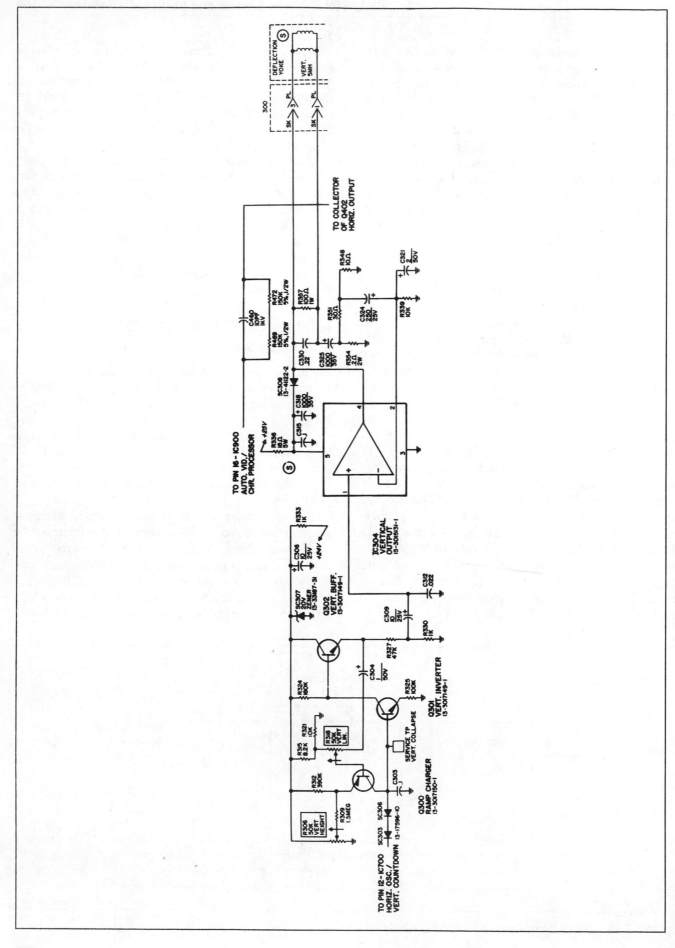

Figure 6-23 Schematic of the Vertical Deflection Circuit (*Courtesy of Philips Electronics N.V.*)

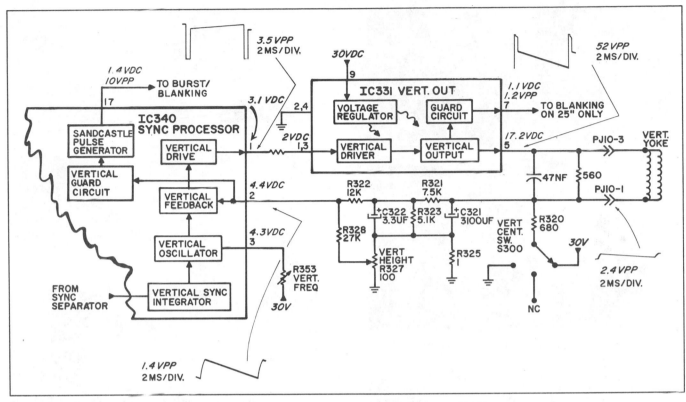

Figure 6-24   IC-Based Vertical Scan System (*Courtesy of Philips Electronics N.V.*)

circuits; the signal is then applied to the cathodes of the CRT.

A peak-to-peak 50-Vdc signal feeds from pin 5 of the vertical output circuit to the vertical deflection coils located in the yoke. Capacitor C21 and resistor R35 provide an ac return path for the yoke along with the vertical centering switch. In addition, the vertical height control adjusts the voltage level in the return circuit and establishes the correct picture height. The vertical frequency varies through the adjustment of R53.

☑ **PROGRESS CHECK**

You have completed Objective 4. Discuss vertical scanning and countdown circuits.

6-6 TROUBLESHOOTING VERTICAL DEFLECTION CIRCUITS

Many times, we divide vertical deflection problems into defects that occur because of problems with either amplitude or frequency. Vertical amplitude problems may cause a complete loss of vertical deflection, a dramatic reduction in the vertical size of the picture, or the loss of vertical linearity. Figures 6-25A, B, and C illustrate each of these three conditions. Problems with

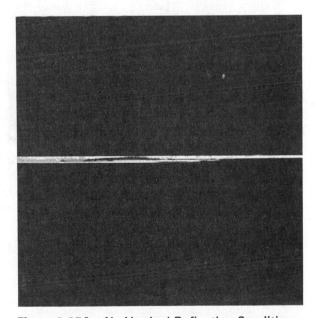

Figure 6-25A   No Vertical Deflection Condition

frequency usually cause the picture to roll in a vertical direction. Figure 6-25D illustrates the symptom caused when the vertical frequency reduces to 20 Hz, resulting in three frames of the same picture.

Another common vertical deflection problem is **vertical foldover**, in which the picture overlaps at either the top or the bottom and distorts another part

Figure 6-25B Poor Vertical Size Condition

Figure 6-25D Reduced Frequency Condition

Figure 6-25C Poor Vertical Linearity Condition

When a receiver has vertical foldover problems, immediate suspects are the vertical output stage, bias circuits, or feedback circuits. In many cases, a leaky or open electrolytic capacitor can cause vertical foldover. In others, either a resistor that has increased in value, a defective transistor, or a faulty integrated circuit can also cause vertical foldover. Often, resistors located in

of the picture. Vertical foldover may take any of four forms:

- Picture has insufficient height with less than an inch foldover in the middle of the picture
- Picture has foldover at the top accompanied by black and white lines
- Picture has excessive height at the bottom and overlapping of the bottom
- Bottom half of the pictures rises and a black area appears at the extreme bottom of the picture

Figures 6-26A, B, C, and D show examples of the four types of vertical foldover.

Figure 6-26A Vertical Foldover in the Middle of the Picture

Figure 6-26B Vertical Foldover at the Top of the Picture

Figure 6-26D Vertical Foldover at the Bottom with Black at the Bottom

Figure 6-26C Vertical Foldover at the Bottom of the Picture

The multimeter also functions as a versatile trouble-shooting tool for vertical foldover problems. When checking possible vertical output transistor problems, always measure the voltages at the transistor terminals to see if the voltages match the values seen on the schematic. If the feedback circuit appears as the likely source of the foldover problem, check the resistance of resistors and capacitors in the feedback loop. Also, check the resistance values of components in the bias circuits of the vertical output transistors.

Test Equipment Primer— The Oscilloscope

Modern oscilloscopes have characteristics such as:

- a measurable sweep rate
- a method for display synchronization
- multiple traces
- vertical sensitivity

Each of the characteristics allows the accurate measurement of frequency, time, peak-to-peak, phase, and voltage. In addition, you can use the oscilloscope to measure the characteristics and distortion of a waveform.

A Measurable Sweep Rate Every oscilloscope has a sweep rate, an input impedance, and a bandwidth. As you learned in chapter five, the horizontal sweep rate of an oscilloscope is the amount of time required to move the CRT beam spot horizontally across one screen division. To measure sweep rate, we calibrate the time per

either the feedback circuit or bias circuits will age and change value. At times, vertical output transistors will become leaky; replacement of both vertical output transistors is the best remedy.

A good troubleshooting procedure for vertical foldover problems involves signal injection, signal tracing, use of the oscilloscope, and use of the multimeter. With IC-based vertical deflection circuits, the injection of a normal signal into the vertical output IC can help determine if a problem exists within the IC or in external feedback circuits. An oscilloscope works as an excellent tool for monitoring the input and output waveforms.

division. Thus, the sweep rate equals a given number of seconds per division. While the oscilloscope exists as an input impedance to the measured signal, the bandwidth specification sets up a reference frequency for the measurement of frequency response. With this, the bandwidth of an oscilloscope can exactly amplify all frequencies of a pulse waveform. If a measured frequency exceeds the upper limit of the oscilloscope bandwidth, the amplitude of the displayed waveform will be lower than the actual amplitude. As an example, a 100-MHz sine wave will have an average value that is 0.707 of the reference frequency amplitude. Displaying the 100-MHz sine wave on the screen of a 100-MHz oscilloscope would show a sine wave that seems to have less than normal amplitude.

Bandwidth also becomes important when measuring the rise time of a waveform. Because a pulse waveform can contain any number of sine wave components, with each component having amplitude and phase, the displayed rise time of a waveform depends on the oscilloscope bandwidth.

Synchronizing the Displayed Waveform   Every oscilloscope display screen has a horizontal and vertical axis. Making a displayed waveform remain stationary on that screen involves synchronizing the internally generated signals of the oscilloscope with the measured signals. To do this, the oscilloscope gathers a pulse waveform from the measured signal or from a part of the signal that corresponds with an internally generated sawtooth waveform. Modern oscilloscopes generate and apply this timing signal, called a trigger, as a synchronizing signal.

Through the use of **triggered sweep**, an oscilloscope can display complicated low- or high-frequency waveforms in the same position. Because the sawtooth waveform period may cover only a small portion of the measured waveform period, the use of a triggered time base allows the oscilloscope to pick out and display a small portion of a complete waveform. Along with that capability, the triggered time base also provides a method for measuring the duration of any part of a waveform.

Multiple Display Traces   As mentioned in chapter five, the multiple channels on oscilloscopes provide the capability to examine two or more waveforms simultaneously. Thus, a technician can observe an input signal to a video amplifier stage while monitoring the amplifier output and make comparisons regarding phase, time, amplitude, and frequency.

Multiple-trace displays utilize two methods for providing real-time analysis, the alternate display mode and the chopping mode. Each method works well for given uses. With the alternate display mode, the first horizontal sweep shows the channel 1 signal while the second horizontal sweep shows the channel 2 signal. This method displays high-frequency waveforms, but only displays repetitive signals that occur during the same time period.

The chopping mode separates the channel 1 and channel 2 signals into pieces. Then the displays are alternately shown during the same horizontal sweep. This mode works with low-speed sweep rates and displays both random and repetitive signals.

Vertical Sensitivity   Like the sweep rate, the sensitivity of an oscilloscope is measured against divisions on the display screen. As you learned earlier, using the vertical axis as a reference, sensitivity measurements occur in volts per division increments. Each vertical division of the CRT screen corresponds with the setting of the vertical sensitivity setting. Therefore, if the sensitivity is set at 1 v/div, a 4-Vdc peak-to-peak signal will cover four screen divisions. The number of divisions covered by a signal with a given sensitivity setting represents the voltage amplitude of the signal and allows voltage measurements with the oscilloscope.

Troubleshooting Vertical Deflection Circuits That Feature Blocking Oscillators

Figure 6-27 repeats the schematic diagram originally shown in Figure 6-14, and shows that the output of Q1 exists as a dividing point for the circuit. The shaded component labels in the drawing indicate possible problem areas. When using an oscilloscope for troubleshooting vertical amplitude problems, you can connect the test equipment to capacitor C8 and ground. A normal sawtooth waveform at this point indicates that the problem exists somewhere after the capacitor. The lack of a waveform at the capacitor indicates that a defect has occurred prior to the test point.

As an example, a common vertical amplitude results in the reduction of the raster to a single horizontal line that extends across the center of the CRT screen. With no waveform at the capacitor, you can use a multimeter to check voltages in the Q1 stage. Identical voltages at the Q1 base and emitter with no voltage at the collector point to a short in the base-emitter junction, a shorted capacitor C3, or an open resistor R14, R15, or inductor L2. A reduced voltage at the Q13 base and a higher than normal voltage at the collector indicates that C22 has opened, or that either inductor L1, L2, or transformer X12 had developed shorted turns.

Moving back to the waveform check at capacitor C8, a normal waveform is an indicator to check the portion of the circuit that follows the oscillator and to check the waveform at the base of the vertical driver. A zero-amplitude waveform at the Q2 base indicates that an

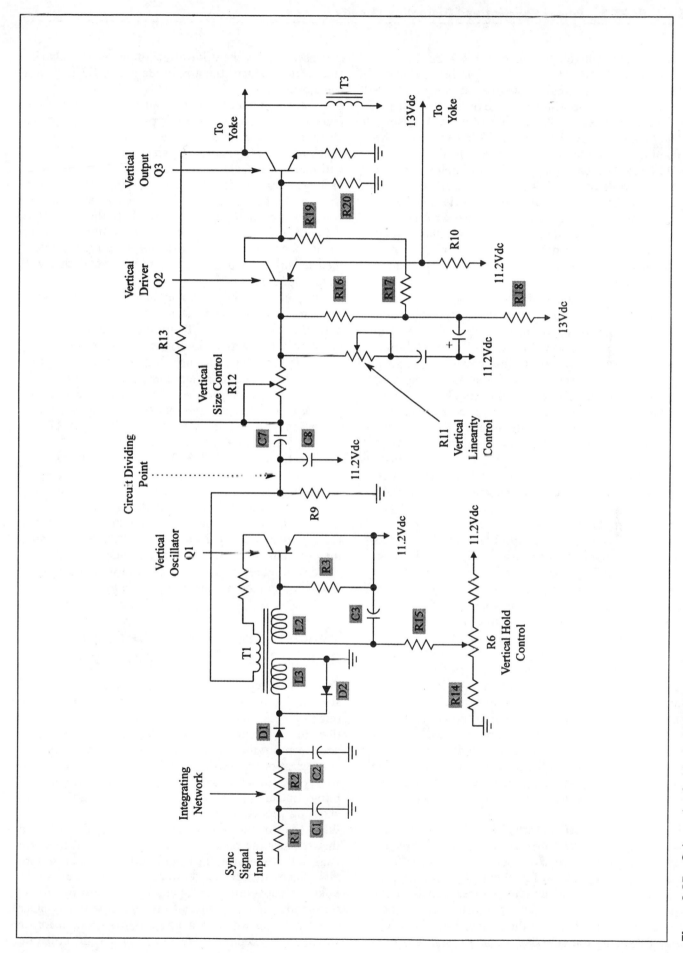

Figure 6-27 Schematic of a Vertical Deflection Circuit with Blocking Oscillator

175

open may have developed in either C7 or R12, or that the vertical driver has developed an internal base-emitter short. A normal waveform at the Q2 base pushes the troubleshooting efforts to the vertical output transistor and a check of the waveform at the Q3 base.

The lack of a waveform at the Q3 base narrows the focus of the search to the area between the Q3 base and the Q2 base. You can again use a multimeter to check for normal voltage and resistance measurements. A dc voltage reading of zero at the vertical driver collector often results from a short within the base-emitter junction of the vertical output transistor. Lower than normal voltage at the vertical driver collector could result in increased resistance in R16, R17, or R18 or an internal problem within Q2.

The oscilloscope check of the waveform at the vertical output transistor base may disclose normal conditions at this point. If so, you need to extend this check to a measurement of the waveforms at collector and emitter. During normal operation, the waveform found at the emitter should have the same shape as the base waveform and a 0.3-Vdc peak-to-peak amplitude. Zero-amplitude waveforms at the collector and emitter of the output transistor point to an open R103 or a defective output transistor. Normal waveforms at the two bases and emitter combined with a zero-amplitude waveform at the collector indicate that either transformer T3 or that a capacitor connected across the yoke coils has shorted.

You can apply the same procedures used for solving the "thin horizontal line" problem to other defects that cause a reduction in the raster height. Rather than look for zero-amplitude waveforms, though, you should check for waveforms that have a lower than normal amplitude. In addition, you can begin to look for components that have changed values rather than open or shorted components. Here the type of symptom—reduced height versus only a thin horizontal line—dictates the type of troubleshooting method used.

If the waveform found at the vertical driver base has a lower than normal amplitude, your search may begin either at a point prior to or following the driver. In this particular circuit, reduced waveforms at the Q2 base result from a decreased C7 capacitance or increased R9 resistance. In addition, a higher than normal resistance value at R18 increases the degenerative effect of that resistor and reduces the amplitude at the vertical driver collector.

A normal waveform at the Q2 base and a lower than normal amplitude waveform at the Q3 base narrow the scope to the area between the two test points and focus on the resistance values of R19 and R20. If R19 has a decreased resistance value, the Q3 base signal-driving current is shunted. A normal waveform at the vertical output base indicates that the defect lies in the circuitry following the Q3 base. If the resistance of R20

increases, the degenerative effect of the resistor also increases and reduces the amplitude of the Q3 collector waveform.

Another common symptom caused by amplitude defects is non-linearity. This symptom turns the search away from checking only for changes in amplitude to checking the waveshape. As an example, a leaky C8 will cause the negative portion of the vertical driver base waveform to flatten. From a symptoms perspective, the bottom of the picture will appear squeezed while the top half remains normal. If the resistance value of R19 decreases, the vertical output transistor cannot conduct normally during the negative portion of the base driving signal, and the top of the picture will appear squeezed.

The picture rolling vertically in one direction always indicates that a problem has occurred in the vertical oscillator circuitry. Even if the problem persists, the adjustment of the vertical hold control can disclose which area of the oscillator circuitry is defective. If adjusting the control in one direction causes the picture to roll down, and adjusting in the other direction causes the picture to roll up, the sync pulses have disappeared. Therefore, problem-solving involves checks of R1, R2, C1, C2, D1, D2, and the L3 winding of T2.

Adjusting the vertical hold control from one extreme to another may cause the picture to roll in the same direction, which indicates that a fault has developed in the oscillator circuitry. This dictates a check of the major frequency-determining components. Decreased C3 capacitance or decreased R3 resistance will cause the oscillator frequency to increase and the picture to roll down. Increased R3 resistance causes the oscillator frequency to increase and the picture to roll up.

Troubleshooting Vertical Deflection Circuits That Feature a Multivibrator

Figure 6-28 repeats the schematic drawing originally seen in Figure 6-16. When you begin troubleshooting this particular vertical deflection circuit, you will notice the differences between the blocking oscillator and the multivibrator. In brief review, the multivibrator configuration establishes the entire circuit as an oscillator, relies on a feedback line, and relies on direct coupling. Because of these three key factors, the troubleshooting procedures will become somewhat different than those in the last section. For example, a defect that reduces the raster to a single horizontal line will cause all waveforms in the circuit to reduce to zero also. There are a few exceptions, however. If either the yoke or capacitor C18 open, the waveforms found throughout the remainder of the circuit will stay close to normal. So we can change our first conclusion to reflect that exception. If the waveform at the collector

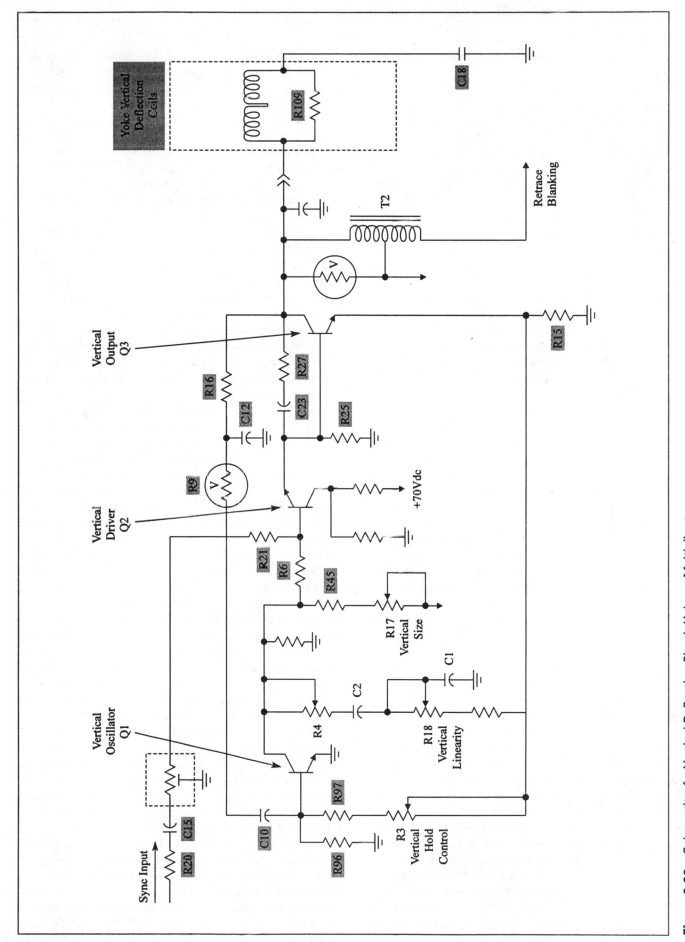

Figure 6-28 Schematic of a Vertical Deflection Circuit Using a Multivibrator

of Q3 is zero, all other waveforms will have zero amplitude.

Because the circuit configuration effectively eliminates waveform analysis for "no vertical deflection" problems, we must turn our attention to voltage and resistance measurements. The direct coupling employed throughout the circuit complicates matters because any defect that prevents the circuit from oscillating will change dc operating voltages by only a small amount. Thus, the dc voltage checks are used to look for drastic differences between the measured values and the normal values seen on a schematic diagram.

In the case of the display of only a horizontal line, an open R6 will cause the operating voltage at the base of Q2 to drop to practically zero. With this single defect, the Q2 emitter current and emitter voltage and the Q3 base voltage will also decrease to zero. As a result, the Q3 emitter current and emitter voltage also drop to zero while the Q3 collector voltage increases to almost 70 Vdc. Along with those changes, the Q1 base voltage drops to zero because of the change at the Q3 emitter. In turn, the Q1 collector current also decreases to zero. This last change causes the Q1 collector voltage to increase to anywhere between 5.6 and 9.6 Vdc. Because R6 has opened, the increased voltage at the Q1 collector does not cause another voltage increase at the Q2 base.

The same type of problem symptom of a horizontal line can also occur if a defect occurs in the feedback line. If C10, R9, or R16 open or if C12 develops a short, the feedback reduces to zero and the signal needed for deflection disappears. With each of those possible defects, the Q1 base voltage increases from -1 to $+0.6$ Vdc. The complete dc loop around the system allows that small increase in dc voltage to affect all the other dc voltages. Voltages at the Q2 collector, Q2 base, Q2 emitter, Q3 base, and Q3 emitter will become slightly less positive.

Measuring the voltages at the Q3 collector and at both sides of R9 provides a quick method for checking the operation of the feedback network. The measurement should show that the three voltages are identical. If not, you can narrow the troubleshooting search significantly by looking at the results. With the voltage found at the left side of R9 less than the voltage on the R9 right side, check for a leaky capacitor C10. Identical voltages at both sides of R9 that are less than the Q3 collector voltage indicate that capacitor C12 is leaky. Zero volts at the right side of R9 tells you that either R16 has opened or C12 has shorted. A voltage ranging between $+56$ and $+62$ Vdc at the right side of R9 combined with $+0.6$ Vdc on the R9 left side points toward either an open R9 or a shorted C10. Sometimes, the feedback voltages will appear normal even though the symptom indicates the lack of vertical deflection. In this case, either C10 or C12 has opened.

As with the blocking oscillator-based vertical deflection circuit, other defects can result in both the shortening and the non-linearity of the picture. In either case, you can rely on your observations of waveforms as a troubleshooting tool, becoming more concerned with the amplitude and the shape of each waveform. The presence of either of the symptoms and normal waveforms throughout the circuit indicates that C18 may have a decrease in capacity, R109 may have a lower than specified resistance, or that the deflection yoke has developed shorted turns.

Smaller than normal or distorted waveforms take us in a different direction. For example, the waveform found across R25 should have a peak-to-peak amplitude of 1 Vdc. At this point in the circuit, the signal represented by the waveform opposes the signal found at the Q3 base. A larger than normal signal at R15 would present greater opposition to the Q3 base signal and result in a picture that covers only a small portion of the raster. The cause for the increased signal found at R25 lies within a resistor that either has developed a larger resistance or has opened.

As you saw earlier in this chapter (Figure 6-17), waveforms found at the Q1 collector, Q2 base and emitter, and the Q3 base should have the same shape and amplitude. In actual practice, the Q2 emitter waveform and Q3 base waveform may be smaller than the Q2 base waveform because of the emitter-follower configuration of Q2. An emitter-follower will always have an output signal smaller than its input signal. Any large difference between the four waveforms usually indicates that problems have occurred in the area following the Q2 stage. For example, R25 and R26 may have a decreased resistance or C23 may have become leaky.

Any decrease in the R6 resistance will cause the Q2 base waveform to have a lower amplitude than the Q1 collector waveform. If the waveforms found at the Q1 collector, Q2 base and emitter, and Q3 base have reduced amplitudes, a defect has developed in the Q1 stage. As an example, an open R45 can cause each of the waveforms to distort and dc voltages at the Q1 collector and Q16 base to decrease.

With vertical frequency defects, you can apply the standard practice of rotating the vertical hold control from one direction to the other and then checking for the direction of the roll. A picture that rolls in both directions because of that control rotation indicates that the circuit has lost the sync pulses. Defects in R20, C15, and R21 can block the sync pulses.

If the picture consistently rolls in the same direction, a problem has developed in the deflection circuitry. Again, the differences between a blocking oscillator circuit and a multivibrator circuit affect the troubleshooting methods. Any defective component in the feedback can cause problems with the vertical frequency. The key components to observe for this particular circuit are C10, R9, C12, and R16. In addition, the failure of R96, R97, R3, and C10—which are located in the Q1 base circuit—can change the vertical frequency.

The waveform found at the Q1 base is particularly important and should have a peak-to-peak pulse amplitude of 2-Vdc. The combination of a good waveform and persistent vertical frequency problems points toward a failed component in the Q1 base circuit. An abnormal peak-to-peak pulse amplitude and a rolling picture indicate that components in the feedback have failed.

Troubleshooting Vertical Scan Circuit Problems

The most valuable tool for troubleshooting a vertical scan circuit is the oscilloscope. Using the combination of an oscilloscope and a multimeter, you can monitor waveshapes and amplitudes as signals flow through the vertical circuits. Keep in mind, however, that some vertical waveforms—such as those found at the output—may not have a stable display. You should always remember to apply basic, logical troubleshooting methods to the vertical circuitry and not jump to conclusions.

Logical troubleshooting methods come into play immediately when checking vertical signal-processing circuits. In this sense, you can limit your initial checks to basic issues such as having the correct input signal and the correct voltage amplitudes. Referring back to the diagram shown in Figure 6-21, the first check should involve the output of the vertical drive found at pin 1. The vertical drive output waveform pictured in Figure 6-29A should have a peak-to-peak amplitude of 3.2-Vdc and should also appear on pins 1 and 3 of the vertical output IC.

Moving to pin 9 of the vertical output IC, we should find a positive 30-Vdc while pins 2 and 4 should remain at chassis ground. Figure 6-29B shows the proper waveform for pin 2. Pin 5 of the same IC should produce the vertical output signal shown in Figure 6-29C if the proper signals and conditions are found at pins 1, 2, 3, and 4. The correct waveform for the bottom of the deflection coils is shown in Figure 6-29D. While it seems easier to limit conclusions to the two integrated circuits, components outside the packages may also cause vertical scan problems.

Capacitor C21 and resistor R25 must have the correct values or the vertical scan will collapse. Many times, a slightly corroded switch or control can cause common symptoms to appear. While any problem within the vertical centering circuit will cause the picture to decenter, an intermittent centering switch is an often overlooked culprit. Vertical size problems often

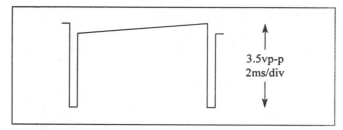

Figure 6-29A   Waveform Taken From Pin 1 of Sync Processor IC (Vertical Drive)

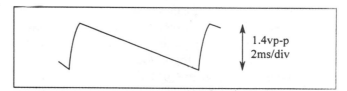

Figure 6-29B   Waveform Taken From Pin 2 of Sync Processor IC (Vertical Feedback)

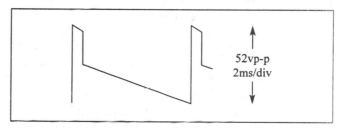

Figure 6-29C   Waveform Taken From Pin 5 of Vertical Output IC (Vertical Output)

Figure 6-29D   Waveform Taken From Bottom of Yoke Vertical Deflection Coils

result from a vertical height control that has dirty contacts. An easy method for checking if the problem exists within the vertical output circuit or the peripheral circuitry is shorting pin 7 of the vertical output IC to ground. If the scan collapses to a single line across the center of the screen, the problem lies within the integrated circuit.

⊃ SERVICE CALL

Mitsubishi Model CS81952R Television has a Short, Distorted Picture

The customer complained that his Mitsubishi television had a "short, distorted picture." From a technical perspective, the raster had foldover at the top and was lacking the proper size. Preliminary checks of the dc voltages showed a higher than normal reading of 42-Vdc at the collector of the second vertical output transistor. The same transistor had a normal voltage reading at the base.

A check of the waveform at the base of the transistor showed the correct dc level and amplitude. However, the check also disclosed that two closely spaced, extremely wide pulses had taken the place of the one normal pulse during each vertical interval. Another waveform check at capacitor C412 showed that the same vertical output transistor was amplifying the incorrect base drive and then applying the voltage waveform to the deflection yoke. With the oscilloscope working as a signal-tracing device, another waveform check at pin 14 of the integrated circuit package that houses a combination of the sync/sweep/chroma/video circuits, or IC201, again showed the presence of the incorrect double-pulse waveform. Additional checks of all other pins of the IC that connected to the vertical circuitry did not reveal any other problems.

Given those findings, the troubleshooting effort moved to the feedback circuitry, the vertical height control, and the vertical linearity control. Again, all voltages and waveforms were normal. An out-of-circuit check of C412 showed that the electrolytic capacitor had become leaky. Replacement of the capacitor, however, only slightly improved the height and had no effect on the foldover problem.

A phone call to Mitsubishi technical support disclosed that the receiver vertical output section has a different configuration than that normally seen in most receivers. With this configuration, the first vertical output transistor operates as a current source for the yoke unless a down-going pulse exists at the base of the second vertical output transistor. A pulse at that location allows the voltages at the Q401 emitter and collector to rise to the same level as the supply voltage.

Accordingly, a higher than normal voltage at the Q402 collector points toward possible problems in the first vertical output transistor circuit. Further inspection of the Q401 circuit showed that the circuit relied on clamping diode D401 to limit the circuit gain and the conduction of the transistor. During normal operation, the diode allows part of the Q401 base current to bypass the transistor.

An out-of-circuit check of Q401 on a transistor tester showed no problems, but a simple check of the clamping diode showed that the diode was open. Replacing the diode restored the proper picture height and eliminated the vertical foldover. The fault in the vertical output transistor circuit caused the oscillator found within IC201 to reset twice during the vertical interval. As a result, the feedback circuitry coupled the double pulse found in the initial troubleshooting checks to the integrated circuit.

➡ **SERVICE CALL** ————————————

Magnavox Model 19C120A Does Not Have a Raster

The Magnavox television receiver had high voltage and audio but no raster when it arrived at the shop. Turning up the screen controls to their maximum settings revealed that the receiver displayed only a thin horizontal line—a sure indicator of a problem in the vertical sweep area. Because of this symptom, initial troubleshooting checks began with the vertical output transistors and worked backwards.

Figure 6-30A shows a schematic of the 19C3 chassis power supply, while Figure 6-30B shows a schematic of the vertical deflection system. Although the dc voltages found at the vertical output transistor terminals seemed abnormal, out-of-circuit checks of the transistors with a transistor check showed no problems. Logic also pointed to possible problems with coupling capacitors C229 and C227. Further checks of the capacitors, however, showed no problems in those components.

The troubleshooting investigation then turned to signal injection and waveform analysis. The injection of a vertical signal at the base of the vertical driver transistor caused the raster to regain vertical deflection and seemed to clear the components following IC200, which contained the vertical oscillator. Injecting the vertical signal and regaining the raster also showed that the video signal was missing.

Moving back to voltage checks, a check of the voltage at the pins of the integrated circuit showed normal readings except for the voltage at pin 18. Instead of having the specified 12-Vdc, the pin had a voltage of only 1.25-Vdc. Voltages at the regulator that supplied the source voltage for the IC were correct. Given the correct voltage at the regulator and the incorrect voltage at the IC200, some component in the line between the regulator and the IC had caused the pin voltage to drop.

Two additional checks pointed toward the faulty component. A visual check of the circuit disclosed that R200 had overheated and was continuing to run hot. An ohmmeter check of zener diode Z200 in the power supply stage that supplied the 12-Vdc source voltage showed that the diode had shorted. Replacing both the diode and the resistor restored the normal vertical deflection and the normal video.

☑ **PROGRESS CHECK**

**You have completed Objective 5.
Troubleshoot vertical signal-processing circuits.**

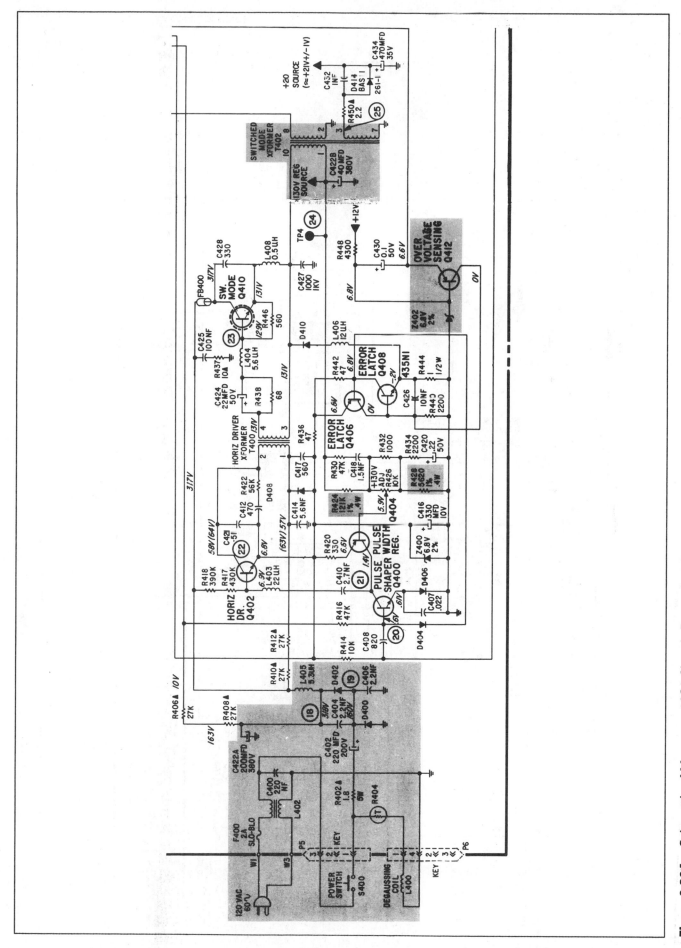

Figure 6-30A Schematic of Magnavox 19C3 Chassis Power Supply

181

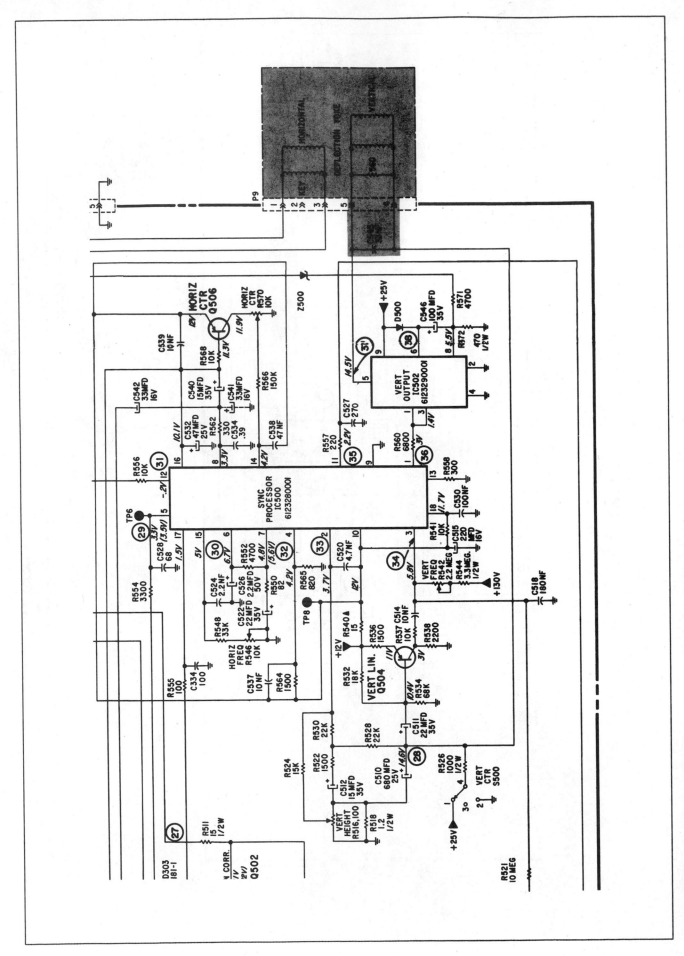

Figure 6-30B Schematic of Magnavox 19C3 Chassis Vertical Deflection System

SUMMARY

Chapter six began by covering basic topics such as deflection and integrator circuits, and moved to an overview of vertical deflection system basics. The discussion of building the vertical sweep voltage used as a natural starting point, the integrator circuits that sit between the sync separator and the vertical system. You learned about the different sections that make up a vertical deflection system, and the various controls that affect the performance of such a system.

The chapter then moved to the operation of vertical oscillator circuits. Older model television receivers rely on either a blocking oscillator or a multivibrator configuration to establish the proper 60-Hz frequency. The chapter provided an in-depth look at the role of feedback in the multivibrator circuit.

Vertical driver and vertical output circuits were covered next, and you learned that waveshaping and ampli-

fication occur in the driver section. The output section provides the drive needed to excite the yoke vertical deflection coils and establish the necessary sawtooth waveform. During your study of the vertical countdown circuit, you found that the vertical and horizontal oscillator functions fit within one integrated circuit package. You also learned about the use of a guard circuit in the IC-based system.

Typical vertical problem symptoms concluded the chapter. Among those symptoms were poor vertical linearity, poor vertical size, lack of a vertical scan, and vertical foldover. Troubleshooting focused on blocking oscillator, multivibrator, output, and IC-based vertical scan circuit problems.

REVIEW QUESTIONS

1. The vertical scan frequency is _____.

2. An integrator is an:
   a. RC filter
   b. LC filter
   c. LR filter

3. Increasing the time constant of an integrator provides better _____.

4. What portion of the vertical sync pulse allows the horizontal deflection circuit to maintain synchronization?

5. The oscillator in the vertical deflection system has a _____ frequency and usually is configured to operate in the _____ state.

6. True or False   A vertical deflection system that has a feedback loop operates as a phase-locked loop.

7. The vertical deflection system includes a waveshaping network that provides a

   _____.

8. How is the trapezoidal waveform required by the vertical deflection system generated?

9. Define the purpose of each of the following controls:
   a. vertical hold
   b. vertical size
   c. vertical linearity

10. True or False   Vertical deflection systems always include some type of large-signal power amplifier in the output stage. During the active scanning period, the amplifier current swings between cut-off and saturation.

11. True or False   The output waveforms from the vertical and horizontal deflection systems travel to the deflection yoke.

12. Why do vertical deflection systems include a vertical driver stage?

13. Name two reasons why a vertical oscillator should be unstable.

14. Refer to Figure 6-13. The waveform found at the collector of the oscillator is classified as

    _____.

15. True or False   An integrating circuit eliminates noise pulses.

16. Is a damped wave desirable in a blocking oscillator circuit?

17. Refer to Figure 6-15D. The shape of the waveform has changed from a _____

    to a _____.

18. A multivibrator pulse generator does not require _____ coupling. Instead,

    the multivibrator system depends on _____.

19. In a multivibrator circuit, a problem in any part of the vertical deflection system:
    a. can stop oscillations and reduce all vertical waveforms to zero
    b. has no effect on the oscillations but reduces all vertical waveforms to zero
    c. is easily found because of the convenient dividing line between the vertical oscillator and amplifier stages
    d. stops oscillations but does not affect most of the vertical waveforms.

20. The waveform shown in Figure 6-17A is produced because of the _____.

21. Refer to Figure 6-15. Why does the same trapezoidal waveform appear at the Q2 emitter and the Q3 base?

22. Deflection output amplifiers increase the level of the signal generated by the _____

    and supply a _____ to the deflection coils.

23. Compared to the collector resistance of a vertical deflection amplifier, vertical deflection coils have a reactance that is:
    a. very low
    b. very high
    c. not noticeable

24. Transistor-based vertical deflection circuits require the addition of an intermediate amplifier, called a

    _____, to produce the large signal needed by the deflection coils. The

    _____ stage fits between the _____

    stage and the _____ stage.

25. Most vertical driver circuits use a _____ configuration because of the
    high-stage gain and the impedance matching given by the configuration.

26. In Figure 6-16, what kind of waveshape is applied to the base of the amplifier?

27. True or False   In the circuit pictured in Figure 6-20, R1 and R2 form a voltage divider that establishes the
                     bias for the driver, while R5 and R6 form a voltage divider that establishes the bias for
                     the amplifier.

28. Refer to Figure 6-20. Potentiometer R4 is the _____ and it controls the
    a. linearity
    b. size
    Potentiometer R7 is the _____ and it controls the:
    a. linearity
    b. size

29. Centering the picture vertically should not disturb either the _____ or the

    _____. Incorrect centering of the raster will obscure either the

    _____ or _____ of the raster.

30. Modern vertical scanning systems apply a _____ pulse to the vertical deflection coils that has the following waveshape:
    a. sawtooth
    b. rectangular
    c. square
    d. trapezoid

31. Name three advantages given by the use of direct coupling when compared to transformer coupling.

32. The circuit pictured in Figure 6-21 does not rely on two power supplies. Why?

33. Describe how the vertical countdown circuit operates.

34. A receiver that utilizes a combination vertical/horizontal oscillator and a countdown circuit will not need:
    a. color and tint controls
    b. vertical and horizontal hold controls
    c. a horizontal pulse tied to the keyed AGC circuitry
    d. vertical linearity and height controls

35. The vertical countdown circuit shown in Figure 6-22A can produce either 525 or 541 counts. Why does the circuit have this capability?

36. What is the purpose of the negative feedback path (through R354) in Figure 6-23?

37. In Figure 6-24, the vertical oscillator is contained in the _____ integrated circuit and provides the _____.

38. True or False   A guard circuit blanks the video signal if a problem occurs within the vertical output stage.

39. Refer to Figure 6-24. To transfer a reference voltage through the luminance circuits to the deflection coils, the circuit uses:
    a. capacitive coupling
    b. direct coupling
    c. transformer coupling
    d. dc coupling
    e. ac coupling

40. Four common vertical signal processing problem symptoms are: _____, _____, _____, _____.

41. List three vertical foldover symptoms.

42. Choose the single best answer. When troubleshooting a non-linearity problem in a blocking oscillator circuit, you should check for the:
    a. proper amplitude
    b. proper waveshape
    c. open output transistor
    d. open resistor in the oscillator circuit
    e. shorted capacitor in the driver circuit

43. True or False   One defect in a multivibrator circuit may cause a series of problem symptoms.

44. A picture that rolls in both directions because of the control rotation indicates that the circuit _____. If the picture consistently rolls in the same direction, a problem has developed in the _____.

45. True or False   You can apply the same troubleshooting logic used with blocking oscillator circuits to multivibrator circuits.

46. True or False   Troubleshooting the IC-based vertical scan circuit shown in Figure 6-24 involves logical methods such as checking for the proper input signals, correct waveshapes, and correct voltage amplitudes.

47. What was the problem symptom in the first Service Call?

48. Refer to the first Service Call. Why did the vertical oscillator reset to a double pulse?

49. What troubleshooting techniques were successfully used in the second Service Call?

50. Refer to the second Service Call. Why did R200 overheat?

51. Describe the operation of a diode sync-separator circuit.

52. Refer to the transistor sync-separator circuit shown in Figure 6-4. The arrival of a sync pulse causes the sync separator transistor to:
    a. become saturated
    b. go into cutoff

53. Refer to Figure 6-4. True or False   The output of the differentiator circuit connects to the horizontal deflection circuit.

54. Refer to Figure 6-4. The integrator circuit consists of a(n) _____ network.

55. Refer to Figure 6-4. What can happen if noise spikes enter a sync-separator circuit?

56. Separation of the sync signals occurs at the _____ stage. The sync signals control the frequency of the _____ in the deflection stages. A sync-separator stage also separates the _____ pulses from the horizontal sync pulses.

57. Sync separation occurs through the _____ of the sync pulses from the composite video signal.

58. True or False   A sync-separator stage amplifies only the horizontal sync, vertical sync, and equalizing pulses.

59. What are the symptoms of the loss of the sync signals?

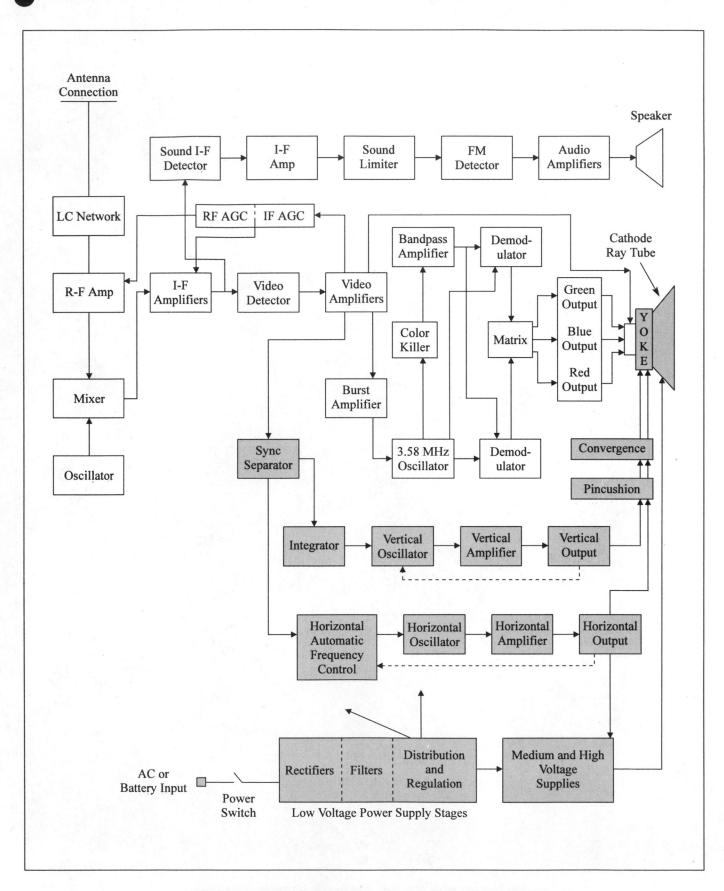

COLOR TELEVISION RECEIVER STAGES AND SIGNAL PATHS

PART 3

Processing Signals

Chapters seven through eleven provide information about circuits that process signals. The order of the subject matter in the chapters—tuners and tuner control systems, I-F/AGG/AFT circuits, sound-processing circuits, luminance signals and video amplifier circuits, chrominance signals and chroma amplifiers—establishes a pattern of input-process-output. Within each chapter, this pattern becomes readily apparent as the text illustrates the principles used to recreate a transmitted image on a television receiver viewing screen. Each of the five chapters also defines and emphasized designs and circuits commonly used in various types of electronic equipment.

As you read through the chapters in part three, remember that the principles used to transmit and receive picture and audio signals have their basis in the way that we see images and hear sounds. For example, our eyes allow us to see both color and monochrome images, different ranges of colors and hues, and to compensate for varying levels of light. In terms of electronics, those sensations translate into brightness, hue, and saturation and the different frequencies found in the electromagnetic spectrum.

The concepts described throughout part three also show how circuits translate the characteristics of an actual live scene into the electrical signals required for transmission, reception, and the efficient processing by electronic equipment. All this becomes apparent through fundamentals such as heterodyning and demodulation. In addition, those concepts illustrate how the circuits process signals for the reproduction of the transmitted image. When we consider the circuits as a system, we can begin to understand how the operations of the power supplies, deflection circuits, and signal-processing circuits combine to produce a viewable image on a display screen.

CHAPTER 7

Tuners and Tuning Control Systems

OBJECTIVES   Upon completion of this chapter, you should be able to:

1. Explain how a MOSFET operates as an R-F amplifier.
2. Discuss tuner oscillators.
3. Understand the use of a cascode amplifier in the tuner mixer stage.
4. Discuss varactor diodes and varactor tuning.
5. Describe how a dedicated microprocessor and associated memory circuits function in an electronic tuner-control system.
6. Describe how a dedicated microprocessor monitors and operates customer controls.
7. Explain the use of frequency synthesis in a tuner-control system.
8. Describe the operation of phase-locked loops.
9. Define typical tuner problems and troubleshooting methods.
10. Determine standard procedures for troubleshooting electronic tuner-control systems.

KEY TERMS

cable-ready tuner

cascode mixer

Colpitts oscillator

hyperband channels

midband channels

mixer stage

radio-frequency (R-F) amplifier

random-access memory

read-only memory

superband channels

tuner oscillator

UHF tuner

varactor diode

VHF tuner

INTRODUCTION

Chapter seven begins a series of chapters that emphasize signal processing. In this chapter, your study of tuners and tuner control systems will lead to the analysis and troubleshooting of amplifier, oscillator, microprocessor control, and frequency synthesis circuits. The opening sections of the chapter break a television down into its individual components and describe how each of those components work together to produce an intermediate frequency. Throughout those discussions, the chapter covers components and circuits such as MOSFETs, cascode configurations, and varactor diodes.

The discussion of tuner-control circuits demonstrates how a dedicated microprocessor operates with associated circuitry such as memory and frequency-synthesis circuits. The chapter also shows the importance of frequency synthesis—and digital electronics as a whole—as it covers the microprocessor control of channel selection, displays, and volume.

As you saw in each of the preceding chapters, troubleshooting involves, among other things, a combination of logical thought, knowledge about the theory of operation, and knowledge about correct application of test equipment. Chapter seven uses this approach as it describes methods for solving common tuner and tuner-control problems. In addition, the chapter introduces the counter and briefly discusses how this measuring device can work with oscillator circuits.

7-1 TUNER BASICS

Televisions take advantage of information carried within R-F or radio-frequency signals to produce sound and pictures. The R-F signal received at the input to a television tuner consists of all the picture and sound information transmitted either through the air or cable systems. As shown in chapter one, the radio-frequency signal band breaks down into divisions called Very High Frequencies (VHF) and Ultra High Frequencies (UHF). Table 7-1 lists the frequencies by channel designations and shows the separations between each frequency band. Figure 7-1 shows how the R-F carrier signal would appear as it becomes the input for the R-F stage of the tuner.

The specific R-F picture and sound signals for each channel have a bandwidth of 6 MHz. Within each 6-MHz band, the picture information carrier wave lies 4.5 MHz below the sound information carrier wave. Figure 7-2 shows the separation of the signals on a frequency response curve. The desired picture signal, which includes both sidebands and sync signals, covers 4 MHz of the bandwidth, while the desired sound carrier and its sidebands cover only 50 kHz.

Table 7-1   R-F Frequencies and Channel Designations

Band	Channels	Frequency (MHz)
VHF (low)	2–6	54–88 MHz
VHF (high)	7–13	174–216 MHz
UHF	14–69	470–806 MHz

Figure 7-1   R-F Carrier Signal

A tuner accepts the selected R-F frequency band, excludes all unwanted signals, and amplifies the desired signal to a level where the demodulation of sound and picture information can occur. A **VHF tuner** can selectively tune to any of the 6-MHz-wide channels from the 54- to 88-MHz band and the 174- to 216-MHz band. A **UHF tuner** selectively tunes to the channels within the 470- to 890-MHz band.

The **midband channels** are designated A to I and include frequencies from 120 to 174 MHz. The **superband channels** range from J to W and include frequencies from 216 to 300 MHz, and the **hyperband channels** range from AA to BBB and cover frequencies from 300 to 776 MHz. **Cable-ready tuners** tune to the channels in both the VHF and UHF bands as well as the mid-, super-, and hyperband channels. During operation, a cable-ready tuner converts the higher cable frequencies to signals within the VHF band. Figure 7-3, located in the Appendix, shows the schematic diagram for a cable-ready tuner used by many manufacturers.

⟳ FUNDAMENTAL REVIEW

Feedback and Phase

Feedback describes a circuit configuration where part of the circuit output signal feeds to the input of the circuit. The signal may take the form of either a voltage or current. In some transistor circuit configurations, the feedback circuit provides the necessary bias voltage.

Examples of this type of feedback bias circuit are collector-feedback and emitter-feedback bias circuits. Negative feedback, which is used in amplifier circuits, occurs when the feedback signal is out-of-phase

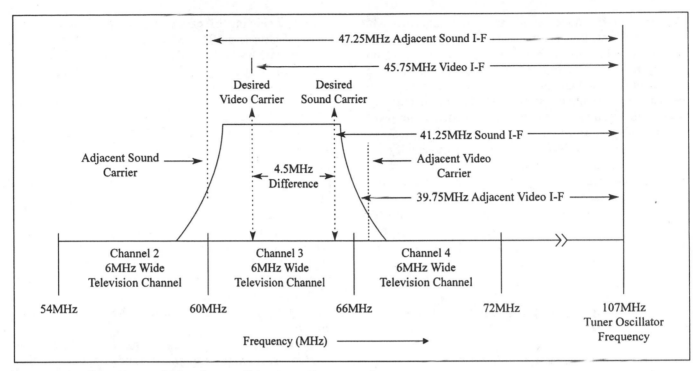

Figure 7-2   Separation of Signals on a Frequency Response Curve

with the input signal. In transistor amplifier circuits, negative feedback offsets lower circuit gain with increased bandwidth, better amplifier stability, and changes in amplifier impedance. When considering the amplifier circuit as a whole, the transistor amplifier produces a 180-degree phase shift. Because the negative feedback circuit does not introduce a phase shift, the feedback signal remains 180 degrees out-of-phase.

Positive feedback is used in oscillator circuits, and occurs when the feedback signal is in-phase with the input signal. In oscillator circuits, a transistor amplifier converts a dc voltage into an ac signal and again produces a 180-degree phase shift. Because the positive feedback circuit also introduces a phase shift, the two signals—input and feedback—remain in phase. Positive feedback in oscillator circuits is also called regenerative feedback because the configuration uses part of the output signal to generate an input to the amplifier, and the output from the amplifier to generate an input for the feedback circuit.

Phase refers to the sinusoidal relationship between two signals. For example, with in-phase signals, an increase in the input signal is matched by an increase in the output signal. With out-of-phase signals, an increase in the input signal is matched by a decrease in the output signal. Phase relationships involve the input voltage and the input current; the input current and the output current; the input voltage and the output current; and the output current and output voltage.

7-2 R-F AMPLIFIER STAGE OPERATION

The **radio-frequency** or **R-F amplifier** provides both selectivity and amplification while increasing the voltage level of an I-F signal applied to its input. Given its flat bandpass, the stage should equally amplify all passing frequencies and reject any signals that lie outside the bandpass. Also, R-F amplifiers should have a good signal-to-noise ratio. With these two characteristics in mind, the R-F amplifier stage design strengthens desired signals while canceling internally or externally generated noise. Internal noise may arise from the mixer or from the conduction of some semiconductor components. External noise is a product of electromagnetic fields generated by appliances, power lines, and unfiltered automobile ignition systems.

Furthermore, the R-F stage should isolate both the oscillator and the mixer from the antenna. Isolating the mixer protects the tuner from any unwanted frequency responses or interference. Isolating the oscillator nearly eliminates any signal radiation or other interference by preventing the oscillator from sending unwanted signals back through the transmission cables. The high-level sine wave signal generated by the oscillator can combine with the cable capacitance to become a receivable signal.

R-F amplifiers used in receivers usually operate in the Class A mode to provide minimum distortion. In addition, an R-F amplifier works as a tuned amplifier.

At the high frequencies of the signals amplified by R-F amplifiers, the internal capacitance of the device produces a low enough reactance that some signal energy feeds back from the output circuit to the input of the amplifier. Depending on the phase relationship between the feedback signal and the input signal, a high-amplitude feedback signal will cause either oscillation or degeneration at the amplifier.

Because of this, R-F amplifier circuits usually employ neutralization. A neutralization circuit consists of an adjustable capacitor, or neutralizing capacitor, which connects from the output circuit to the input circuit. The circuit feeds back a small amount of 180-degree out-of-phase energy. This small amount of feedback energy cancels the original feedback signal that travels through the amplifier.

A Dual-Gate MOSFET As an R-F Amplifier

The need for a combination of amplification, selectivity, and isolation has changed R-F stage designs because of the availability of new, more efficient components. Throughout the past thirty years, vacuum tubes, transistors, field-effect transistors (FETs), metal-oxide semiconductor FETs, (MOSFETs), and integrated circuits have served as R-F amplifiers in tuners. For our purposes, we will concentrate on a recent, popular design that uses a dual-gate MOSFET as an R-F amplifier.

Looking at Figure 7-4, the received R-F channel signal transfers to gate 1 of the MOSFET through a tuned circuit consisting of capacitor C6, the input capacitance of gate 1, a 101.5-MHz tuned coil, and another capacitor. (The last-mentioned coil and capacitor are not shown on the diagram.) Given the operating characteristics of the dual-gate MOSFET, the circuit requires very little current from the antenna and generates very little internal noise.

Because the MOSFET source ties to ground through a simple RC circuit, gate 1 and the source form a common-source input of the amplifier. With the common-source input, the connection of the AGC voltage through a resistor allows the 12 Vdc at gate 2 to change with variations of the AGC voltage. Feeding through resistor R5 and into gate 2 of the MOSFET, the AGC voltage is directly proportional to the received VHF signal. With a strong signal at the antenna, the AGC voltage increases and reduces the gain of the MOSFET. With a weak signal at the antenna, the AGC voltage decreases and allows the gain at the MOSFET to increase.

Grounding gate 2 through bypass capacitor C5 allows the combination of the grounded gate, the channel, and the drain to form a common-gate circuit and sets up a high impedance at the amplifier output. The output signal at the MOSFET drain becomes the input signal for the mixer stage. Given this configuration, the circuit offers several advantages. The common gate prevents the entire circuit from oscillating and allows any change of bias for gate 2 to effectively change the drain current of both sections of the amplifier. In addition, the common-gate configuration provides a low-input impedance, a high-output impedance, and a voltage gain greater than one.

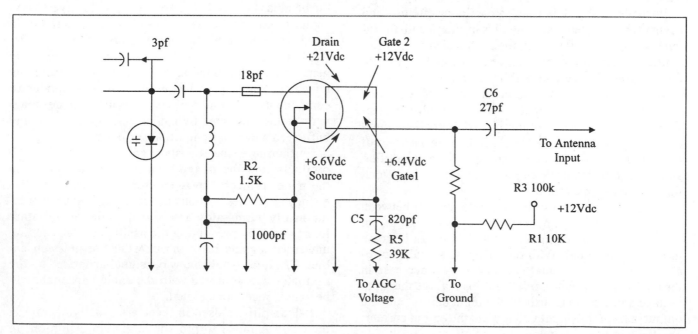

Figure 7-4   Schematic of a Dual-Gate MOSFET-Based Tuner R-F Amplifier

FUNDAMENTAL REVIEW

Capacitance

Capacitance occurs with the separation of two or more conducting materials by an insulating material. We measure capacitance with basic units called the Farad (1 Farad), the milliFarad (.001 Farad), the MicroFarad (.0000001), and the PicoFarad (0000000000001 Farad or pF). The capacitance value depends on the amount of total surface area taken by the conducting materials, and the amount of spacing between the conducting materials, and the thickness and the type of insulating material. Capacitors block dc current, allow ac current to pass, and store voltage. Like inductors, capacitors also have reactance and impedance values.

FUNDAMENTAL REVIEW

Field-Effect Transistors

As discussed in chapter four, a field-effect transistor is a voltage-controlled semiconductor that consists of a single n-type material and a single p-type material. Physically, FETs resemble bipolar junction transistors. However, while the transistor has an emitter, a collector, and a base terminal, the FET has a source, a gate, and a drain terminal. The source of the FET is connected to the drain by n- or p-type material called a channel. The FET operates by using the width of the

gate to directly change the amount of current flowing through the drain. Correctly biasing the gate causes the depletion layer at the gate to either widen or narrow.

FETs can be configured as common-source, common-gate, and common-drain amplifiers. Amplifier circuits based on an FET provide high-input impedance and very low interference with any transmitted signal.

A MOSFET is an improvement on the FET design and uses an input signal to increase the size of the channel. Like the FET, the MOSFET has a source, gate, and drain. The MOSFET differs in construction regarding the channel and the addition of substrate material. Depletion-type MOSFETs have a physical channel while enhancement-type MOSFETs rely on the gate voltage for the formation of a channel. The substrate material consists of either p- or n-type material and connects to the source. MOSFET amplifier circuits operate with an extremely low input current.

PROGRESS CHECK

You have completed Objective 1. Explain how a MOSFET operates as an R-F amplifier.

7-3 OSCILLATOR STAGE OPERATION

Tuner oscillators produce a sine-wave frequency that has a much higher level than the incoming R-F signal.

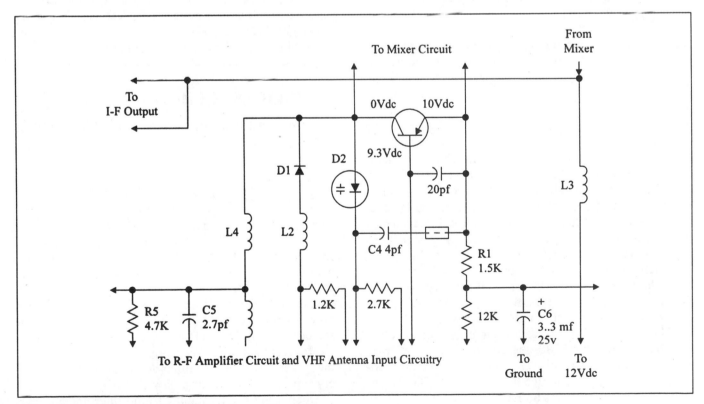

Figure 7-5 Schematic of a Colpitts Tuner Oscillator

With the output of the oscillator stage connected to the input of the mixer stage, the high-frequency, high-amplitude oscillator output is injected into the mixer. Using Figure 7-5 as a reference, the transistor oscillator is configured as a **Colpitts oscillator**, providing the regenerative feedback needed to produce oscillations. When you study the diagram, remember that the transistor is configured as an emitter follower. Because of this configuration, the transistor produces a 180-degree phase shift from the base to the collector. To produce regenerative feedback, the two tapped capacitors and the inductor also produce a 180-degree phase shift.

In this single-transistor oscillator circuit, a network of capacitors and AFC components make up the major tuning components. The combination of inductor L4, resistor R4, capacitors C5 and C4, and the varactor diode D2 in the collector circuit of the transistor develop the regenerative feedback signal and provide forward bias for the oscillator. In addition, these feedback components work as a resonant tuned circuit and establish the operating frequency of the Colpitts oscillator. Two other capacitors in the same circuit establish a capacitor voltage divider and determine the amplitude of the oscillator signal feeding to the base of the mixer. Diode D1 and the AFT circuit provide automatic fine-tuning of the oscillator signal.

↻ FUNDAMENTAL REVIEW

Oscillators

An oscillator is a circuit that produces an output waveform but does not have an external signal source. Oscillator circuits always include a positive-feedback circuit that generates an input to an amplifier. In turn, the oscillator generates an input signal for the feedback circuit. An external signal works as a trigger for the oscillator action.

Oscillators vary in stability, or the ability to maintain an output that has a constant frequency and amplitude. Well-designed oscillator circuits maintain a constant rate of oscillation over a long period of time. The loss of amplitude seen for progressive cycles of oscillation is called damping. Different circuit configurations and oscillator types affect stability. Of the different types, crystal-controlled oscillators provide the most stability. Other oscillator types include bipolar junction transistors, field-effect transistors, operational amplifiers, and multivibrators.

The popular Colpitts oscillator uses only positive feedback to achieve a 180-degree phase shift. The operating frequency of a Colpitts oscillator equals the resonant frequency of the feedback circuit. The Wien-Bridge oscillator utilizes both positive feedback to control the operating frequency of the circuit and negative

feedback to control gain. Wien-Bridge oscillators have no phase shift at the resonant frequency.

Hartley, Clapp, and Armstrong oscillators also achieve a 180-degree phase shift, but have different circuit configurations. While the Colpitts oscillator feedback network features tapped capacitors and an inductor, a Hartley oscillator feedback network has tapped inductors and a capacitor. Clapp oscillator circuits have a capacitor in series with an inductor. The feedback circuit of an Armstrong oscillator will have a single transformer in parallel with a capacitor.

Crystal-controlled oscillators have the highest stability of all oscillator types and work well for communications applications. In these oscillator circuits, a quartz crystal controls the operating frequency. Operation of a crystal-controlled oscillator is based on the piezoelectric effect, or the fact that crystals vibrate at constant rates when exposed to an electric field.

↻ FUNDAMENTAL REVIEW

Resonance

Resonance occurs when a specific frequency causes the inductances and capacitances in either a series or parallel ac circuit to exactly oppose one another. With resonance, a single particular frequency is the resonant frequency, and three basic rules emerge. First, capacitors and inductors with larger values have lower resonant frequencies. Second, capacitors and inductors in series have a low impedance at the resonant frequency. Finally, capacitors and inductors in a parallel circuit have a high impedance at the resonant frequency.

☑ PROGRESS CHECK

You have completed Objective 2. Discuss tuner oscillators.

7-4   MIXER STAGE OPERATION

The **mixer stage** converts the selected VHF television channel frequency into the lower, intermediate-frequency signal. This change occurs through the heterodyning, or mixing, of the high-frequency oscillator output signal and the radio-frequency signal associated with the selected channel. With this operation, the tuner works with an R-F signal that changes with the selection of a channel and produces an intermediate-frequency signal that never changes. As a result, the stages following the tuner do not need to retune every time the user selects a new channel.

Decreasing the frequency to the intermediate level allows an increase in amplifier gain, provides better amplifier stability, and improves the frequency response of the circuit. The lower frequency is the difference between the R-F signal and the local oscillator frequency.

Figure 7-6 shows the mixer stage from the same tuner schematic shown in Figures 7-2 and 7-3. This particular stage consists of two transistors that have direct coupling from the collector of one to the emitter of the other. Defined as a **cascode mixer**, this configuration provides good mixer stability at high frequencies, high input impedance, and low overall capacitance.

In the figure, the input circuit consisting of C1, C2, C3 and inductors found in the R-F amplifier circuit tunes the common emitter configured Q2 to the R-F signal frequency. Any change of the channel selection switch by the customer varies the inductance seen in the R-F amplifier circuit. As Figure 7-6 shows, capacitor C4, along with other capacitors in the oscillator circuit, couples the oscillator signals to the base or input of Q2. While this capacitor provides capacitive coupling and impedance matching, the values of the capacitor and the oscillator circuit components also determine the input signal levels. Resistor networks in both transistor circuits act as voltage dividers and set up the correct voltage on the base of Q2.

During normal operation, the mixer circuit changes the channel frequency into an I-F frequency because of the configuration of the transistor. Again looking at Figure 7-6, the common-base configuration of Q1 yields non-linear operating characteristics. Voltage changes across a linear device, such as a resistor, produce directly proportionate changes in current across the same device. The same changing voltage across a non-linear device does not produce a directly propor-

tionate change in current and allows the incoming R-F signal to "beat" against the oscillator signal. While the I-F signal develops across the collector circuit of the common base configured Q1, a combination of two capacitors and two inductors tune the amplified signal to the proper I-F frequency.

The mixer output signal contains the sum and the difference of the R-F amplifier and oscillator output signals. As an example, the mixing of the local oscillator output frequency for channel 4 (113 MHz) with the R-F signal for channel 4 yields 44 and 182 MHz. The difference frequency (44 MHz) becomes the I-F output signal, containing 41.25-MHz sound and 45.75-MHz picture carriers.

↻ FUNDAMENTAL REVIEW

Amplification and Gain

Amplification is the process of increasing the power of an ac signal. Amplifier circuits produce gain or a multiplication characteristic between the input and output of a circuit. As an example, if an amplifier circuit has a gain characteristic of 50, the signal at the output is 50 times larger than the signal found at the input. The gain of a circuit is controlled by component values. Gain is quantified in terms of current, voltage, and power.

☑ PROGRESS CHECK

You have completed Objective 3. Understand the use of a cascode amplifier in the tuner mixer stage.

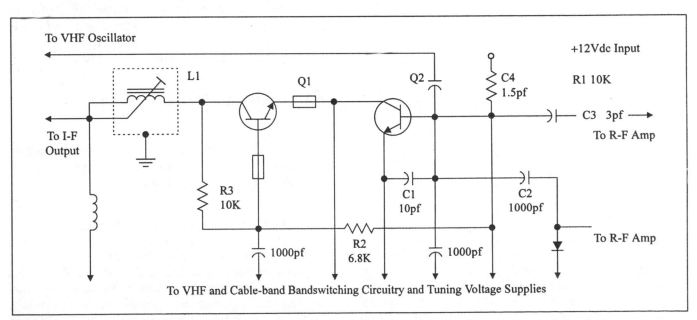

Figure 7-6   Schematic of a Tuner Mixer Stage Configured as a Cascode Amplifier

7-5 VARACTOR DIODE TUNING

Each stage of the tuner is tuned simultaneously through inductance and capacitance with the selection of a new channel. While mechanical tuners rotated different sets of inductors and capacitors in series with the stages, electronic tuners often rely on a combination of varying dc tuning voltages and the characteristics of varactor diodes. This variable combination determines the operating frequencies of the tuner stages.

Varactor Diode Principles

A **varactor diode** is a P-N junction diode specifically designed to have a variable capacitance when its reverse-bias voltage varies. This change in capacitance allows the varactor diode to tune across the low VHF, the high VHF, and the UHF bands. Figure 7-7 shows a simple varactor diode circuit and the schematic symbol for a varactor diode.

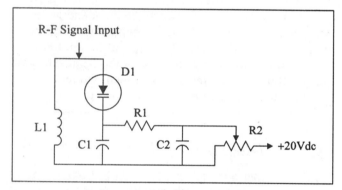

Figure 7-7   Simple Varactor Diode Circuit

When considering how the circuit functions, some knowledge about capacitance theory becomes useful. In brief review, the total circuit capacitance of a series circuit containing one large capacitance and several smaller capacitors effectively equals the capacitance of the smaller capacitor. Two formulas for capacitors in series exist. If the circuit has only two capacitors in series, the formula is:

$$C_T = C_1 C_2/C_1 + C_2$$

The formula for a series circuit with more than two capacitors is:

$$1/C_T = 1/C_1 + 1/C_2 + 1/C_3 + \ldots$$

In Figure 7-7, C2 has a capacitance of 1000 picofarads. As we use resistor R2 to change a variable dc tuning voltage, and then apply that voltage to the cathode of varactor diode D1 through resistor R1, a smaller capacitance develops across the diode. As the dc tuning voltage changes, the capacitance of the diode ranges from 5 to 11.25 pF. If we place the values for C2 and the diode capacitance into the first series capacitance formula, we find that the total circuit capacitance for each end value of the diode capacitance equals:

$$C_T = C_1 C_2/C_1 + C_2 \text{ or } C_T$$
$$= 1000 \text{ pF} * 11.25 \text{ pF}/1000 \text{ pF} + 11.25 \text{ pF}$$
$$\text{or } 11.14 \text{ pF}$$

$$C_T = C_1 C_2/C_1 + C_2 \text{ or } C_T$$
$$= 1000 \text{ pF} * 5 \text{ pF}/1000 \text{ pF} + 5 \text{ pF}$$
$$\text{or } 4.98 \text{ pF}$$

Moving back to the figure, C2, R2, and C3 complete two functions. First, the three components place the cathode of the varactor diode at signal ground while preventing the dc voltage from shorting to ground. As a result, the R-F signal "sees" that the varactor diode is in parallel with the coil. The significance of the circuit values and the placement of the components become apparent when we extend our analysis to cover the resonant frequency of the circuit. As you know, resonance occurs when the effects of inductance and capacitance exactly oppose each other. Resonance occurs at a single frequency, which is defined with the formula:

$$f = 1/(2?LC)$$

with L and C representing the circuit inductance and capacitance. If we substitute the two end capacitance values for C and the value for L_1 for L, we find that the resonant frequency varies from 90 MHz to 60 MHz.

Varactor tuners use the change in capacitance produced by the voltage characteristics of the varactor diodes to tune across the VHF, UHF, and cable bands. While varactor diodes offer exact tuning, that characteristic may be offset by low circuit Q and some noise degradation within the circuit. Because the Q of a circuit determines the selectivity, phase angle, and power factor of a tuned circuit, every tuner design provides additional amplification to offset any losses of Q.

↻ FUNDAMENTAL REVIEW
Circuit Q

The Q of a circuit describes the qualities of a resonant circuit and compares the amount of energy stored versus the amount of energy dissipated.

Figure 7-8 shows a functional diagram of the same tuning system shown in Figures 7-1, 7-2, and 7-3, but emphasizes the varactor diodes rather than the R-F amplifier, oscillator, or mixer. The figure shows that varactor diodes control R-F selectivity, interstage tuning, and the frequency of the local oscillator. Looking at the interstage tuning, the different levels of the dc tuning voltages correspond to the range of channels

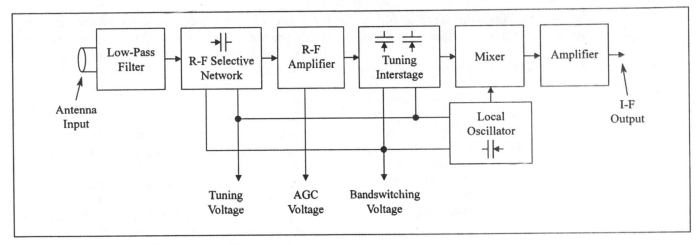

Figure 7-8 Functional Diagram of a Varactor Tuner

that make up the available bands. In the diagram, four varactor diodes control the tuning voltages for the three stages. When the consumer selects a channel, the level of dc tuning voltage applied across all four diodes tunes each stage of the tuner to the correct channel.

☑ **PROGRESS CHECK**

You have completed the Objective 4. Discuss varactor diodes and varactor tuning.

7-6 MICROPROCESSOR CONTROL SYSTEMS

Every tuner system requires some method for allowing the customer to select a channel, control the tuning, and see the channel display. Because of the evolution of video products, tuner-control systems have undergone many changes. Older-style, mechanically controlled tuners have a thirteen-position rotary switch that divides into sections and covers only the VHF band or the UHF band. Each section contains tuning coils connected to the R-F, mixer, and local-oscillator circuits.

Recently, tuner-control systems have evolved from mechanical tuners, first to a hybrid approach that used potentiometers and varactor diodes, and finally to designs based on frequency synthesis. With frequency synthesis, electronically controlled tuners combine a control assembly with a tuning system, control processor, memory, frequency synthesizer control units, and different types of keyboard input. As you saw in section 7-5, the electronic tuning-control system uses a varying dc control voltage to change the capacitance of varactor diodes and determine the operating frequency of the tuner.

Although you may encounter the older-style tuners, the television industry relies today on electronically controlled tuners that can tune across the VHF, UHF,

and cable-bands. Because of this, our effort will concentrate on the newer tuners.

Electronic Tuner-Control System Basics

Figure 7-9, located at the end of this book, provides a schematic diagram of a 125-channel, electronic controlled tuner-control system. At first glance, the schematic seems extremely complex. However, we can divide the design into four sections that set up the functions of the tuner R-F amplifier, mixer, and oscillator. This combination provides exact tuning of channels, automatic fine-tuning, automatic channel search capabilities, and easy push-button control.

By dividing the complete design of the electronic tuning-control system into sections, we can gain a better understanding about how the control system functions and how to apply standard troubleshooting methods. The four sections are:

- dedicated microprocessor and system memory section to issue and store operating commands
- customer preference controls and displays
- a frequency synthesis section designed for exact channel selection and continuous frequency control
- an electronic bandswitching system

↻ **FUNDAMENTAL REVIEW**

Binary and BCD Number Systems

Computer systems work with the binary number system, a system that allows only two values—zero and one. If we look at the digits of a binary number, each value in the columns equals a value based on the powers of 2. The simplicity of the binary system allows computer systems to move numbers from one part of a system to another and to work with large numbers. Each binary position is called a bit while a group of 8 bits is called a byte.

Keyboards, LED displays, and switches rely on a variation of the binary system called binary-coded decimal, or BCD. The BCD system uses 4 bits to represent each digit of a decimal number. As an example, decimal number 759 uses 0111 for the 7, 0101 for the 5, and 1001 for the 9 and appears as 0111 0101 1001.

In a binary system, all numbers are represented by high signals and low signals. Each high and low signal is separated by an area of voltage that has no binary meaning. While a high signal has a value of 3 to 5 Vdc, a low signal has a value of 0 to 1 Vdc.

To accomplish the tasks listed above, the microprocessor interfaces with a keypad or remote-control receiver, the system memory, and logic circuitry, and follows a simple input/process/output routine. As Figure 7-9 shows, data travels from either the keyboard—which contains customer and channel selection controls—or the remote-control interface, or a memory section into IC1000, the microprocessor, and into an input section.

The microprocessor response includes storing data in the tuner memory and sending the channel selection information to the frequency-synthesis circuitry. After leaving the central processing unit or CPU, the information travels either into memory, IC1050, or along an output path to either the channel display, IC1030, or the switching circuit, or the frequency-synthesis circuit, IC1200. Data travels to the CPU in the form of binary numbers. When studying the diagram, note that eight arrows point into the microprocessor from the input devices interface and out of the microprocessor into the output devices interface. The eight arrows indicate that the tuner uses 8-bit data commands.

The binary number system provides a convenient method for performing low-level instructions and for both temporarily placing information into and retrieving information from the **random-access memory** or RAM of the tuner. Another set of memory, referred to as **read-only memory** or ROM, holds preset, permanent instructions for the microprocessor. Whether configured as read-only or random-access memory, the memory devices store both channel selection information and most-often needed instruction sets. The instructions take the form of a preprogrammed set that the microprocessor follows when prompted by some type of signal. For example, designating channel 19 as a cable channel and placing that channel into memory as a recognized channel retrieves a specific instruction set from the read-only memory.

Along with using memory for storing data, the microprocessor also temporarily stores data in a register. When the microprocessor issues an instruction routine, a sequence in that routine may use the data group during a routine operation or may use the group to latch desired conditions. Because the register offers only temporary storage, the system discards the stored data after completing the operation.

Microprocessor Control   Figure 7-10 provides a block diagram of a tuning system built around an 8-bit 6502-type microprocessor. With normal ac power applied to the receiver, the microprocessor has +5 Vdc applied to pin 3. As the microprocessor internal circuitry charges capacitor C1 to +5 Vdc, a RESET operation occurs and prepares the microprocessor for operation. At the same time, the receiver power supply provides the power on/off relay circuit with +18 Vdc.

The microprocessor communicates with the keyboard, the remote control preamplifier circuit, an LED (light-emitting diode) display drive circuit, the volume control circuit, the system clock, and a favorite station circuit through interface circuits. With power applied to the system, the microprocessor begins to send scan pulses to the keyboard. When the customer closes a key, the scan pulses return to the microprocessor and initiate an operation.

As an example, the closing of the "power on" key causes pin 22 of the microprocessor to read a logic 1 from its pin 23. Given that signal level at pin 22, the microprocessor internal circuitry sends a low scan pulse to the same pin and detects the pressing of the power on button. In addition, the internal circuitry sends a logic 1 signal to pin 14 of the microprocessor. This signal activates the on/off relay driver and applies power to the receiver.

✓ PROGRESS CHECK

You have completed Objective 5. Describe how a dedicated microprocessor and associated memory circuits function in an electronic tuner-control system.

Customer Preference Displays and Controls

Control of any tuner begins with the customer interface. In the past, a consumer working with a mechanical tuner used a knob attached to a shaft to select and fine-tune channels. With electronic tuners, the customer interface and the control system begin with some type of keypad or remote-control device. In addition, the customer interface includes either a channel display based on light-emitting diodes or an on-screen display.

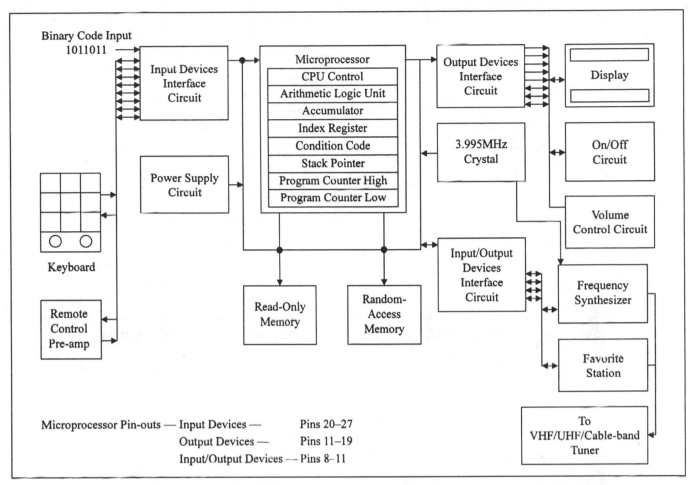

Binary Code Input
1011011

Figure 7-10   Microprocessor-Controlled Tuning System (*Courtesy of Philips Electronics N.V.*)

Obviously, channel selection and tuning begin when the customer uses either a push-button, a keypad, or a remote-control device to start instruction sets into motion. In addition to working with receiver controls, the customer also needs some method for finding which channel is selected. While the older mechanical tuners relied on a light bulb that illuminated the numbers on a rotating wheel, modern tuners use either displays based on light-emitting diodes or on-screen displays.

LED Channel Displays   Figure 7-11 shows a block diagram for a channel display using light-emitting diodes and data from a microprocessor. When the customer chooses a channel by either touching the front-panel keyboard or using the remote control, the microprocessor senses the key closure. Then the processor sends data, clock, and enable signals to the channel display circuit through three output lines. The display circuitry—which includes a combination of latches, buffers, and a shift register—decodes and stores the serial data. The proper segments of the channel display are driven to light by the data. In addition, three voltage sources—+5.4 Vdc for the display circuitry, +3.2 Vdc

for the LED, and +4.3 Vdc for the LED segment brightness—power the display assembly.

On-Screen Channel and Menu Displays   Many televisions, VCRs and satellite TV receivers use on-screen displays to show everything from the time, volume settings, and current channel to customer preferences and menu selections. In all cases, a character generator enclosed in an integrated circuit produces alphanumeric characters that can be positioned in different areas of the screen. Generally, however, the current channel number will display in the upper right of the screen, and the time will display in the lower left.

The display circuit takes advantage of the red, blue, and green output signals found at the chrominance/luminance-control integrated circuit. For example, the selection of a new channel causes the on-screen display IC to send a blanking signal to the character-blanking transistor Q1. The transistor shorts the output signals from the chrominance/luminance IC to ground, while the data from the on-screen display IC output goes to the red, blue, and green buffer transistors. As Figure 7-12 shows, the outputs from the buffer transistors connect to the CRT.

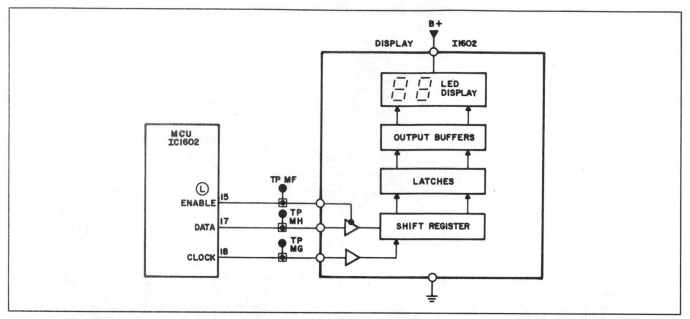

Figure 7-11 Schematic Diagram of an LED Channel Display (*Courtesy of Philips Electronics N.V.*)

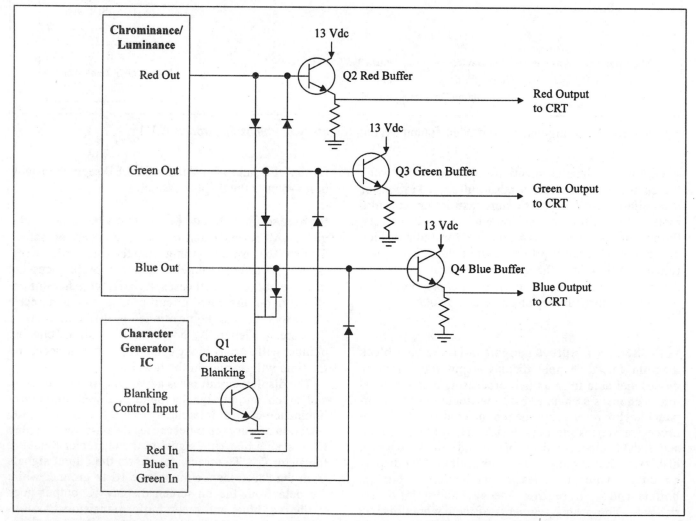

Figure 7-12 Schematic of the On-Screen Channel Display Character-Generator Circuit (*Courtesy of Philips Electronics N.V.*)

Remote Controls and Remote Receivers   Figure 7-13A shows the schematic for an infrared remote transmitter. The transmitter sends a 14-bit signal containing channel and volume information to the remote circuit shown in Figure 7-13B. During operation, the remote preamplifier applies the 14-bit signal to pin 9 of the microprocessor. When the first bit of the data string reaches pin 9, the digital level at the pin goes high, and the microprocessor releases digital low-enable pulses at its pin 8. The presence of the enable pulses allows the remote preamplifier to send the remainder of the data to the microprocessor.

Microprocessor Control of Volume   Like the LED display drive circuit, the volume-control circuit shown in Figure 7-14 also receives serial data from the microprocessor. It converts the serial data to a variable dc voltage, which controls an audio IC.

Going back to the keyboard for a moment, the microprocessor senses the closing of either the volume up or volume down control. If the customer pushes the volume up key, the microprocessor internal circuitry sends a logic-high signal to pin 11 of the microprocessor. From there, an enable signal travels to pin 2 of a shift register—an integrated circuit that accepts and holds serial data until the customer changes the volume. When the customer turns the receiver off, the micro-

processor uses an instruction set to store the serial data representing the current volume level into the system random-access memory. Turning the receiver on causes the microprocessor to issue another instruction set that outputs the stored serial data back to the volume-control circuit. As a result, the receiver has the same volume level heard before the customer turned the power off.

An almost identical process occurs when the customer uses the mute button or selects a new channel. Again, the microprocessor issues an instruction set that stores serial data representing a volume level into the system RAM. The most important difference in the process occurs when the microprocessor also sends a string of serial data 1s to the volume control. The string of 1s reduces the volume level to its lowest level during the mute or channel select operation. When the customer completes either operation, the microprocessor uses an instruction set to replace the serial data 1s with the stored data from the system RAM.

Favorite Station Selection   When the customer decides to use the favorite-station option provided in many receivers, another set of integrated circuits goes into operation. Shown in Figure 7-15, the favorite-station section consists of a shift register IC and a memory IC that also contains an address decoder.

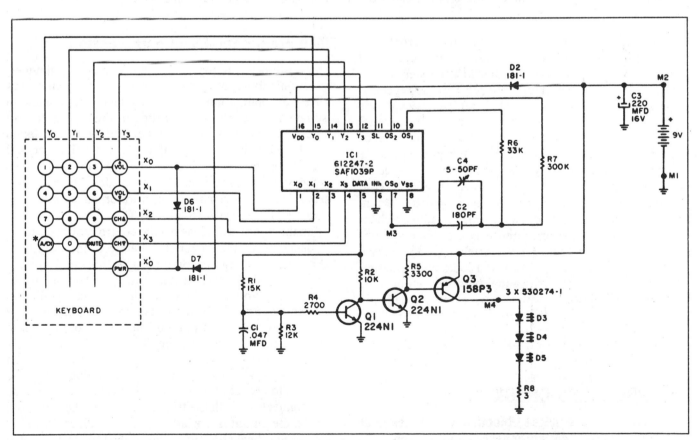

Figure 7-13A   Schematic of an Infrared Remote Transmitter *(Courtesy of Philips Electonics N.V.)*

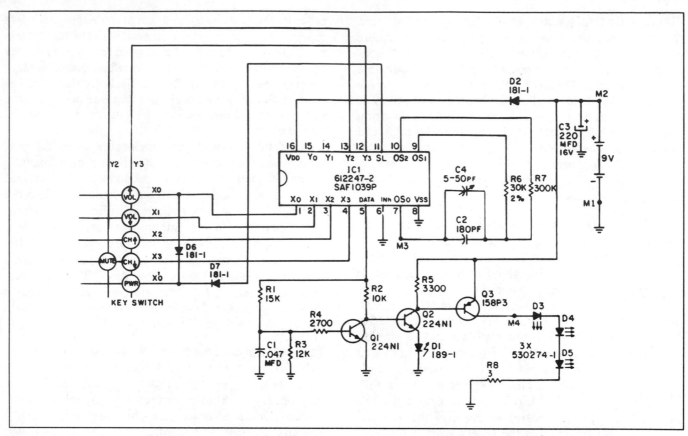

Figure 7-13B   Schematic of a Remote Preamplifier Circuit (*Courtesy of Philips Electronics N.V.*)

Selection of a favorite station causes the microprocessor to send a data stream representing the channel through its pin 17 to pin 2 of the shift register, and an enable command from pin 12 to pin 1 of the register.

For this application, the shift register accepts the 8-bit serial data but outputs two groups of 4-bit parallel data. This parallel data travels to the address decoder of the memory IC and selects one of eighty-two possible addresses. Each memory address represents a specific channel and contains one bit of data. If the address contains a 1, the channel is selected as a favorite station, while a 0 represents the default state of an unused channel. Selection of a favorite station causes the microprocessor to send an instruction that changes the 0 for a particular address to a 1. When the customer pushes either the channel scan or channel up/down buttons, the microprocessor uses a connection from its pin 13 to pin 13 of the memory IC to scan for channels that are represented by a 1.

☑ PROGRESS CHECK

You have completed Objective 6. Describe how a dedicated microprocessor monitors and operates customer controls.

The Frequency-Synthesis Section

In addition to microprocessors and memory devices, electronically-controlled tuners rely on frequency synthesis for maintaining the exact selected frequency. In some instances, the VCO portion of the tuning-control circuit consists of two emitter-follower transistors combined with a differential amplifier; a fixed crystal that resonates at approximately 4 MHz; and a frequency-adjust capacitor. Most tuner designs, however, use a 555-based astable multivibrator, such as the one depicted in Figure 7-16, as a VCO.

In the figure, a decrease in the control voltage causes a decrease in the difference between the oscillator threshold and trigger voltages. The threshold voltage is a value that switches the multivibrator on, while the trigger voltage is a value that causes oscillation to occur. Varying R1 can either increase or decrease the control voltage at pin 4 of the multivibrator, while the internal circuitry of the IC controls the charge-discharge of C1. Thus, reducing the difference between the trigger and threshold voltages also reduces the time needed to charge and discharge the capacitor. Because C1 connects to both the trigger input and the threshold input, the output is a steady stream of pulses. With the decreased discharge time, the circuit cycle time also decreases and causes the output frequency to increase.

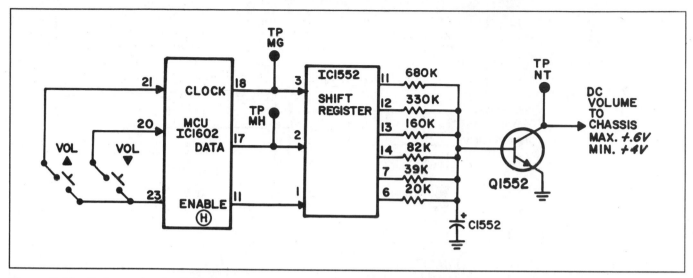

Figure 7-14   Schematic of a Microprocessor-Controlled Volume-Control Circuit (*Courtesy of Philips Electonics N.V.*)

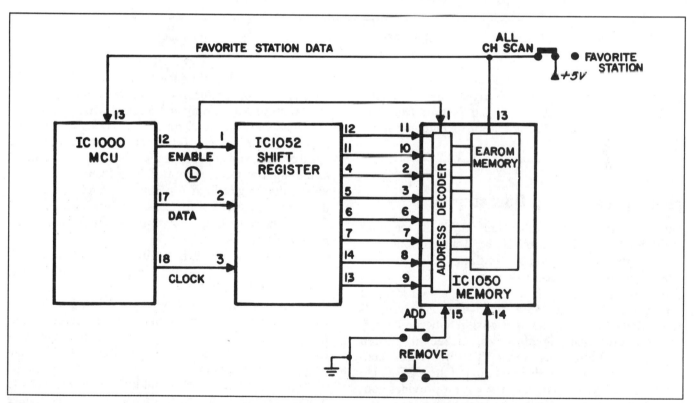

Figure 7-15   Schematic of the Favorite-Station Selection Circuits (*Courtesy of Philips Electronics N.V.*)

☑ PROGRESS CHECK

You have completed Objective 7. Explain the use of frequency synthesis in a tuner-control system.

↻ FUNDAMENTAL REVIEW

Multivibrators

Most VCO circuits are based on multivibrators. When a circuit is designed to have zero, one, or two stable output states, the circuit is a multivibrator. An astable multivibrator has no stable output state and is a square-wave oscillator. Monostable multivibrators have one

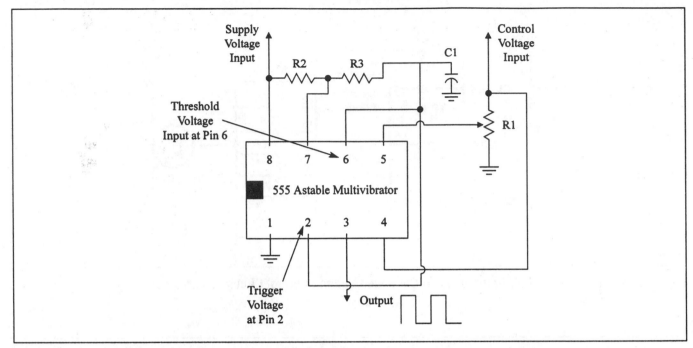

Figure 7-16   555-Based Astable Multivibrator Operating as a VCO (*Courtesy of Philips Electronics N.V.*)

stable state and generate a single output pulse with the application of an input trigger signal. Bistable multivibrators have two stable states and switch from one state to the other with the application of an input trigger signal.

Phase-Locked Loop Operation

The PLL circuit shown in Figure 7-17 locks the tuner oscillator to either a harmonic or a subharmonic frequency given by a reference oscillator. To do this, the circuit compares a feedback signal derived from the output of a VCO with a reference signal derived from the output of a stable, fixed-frequency, crystal-controlled oscillator. The relationship between the reference and feedback signals forms the checking portion of the PLL circuit. Because the PLL circuit uses feedback signals to monitor the operation of the VCO and to produce a dc tuning voltage, the circuit provides continuous protection against frequency drift.

⚓ FUNDAMENTAL REVIEW ————
PLL Circuits

A phase-locked loop contains a voltage-controlled oscillator, prescaler and divider circuits, a comparator, and a quartz crystal. The PLL receives an input signal and then compares that signal with the feedback of an internal clock signal generated by the VCO.

We can trace the reference signal from a local oscillator to the comparator. As with most tuner designs, a crystal oscillator running at 3.581055 MHz operates as the reference oscillator. This 3.58-MHz signal from the crystal travels along a reference path and through a fixed divider, and appears as the reference signal found at the input of the comparator. The fixed divider provides a dividing ratio of 1 to 3667. Given this ratio, the divider establishes a 976.5625-Hz reference signal that fits into the same frequency range as the feedback signal taken from the VCO.

Operation of the PLL begins when the customer selects a new channel and the microprocessor internal circuitry sends digital low pulses to pins 16 and 19 of the processor. The divide-by-n factor found within the PLL numerically matches the selected channel frequency. For example, the selection of VHF channel 13 (210 MHz) establishes a divide-by factor of 210 at the programmable divider.

At pin 16, the channel signal information leaves the microprocessor in the form of a 1- to 4-MHz square wave. In this scaled-down form, the square wave information represents the frequency of the desired channel. While the low at pin 19 disables the AFC circuit, the low at pin 16 enables the frequency-synthesizer IC. Once the synthesizer is enabled, the microprocessor sends 12 bits of serial data to pin 17 of the synthesizer. The 12-bit data stream represents the frequency for the selected channel.

The connection at pin 17 also ties the tuning data to a programmable divider that is part of the phase-locked loop circuit. At this point, the programmable divider

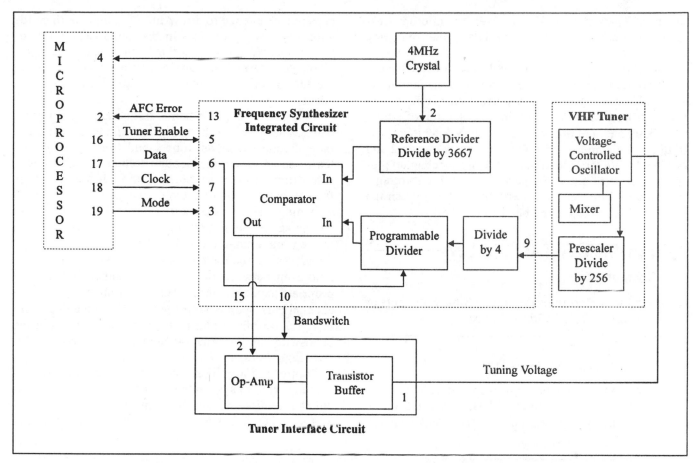

Figure 7-17   Block Diagram of a Tuner-Control PLL Circuit (*Courtesy of Philips Electronics N.V.*)

selects the correct division factor for the selected channel. Once the division factor is converted to a binary number and stored in the microprocessor memory, it becomes the address for a specific channel. Because the divider is programmable, the dividing ratio can adjust to any channel frequency. Thus, despite the local changes in oscillator frequency caused by channel selection, the local oscillator frequency stays at a level suitable for comparison with the reference.

↻ FUNDAMENTAL REVIEW

Dividers

A divider divides input pulses by an output pulse.

In addition, tuning data—in the form of square-wave input pulses—for the selected channel feeds to the VCO. As the Figure 7-17 shows, a prescaler in the tuner divides the frequency at the output of the VCO by 256 and then applies the divided frequency to the combination of a fixed and a programmable divider. With this input, the divider set provides a low-frequency feedback signal that corresponds to a specific channel. With the feedback signal at one input, the comparator then compares the phase and frequency of the sample output from the VCO with the phase and frequency of a reference signal.

The comparator compares both the frequency and the phase of the two signals found at its inputs. If the VCO is producing the correct frequency for the selected channel, the feedback signal and the reference signal will match in both frequency and phase. If the frequency produced by the VCO does not correspond with the selected channel frequency, the microprocessor issues an instruction set that causes a train of error pulses to appear at the output of the comparator and pin 2 of the frequency-synthesizer IC.

Within the tuner-interface circuit, the integrating op-amp integrates this train of error pulses and produces a low-level dc voltage, which becomes a correction voltage for the voltage-controlled oscillator. In turn, the correction voltage causes the VCO to tune 1 MHz lower in frequency in an attempt to find the correct frequency. If retuning by 1 MHz does not correct the off-tune condition for the selected channel, the microprocessor issues another command that causes the tuner to tune 1 MHz higher in frequency in a similar attempt. At times, however, the ±1MHz swing will not find the proper frequency needed to tune the selected channel. When this condition arises, the microprocessor sends a new string of tuning data to the frequency synthesizer.

The correction, or tuning, voltage will not have a value higher than 24 Vdc. Selection of a channel in the low VHF band produces a correction, voltage that ranges from 2 to 20 Vdc, while the selection of a high VHF band channel produces a 7 to 20 Vdc voltage. For UHF band channels, the voltage ranges between 2 and 24 Vdc. While the selection of each higher channel produces a proportionate increase in the tuning voltage, the voltage drops to its lowest "band" level when the selection moves from one band to another.

↺ FUNDAMENTAL REVIEW ⸺
Comparators

A comparator determines the sign of binary numbers or compares the polarity of dc voltages.

Electronic Bandswitching

Varactor diodes have enough capacitance change to tune across either the individual VHF low bands or the VHF high bands. They do not, however, have enough

capacitance change to allow tuning across both bands with the same set of coils in the tuned LC circuits of a tuner. In addition, during the discussion of PLL operation, you learned that some overlap occurs between the tuning voltage ranges of the different television bands.

To avoid tuning the same frequency for different bands, and to allow the reception of several different radio-frequency bands, tuners and tuner-control systems use an electronic bandswitching circuit to activate different tuner sections. Although these circuits feature basic switching techniques, electronic bandswitching also takes advantage of the PLL operation. As the sample signal travels along the feedback path to the programmable divider, the signal goes through the prescaler, and the prescaler sets up a frequency relationship between the selected channel and the appropriate band. The local oscillator frequency for each channel corresponds with the tuned frequencies. With this relationship in place, the microprocessor and the bandswitching circuitry generate the correct bandswitching voltage.

Figure 7-18 highlights the bandswitching portion of the tuner-control system that we have analyzed throughout this section. In the diagram, each of the transistors switches to a specific band. When the customer

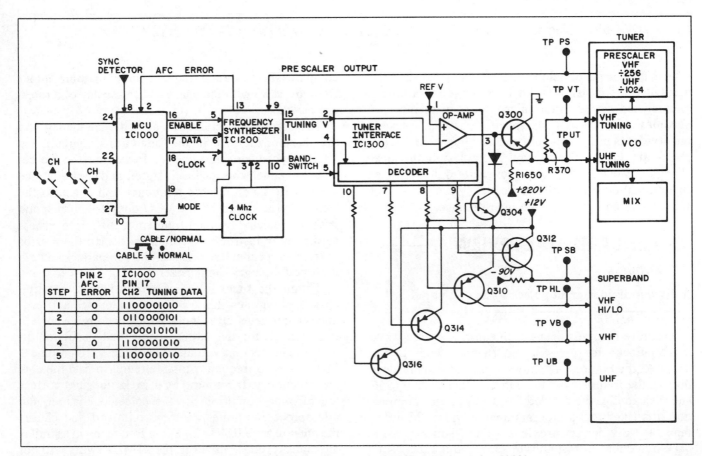

Figure 7-18 Tuner-Control Bandswitching Circuit (*Courtesy of Philips Electronics N.V.*)

chooses channel 4, the microprocessor disables the AFC circuit, enables the frequency synthesizer, and sends 12 bits of serial data to pin 17 of the synthesizer, IC1200. While the entire data stream represents a specific channel, the first two bits of the data represent the band information for the selected channel.

The frequency synthesizer outputs the two bits of bandswitching data to the tuner interface IC for decoding. Then the decoded bandswitch data travels from the tuner interface to the bandswitch transistors, Q310, Q312, Q314, and Q316. Switching occurs when the data in the form of a square-wave input drives the transistor back and forth between saturation and cut-off and produces a square-wave output. A negative dc input voltage to the transistor causes the emitter-base junction of the transistor to bias off, while a positive dc input voltage causes the transistor to saturate. When biased off, the transistor is an open switch; when saturated, the transistor is a closed switch.

☑ PROGRESS CHECK

You have completed Objective 8. Describe the operation of phase-locked loops.

7-7 TROUBLESHOOTING TUNER SYSTEMS

Whether working with televisions, VCRs, or satellite television receivers, you can see the symptoms of a defective tuner by watching the reproduced picture on the CRT screen. In addition, you can draw a few conclusions about the source of the problem by remembering the functions individually provided by the tuner stages. You know that a certain type of signal enters the tuner, that the R-F amplifier, oscillator, and mixer process the signal in a certain way, and that a certain type of signal exits from the tuner.

Yet, as Table 7-2 indicates, one symptom can point toward several sections. Therefore, while looking at the symptom, you also need to use troubleshooting methods such as signal tracing to find the source of the

problem. In addition to those problems, misalignment of any of the stages can cause the intermittent operation of some channels and poor or weak color reproduction. Problems with the automatic fine-tuning circuit or the fine-tuning of the tuner can also cause the loss of sound, the loss of color, or a weak and snowy picture.

Given the modular construction of tuners, you could remove the entire tuner and exchange the unit for a new or factory-serviced tuner. Several repair depots specialize in the repair and replacement of inoperable tuners. Because troubleshooting some tuner problems—such as those related to the oscillator—may take time, the use of a repair depot may stand as the most cost- and time-effective option. However, because the I-F section may cause the receiver to exhibit some of the same symptoms, it may also be useful to consider checking the tuner for some obvious faults before replacing the unit.

Safety First!

Rather than attempting to squeeze a test probe between components and risk shorting two pins of a microprocessor, use available clip-on adapters for the testing of integrated circuits. The clip-on adapters fit dual in-line packages or DIPS, plastic leaded chip carriers or PLCCs, small outline integrated circuits or SOICs, or plastic quad flat packs or PQFPs. With no power applied and using a pin locater, align the adapter with the IC and gently slide it over the integrated circuit. The test clips have labels for each pin number and offer an easy method for checking readings at the correct access point. Finally, remember to observe the proper antistatic precautions.

Test Equipment Primer: Counters

A counter measures frequencies, periods, frequency ratios, and time intervals, and is especially useful for measuring the frequency of the local oscillator in a television, the clock crystal in a microcomputer, or the 3.58-MHz master clock in a VCR. When used for

Table 7-2 Tuner Problem Symptoms

R-F Amplifier	Oscillator	Mixer
Weak and snowy picture on all channels	Raster but no picture, or sound on all channels	Interference such as herringbone in picture
Raster but no picture, or sound on some channels	Weak sound and snowy picture on all channels	Raster but no picture, or sound on some channels
Interference such as herringbone in picture	Raster but no picture, or sound on some channels	No I-F output signal

measuring frequencies, a counter counts the number of complete cycles per second over a precise period of time. With the measurement of frequency ratio, a counter compares the measurements of two different input signals applied to different input channels.

When purchasing a frequency counter, consider:

- Frequency range
- Resolution
- Time-base stability
- Sensitivity
- Ruggedness
- Shielding against R-F signal interference

Most general-purpose counters have ranges that extend from 10 Hz to 10 MHz, while the range of counters intended for specialized duties may extend up to 100 MHz. The attachment of a prescaler, or a frequency-divider circuit, extends the high-frequency measurement capability of a counter.

When purchasing a counter, check for selectable resolution and verify the stability of the internal time base. Along with time-base stability, the resolution of a counter establishes the accuracy of a measurement. Resolution is the smallest increment of change displayed on a counter, while stability is measured in parts per million. A good stability measure will show little or no changes in the quartz crystal oscillator frequency if the counter undergoes any variations in temperature or operating voltage. Measured in fractions of a volt, the sensitivity of a counter shows the lowest signal strength that the counter will measure.

R-F Amplifier Problems, Symptoms, and Solutions

Using the same tuner depicted throughout this section, we will take a look at possible problems that can occur with each stage. The characteristics of the MOSFET used for the R-F amplifier suggest that fairly basic problems can develop. If the R-F amplifier does not seem to have an output, check that an input signal appears at input circuitry leading to gate 1. In addition, verify that the correct voltage is at the drain and that the amplifier has good ground connections.

After verifying the presence of an input signal, begin inspecting the components that surround the MOSFET and look for possible open conditions. Signal tracing is one method for finding open coupling and bypass capacitors in a circuit. To find an open capacitor through signal tracing, connect a signal generator to the input of the circuit. While some signal attenuation may occur at the output, you should find approximately the same signal amplitude on both sides of any coupling capacitor. No signal at the output side of a coupling

capacitor indicates that the capacitor is open. Because bypass capacitors pass a signal to ground, no signal should exist across the capacitor. If the bypass capacitor has opened, then a large signal will appear across the capacitor.

In the R-F amplifier circuit, an open C6 (a coupling capacitor) causes the MOSFET to lose the input R-F signal. When either R1 or R3 opens, the MOSFET loses its forward-bias voltage and, in effect, becomes self-biased by the drain voltage. If this condition occurs, the voltage at gate 2 drops to zero, and the amplifier will begin to clip the output signal. With an open C5, gate 2 loses its connection to ground and equal voltages appear at the drain and the source.

MOSFETs display specific symptoms if the device has an internal short or an open. A MOSFET with a shorted gate/source junction will measure 0 Vdc at the junction, and the current flowing through the drain circuit will increase. Under normal conditions, the current flowing through the drain and source will be approximately equal. If the gate/source junction opens and the depletion layer cannot form, the voltage at the junction will increase beyond its normal rated value.

Oscillator Faults, Symptoms, and Solutions

Oscillator circuits have two basic characteristics that can aid your troubleshooting efforts. First, the only input to an oscillator circuit is the dc supply voltage. Second, the circuit pictured in Figure 7-5 uses regenerative feedback to provide forward bias for the amplifier and does not have an external signal source.

Start the troubleshooting process by verifying that both the circuit and the oscillator have the correct supply voltages. Compare the measurements at the oscillator emitter, base, and collector with those shown on the schematic. Because the transistor is configured as an emitter follower, the circuit will have very little voltage gain between the input and the output, and the emitter and base will have almost equal voltages. If the collector and emitter have nearly equal voltages, the oscillator has become cut-off. As a result, the base circuit may have an open or the oscillator may have an internal defect. An open R1 in the emitter circuit can cause the voltage at the emitter to range close to the +12 Vdc value of the supply voltage. If this occurs, check for the presence of the input signal at the base of the oscillator. The presence of a signal points to a defective transistor.

After checking the supply voltages, check the values of R4, C4, and C5 through the use of a multimeter and an in-circuit capacitance meter. If those components have values within the rated tolerance, the oscillator probably has an internal defect. Also, check varactor diode D2 for a shorted condition. A shorted AFT diode can ruin the oscillator transistor.

Mixer Faults, Symptoms, and Solutions

As with the R-F amplifier and the oscillator, you need to know both the function of the mixer and how the configuration of the individual components in the mixer circuit allow this function to occur. A mixer circuit changes an R-F frequency to an I-F frequency and then amplifies the I-F signal. Referring to the mixer circuit in Figure 7-8, one stage of the circuit changes the frequency while the other provides the I-F amplification. By tracing the signal from the output of the mixer back to the input, you have a better chance of finding defective components. If the circuit does not produce an I-F signal at the I-F output, use an oscilloscope to check for the presence of the I-F output signal at both sides of tuned coil L1.

After making these basic tests, you can divide the cascode mixer into the common-emitter circuit of Q2 and the common-base circuit of Q1 and trace the signal from input to output. The common-emitter circuit changes the frequency, while the common-base circuit works as the amplifier. With the voltage-divider bias circuit consisting of R1, R2, and R3 setting the value of the input voltages for both circuits, an open in any one of these resistors stops the flow of current and the development of any bias voltage.

Each of the amplifier circuits should produce signal gain from input to output. Looking first at the common-emitter circuit, the input signal travels to the base and the output signal exits at the collector. The common-emitter amplifier produces high gain from input to output and a 180-degree phase change. Because Q1 amplifies the I-F signal, use an oscilloscope to check both the output and the input of the transistor. Given the common-emitter characteristic of producing high input-to-output gain, remember to change the volts/division setting on the oscilloscope.

With no I-F signal at the emitter, you can begin checking the Q2 common-base circuit, where the input signal enters at the emitter and the output signal exits at the collector. The common-base amplifier produces no phase change and midlevel signal gain. While an open C3 prevents the coupling of the R-F signal to the circuit, an open C4 eliminates the oscillator signal as an input signal. If there is an I-F signal at the emitter of Q1, you know that the Q2 circuit is functioning correctly.

✓ PROGRESS CHECK

You have completed Objective 9. Define typical tuner problems and troubleshooting methods.

Tuner-Control Problems

Before we begin to look at specific problems and solutions, it may be helpful to divide any electronic tuner system into different categories and examine the linkages between the categories and the problems. All this establishes a logical troubleshooting method and avoids a scattered shotgun approach to repairing the unit. When troubleshooting a microprocessor control system, use a multimeter to check for the correct dc supply voltages, an oscilloscope to check for the correct logic levels and data content at key test points, a frequency counter to monitor the reference and local oscillator frequencies, and an isolation transformer for ground isolation.

Troubleshooting Microprocessor Control Problems

LED Display Problems In some instances, the LED display for a tuner system will fail to light. Before ordering a new LED assembly for replacement, you should check that the failure of other components has not affected the operation of the LED display. First, find if the LED has any illumination and whether an operation such as "Add a Favorite Station" works as it should. If the LED is slightly illuminated and "Add a Favorite Station" works, check the +24-Vdc voltage supply for the display. Many times, a resistor will open in the voltage supply and prevent the proper voltage from reaching the display. With the proper voltages and a dimly illuminated LED, you can conclude that the LED display assembly is defective.

Another common condition occurs with the part of the display on or off at all times. With this condition, check to see if the display changes with every key sequence. If you find that the display does not change, then the LED display assembly is defective. A changing but constantly illuminated display means that the microprocessor has an internal defect.

If a customer reports that the remote transmitter no longer works, consider which parts of the system affect the transmitter operation. The operation of the remote transmitter involves the transmitter, the remote preamplifier circuit, and the microprocessor. You also know that pressing a key on the remote should cause the signal at pin 8 to reach the +3 to 5-Vdc level; the signal level at pin 9 should stay between 0 and +1 Vdc.

Given your knowledge of the system, you should first check for the presence and status of the signals at pins 8 and 9 of the microprocessor. A condition of having the remote key pressed, good signals at the pins, and no remote operation indicates that the microprocessor has developed an internal defect. If no signals are present at the pins, disconnect the remote-control preamplifier from the system and again check the status of the signals at the two microprocessor pins. While the presence of the signals with a disconnected remote preamplifier tells you that the preamplifier is defective, the absence of signals indicates either a defective remote control or a defective microprocessor.

Volume Control Problems   Customers may report that they can no longer control the volume by pressing on the volume up or down buttons. As with the other microprocessor control subsystems, check the dc supply voltages, logic conditions, and data streams. When working with the circuit shown in Figure 7-15, check for the +24-Vdc supply voltage at collector of Q1 before assuming that a component failure has caused the problem. Also, use the multimeter to check for the proper dc voltage at the Q1 emitter. Because the Q1 emitter ties to the dc volume-control voltage found at the chassis, the voltage should vary from 0.6 Vdc at maximum volume, to 3.4 Vdc at a typical volume, and to 4 Vdc at a minimum volume.

From our earlier discussion, you know that the microprocessor should sense the pressing of either the volume up or down control and send a logic-high signal to its pin 11 and a stream of serial data to its pin 2. Using an oscilloscope, check for a pulse of +3 to 5 Vdc at pin 11 and a series of pulses at pin 2. When pressing the volume up key, the serial data should count down with all logic 0s showing at maximum volume. Pressing the volume down key should cause the serial data to count up, with the data at a typical volume setting appearing as 101101. At minimum volume, the serial data should appear as all logic 1s.

You can perform the same type of tests at the data input pins of the shift register. Referring to Figure 7-15 those pins are shown in Table 7-3.

Table 7-3   Volume Control Shift Register Pins and Data Readings

Pins	6	7	14	13	12	11
Maximum Volume	0	0	0	0	0	0
Typical Volume	1	0	1	1	0	1
Minimum Volume	1	1	1	1	1	1

Frequency Synthesis Problems

One of the most complex problems encountered by technicians involves the improper tuning of channels by the control system. Efforts to pinpoint the problem lead to the bandswitching circuits, the frequency synthesizer, the microprocessor, peripheral circuitry, and the tuner.

Improper Tuning   Before attempting to solve the problem, you need to know whether the system correctly tunes across the low VHF band, the high VHF band, the UHF band, and the cable bands. If the system tunes across some but not all bands, the process of elimination takes you to the bandswitching circuitry and then to the frequency synthesizer circuitry. Improper tuning across all bands leads to the frequency synthesizer and the tuner.

If the system does not tune across the low VHF band, check for the proper bandswitching voltages at the collectors of Q10, the superband switch; Q12, the high/low bandswitch; Q14, the VHF bandswitch; and Q16, the UHF bandswitch. Figure 7-18 showed the transistors and the voltages. Having the correct voltages at those test points indicates that a problem exists either within the tuner-interface IC, the frequency-synthesizer IC, or within the tuner. To further narrow the search, you can check for the presence of the VHF tuning voltage at pin 16 of the tuner interface and verify that the correct frequency exists at the point where the tuner prescaler connects to the remainder of the system. If you find the wrong frequency at this point, you know that a defect exists within the tuner.

To check the prescaler frequency, calculate the local oscillator frequency for a given channel by adding 44 MHz to the center of the selected channel frequency. With the prescaler dividing by 256, you can determine the correct divider output frequency for any channel. Then use a frequency counter to measure the actual divider-output frequency and compare the reading with the calculated value.

The correct frequency at the prescaler takes you back to the tuner-interface IC. Here you can use an oscilloscope to check for logic 1s at pins 7, 8, and 10 and a logic 0 at pin 9. With the proper logic levels in place, you can turn your attention to the +24-Vdc and −40-Vdc voltage sources in the bandswitching circuit. If the correct levels do not exist at the tuner interface, you can check for logic 0s at pins 4 and 5 of the frequency synthesizer. The presence of logic 0s at those pins verifies that the tuner interface IC is defective. Logic 1s at either of those pins indicate a defective frequency synthesizer or microprocessor. Again using an oscilloscope, you can check for proper microprocessor operation by examining and comparing the waveforms at pins 3 and 12 of the synthesizer to the waveforms shown in the schematic of Figure 7-9.

If a tuner system will tune some channels properly while not tuning others, check the logic transitions at pin 15 of the synthesizer. Selecting a channel should cause the logic state at the output of the comparator to change as the tuned frequency goes above and below the center frequency of the channel. Each transition should occur at the center frequency of the selected channel.

With a few exceptions, you can follow the same procedures for solving problems with the high VHF band, the UHF band, and the cable bands. In each case, check the bandswitching voltages and logic levels at pins 7, 8, 9, and 10 of the tuner interface and pins 4 and 5 of the frequency synthesizer and follow the same troubleshooting path. There are some differences, however. The logic levels at the tuner-interface IC pins and frequency-synthesizer pins change with the selection of the UHF, low and high VHF, and cable bands. The selection of different bands also means that the test point for the tuning voltage changes from VHF to UHF and to cable band.

Table 7-4 lists the logic levels and test points for each band. The test points refer to the schematic shown in Figure 7-10.

☑ PROGRESS CHECK

You have completed Objective 10. Determine standard procedures for troubleshooting electronic tuner-control systems.

Table 7-4 Tuning System Logic Levels and Test Points

	Tuner-Interface Logic				Frequency-Synthesizer Logic			Test Point
Pins	7	8	9	10	Pins	4	5	
Low VHF	0	1	1	1		1	1	ME
High VHF	0	0	1	1		0	1	ME
UHF	1	0	1	0		0	0	MV
Cable Band	0	0	0	1		1	0	ME

SUMMARY

Chapter seven used discussions of tuner stages and electronic tuner-control systems to increase your knowledge of components and circuit configurations. The examination of the R-F amplifier stage allowed us to explore the operation of a dual-gate MOSFET. You learned about different types of oscillators in the tuner oscillator stage and saw an application of a cascode amplifier in the mixer stage. The analysis of the electronic tuner-control system included study of dedicated microprocessors, storage memory, registers, dividers, and the operation of a PLL circuit, and reviewed the applications of binary logic throughout the circuits.

Chapter seven presented troubleshooting methods that you will apply throughout this text. The examination of tuner problems, included basic points and needs of discrete amplifier operation, as well as an analysis of integrated circuit operation.

REVIEW QUESTIONS

1. Tuners _____ all unwanted signals and _____ the desired signal to a level where the _____ of sound and picture information can occur.

2. The frequency range for the low VHF band is _____, for the VHF high band is _____, for the UHF band is _____, for the superband is _____, for the Hyperband is _____.

3. Why does the R-F amplifier stage strengthen desired signals and cancel internally or externally generated noise?

4. The R-F amplifier isolates the _____ and the _____

   because _____.

5. Why is a dual-gate MOSFET a better choice for an R-F amplifier than a transistor?

6. Refer to Figure 7-4. Capacitor C6 in that figure is what kind of capacitor?
   a. coupling
   b. bypass
   c. filter
   d. tuning

7. Refer to Figure 7-4. Capacitor C5 in that figure is what kind of capacitor?
   a. coupling
   b. bypass
   c. filter
   d. tuning

8. In Figure 7-4, the voltage applied to resistors R1 and R3 provides _____.

9. List three common tuner problem symptoms.

10. Refer to the circuit shown in Figure 7-4. The customer has reported that the receiver cannot tune VHF channels. To this point, you have found the following voltages at the MOSFET terminals: gate 1 (6.4 Vdc), gate 2 (0 Vdc), drain (21 Vdc), source (6.6 Vdc). Show how you would find the circuit problem that causes the reported problem. List any other symptoms that you may find.

11. True or False   The output of a tuner local oscillator is a low-frequency sine wave.

12. What are the characteristics of a Colpitts oscillator?

13. Describe the cascode configuration and why it is a good choice for a mixer stage.

14. In the circuit shown in Figure 7-6, Q1 provides _____. Q2 provides

    _____. The input for Q1 is the _____.

    The output for Q2 is the _____.

15. True or False   Q1 is configured as a common-base amplifier and Q2 is configured as a common emitter amplifier.

16. Refer to the circuit shown in Figure 7-6. R2 has opened, yet replacing R2 does not restore the I-F signal at the output of the mixer circuit. What would you do to track down the source of the problem?

17. Draw a simple varactor diode circuit. What causes the capacitance of the varactor diode to change?

18. The _____ of a circuit describes the qualities of a resonant circuit and compares the amount of energy stored versus the amount of dissipated energy.

19. What is the capacitance theory that illustrates varactor diode operation?

20. In the block diagram of a varactor tuner shown in Figure 7-8, the varactor diodes control

    _____, _____, and

    _____.

21. What are the four key subsystems that make up an electronically controlled tuner?

22. The dedicated microprocessor found in an electronically controlled tuner _____;

    _____; and _____.

23. Refer to Figure 7-14. When data moves from the input circuits to the microprocessor, the data is in the form of binary code with each 1 and 0 representing a value. What are those values?

24. True or False   Memory devices in an electronically controlled tuner may be either read-only or random-access devices. What is the difference between read-only and random-access memory?

25. Describe the purpose and operation of each of the listed components?
    a. voltage-controlled oscillator
    b. phase comparator
    c. programmable divider
    d. crystal oscillator
    e. latch
    f. register
    g. microprocessor
    h. RAM and ROM
    i. prescaler

26. In a frequency-synthesis tuner, the channel selection information moves from the microprocessor to a programmable divider. What is the form and frequency of that information?

27. Draw a simple diagram of a phase-locked loop. What is the purpose of a PLL? How does a PLL work?

28. The output of the phase comparator in a frequency-synthesis tuner is _____

    and is used to _____.

29. The output of the reference oscillator in a PLL is fixed at _____ MHz.

30. When troubleshooting microprocessor control systems, you should check _____,

    _____, and _____ before replacing components.

31. What test equipment would you use to check a microprocessor control system?

32. When checking the volume-control circuit shown in Figure 7-15, the voltage at the

    _____ of Q1 should vary with different volume settings.

33. Refer to Figure 7-17. If the tuner does not tune all channels properly, you can check the logic condition at pins 4 and 5 of the frequency synthesizer. The presence of logic 1s at those pins verifies that the tuner

    interface IC is _____.

34. Why would you check the tuning voltage when troubleshooting a tuner control problem?

35. True or False   The correct voltage for a logic-high state is +0 to 1 Vdc, and the correct voltage for a logic-low state is +3 to 5 Vdc.

36. What is the difference between a local oscillator and a reference oscillator?

37. One method for checking the operation of a frequency-synthesis tuner is checking the output of the comparator during channel changes. How would you check the output and what would you expect to find?

38. An electronic tuner-control system has problems with tuning the superband channels. Describe the procedure for finding the source of the problem.

CHAPTER

8

Intermediate Frequency, AGC, and AFT Circuits

OBJECTIVES Upon completion of this chapter, you should be able to:

1. Define intermediate-frequency signals and automatic fine-tuning.
2. Discuss circuit Q.
3. Understand how wave traps and bandpass filters affect selectivity and attenuation.
4. Explain amplifier stage gain, coupling, and neutralization.
5. Discuss automatic gain control circuits.
6. Discuss stagger-tuned I-F circuits and SAW filters.
7. Explain the integration of the I-F amplifier circuit, AFT circuit, and AGC circuit into one device.
8. Explain sound and video I-F signal extraction, demodulation of the composite video signal, and video detector circuits.
9. Identify I-F amplifier, AFT, and AGC circuits troubleshooting methods.

KEY TERMS

absorption trap

adjacent channel

adjacent channel traps

amplified AGC

automatic fine-tuning (AFT) circuit

automatic gain control (AGC) circuit

bridged-T trap

capacitance meter

critical coupling

delayed AGC

discriminator

forward AGC

frequency response curve

gated AGC

gated pulse

intercarrier approach

keyed AGC

keyed pulse

mutual coupling impedance

notch trap

overcoupling

parasitic oscillation

reverse AGC

selectivity

sound intermediate frequency

split-sound approach

stagger tuning

surface acoustic wave (SAW) filter

transistor tester

vestigial transmission

video detector

INTRODUCTION

In the last chapter, you learned that the mixer stage of the tuner outputs an intermediate frequency signal to the I-F section. Although the terms "intermediate frequency signal" may seem to describe one specific frequency, the mixer output actually contains signals with various frequencies. From those signals, only two—the video I-F signal and its sidebands and the sound I-F signal and its sidebands—are desirable. The video I-F signal contains picture information while the sound I-F signals contain the audio information. Any other signals at the mixer output are unneeded by the remainder of the video receiver.

Because of this, the I-F section of the receiver must complete two tasks. First, the section must eliminate the unneeded signals found at the mixer output. Also, the I-F portion uses tuned circuits to provide not only selectivity but also the proper signal ratios. Any unwanted I-F frequencies can cause interference in the reproduced picture. Second, the I-F section must provide sufficient amplification for the desired video and sound I-F signals.

Chapter eight explores the how electronic circuits eliminate the unneeded signals, provide selectivity, and amplify the desired video and sound I-F signals. The chapter illustrates the relationships between the tuner, I-F section, automatic fine-tuning, and automatic gain control circuits used in television receivers. You will also learn more about attentuation, frequency response, and circuit Q, which in turn provide a basis for the discussion of filters and wave traps.

The chapter builds on these topics with detailed studies of different types of circuits used in video electronic systems, including stagger-tuned I-F circuits, surface acoustic wave filters, and single-IC devices that contain all I-F, AFT, and AGC functions within one package. Finally, the troubleshooting section shows how to pinpoint problems with I-F, AFT, and AGC circuits.

8-1 PROCESSING INTERMEDIATE FREQUENCY SIGNALS

Three different signal relationships exist between the I-F section, the tuner, and the remainder of the receiver. First, the frequency-modulated sound I-F and amplitude-modulated picture I-F signals travel from the mixer to the I-F section through a transmission cable. At the input to the I-F section, the I-F signal contains the desired as well as adjacent channel picture and sound frequencies. After the signals are shaped to eliminate the undesired signals and then amplified, demodulation of the signals begins the development of the sound and composite video signals. Second, automatic

gain control circuitry uses signals from the I-F section to monitor and correct the overall gain of signals processed within the tuner and the I-F section. Lastly, automatic fine-tuning circuitry uses signals from one of the I-F amplifiers to monitor and correct any video I-F signal changes by applying a dc correction voltage to the tuner.

Intermediate-Frequency Signals

Although a video receiver could use any set of picture and sound frequencies, all television receivers utilize a **video intermediate frequency** of 45.75 MHz and a sound intermediate frequency of 41.25 MHz. This standard approach is necessary because of the design of other receiver sections. The use of lower intermediate frequencies offers higher amplifier stage gain and good stability. Yet the use of a lower intermediate frequency pushes the tuner oscillator frequency closer to the R-F input signal and allows the I-F signal to become a broadcast signal.

The transmitted sound carrier is 4.5 MHz higher than the picture carrier. Because of this separation at transmission, the video and sound I-F signals found at the output of the mixer also have a 4.5 MHz separation. Unlike the transmitted signals, however, the sound I-F has a frequency 4.5 MHz lower than the video I-F. Heterodyning within the mixer stage causes this change to occur.

Both the 41.75-MHz sound carrier and the 45.25-MHz picture carrier travel from the mixer to the I-F section. Although older televisions utilize a technique called the intercarrier method for processing the I-F signals, newer designs use a split-sound approach. With the **intercarrier method**, one strip of amplifiers amplifies both the video and sound I-F signals. At the end of the I-F strip, sound and video detectors separate the two signals for additional processing and amplification.

The **split-sound method** uses a separate sound stage to process and develop the audio output signal. While the sound carrier passes through a portion of the video I-F section, there is a sound take-off point in the third video I-F amplifier stage. Once the sound carrier enters the separate sound stages, additional I-F amplifiers shape the signal. Then a detector demodulates an audio signal for further development and amplification in the audio output section.

Along with the sound and video carriers, the I-F section must also allow other signals to pass to the next sections. The color sync signal, or color burst, has a frequency located 3.579545 MHz away from the video carrier. Color sidebands—the chrominance signal—extend approximately 0.5 MHz above and below the 3.58-MHz color burst and have a bandwidth of 1 MHz. Because of the presence of these signals, the I-F signal must have the capability to provide proper amplification across a broad bandwidth.

Figure 8-1 shows a **frequency response curve** for the intermediate frequencies taken from the mixer. Markers on the curve represent the 41.25-MHz sound and 45.75-MHz picture carriers, adjacent channel signals, the 42.17 MHz color subcarrier, and the 42.57-MHz upper and 41.67 MHz lower color sidebands. During an alignment of the I-F amplifier section, a technician can use a combination of a sweep-marker generator and an oscilloscope to display the frequency response curve and the markers.

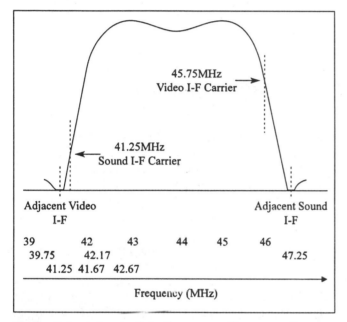

Figure 8-1   Intermediate Frequency Response Curve

In itself, the frequency response curve shows the amount of gain found at given frequencies. While the baseline of the frequency curve represents zero percent gain, the width of the curve at the 50 percent gain mark represents the maximum bandwidth of highest frequencies. A frequency response curve also shows whether an amplifier maintains constant gain when operating within its bandwidth, and shows the cut-off frequencies—f_1 and f_2—of the amplifier. On the curve, the cut-off frequencies occur at the points where the amplifier power gain drops to 50 percent. The center frequency of the bandwidth equals the geometric average of the cut-off frequencies, or:

$$f_O = f_1 f_2$$

↻ FUNDAMENTAL REVIEW

Carriers and Sidebands

A carrier is a single, unmodulated radio-frequency signal. Either amplitude modulation or frequency modulation superimposes information onto the carrier and produces side-carrier, or sideband, frequencies that extend above and below the original frequency. When modulated, the carrier frequency is the center frequency of the modulated wave. Carrier waves are equal to the sum of the unmodulated R-F carrier and two R-F sideband frequencies.

Signals with a high-modulation frequency lie farther from the carrier than signals with low-modulation frequencies. Consequently, a signal containing both low and high frequencies—such as a broadcast television channel—requires a greater bandwidth than a signal containing only low frequencies.

While the picture signal is an amplitude-modulated carrier wave that has symmetrical amplitude variations above and below the carrier, the sound signal is frequency-modulated by audio frequencies ranging from 50 to 15,000 Hz. The video signal—which includes luminance, chrominance, and sync information—modulates the picture carrier.

With a 6-MHz television channel, the picture carrier frequency is always 1.25 MHz above the low end of the channel. The sound carrier frequency is always 4.5 MHz above the picture frequency, or 0.25 MHz below the upper end of the channel. As an example, the picture carrier for channel 3, which is broadcast from 60 to 66 MHz, is 60 MHz + 1.25 MHz or 61.25 MHz. The sound carrier frequency is 66 MHz − 0.25 MHz or 65.75 MHz.

Vestigial Sideband Compensation

Several factors affect the transmission and reception of a video signal. Reproducing the desired amount of detail seen in the original picture requires modulation frequencies up to approximately 4 MHz. Therefore, double-sideband transmission of those frequencies would require an 8-MHz band for the carrier and both sidebands. In addition to the modulated video carrier, a television channel must also have space for the modulated audio carrier. Yet, as we have discussed, the bandwidth for each allocated television channel is only 6 MHz.

This difficulty is overcome by vestigial transmission. **Vestigial transmission** is the suppression through filtering of a major portion of the lower video sideband. The suppression allows the entire sound and picture signal to fit within the 6-MHz bandwidth. Although a filter completely attenuates all frequencies more than 1.25 MHz below the carrier, attenuation of the lower transmitted sideband does not occur until a point 0.75 MHz below the carrier.

Without vestigial compensation, out-of-proportion signals will appear at the output of the video detector. Video signals below 0.75 MHz will have an amplitude twice as great as the higher-frequency signals. Any signals from 0 to 750 kHz would receive full double-sideband amplification, while signals above 1.25 MHz would receive only single-sideband amplification. Thus, large areas of the reproduced picture would have greater contrast than smaller areas of the picture.

When we look at the uncompensated frequency response curve shown in Figure 8-2, the reasons for applying vestigial compensation become evident. With the 42.17-MHz chrominance marker positioned at the midpoint of the frequency response curve, only upper sideband signals receive sufficient amplifications. As you know, the use of the vestigial technique lessens the amplification for the lower sideband. Circuits in the television correct the unbalance in the sideband amplification.

Properly tuned I-F circuits equalize the vestigial sideband effect by reducing the gain response of the video I-F carrier by 50 percent as compared to the gain response of the single-sideband frequencies. The compensated frequency response curve shown in Figure 8-3 provides a flat detected output from 0 Hz to approximately 4 MHz at the output of the video detector. With this response, the original low, medium, and high picture frequencies receive equal amplification and the reproduced picture has excellent overall quality.

Automatic Fine-Tuning

Properly operating tuner and I-F sections require the precise alignment of the signals processed throughout the circuits. With the correct local oscillator frequency, the I-F frequency response curve closely resembles the curve shown in Figure 8-1. In that figure, the 45.75-MHz picture carrier and the 42.17-MHz chroma carrier have nearly equal amplitudes. If the local oscillator frequency deviates from its center point, problem symptoms, including beat interference, no color, excessive color, and weak sound, will occur. A frequency shift above 107 MHz allows the

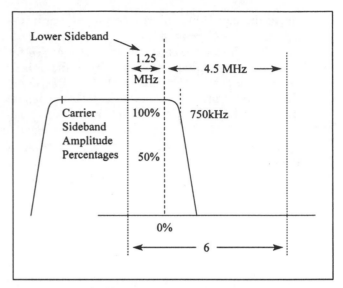

Figure 8-3   Compensated Frequency Response Curve

sound carrier and the color burst signal to heterodyne, causing a herringbone-like pattern to cover the picture. A frequency shift below 107 MHz reduces the amplitude of the color burst and causes a loss of color. In each case, the shifting of the oscillator frequency also causes the picture I-F, sound I-F, and color subcarrier signals to shift.

Figure 8-4 shows a block diagram of an **automatic fine-tuning** or **AFT circuit** that uses a frequency discriminator to control the local oscillator frequency. Tied to the third I-F amplifier, the AFT circuit uses phase detection to monitor the 45.75-MHz video carrier for any positive or negative changes in frequency. When a frequency change occurs, the AFT circuit generates and feeds either a proportional positive or a negative dc correction voltage back to the tuner local oscillator. This correction voltage slightly changes the bias of the oscillator and forces it back to the original frequency.

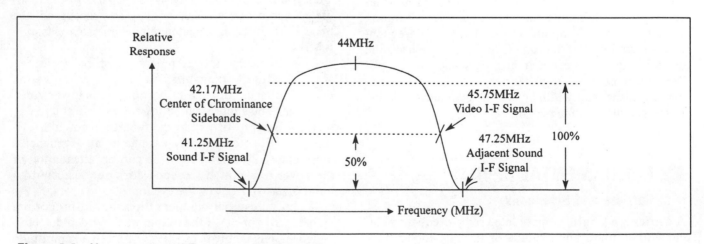

Figure 8-2   Uncompensated Frequency Response Curve

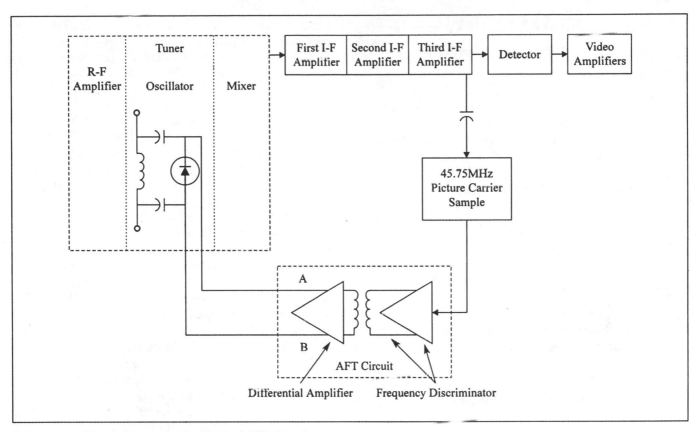

Figure 8-4   Block Diagram of a Typical AFT Circuit

The **discriminator** converts the 45.75-MHz ac signal found at the output of the I-F amplifier to a dc control voltage, while the changing capacitance of the varactor corrects the oscillator frequency. The amount and polarity of the dc control voltage corresponds to the amount and direction of the picture carrier deviation from the 45.75-MHz center frequency. As the figure shows, the discriminator output travels through a differential amplifier and to a varactor diode located in the oscillator circuit. Going back to our discussion about tuners, the varactor forms part of a tank circuit for the local oscillator.

Correct AFT operation produces two equal dc voltages at points A and B of the differential amplifier output and causes no changes in the varactor diode capacitance. If the local oscillator frequency tunes too low, the picture carrier amplitude increases and the chroma carrier amplitude decreases. As a result, the reproduced picture has either very weak color or a loss of color.

The discriminator responds to the low local oscillator frequency by using an increase in the differential amplifier point A output voltage, and a decrease in the point B output voltage, to apply an increased reverse bias to the varactor diode. With an increased reverse bias, the capacitance of the varactor diode decreases and causes the local oscillator frequency to shift higher

and back to its original frequency. A high local oscillator frequency causes the opposite conditions to occur at the output of the differential amplifier: the point A voltage decreases while the point B voltage increases. As a result, the reverse bias of the varactor diode decreases and causes the diode capacitance to increase. The local oscillator frequency then shifts down and back to its original operating frequency.

Earlier AFT circuit designs were packaged in a single integrated circuit and utilized two external, adjustable coils. More recent designs use a synchronous demodulator for automatic fine-tuning and place the entire circuit within the same IC that houses the I-F and automatic gain control sections of the receiver. For this application, the synchronous demodulator compares a pure 45.75-MHz unmodulated carrier with a sample of the modulated video I-F signal. In most designs, the synchronous demodulator takes the form of a phase-locked loop and a local oscillator locked to 45.75 MHz. This approach for the AFT circuit:

- maintains the circuit gain
- reduces the chance for I-F harmonics through low-level switching
- cuts distortion

In later sections of this chapter, you will learn how the AFT operation fits within a single IC package.

FUNDAMENTAL REVIEW ──────

Bandwidth and Frequency Response

Amplifiers have constant gain over a specific range of frequencies called a band. Because this range fits within upper and lower limits, the frequencies have a bandwidth. The upper and lower limits of the bandwidth are set by the cut-off frequencies of the amplifier. For example, an amplifier that has constant gain for a range of frequencies extending from 40 to 44 MHz has a bandwidth of 4 MHz. For the purposes of this example, the cut-off frequencies are 40 and 44 MHz. The specific values for the cut-off frequencies vary from amplifier to amplifier.

─────────────────────────────

☑ PROGRESS CHECK

You have completed Objective 1. Define intermediate-frequency signals and automatic fine-tuning.

8-2 SELECTIVITY AND ATTENUATION

In chapter seven, we discussed the capability of the tuner to receive channels across a broad bandwidth that included VHF, UHF, hyperband, and superband channels. Although having a broad bandwidth extends the capability of the tuner, it also increases the potential for interference between adjacent channels. Interference

occurs because of the overall amplification of the adjacent channel frequencies as well as the desired signals. The I-F section uses tuned circuits to provide the selectivity needed to eliminate any possible adjacent channel interference.

When considering adjacent channels, frequencies rather than channel numbers provide the best definition. An **adjacent channel** always has a frequency that overlaps with the frequency of the desired channel. For example, channels 2 (54 to 60 MHz) and 4 (66 to 72 MHz) are adjacent channels for channel 3 (60 to 66 MHz). However, channel 4 is not an adjacent channel for channel 5 (76 to 82 MHz), and channel 6 (82 to 88 MHz) is not an adjacent channel for channel 7 (174 to 180 MHz).

Figure 8-5 shows the relationship of a selected station, such as channel 3, to its adjacent carrier frequencies. The response curve has markers for the desired channel carriers and the adjacent channel carriers. With the standard 45.75-MHz video I-F and 41.25-MHz sound I-F, the local oscillator frequency for channel 3 is 107 MHz. When looking at the response curve, note that the sound carrier for the channel below and the video carrier for the channel above the desired channel have the most potential for interference. Each of these adjacent signals appears within the indicated tuner response curve.

With the receiver set to channel 3, any channel 2 adjacent sound carrier has a frequency of 59.75 MHz. When the adjacent sound carrier heterodynes with the 107-MHz channel 3 frequency, the mixing produces an

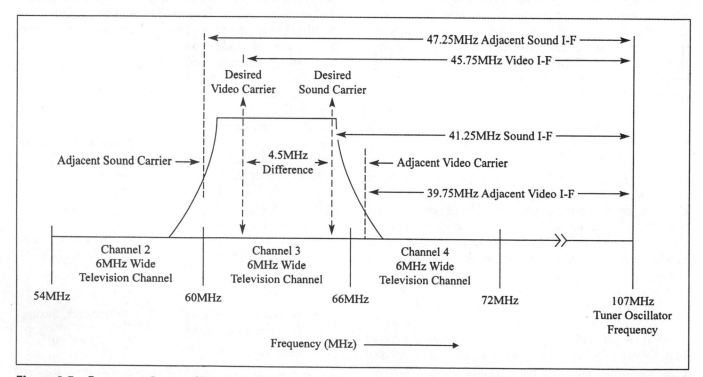

Figure 8-5 Response Curves Showing a Desired Channel 3 Signal and Adjacent Channel 2 and Channel 4 Signals

adjacent I-F difference of 47.25 MHz. Any channel 4 adjacent video carrier with a frequency of 67.25 MHz beats with the 107-MHz signal and produces an adjacent I-F difference of 39.75 MHz. Through a process called **selectivity**, the receiver I-F section discriminates against the undesired 47.25-MHz sound carrier and the 39.75-MHz video carrier by having a broad enough bandwidth to allow the amplification of the desired 41.25-MHz and 45.75-MHz I-F signals. To accomplish this, the I-F section has a bandwidth of 3 to 4 MHz and enough attenuation beyond the bandwidth limits to eliminate any adjacent channel interference. When such interference occurs, picture information from the interfering station becomes superimposed over the desired picture.

Circuit Q

We can measure selectivity in terms of circuit quality, or Q. When defining circuit Q, we consider the amount of total energy in the circuit versus the amount of energy dissipated throughout the circuit. Thus, some type of source must replace the dissipated energy. A tuned circuit that has a large Q offers high selectivity, a sharp response curve, and a large developed voltage. Tuned circuits with a low Q have a poor selectivity, a broad response curve, and a small developed voltage. In mathematical terms, circuit Q equals:

$$Q = X_L/2$$

Because the stages of an I-F section need to pass a 6-MHz band of frequencies, the circuit Q must remain low. With high-frequency carriers, capacitance factors reduce L/C ratios and keep the circuit Q low. With low-frequency carriers, however, different circuit parameters are used to maintain a low Q.

One method for stabilizing the Q relies on placing a shunt resistance in parallel with a tuned circuit. The value of the parallel resistance is the maximum impedance that the tuned circuit can present under any condition. Here, the selectivity of the circuit broadens symmetrically, and the bandwidth has an inversely proportional relationship to the shunting resistance. Therefore, the Q remains low while the bandwidth becomes broader.

☑ PROGRESS CHECK

You have completed Objective 2. Discuss circuit Q.

Attenuation

Depending on the type of I-F stage used in the video receiver, attenuation occurs through either a set of traps and filters or through a **surface acoustical wave** or **SAW filter**. Both designs satisfy the gain and selectivity requirements of the I-F section. Earlier receivers take advantage of bandpass filters, or stagger-tuned, single coils, or a combination of filters and coils to provide interstage coupling. Rather than rely on interstage coupling, newer receivers use a SAW filter to provide the proper I-F response. A SAW filter acts as a tuned circuit, establishes the intermediate frequency bandwidth, and provides waveshaping. Along with attenuating the adjacent channel frequencies, the circuit must also limit the desired sound I-F carrier to a level much lower than the video I-F carrier. Without attenuation, the video carrier sidebands can interfere with the audio carrier sidebands.

Wave Traps and Bandpass Filters

In some I-F section designs, attenuation of the undesired signals occurs through the use of sharply tuned circuits called **wave traps**. These series- or parallel-tuned circuits work within the I-F interstage coupling circuits and either reject, absorb, or provide a path to ground for a particular signal. The tuned signal may be an adjacent sound I-F signal, an adjacent video I-F signal, or—in the video I-F section—the desired sound I-F signal. Depending on the requirements, wave traps fall into four categories:

1. Series-resonant traps
2. Parallel-resonant traps
3. Absorption traps
4. Bridged-T or notch traps

Attenuation of undesired signals occurs because of the reduction of I-F signal gain at the trap frequencies. Because of the rejection of some part of the I-F signal, the tuning of the traps affects the shape of the frequency response curve. Each trap affects the I-F response at the edges of the passband by cutting into the skirt of the response curve.

Adjacent channel traps reject any 39.75-MHz or 47.25-MHz signals for any selected channel. Usually, adjacent channel traps are located between the mixer and the input to the first I-F amplifier. 41.25-MHz sound I-F carrier traps maintain a constant sound I-F carrier level throughout the I-F section. An overly strong 41.25-MHz sound carrier can cause the 4.5-MHz sound I-F carrier to heterodyne with the color burst signal. As a result, the amplitude-modulated video signal modulates the 4.5-MHz sound carrier. Under the proper operating conditions, a frequency modulated signal modulates the sound carrier. 4.5-MHz traps prevent the sound I-F signal from reaching the video amplifier circuits and causing interference to appear within the reproduced picture. Usually, I-F sections will feature 4.5 MHz traps at the video detector and following the video amplifier stages.

Wave Trap Configurations   In Figure 8-6A, L1 and C1 make up a series-resonant wave trap that shunts the 39.75-MHz adjacent video signal to ground. This shunting effect occurs because of the low impedance presented by the components at a resonant frequency of 39.75-MHz. In Figure 8-6B, L2 and C2 make up a parallel-resonant trap that also couples the I-F signal to the second I-F stage. In this configuration, the components present a high impedance to any unwanted signals and prevent those signals from traveling to the amplifier.

As the name implies, an **absorption trap** absorbs energy at a resonant frequency. An absorption trap is a parallel-resonant circuit that is inductively coupled to a tank circuit. During operation, the absorption trap absorbs energy from the tank circuit. Figure 8-6C shows the schematic diagram for an absorption trap. The trap is coupled to an I-F coil found in the input circuit of the first I-F amplifier. With the trap wound on the same form as the I-F coil, the absorption trap absorbs any energy found at the resonant frequency of 41.25 MHz. As a result, the trap attenuates the desired sound I-F signal.

A **bridged-T Trap** blocks all frequencies that fall within its bandwidth. Figures 8-7A and B show a block diagram of a bridged-T trap and a corresponding fre-

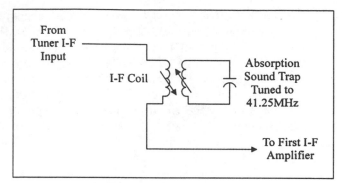

Figure 8-6C   Schematic Drawing of an Absorption Trap

quency response curve. The **notch trap** consists of a low-pass filter, a high-pass filter, and a summing amplifier. While the high-pass filter establishes the lower cut-off frequency, or f_1, for the complete filter, the low-pass sets the upper cut-off frequency, or f_2, for the complete filter. The summing amplifier produces an output voltage proportional to its input voltages.

Moving to the response curve, the gap between the two cut-off frequencies sets the bandwidth for the filter. With an input frequency lower than f_1, the input voltage produced at the output of the high-pass filter equals zero, while a positive voltage appears at

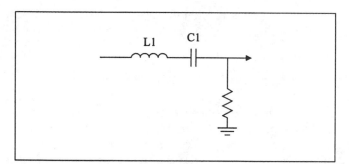

Figure 8-6A   Schematic Drawing of a Series-Resonant Wave Trap

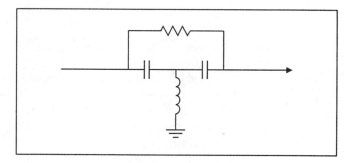

Figure 8-7A   Schematic Diagram of a Bridged-T Trap

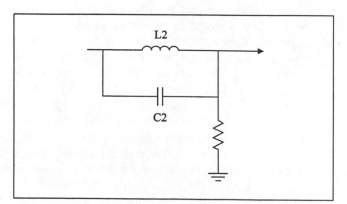

Figure 8-6B   Schematic Drawing of a Parallel-Resonant Wave Trap

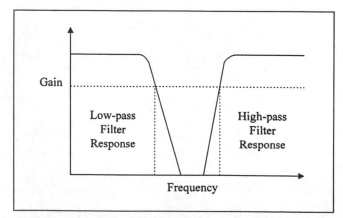

Figure 8-7B   Frequency Response Curve for Bridged-T Trap

the output of the low-pass filter. With an input frequency higher than f_2, the input voltage produced at the output of the low-pass filter also equals zero, while a positive voltage appears at the output of the high-pass filter.

Thus, with either an input frequency below f_1 or higher than f_2, the signal passes through to the summing amplifier, and the amplifier has an output signal. However, when the input frequency fits anywhere in between the upper and lower cut-off frequencies, neither the low-pass filter, the high-pass filter, or the summing amplifier has an output. As Figure 8-7B shows, the notching action of the complete filter circuit either reduces or rejects any frequency fitting within the circuit bandwidth.

Bandpass Filters   In addition to wave traps, the I-F circuits also utilize bandpass filters that attenuate any frequency outside of established band limits. By employing bandpass filters, the circuit assures the maximum response to a given frequency. Like the notch filter, a bandpass filter consists of a high-pass and a low-pass filter. Unlike the notch filter, the bandpass filter allows any frequency below the value of f_2 and above the value of f_1 to pass, blocking any frequencies above f_2 and below f_1.

Shown in Figure 8-8, the most simple bandpass filter consists of several tuned circuits and a terminating resistor. With this basic design, the frequency limits of the passband and a given undesired frequency lying outside the band limits determine the amount of induc-

tance, the amount of capacitance, and the value of the terminating resistor.

We can also use the filter circuit shown in the figure to make a transition from purely trapping circuits to circuits that provide both attenuation and coupling. Because the bandpass filter consists of two tuned circuits coupled in such a way that the circuit has a maximum response at two frequencies, it is also called a coupled filter. In the figure, both the primary and the secondary of a double-tuned coupling transformer tune to the same frequency. As you will see in the next section, the close coupling of the two coils provides the desired frequency response.

☑ PROGRESS CHECK

You have completed Objective 3. Understand how wave traps and bandpass filters affect selectivity and attenuation.

8-3 I-F AMPLIFIER STAGE GAIN, COUPLING, AND NEUTRALIZATION

In a television, the I-F stages have an overall gain that ranges from 500 to more than 10,000; they produce most of the video signal amplification. Whether based on transistors or an integrated circuit, most I-F sections contain at least three and sometimes four amplifiers. The net gain of the I-F section is derived from the product of the individual stage gains. As such, an I-F section

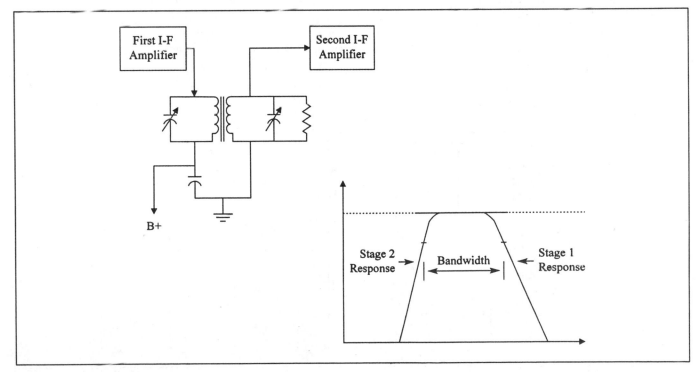

Figure 8-8 Bandpass Filter

that has four amplifier stages each with a stage gain of 15 will have an overall gain of 15 × 15 × 15 × 15 or 50,625.

Coupling

Figure 8-9 shows the schematic of a typical transistor-based I-F amplifier circuit. Note that the circuit features more than one amplifier. Instead, a group of amplifiers connected in series, or cascaded, handles the amplification of the I-F signal. With cascading, each amplifier stage multiplies the signal gain. While the cascaded amplifiers form a multistage amplifier, coupling capacitors connect the stages to one another, allow the signal to pass from stage to stage, and provide dc isolation from stage to stage. Each coupling capacitor connects in series from the collector of one transistor to the base of the next transistor.

↻ FUNDAMENTAL REVIEW

Coupling

As you learned in chapter one, coupling is a method for transferring energy from one circuit to another. Communications products use several different types of coupling, including direct coupling, resistance-capacitance coupling, impedance coupling, and transformer coupling. Each type of coupling works well for specific types of applications.

Direct coupling is used in circuits that operate at low frequencies; one amplifier is directly connected to another. Also good for low frequency applications is capacitance coupling. This type of coupling allows an ac signal to pass through a capacitor from the output of one amplifier to the input of another. Any dc component is blocked by the capacitor and does not affect any dc voltages at the second amplifier.

Impedance coupling is used with amplifier circuits that must have a wide bandwidth. Rather than use a resistor as the load, an amplifier circuit using impedance coupling uses the inductive reactance of a coil as a load. At high frequencies, the impedance presented by the coil increases and causes the circuit gain to also increase. Transformer coupling is also used in amplifier circuits that operate with high frequencies. It involves the use the primary and secondary of a transformer for the transfer of energy. An ac signal at the output of one amplifier transfers from the primary winding of a transformer to its secondary winding, which is connected to the input of the next amplifier.

Neutralization

In addition to coupling capacitors, the cascaded amplifier also features components that provide neutralization. Without neutralization, an I-F amplifier will begin to oscillate at a high frequency. The consequences of

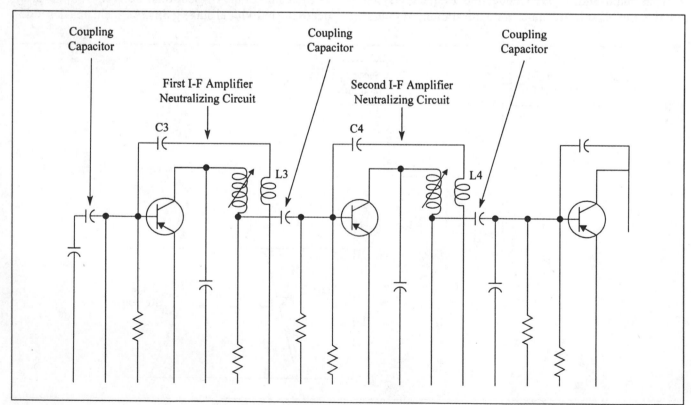

Figure 8-9 Illustration of Cascaded Amplifiers and Coupling Capacitors

these **parasitic oscillations** is the overheating of the amplifier, a loss of gain, and noise distortion.

Returning to Figure 8-9, the secondary windings L3 and L4 supply small neutralizing voltages to the bases of the first and second I-F amplifiers through capacitors C3 and C4. The neutralizing signal has the opposite phase of the signal found at the output of the amplifiers.

⟳ FUNDAMENTAL REVIEW

Transistor Characteristics and Signals

Capacitance: Applying R-F and I-F signals to a transistor amplifier affects the performance of the semiconductor In several ways. For example, the stray capacitances between the transistor and its circuit connections or within the transistor can be considered as part of the amplifier circuit. These capacitances can affect the establishment of a resonant frequency for a tuned circuit and may place a limit on the inductance/capacitance ratio found in the amplifier circuit. As a result, the circuit Q decreases and limits the amplifier gain for a given bandwidth.

Transit Time: Any change in emitter current caused by an input signal causes a corresponding change in collector current. At low radio frequencies, the change of emitter and collector currents occurs at a rate that matches the change in signal because of the slower signal cycle time. However, high radio frequencies that have a faster signal cycle time require a much faster change in the emitter and collector currents. Thus, transistors that amplify high frequencies must have a short transit time.

Inductance: Very high frequencies also cause leads and connections to exhibit inductive properties and to have reactance. Any internal inductance adds to external inductances found in the amplifier circuit, affecting tuning. In addition, the collector current develops an induced voltage across the inductance presented by the emitter lead. The induced voltage feeds in-phase with the base voltage, decreases the total input resistance of the amplifier, and reduces the circuit Q.

Noise: The internal movement of electrons within a transistor generates noise that can interfere with signals. Resistance noise develops because of the movement of electrons across a conductor. The moving electrons, which result from thermal action, develop a voltage across the ends of the conductor. Noise occurs because of the random variations of the voltage. Random variations in collector current cause shot-effect noise. When electrons leave the emitter of an NPN transistor or holes leave the emitter of a PNP transistor at random frequencies, slight changes in collector current

occur. These slight changes produce noise in the collector circuit.

Transformer Coupling The coupling capacitors highlighted in Figure 8-9 show one method for transferring energy from one circuit to another. Coupling also occurs if two circuits share inductances. With transformer coupling, both the primary and the secondary are tuned to the same frequency. The transistor-based I-F circuits that we have discussed use coupling to obtain the maximum response at two frequencies. Close coupling of the coils provides the double-peaked response curve and wide bandwidth that we examined in Figure 8-1.

Figures 8-10A, B, and C show how varying the degree of coupling between coils affects the shape of a frequency response curve and the amount of bandwidth. The relationship between the amount of coupling, shape, and bandwidth is graphed against the amount of secondary voltage. In Figure 8-10A, tuning two loosely-coupled coils to the same resonant frequency establishes the narrow, sharply-peaked response. Very little coupling exists with a wide space between the coils.

Bringing the two coils slightly together increases the amount of coupling and sets up the broad response curve with steep sides, as shown in Figure 8-10B. The sides of the response curve have a steeper angle as it becomes larger and broader. The curve continues to broaden until the secondary voltage reaches its maximum value. When this balance of response broadening

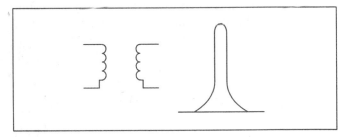

Figure 8-10A   Effect of Loose Coupling on Frequency Response and Bandwidth

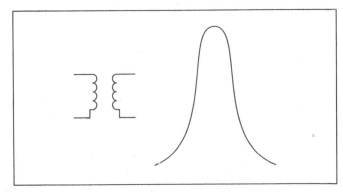

Figure 8-10B   Effect of Critical Coupling on Frequency Response and Bandwidth

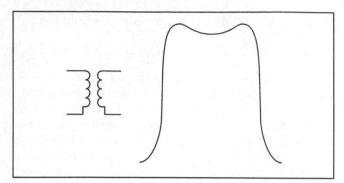

Figure 8-10C   Effect of Overcoupling on Frequency Response and Bandwidth

versus maximum value of secondary voltage occurs, the coupling reaches a level called **critical coupling**.

Increasing the coupling to the point of **overcoupling** spreads the emphasized response humps farther apart as shown with in Figure 8-10C. With overcoupling, the sound and picture carriers plus the sidebands receive uniform amplification. As curve C shows, overcoupling provides a frequency response that matches the desired curve shown in Figure 8-1.

All this occurs because of a property common to all coupled filters called **mutual coupling impedance**. Because mutual coupling impedance is common to both halves of a coupled filter, it provides the maximum amount of energy transfer. Depending on the circuit configuration of the coupled filter, we can measure the mutual coupling impedance in terms of inductance, capacitance, or resistance.

Figure 8-11 shows a loosely coupled bandpass filter connected to another coil. Although little or no mag-

netic coupling occurs between the tuned coils, coil L_m provides the coupling impedance. In the output circuit of the first I-F amplifier, L1 and L_m form an inductive component, and the output capacitance of the amplifier forms a capacitive component of the primary tuned circuit. In the input circuit of the second I-F amplifier, L2 and L_m form an inductive component, and the input capacitance of the amplifier forms a capacitive component of the secondary tuned circuit.

During operation, signal currents in the primary circuit produce an electromagnetic force across coil L_m. In turn, this small emf causes currents to induce larger signal voltages across the tuned secondary circuit. At the resonant frequency of the coupled circuit, the high-frequency response leads to the maximum transfer of energy between the primary and the secondary.

FUNDAMENTAL REVIEW
Gain

Gain can be measured in terms of current, voltage, or power. Because amplifiers should have predictable output values, the value of amplifier gain should remain stable under normal operating conditions.

Current gain, or A_I, occurs when the amount of ac current flowing through an amplifier from input to output increases. The equation for finding the value of current gain is:

$$A_I = I_{out}/I_{in}$$

where I_{out} equals the current at the output of the amplifier and I_{in} equals the current at the input of the

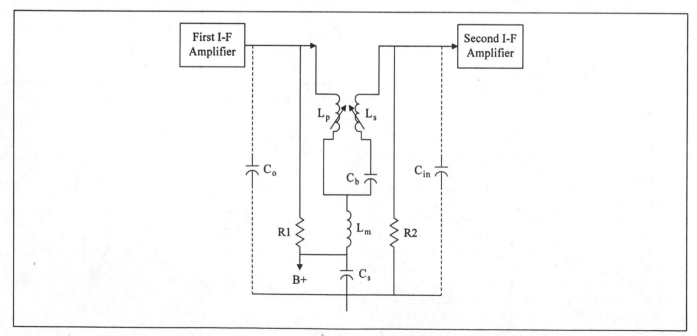

Figure 8-11   Schematic of a Loosely-Coupled Bandpass Filter Connected to a Second Coil

amplifier. The current gain of a multistage amplifier is the product of the individual-stage current gain values.

Voltage gain, or A_V, occurs when the amount of ac signal voltage increases from the input to the output of an amplifier. The equation for finding the value of voltage gain is:

$$A_V = V_{out}/V_{in}$$

or the value of the voltage at the output of the amplifier divided by the value of the voltage at the amplifier input. The voltage gain of a multistage amplifier is the product of the individual-stage voltage gain values.

Power gain, or A_P, occurs when the ac signal power increases from the input of the amplifier to the output. The equation for finding the value of power gain is:

$$A_P = A_I A_V$$

or the value of the current gain multiplied by the value of the voltage gain. The power gain of a multistage amplifier is the product of the values of the multistage voltage gain and the multistage current gain.

☑ PROGRESS CHECK

You have completed Objective 4. Explain amplifier stage gain, coupling, and neutralization.

8-4 AUTOMATIC GAIN CONTROL

Any type of communications equipment includes an **automatic gain control** or **AGC circuit** for controlling the gain of a signal amplifier. In video products such as televisions, the R-F and I-F amplifiers connect to an AGC line and are forward biased at all times. AGC circuits take advantage of the forward bias characteristics of the amplifiers and either change or maintain amplifier gain by adjusting the operating con- ditions of the amplifier. Because of this, the dc control voltage pro-

duced by the AGC circuit may push the amplifier towards either saturation or cut-off.

The AGC circuit works as part of a closed-feedback loop that includes a detector, a filter circuit, a feedback path to the amplifier, and the amplifier stage. When the input signal varies in amplitude, a dc correction voltage feeds from the AGC circuit to the amplifier and maintains a constant amplifier gain by controlling the amplifier forward bias. Figure 8-12 shows a block diagram of the AGC operation.

If the input signal amplitude increases, the AGC circuit prevents the output of the amplifier from increasing by producing a higher dc bias control voltage. The increased bias at the amplifier reduces the gain of the amplifier stage. A decrease in the input signal amplitude causes the AGC circuit to produce a lower dc bias control voltage. Here, the decreased bias at the amplifier causes the amplifier stage gain to increase.

Forward and Reverse AGC

As mentioned, AGC circuits use the operating characteristics of an amplifier to control amplifier gain. With **forward AGC**, the automatic gain control circuit begins with a no-signal bias that produces the maximum gain of the amplifier. Increasing the bias in a forward direction causes an increased amplifier current, an increased voltage drop across a series load, and a decreased gain value at the amplifier output. In this case, the forward AGC circuit pushes the amplifier toward saturation with a control voltage that has the same polarity as the bias voltage, and thus controls gain.

Again starting with a no-signal bias that produces maximum amplifier gain, an AGC circuit using **reverse AGC** decreases the forward bias of the amplifier circuit. With a decrease in forward bias, a decrease in current through the amplifier also decreases the gain at the amplifier output. With reverse AGC, the AGC circuit uses a control voltage that has a different polarity from the bias voltage and pushes the amplifier toward cut-off to decrease gain.

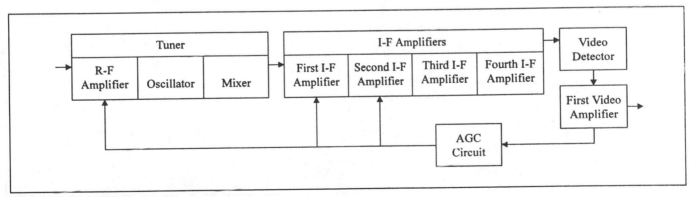

Figure 8-12   Block Diagram of AGC Operation

R-F and I-F AGC Circuits

Although an obvious link exists between the gain of an R-F amplifier and the gain of I-F amplifier stages, the two sections have different operating requirements regarding signal strength. The AGC circuit controlling I-F stage gain controls strong signals so that the gain produced by the first and second I-F amplifiers will not overload the stages that follow. In addition, the AGC circuit increases I-F amplifier gain under weak signal conditions.

In a tuner, driving the mixer with the strongest possible signal maintains a high signal-to-noise ratio. As a result, any gain control at the R-F amplifier during weak signal conditions is undesirable. Under those conditions, the developed AGC voltage would further weaken the signal. Despite the need to drive the mixer with a strong signal, the use of an AGC voltage during strong signal conditions prevents the R-F amplifier from overloading the mixer stage. Without a dc control voltage reducing the amplifier gain, overload condi-

tions at the mixer would allow crosstalk between adjacent signal frequencies to occur.

To provide the best signal-to-noise ratio, the R-F amplifier stage of the tuner operates at the maximum gain for R-F signal levels that have a threshold level up to 1 millivolt. However, uncontrolled gain at this point can overload the I-F amplifiers that follow the tuner. To provide incremental control over a wide range of signal levels, an R-F AGC circuit connects to the first and second I-F amplifier stages and the tuner R-F stage and compares the video sync tip level with a fixed dc reference value. Any signal increase above the reference level causes the AGC circuit to apply a dc correction voltage to the amplifier stages and restore the proper sync tip amplitude.

Keyed AGC   Figure 8-13 shows a schematic diagram of a **keyed** or **gated AGC** circuit. This system that uses a **keyed** or **gated pulse** to control the amount of developed AGC voltage. The AGC system compares the video sync tip level with a reference voltage derived

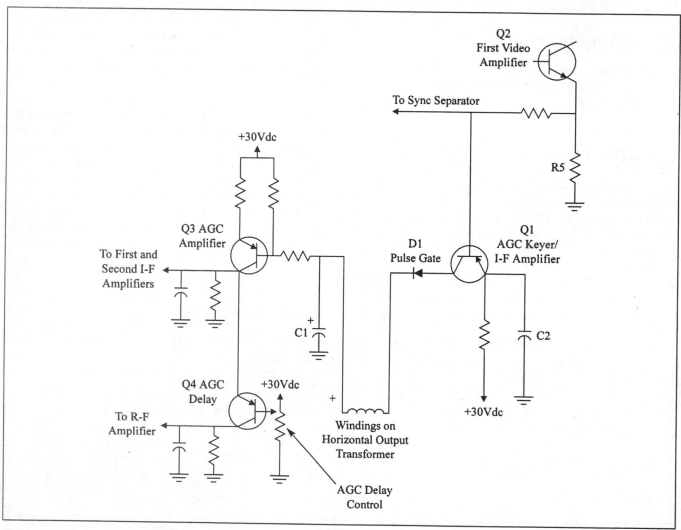

Figure 8-13 Schematic Diagram of a Keyed AGC Circuit

from a stage that operates only when a sync signal is transmitted. This reference, or keyed, pulse consists of a burst of signal from the horizontal output section at the 15,750-Hz horizontal frequency rate.

As the figure shows, both the video sync and the keyed pulse are inputs to the AGC system. Transistor Q1 operates as both an I-F amplifier and the AGC keyer, while transistor Q2 amplifies the AGC signal. At the output of the circuit, the AGC signal travels to the tuner and the I-F amplifiers at resistors R6 and R7. Under no-signal conditions, Q1 remains biased into cut-off by a reverse-bias voltage produced by the power supply, and no AGC voltage is produced. When the video sync pulse reaches a specified amplitude, it becomes part of a bias voltage for the AGC keyer.

The keying pulse from the horizontal output section travels to the collector of Q1 through D1, a pulse gate diode, and produces a collector voltage of 1 Vdc. The pulse gate diode rectifies the keyed pulse; prevents the voltage at the base-emitter junction of the transistor from shorting the AGC voltage; and prevents the forward biasing of the transistor during intervals that occur between the pulses. The gating pulse intervals coincide with the horizontal scan period. Combined with the video sync pulse, the keyed pulse causes the transistor to conduct at the horizontal frequency rate. When Q1 conducts, samples of the I-F signal pass through transformer T1, a coupling transformer tuned to 42 MHz, and become rectified by diode D2.

During this period, C1 charges to a level that matches the amplitude of the peak value of the video sync pulses and supplies the AGC voltage. In turn, this voltage biases Q2. Because the AGC amplifier is configured as an emitter follower, the AGC line returns to the emitter. Any variations in the I-F signal also change the forward bias of the AGC amplifier, the voltage at the emitter of Q2, and the AGC voltage. During the gating pulse interval, the capacitor discharges slowly through R3 and R4 and maintains a constant AGC voltage.

Amplified AGC Transistor Q3 functions as a dc amplifier, amplifying the AGC voltage for the R-F amplifier. Under some conditions, the dc control voltage supplied by the keyed AGC circuit may be too weak to control the strong signals. **Amplified AGC** offsets those conditions.

Delayed AGC Although not discussed in the analysis of Figure 8-13, the diagram also introduces the concept of delayed AGC with the circuit that includes transistor Q4 and the AGC delay control. Figure 8-14 shows the connecting points and the feedback paths for an R-F AGC circuit. This incremental control, or **delayed AGC**, senses increased R-F signal levels and reduces the gain of the I-F amplifiers in proportion to the R-F signal level increases. When the R-F signal increases to the limit of the second I-F stage gain reduction, the AGC system reduces the gain of the second I-F amplifier stage. Further increases cause the AGC system to reduce the gain at the first I-F amplifier and finally at the tuner R-F amplifier stage.

Figure 8-15 shows a typical AGC delay circuit for two transistors operating as an R-F amplifier and an I-F amplifier, and another transistor providing an AGC delay voltage. R3, an AGC delay control, allows some adjustment of the AGC delay transistor turn-on bias. With weak or no signal conditions, a dc voltage applied from the AGC amplifier and through a voltage-divider network consisting of resistors R1 and R2 biases the AGC delay transistor into cut-off; it also reduces the AGC voltage at the R-F and I-F amplifier stages. As a result, the gain at the R-F amplifier remains unchanged, and the gain at the I-F stages increases under weak signal conditions.

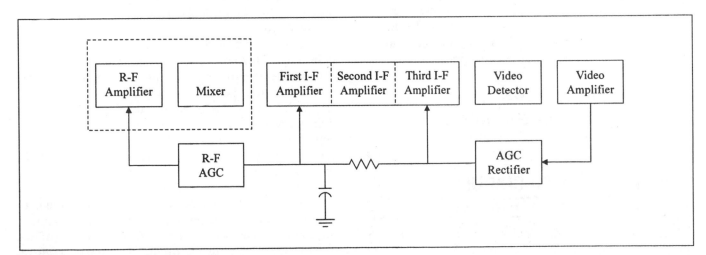

Figure 8-14 Block Diagram of a Delayed AGC Circuit

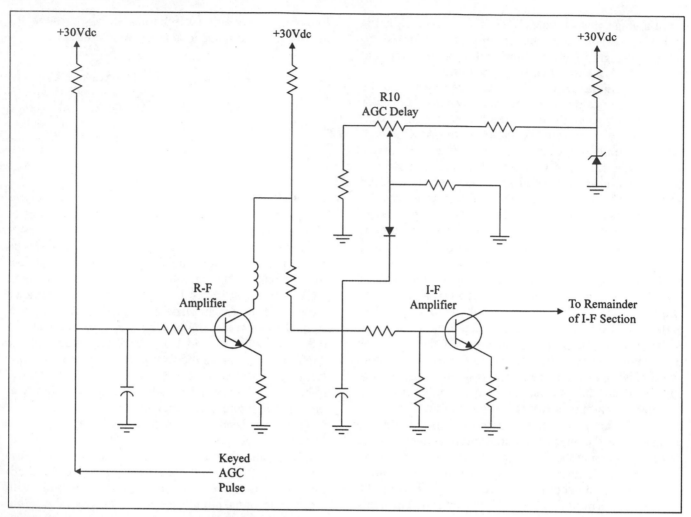

Figure 8-15   Schematic of an AGC Delay Circuit

Strong signal conditions cause the AGC voltage to increase and reduce the gain at the I-F stages. When the received signal increases to approximately 4 to 5 Vdc, the increased AGC voltage overcomes the reverse bias of the AGC delay transistor and allows the transistor to conduct. The conduction of the transistor increases the voltage at the collector of T1 and drives the R-F transistor toward saturation. Thus, gain reduction occurs at all the stages during strong signal conditions.

✓ PROGRESS CHECK

You have completed Objective 5. Discuss automatic gain control circuits.

8-5   I-F AMPLIFIER CIRCUIT CHARACTERISTICS

When considering individual tuned circuits, we also find several key relationships between bandwidth, amplitude, and gain. The top and bottom extremes of any

bandwidth are at equal distances above and below resonance, and occur when the amplitude equals 70.7 percent of the maximum resonant frequency. The overall bandwidth of an I-F stage is 70.7 percent of the maximum overall response. Therefore, the combined I-F amplifier stages have a lower bandwidth than any individual stage.

Again looking at the two points where the amplitude equals 70.7 percent of the maximum resonant frequency, the gain of the individual stage equals the gain at those points. For an I-F amplifier strip, the overall amplification equals the product of the amplification characteristics of the individual stages. To see how these relationships develop, we can use the two I-F stages shown in Figure 8-16 as an example. There, two identical I-F amplifier stages are tuned to the same frequency. Because of this, the total gain of the complete circuit—or the circuit gain at 70.7 percent of the overall maximum response—equals the square of the gain of one stage. The overall response of the circuit equals the square of the individual stage response, or $(.707)^2$. With each individual circuit having

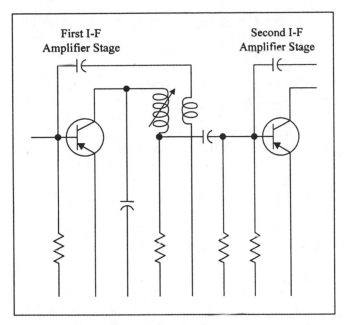

Figure 8-16 Two I-F Amplifier Stages

an identical resonant frequency, points for the 70.7 percent maximum resonant frequency lie closer to one another on the graph shown in Figure 8-17.

With the two points moving closer together, the corresponding bandwidth shrinks to approximately 64.4 percent of the single-stage bandwidth. Thus, if each stage has a gain of 10, a bandwidth of 4 MHz, and remains tuned to the same frequency, the overall circuit gain equals 10^2 or 100, and the overall circuit bandwidth equals 0.644×4 MHz or 2.58 MHz. Unfortunately, reducing the bandwidth from 4 to 2.58 MHz degrades the signal and causes a loss of picture detail.

We can maintain a 4-MHz bandwidth while increasing the number of stages by decreasing the Q of the individual stages. Again, reducing the resistance of the shunting, or loading, resistor—the maximum impedance of the tuned circuit under any conditions—also decreases the Q of the individual stages. However, the reduction in individual circuit Q has a cost measured in the loss of amplifier gain per stage.

As an example, the desired center frequency, or f_r, of the circuit is 44 MHz. Each amplifier stage shown in Figure 8-16 has a voltage gain of 10 and a bandwidth of 4 MHz. Thus, the overall bandwidth equals 1.4 percent of the single-stage bandwidth or 5.6 MHz, and the overall gain equals $.707 \times 10^2$ or 71. One stage is tuned to the center frequency plus one-half of the desired bandwidth, while the other is tuned to the center frequency minus one-half of the desired bandwidth. If we place this into equation form, we find that:

$$f = 1.4 \times 4 \text{ MHz} = 5.6 \text{ MHz}$$

$$f_r = 44 \text{ MHz}$$

$$f_1 = f_r + .5\, f = 44 \text{ MHz} + (.5 \times 5.6 \text{ MHz})$$
$$= 46.8 \text{ MHz}$$

$$f_2 = f_r - .5\, f = 44 \text{ MHz} - (.5 \times 5.6 \text{ MHz})$$
$$= 41.2 \text{ MHz}$$

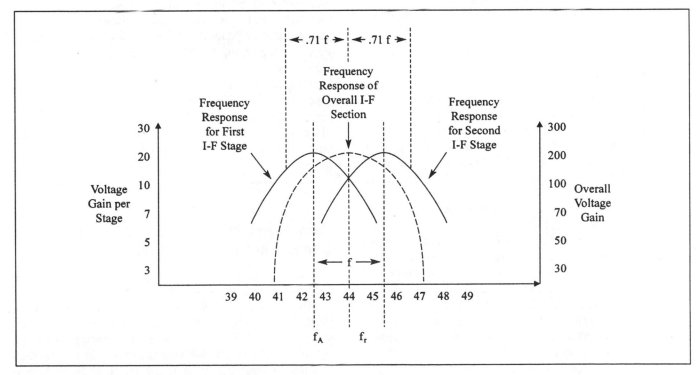

Figure 8-17 Frequency Response Curve for Two Identical I-F Stages

A separation of 2.8 MHz exists between each frequency and the center frequency. The frequencies f_1 and f_2 represent the resonant frequencies for each of the tuned circuits.

If we consider the rejection requirements of most video receivers, we find that we must reduce the overall bandwidth from 5.6 to 4 MHz. That is, we need to pass the 41.25-MHz sound I-F signal, the 45.75-MHz video I-F signal, the 42.17-MHz color sync signal, and the color signal sidebands while rejecting the 39.75-MHz adjacent channel video I-F signal and the 47.25-MHz adjacent channel sound I-F signal. To accomplish this, we simply decrease the bandwidth of each of the tuned circuits by 29 percent. With that reduction for each stage, the respective resonant frequencies equal:

$$f_1 = f_r + .5(.71 \, \pi \, f)$$
$$f_2 = f_r - .5(.71 \, \pi \, f)$$

As Figure 8-17 shows, the combination of the individual tuned circuit response curves establishes the desired I-F response curve for the entire circuit. Even though the bandwidth of each stage has narrowed to 2.84 MHz, the combined stages provide the proper response. In addition, each stage has a greater gain than is possible if the individual stages had a bandpass of 4 MHz.

Greater stage gain occurs through an increase in circuit Q. The narrowed response of the individual stages increases the individual circuit Q by a factor of 1.4. This increase in Q is accompanied by a corresponding increase in stage gain, and the overall gain equals $.707 \times (1.4 \times 10)^2$ or 140.

Stagger Tuning

To overcome the gain loss, we could tack additional amplifier stages onto the circuit, but that would take us back to the problem of reduced bandwidth. To counter both of these factors, transistor-based I-F amplifier sections use a technique called **stagger tuning**. With this arrangement, a single-tuned circuit works as the common coupling impedance between transistor amplifier stages. Television receivers that rely on stagger tuning use several single-tuned stages.

Tuning the alternate stages of the section to frequencies slightly above and below the center frequency allows the circuit to maintain both bandwidth and gain. Stagger tuning of the individual stages occurs through the tuning of various coils to different frequencies. With this approach to tuning, the complete circuit produces a desired overall response that matches the frequency response curve originally shown in Figure 8-1.

Figure 8-18 depicts a typical stagger-tuned I-F section. In the circuit, each single-tuned stage is adjusted to a different frequency within the desired bandwidth. Using the center frequency of the bandwidth as a reference point, the two frequencies set at equal distances above and below the reference. The amount of separation between the two frequencies equals the desired bandwidth. The entire circuit has a bandwidth 1.4 times greater than the bandwidth of one stage.

Figure 8-18 also breaks the stagger-tuned circuit down into individual tuned and trap circuits. Coils L1 and L2 are tuned to the center of the I-F response curve, 44 MHz. Moving from left to right, a pair of LC traps resonate at the 47.25-MHz adjacent sound and 39.75-MHz adjacent video frequencies. Resistors in the base circuits of the first and second I-F amplifiers form voltage dividers for the adjacent channel traps.

The two secondary windings L6 and L8 supply neutralizing voltages to the bases of the first and second I-F amplifiers. At the lower right-hand side of the figure, two resistors in the emitter circuit operate as a voltage divider and supply a neutralizing voltage for the third I-F amplifier. Resistors placed in the emitter circuit of each I-F amplifier are bypassed to ground and prevent the overheating of the transistor.

⟳ FUNDAMENTAL REVIEW
Tuned Circuits

If we connect a capacitor in parallel with a series circuit consisting of a resistance and an inductance, the complete tuned circuit provides different impedances for different frequencies. In most cases, the resistance is the resistance of the coil. The impedance of a tuned circuit is directly proportional to the inductance and inversely proportional to the capacitance and the series resistance.

A tuned circuit will yield a large impedance with a given resistance if the inductance is large and the capacitance is small. Because the tuned circuit voltage is directly proportional to the impedance, a large inductance develops a larger voltage than a large capacitance.

Using SAW Filters and Integrated Circuits in the I-F Section

The introduction of SAW filters into I-F circuits allowed circuit designs to consolidate all the I-F section functions into one integrated circuit package. However, the knowledge that you gained from studying stagger-tuned I-F circuits, wave traps, bandpass filters, coupling circuits, and I-F signal shaping is useful when applied to your study of these modern designs. Enclosing the functions into a single IC package:

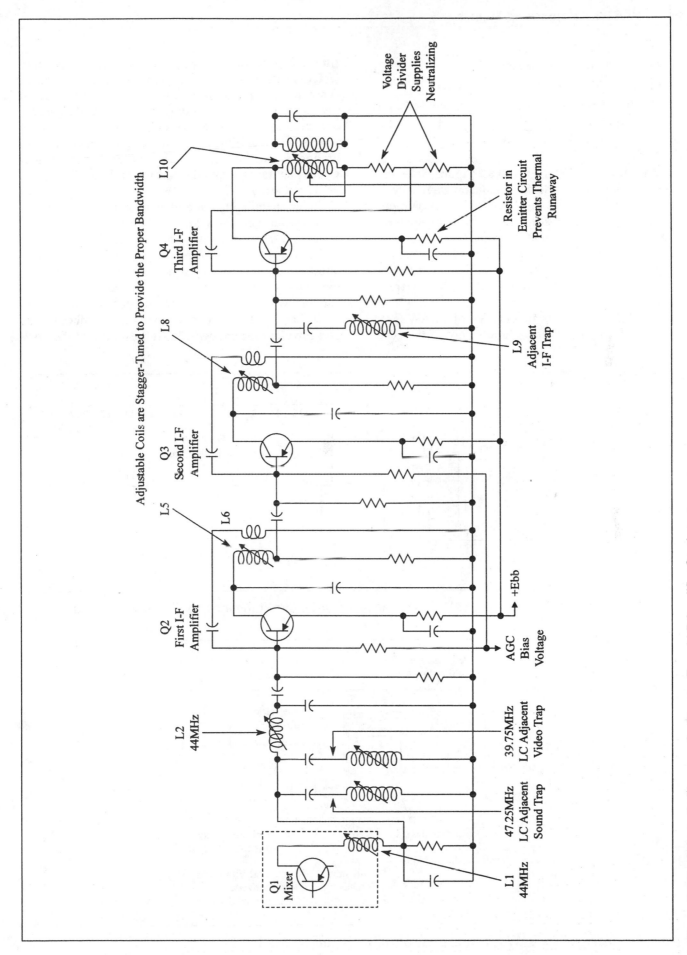

Figure 8-18 Schematic Diagram of a Stagger-Tuned I-F Amplifier Section

- increases circuit efficiency
- decreases the chances for signal losses
- eliminates the need for wave traps, tuned circuits, and alignment
- cuts the need for troubleshooting

Surface Acoustic Wave Filters   Figure 8-19 is a schematic diagram of a surface acoustic wave filter in an I-F/AFT/AGC circuit. The SAW filter is constructed from a piezoelectric substrate that allows the device to function like a crystal in that it resonates at specific frequencies. It also works as a double transducer because it has two sets of metal electrodes of different lengths extending across the substrate that allow the two transfers of energy.

Figure 8-20 shows a basic diagram of a SAW filter. The I-F signal appears at the input array of the SAW filter as an ac signal. The array is called amplitude-weighted because the varied length fingers provide a sharp cut-off for a particular response. Because of the piezoelectric effect, the twisting or bending of the crystal substrate converts the electrical signal into a mechanical vibration called an acoustic wave. The vibration occurs at the signal frequency and creates waves that travel across the surface of the SAW filter to its output array.

Given specific crystal and array parameters, the filter uses the vibrations to establish the desired bandwidth and frequency response characteristics. Various types of crystal materials provide a better response and allow less signal loss at desired I-F signals. The weighting of the input and output arrays can affect selectivity and bandwidth. Increasing the number of array fingers limits bandwidth, while decreasing the number of fingers broadens the bandwidth. At the output of the SAW

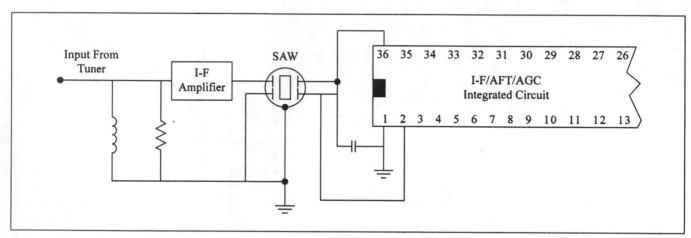

Figure 8-19   Schematic Diagram of a SAW Filter in an I-F/AFT/AGC Circuit (*Courtesy of Philips Electronics N.V.*)

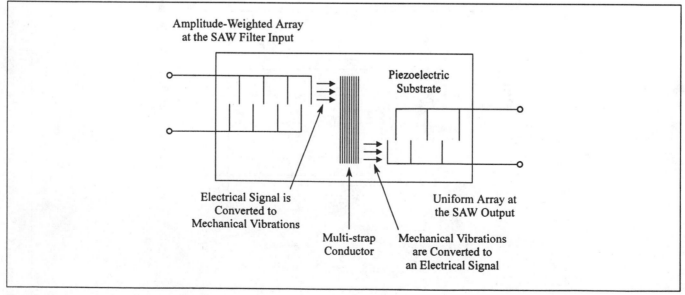

Figure 8-20   Diagram of a SAW Filter (*Courtesy of Philips Electronics N.V.*)

filter, a uniform array consisting of a specified number of uniform fingers maintains the bandwidth within the desired 6-MHz range.

FUNDAMENTAL REVIEW

Transducers

A transducer connects the physical and electronic quantities together for the purpose of producing some type of energy. Transducers respond to changes in temperature, light, strain, movement, and position. In electronics, transducers convert a quantity, such as mechanical vibrations, to an electronic signal.

☑ PROGRESS CHECK

You have completed Objective 6. Discuss stagger-tuned I-F circuits and SAW filters.

The Integrated Circuit I-F System

With Figure 8-21, two integrated circuits handle all the I-F amplification duties. Preceding the two integrated circuits, the complete I-F signal from the tuner feeds into the input of Q1, a transistor amplifier, and into a SAW filter. In part, the transistor establishes impedance matching between the tuner and the SAW filter. In addition, Q1 provides some isolation between the tuner and the filter, preventing the signals from the tuner from changing the characteristics of the filter. The SAW filter attenuates any undesired adjacent channel signals.

IC1—an untuned, broad-band, high-gain amplifier—provides most of the gain for the complete I-F system. Along with containing the first two stages of I-F signal amplification, the IC also includes all the AGC circuitry as well as the sync-separator stages. IC2 contains the third I-F amplifier, the AFT circuitry and associated take-off points, the sound I-F circuitry, and the sound I-F take-off points. After amplification by IC1, the video and sound I-F signals travel to IC2. Fur-

ther amplification in IC2, extraction of the sound and video signals, and the demodulation of the video signal by an internal detector produce the composite video signal.

Figure 8-22 shows the schematic of a single integrated circuit I-F/AFT/AGC package. Although the schematic does not depict any of the internal components that make up the IC, the package includes all three stages of I-F amplification, a synchronous video detector, all the AFT circuitry, and all the AGC circuitry. Figure 8-23 uses a block diagram to illustrate the functions enclosed within the integrated circuit.

Connected to the input of the IC, a single transistor operates as an I-F amplifier and provides compensation for any insertion loss seen at the input of the SAW filter. Most SAW filters introduce signal loss into the circuit because of the need to remove any transient responses or other spurious signals. A 47.25-MHz trap in the base circuit of the amplifier eliminates the adjacent sound carrier. Source voltages for the IC arrive from an external 11.4-Vdc regulated power supply. Pins 1 and 2 of the IC provide the 6.4-Vdc AFT voltage for the tuner.

Once the SAW filter properly shapes the I-F signal, the signal travels through differential inputs into the integrated circuit. After passing through three stages of amplification, the signal passes into a synchronous detector for demodulation of the composite video signal. The demodulated video signal containing the video, chrominance, and sound carrier information then couples from the IC to a 4.5-MHz trap for elimination of the sound information from the luminance and chrominance signals. Pin 12 of the IC couples the sound carrier through a 4.5-MHz bandpass circuit and supplies the sound carrier information to the audio circuits.

As we know, the I-F AGC circuitry varies the gain of the first, second, and third I-F stages, while the R-F AGC circuit develops an R-F AGC correction voltage. Unlike AGC designs previously discussed, the R-F AGC circuit enclosed in the IC does not require keying pulses from the horizontal output section. Instead, the

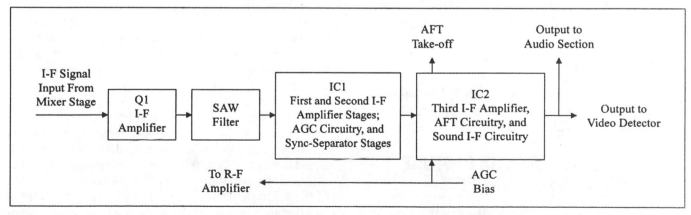

Figure 8-21   Block Diagram of an IC-Based IF/AFT/AGC System (*Courtesy of Philips Electronics N.V.*)

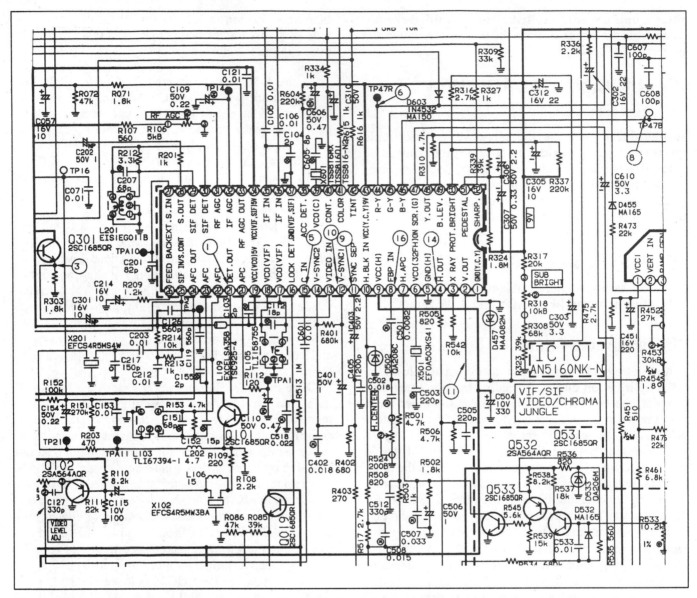

Figure 8-22   Schematic Drawing of a Single IC I-F/AFT/AGC Package (*Courtesy of Matsushita Electronics*)

design prevents any possible phasing problems between the keying pulse and horizontal sync signals by using noise limiting circuits in the AGC. The noise-limiting circuits eliminate any need for a keyed AGC pulse.

☑ PROGRESS CHECK

You have completed Objective 7. Explain the integration of the I-F amplifier circuit, AFT circuit, and AGC circuit into one device.

8-6 DETECTION OF THE PICTURE AND SOUND CARRIERS

Figure 8-24 uses a block diagram to show the sound and video carrier extraction points for a television. One of

the most complex problems encountered by technicians involves the improper tuning of channels by the control system. Efforts to pinpoint the problem lead to the bandswitching circuits, the frequency synthesizer, the microprocessor, peripheral circuitry, and the tuner. The diagram illustrates that the take-off and demodulation of the sound carrier occur before the take-off and demodulation of the video carrier.

With each of the I-F sections shown in this chapter, the extraction of the sound signal from the complete intermediate frequency occurs before the signal reaches the **video detector**, which demodulates the video I-F signal and recovers the composite video signal. This location is important because of the heterodyning action of the video detector. The detector heterodynes all the I-F signals together to produce a video signal, so it would heterodyne the 42.17-MHz chrominance signal

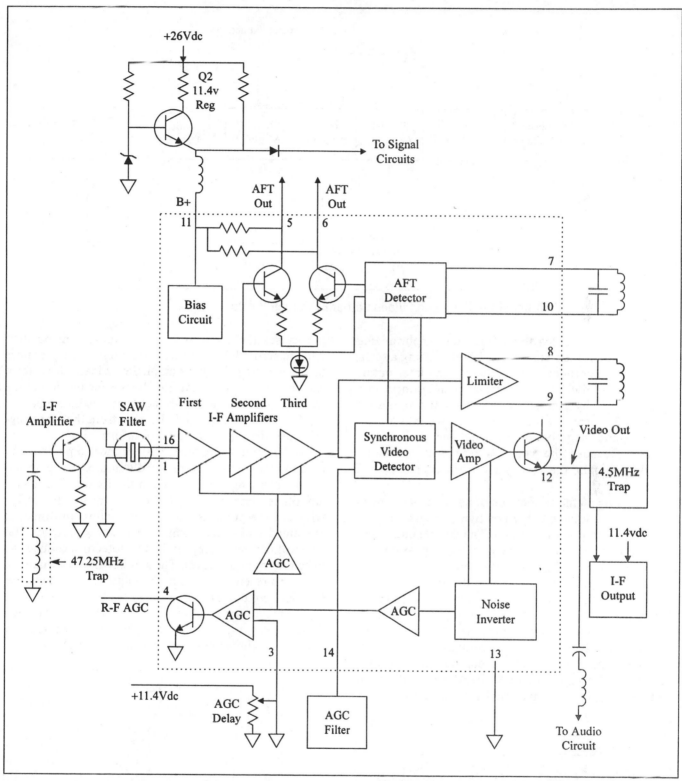

Figure 8-23   Internal Block Diagram of the Single IC I-F/AFT/AGC Package (*Courtesy of Philips Electronics N.V.*)

with 41.25-MHz sound carrier. The resulting 920-kHz beat pattern would show as annoying interference in the reproduced picture. Traps following the sound take-off point attenuate any remaining 41.25-MHz sound signal.

Demodulating the Video I-F Signal

A video detector has at its output three types of signals—the video, blanking, and synchronizing signals. Each has a proportional amplitude and exactly the same

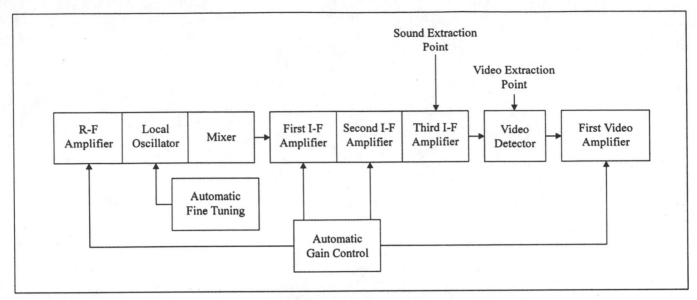

Figure 8-24   Block Diagram of the Sound and Video Carrier Extraction Points

frequency as the video, blanking, and synchronizing signals found at the transmitter. Whether using a rectifier or the synchronous demodulator seen later in this section, detector circuits remove sideband information from a carrier by converting the video and sound I-F signals into current. As the current flows into a load, positive- and negative-polarity ac and dc voltages develop. Because the voltages represent constantly changing signals, the voltages vary in frequency and amplitude.

Depending on the connection of its elements, a video detector demodulates either positive or negative alterations of the applied I-F signal. The demodulation polarity determines the polarity of the composite video signal. In Figure 8-25, the amplified I-F signal is applied to the input of a series video-detector circuit and the anode of the diode. Given the configuration of the circuit, the diode conducts only during the positive alternations of the I-F signal. With the diode conducting from cathode to anode, a closed-signal loop moves from the diode cathode to its anode, through adjustable coil L1, and through resistor R1 back to the cathode of the diode. Applying the pulsing I-F signal across the in-

put causes the development of a maximum positive voltage across R1. The filtering action of C1 produces the negative-polarity composite video waveform shown in Figure 8-26. Although the sync portion of the waveform points in the direction of positive polarity, the "negative polarity" label comes from the more negative positioning of the picture information.

At the point where the signal reaches 25 percent less than the maximum, the reduced output voltage produces the blanking pulse shown in the figure. The lower-amplitude time period of the waveform—identified with a P—represents the actual picture information contained within the composite video waveform. Because of the lower amplitude, the detector produces an even lower voltage level. The subtle voltage variations, which represent changes in the picture, occur between the blanking pulses.

The maximum amplitude of the carrier signal occurs during the sync-pulse period. At this point, the detector also has maximum conduction and develops the highest

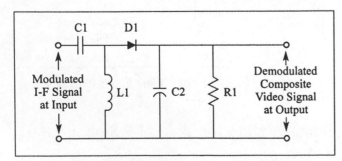

Figure 8-25   Video-Detector Circuit Based on a Single-Rectifier Diode

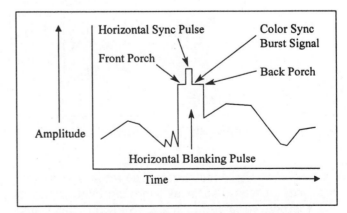

Figure 8-26   Negative-Polarity Composite Video Waveform

voltage seen at the detector. This maximum voltage translates into a positive sync pulse.

Reversing the polarity of the diode applies the I-F input signal across the cathode of the detector and causes conduction to occur on only the negative alternations of the I-F signal. With this configuration, the closed-signal loop moves from the diode cathode to its anode, through load resistor R1, through the adjustable coil, and back to the cathode of the diode. The maximum negative alternations of the I-F signal that occur during the sync-pulse period cause the diode conduction and produce the positive-polarity composite video waveform shown in Figure 8-27.

As Figures 8-26 and 8-27 show, the picture information may appear in either the positive or negative portion of the detected waveform. Although the detector may develop either positive or negative polarity voltages at its output, the polarity of the voltage must match the requirements of the video amplifier section. In turn, the video amplifiers must apply the proper polarity video signal to the input of the CRT.

Although older televisions use the rectifier diode circuit for video demodulation, newer designs rely on a synchronous demodulator packaged within an integrated circuit. As with the AFT circuit seen earlier, a synchronous demodulator provides linear detection at all signal levels, some amplification, and trapping. Given these characteristics, demodulator circuits built around a synchronous demodulator require less amplification and few external traps.

Figure 8-28 uses a block diagram to show a video demodulator circuit using a synchronous demodulator. During operation, the demodulator compares the transmitted video I-F signal with a 45.75-MHz reference signal and uses the sum and difference of the two frequencies to produce a composite video signal. The circuit first eliminates the carrier frequencies. While filters trap the 90-MHz carrier sum frequency, the carrier difference frequency equals zero. The difference of the remaining portion of the video I-F signal—sidebands containing picture information—becomes the composite video signal.

FUNDAMENTAL REVIEW

Demodulation

Demodulation is the extraction of sideband information from a carrier wave. In video products, video and sound detectors recover transmitted picture and audio information from the video and sound carriers.

Composite Video Signal Phase

The composite video signal may have either a positive or negative phase. Looking at Figures 8-26 and 8-27, we can compare the differences between a positive- and negative-phase composite video signal. As Figure 8-27 shows, a composite video signal with the sync tips pointing downward has a positive phase. The portion of the signal that produces white in the reproduced picture is in the positive direction. Figure 8-26 shows a negative-phase video signal that has positive-polarity sync pulses and negative-polarity picture information.

The phase of the composite video signal found at the output of the video amplifier stage depends on whether the system uses grid or cathode drive. Feeding the signal to the grid of the CRT requires a positive-polarity composite video signal at the amplifier output. Applying the video signal to the cathode of the

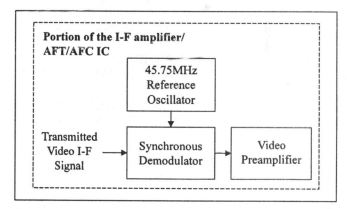

Figure 8-28   Video Detection Circuit Using a Synchronous Demodulator

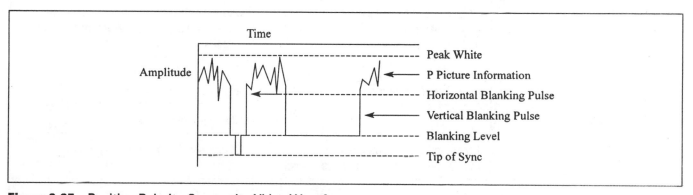

Figure 8-27   Positive-Polarity Composite Video Waveform

CRT requires a negative-phase composite video signal with positive-polarity sync pulses and negative-polarity picture information.

The sync and blanking portions of the composite video signal provide the bias needed to drive the CRT into cut-off. The blanking portion is often called the black portion of the signal. During operation with normal bias, the amplitude of the blanking portion equals the amplitude needed for cut-off. The sync-pulse level exceeds the blanking amplitude and enters a region called "blacker than black." These two signal levels establish a reference for the fluctuating picture signal as its level changes the amount of light given by the CRT over a range from white to black.

☑ PROGRESS CHECK

You have completed Objective 8. Explain sound and video I-F signal extraction, demodulation of the composite video signal, and video detector circuits.

8-7 I-F SECTION TROUBLESHOOTING

You have learned how the tuner, the I-F amplifiers, and the AFT and AGC circuitry are linked to one another. Because of the nature of those links and the types of signals moving back and forth between the tuner and the I-F section, many of the same problem symptoms appear for both sections. All this makes finding the cause and solutions for a problem more difficult.

As with tuner problems, troubleshooting I-F section problems involves an analysis of the problem, signal substitution, signal tracing, signal analysis, and voltage checks. Signal substitution can play a key role in determining the source of the problem because of the relationship between the tuner and the I-F section. Tables 8-1 to 8-6 illustrate the thought processes that go into troubleshooting the I-F section.

Table 8-1   Selectivity and Attenuation Problems

Problem	Symptom	Solution
Adjacent channel interference	Images from adjacent channel float through picture	Check alignment of tuned circuits.
Noise	Weak picture with snow; light contrast	Check adjustment of traps. Check alignment of tuned circuits.
I-F oscillations	Streaks in picture and poor sound quality	Check adjustment of traps. Check alignment of tuned circuits.
Ringing	Images in picture have shadows on the right side; good raster and sound	Regenerative feedback in I-F amplifier loop. Check for open bypass capacitors in I-F amplifier circuit, for open resistors in I-F amplifier line, and for open in neutralizing circuit.
Excessive peaking	Fine details in picture have ghosts	Check peaking coils. Check alignment of I-F amplifier and tuned circuits.
Video overload	Buzz in the sound; very black contrast; picture has poor horizontal or vertical locking	Check filter and bypass capacitors in tuned circuits. Check alignment of tuned circuits.
	Poor black-and-white color picture quality; good raster and good sound	Check adjustment of traps. Check alignment of tuned circuits.
No or poor color		Check alignment of tuned circuits.
Audio crosstalk	Black-and-white bars bloat through picture and vary with changes in sound	Check alignment of tuned circuits.
Smearing	Distorted picture; no detail in picture	Check alignment of tuned circuits.

Table 8-2   I-F Amplification Problems

Problem	Symptom	Solution
Video overload	Buzz in the sound; very black contrast; picture has poor horizontal or vertical locking	Check I-F amplifiers. Check I-F amplifier supply voltages. Check filter capacitors in I-F amplifier power supply line.
Smearing	Distorted picture; no detail in picture	Check I-F amplifier supply voltages.
	Good picture, good raster, no or weak sound	Check I-F amplifier supply voltages. Check for presence of sound carrier at input to I-F section.
	No picture, good raster, good sound	Check I-F amplifier supply voltages. Check for video signal at input to I-F section and at output of each I-F amplifier. Replace I-F amplifiers(s).
	No picture, good raster, no sound	Check first and second I-F amplifiers. Check AGC for proper bias of I-F section. Check connections and supply voltages.
	Negative out-of-sync picture, good raster	AGC circuit is overdriving third I-F amplifier.
	Buzz in sound, hum, bars floating through picture	Check filter capacitors in I-F amplifier supply.

Table 8-3   AGC Problems

Problem	Symptom	Solution
Video overload	Buzz in sound; very black contrast; picture has poor horizontal or vertical locking	R-F AGC circuitry is not producing any negative AGC voltages. Check AGC negative voltage sources. Check AGC keyer circuitry for shorted rectifier diodes and clamp diodes, and open filter capacitors.
Noise	Weak picture with snow; light contrast	Check AGC delay control. Low gain or AGC negative-voltage problem. Check for open resistors and bypass capacitors in R-F AGC line. Check tuners and AFT circuitry.
Ringing	Images in picture have shadows on the right side; good raster and sound	Replace open capacitor in AGC circuit.
No picture or sound, good raster		Misadjusted AGC delay control. Check AGC line for open bypass and filter capacitors.
Adjacent channel interference	Other channel images float through picture	Check R-F AGC voltages during channel change. Check for open bypass components in R-F AGC line.
Flutter	Contrast pulses from too dark to washed out	Replace open bypass capacitor in AGC keyer circuit. Check AGC keyer and AGC amplifier.

Table 8-4   SAW Filter Problems

Problem	Symptom	Solution
Reflection from output to input of SAW filter	Ghosts/images float through picture	Add multistage coupler to SAW filter, replace SAW filter.
Reflection from crystal portion of SAW filter	Ghosts/images float through picture	Replace SAW filter.
Adjacent channel interference	Images from adjacent channel float through picture	Replace SAW filter.
Weak picture with poor contrast		Check output from I-F preamp. Replace I-F preamp. Replace SAW filter.
No picture, good raster, no sound		Check signal at input to SAW filter. Replace SAW filter.

Table 8-5   Detector Problems

Problem	Symptom	Solution
920-kHz beat interference	Herringbone in picture	Replace detector.
	Weak picture with poor contrast	Replace detector.

Table 8-6   AFT Problems

Problem	Symptom	Solution
	No or poor color	
	Excessive color	
	Weak sound	
920-kHz beat interference	Herringbone in picture	
Very sensitive fine tuning	Picture drifts from good signal to poor signal	Alignment

Certainly, the large-scale integration of I-F/AFT/AGC circuitry has eliminated many of the problems associated with selectivity and attenuation. You have probably noticed that a large number of I-F amplifier troubleshooting solutions involve the realignment of the tuned circuits. Before you begin to study the methods and equipment used to accomplish this realignment, it is important to emphasize that realigning the tuned circuits should be the last option for solving an I-F section problem. Too many technicians jump to the conclusion that the alignment has changed and

make unnecessary adjustments. The resulting problems that may be caused by those adjustments can sometimes cover the actual fault and further complicate the repair.

For the most part, SAW filters and integrated circuits have eliminated the need for alignment. Although the modern I-F system retains a few tuned circuits, alignment is rarely necessary. But for systems that utilize the many tuned circuits associated with stagger tuning, then there is a potential need for realigning the system.

Test Equipment Primer: Component Testers

Component testers apply a small ac voltage across the device under test and measure the resulting ac current. Devices covered by these tests include resistors, capacitors, inductors, diodes, transistors, and integrated circuits. Each voltage value and the type of component match to produce individual "signatures" that disclose the condition of the device. Because component testers work with unpowered devices and apply only an ac signal, the test may occur either in-circuit or out-of-circuit.

A **transistor tester** allows for the in-circuit or out-of-circuit testing of transistors, diodes, and field-effect transistors. When purchasing a transistor tester, verify that the tester identifies the leads and type of a device, that it measures leakage and gain, and that it offers either an audible or visual good/bad indicator. **Capacitance meters** check only the value of capacitance and should offer measurements from 0.1 picofarads to 999.90 microfarads. The meter also becomes useful for checking tolerance and sorting unmarked capacitors.

⊃ SERVICE CALL

Receiver Would Not Lock onto Medium-to-Strong Signals

Problems within the tuner, I-F amplifier, AGC, or sync stages can prevent a receiver from locking onto medium-to-strong signals. Without locking, the picture will roll either horizontally or vertically. To find the cause of the problem, we can video-inject signals at different stage locations and work by the process of elimination. For this example, we will inject a signal into the sync-separator base and the sync-amplifier base. The sync stages locked vertically and horizontally at all signal levels. From this we can conclude that the sync separator and vertical circuits work properly.

Moving from the sync section, we then repeat the process for the first video amplifier. Locking occurred at all signal levels when we injected the video signal into the video-amplifier base. By using a tuner subber, we can check to see if changing the tuner solves the medium-to-strong signal-locking problem. As with the other tested stages, we find that the problem persists.

Having eliminated the sync stages, first video amplifier, and the tuner from consideration as problem areas, we are left with the AGC section. With normal signal conditions applied to the tuner input, we can ground the R-F AGC at the tuner and check for a good signal lock. Seeing a normal picture tells us that a problem exists in the R-F AGC circuit. Because the section involves amplified AGC, delayed AGC, and keyed AGC, we need to look for specific problem spots. First, we will monitor the relationships between the R-F and I-F AGC voltages for different signal levels. If the proper relationships do not exist, the next step is either to adjust the AGC control or inject a dc voltage at the R-F AGC point.

If neither the adjustment or the dc voltage injection eliminated the rolling, we need to consider which components in the AGC section could cause a problem. For example, we can recheck the tuner for a leaky MOSFET in the R-F amplifier stage for the correct voltages at the source and drain. In addition, we can check the emitter, base, and collector voltages at the AGC amplifier and the AGC keyer. We find that while the voltages at the R-F amplifier and AGC check within tolerance, the voltages at the AGC keyer are wrong. However, an out-of-circuit check of the AGC keyer transistor proves that the transistor is good.

Therefore, we can begin to look for components that supply signals to the AGC keyer stage. Figure 8-29 shows a schematic of the circuit area that seems to have the problem. Of the components shown in the stage schematic, C5, an electrolytic capacitor, bypasses several signals and circuits and connects directly to the AGC keyer emitter. With an open capacitor at this point, distorted horizontal pulses feed into the R-F and I-F AGC voltages and into the base circuit of the first video amplifier. Mixing those pulses with the AGC and video signals caused the receiver to lose any locking at medium-to-strong signals. Replacing the open electrolytic capacitor returned the receiver to its normal operating state.

✓ PROGRESS CHECK

You have completed Objective 9. Identify I-F amplifier, AFT, and AGC circuits troubleshooting methods.

SUMMARY

Chapter eight covered three sections that set up the operating conditions for almost all the stages covered in the remainder of this text. After the tuner converts and reduces the received R-F signal into the intermediate frequencies, the I-F section shapes and amplifies the signal in preparation for the recovery of the picture and sound information. By controlling the gain, selectivity, and response of the intermediate frequency

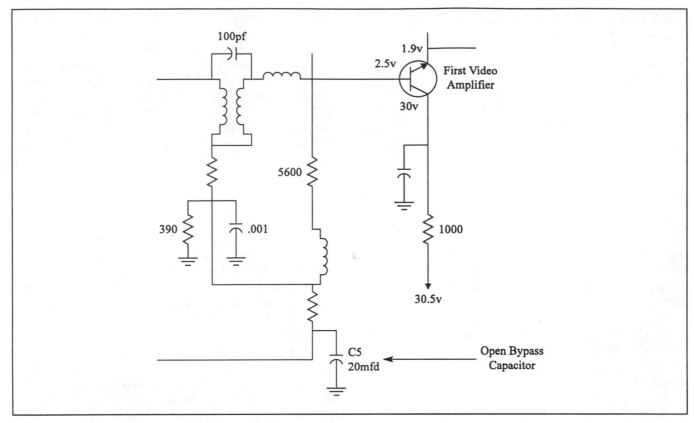

100pf

5600

390

.001

1.9v

2.5v

First Video
Amplifier

30v

1000

30.5v

C5
20mfd

Open Bypass
Capacitor

Figure 8-29   Schematic of the Service Call I-F Circuit

signals, the I-F amplifier, AFT, and AGC sections affect the overall reproduced picture and sound. One glance at the troubleshooting charts shown at the end of the chapter discloses that problems within the sections affect the stability and clarity of the picture, the quality of the sound, the accuracy of reproduced colors, and the ability of the receiver to maintain consistent tuning.

The AFT and AGC circuits link the I-F section back to the tuner and—in the case of the AGC circuit—into the video amplifier section and the horizontal output section. Automatic fine-tuning provides a method for continuously controlling the oscillator frequency by monitoring the I-F frequencies. Any deviation at the I-F level causes a corrective action to occur at the tuner oscillator stage. The automatic gain control section monitors and controls both the R-F and I-F ampli-

fier signal gains and prevents the overdriving of the amplifiers.

Modulation of the picture and sound carriers provides an efficient method for transferring transmitted information to a receiver. Before demodulation can occur, though, both the sound and picture I-F signals are extracted at specific points within the I-F section. While the sound extraction point takes the sound signal to the sound circuits for additional processing, video extraction couples the video I-F signal to the video demodulator. At this point, the demodulation of the video signal yields the sync, blanking, and picture information that makes up the composite video signal.

The troubleshooting section took a look at solving problems that occur in both stagger-tuned systems that use discrete components and I-F systems that utilize one integrated circuit as a package for all the functions.

REVIEW QUESTIONS

1. True or False   The output of the mixer contains only two signals.

2. What is a carrier? What are upper and lower sidebands?

3. High frequency signals contain more _____ than low frequency signals.

4. The bandwidth for a signal containing low and high frequencies would be (narrow) (broad).

5. What is vestigial sideband compensation and why is it necessary?

6. Desired I-F signals are the _____ MHz _____ and the _____ MHz

   _____.

7. The sound carrier is 4.5 MHz (below) (above) the video carrier at the output of the mixer.

8. Why are lower intermediate frequencies desirable?

9. The color burst signal is located _____ MHz from the _____ signal.

   The color signal sidebands, or _____ signal, are _____ above and
   below the color burst signal.

10. The I-F section must accomplish two tasks. Describe what those tasks are.

11. In a color television, the I-F signal also contains the _____ and the

    _____.

12. True or False   An AFT circuit controls the frequency of the R-F amplifier.

13. In an I-F section, the sound take-off point (precedes) (follows) the video detector.

14. True or False   The intercarrier approach for transferring sound and video I-F signals is used in modern
    video receivers.

15. Describe how an AFT circuit works.

16. The amplification of _____ frequencies can cause interference in a
    reproduce picture.

17. What is circuit Q?

18. In a high-frequency circuit, you would want (high) (low) circuit Q.

19. True or False   Shunt resistances are used to stabilize the circuit Q.

20. The 41.25-MHz sound I-F carrier is a desired signal in I-F systems. Why do circuit designs feature 41.25-MHz
    traps?

21. Traps are tuned to _____ frequencies and _____,

    _____, or _____ those frequencies.

22. What is the main purpose of wave traps?

23. Draw and compare the frequency response curves for a bridged-T trap and a bandpass filter.

24. An absorption trap _____.

25. What are the differences between a series resonant trap and a parallel resonant trap?

26. True or False   In a television, the I-F stages have an overall gain that ranges from 500 to more than 10,000
    and produce most of the video signal amplification.

27. Where do coupling capacitors normally appear in a transistor-based amplifier circuit? Provide a rough
    drawing of three transistor-amplifier stages coupled with capacitors.

28. An I-F section that has three amplifier stages that have an individual stage gain of 18 will have an overall

    gain of _____.

29. Why is neutralization necessary in a tuned circuit?

30. Mutual coupling impedance provides the (maximum) (minimum) amount of energy transformer.

31. What is the difference between "loose" and "tight" coupling? What effect does loose coupling have on
    selectivity and bandwidth? What effect does tight coupling have on selectivity and bandwidth?

32. Draw the frequency response curve for an overcoupled circuit.

33. Why do parasitic oscillations occur?

34. Parasitic oscillations in an amplifier can cause the _____,
_____, and _____.

35. True or False   Whether based on transistors or an integrated circuit, all I-F sections contain four
    amplifiers.

36. Mutual coupling impedance is measured in terms of:
    a. resistance and capacitance
    b. inductance and capacitance
    c. resistance, capacitance, and inductance
    d. voltage, capacitance, and resistance

37. Draw the frequency response curve for critical coupling.

38. Under strong signal conditions, should the gain at I-F amplifiers increase or decrease? Under weak signal
    conditions should the gain at I-F amplifiers increase or decrease?

39. What is the purpose of an AGC circuit?

40. What happens if an R-F amplifier overloads the mixer?

41. An AGC system using forward AGC biases an amplifier towards:
    a. cutoff
    b. saturation

    An AGC system using reverse AGC biases an amplifier towards:
    a. cutoff
    b. saturation

42. What does a pulse gate diode provide?

43. True or False   The gain of an R-F amplifier should be highly controlled under weak signal conditions.

44. Why do some circuits use amplified AGC?

45. The keyed or gated pulse in a keyed AGC system comes from the _____
    section and has a frequency of _____.

46. Refer to Figure 8-13. When Capacitor C1 charges and discharges, it provides _____
    _____.

47. With keyed AGC, the AGC circuit uses a _____ pulse and a
    _____ as input signals. The output signal is _____.

48. Why does an interval exist between the keying pulses in a keyed AGC system?

49. Why does a keyed AGC system derive the keyed pulse from the horizontal output section?

50. How does a delayed AGC system operate?

51. With a keyed AGC system, the reference voltage is taken from a stage that _____
    _____. A _____ pulse is the keyed pulse.

52. Why does the circuit shown in Figure 8-13 include an AGC amplifier?

53. What is the difference between forward and reverse AGC? Why are both methods used?

54. In Figure 8-21, two identical I-F amplifier stages are tuned to the same frequency. Because of this, the total
    gain of the complete circuit _____. The overall response of the circuit
    equals_____.

55. The overall bandwidth of an I-F stage is 1.4 percent of the _____.

56. Stagger tuning is accomplished through _____.

57. What is the advantage of stagger tuning?

58. Bandwidth is a measurement of a _____ between two points.

59. The integration of the I-F amplifiers, AFT circuit, AGC circuit, and demodulator into one package provides benefits such as _____.

60. What is the purpose of the I-F amplifier that precedes the SAW filter in Figure 8-22?

61. Describe the construction and operation of a SAW filter. Are SAW filters used in modern television receivers?

62. A 45.25-MHz trap eliminates the
    a. video I-F carrier
    b. sound I-F carrier
    c. adjacent video I-F carrier
    d. adjacent sound I-F carrier

63. The I-F amplifier/AFT/AGC circuits shown in Figures 8-21 and 8-22 do not use keyed AGC. Why not?

64. Demodulation of the composite video signal in Figure 8-23 is accomplished through the use of a

    _____.

65. True or False   The bias voltages for the circuit shown in Figure 8-23 are provided by a 11.6-Vdc regulator.

66. Demodulation is _____.

67. Does the detection of the sound I-F and video I-F signals occur at the same point in the I-F circuit? If so, describe why. If not, describe why not.

68. True or False   The sound I-F signal is amplitude modulated and the picture I-F signal is frequency modulated.

69. Match the following labels to the correct portion of the waveform.
    a. sync pulse
    b. blanking pulse
    c. picture information

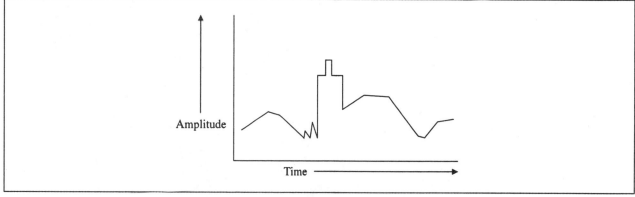

70. For the waveform portions listed in the previous problem, which produces the highest voltages? The mid-level voltages? The lowest dc voltages?

71. A synchronous demodulator offers key advantages over a simple detector diode. What are those advantages?

72. True or False   Either a negative- or positive-polarity composite video signal may be found at the output of a video detector.

73. Name the signals found in the composite video signal.

74. A demodulator circuit built around a synchronous demodulator requires less _____

    _____ and fewer _____.

75. What do the voltage variations that occur between blanking pulses represent?

76. **True or False**   If you suspect that a problem exists in the I-F amplifier section, you should realign the section first and then check for component failures.

77. **True or False**   Circuits using SAW filters require more alignment than stagger-tuned amplifier circuits.

78. What are the symptoms of the following problems?
    a. adjacent channel interference
    b. noise
    c. I-F oscillations
    d. audio crosstalk

79. What are the symptoms of the following problems?
    a. ringing
    b. excessive peaking
    c. video overload
    d. smearing

80. Can a problem in the AGC circuit affect the I-F amplifiers?

81. What is a probable cause of 920-kHz beat interference?

82. What condition will cause oscillations to occur in an I-F amplifier circuit?

83. Refer to the Service Call at the end of the section. What did you learn about troubleshooting methods from the example? Why did an open bypass capacitor in the AGC keyer stage cause the problem?

84. In modern televisions, one integrated circuit contains the I-F amplifier, AFT, and AGC circuits. Would you expect to find more or less problems with the single IC approach when compared to stagger tuning? List three reasons for your answer.

CHAPTER 9

Processing the Sound Signal

OBJECTIVES   Upon completion of this chapter, you should be able to:

1. Discuss pre-emphasis, de-emphasis, and sound I-F detection, amplification and limiting.
2. Discuss sound I-F demodulation.
3. Explain discrete and active components in audio amplifier circuits.
4. Discuss how monaural sound circuits function.
5. Discuss how stereo sound circuits function.
6. Describe special-purpose sound circuits.
7. Troubleshoot different types of sound-circuit problems.

KEY TERMS
cross-over distortion
cross-over network
de-emphasis
differential peak detector
figure of merit
Foster-Seeley discriminator
monaural sound
multichannel sound
pre-emphasis

quadrature detector
quasi-parallel sound method
ratio detector
separate audio programming
split-supply Class AB amplifier
stereo sound
synthesized sound
thermal runaway

INTRODUCTION

During our study of the I-F section, we considered circuits that, in many ways, resemble a busy set of intersecting highways. While a complete I-F signal enters the circuit, the signal separates at different extraction points. Portions of the signal go to the video amplifier section while other portions go into the sound section. As you will find in a later chapter, other samples of the same signal become part of the sync-separation signal.

Along with those signals, AGC and AFT signals control parts of the entire operation.

Chapter nine takes us from the sound I-F extraction point in the I-F section and leads us through the detection, processing, and demodulation of the sound I-F signal, and the amplification of the audio output signal. Each section of the chapter describes the type of circuits used to accomplish those tasks. In addition, the chapter illustrates how sound sections have evolved from the common one-channel circuits to stereo sections that reproduce high-fidelity sound.

As you read through chapter nine, you will encounter phrases such as intercarrier, split-sound, and quasi-parallel sound. Each phrase describes methods for transporting the video I-F and sound I-F signals and then separating the two signals at the appropriate extraction points. Figures 9-1A, B, and C use block diagrams to illustrate the three methods.

Chapter eight concentrated on the intercarrier method, shown in Figure 9-1A, which uses the I-F section to amplify both the video and sound carriers of the desired channel. Mixing the signals at the video detector produces a 4.5-MHz difference frequency that contains sound-carrier frequency modulation and serves as the sound I-F signal. Attenuation before the video demodulator prevents the sound signal from heterodyning with the color subcarrier and protects the system from beat interference.

The intercarrier system offers several advantages. Because a frequency difference between the video and sound carriers is established at the transmission point, the frequency and demodulation of the sound carrier remains independent of any receiver tuning. Despite the advantages, the intercarrier system is susceptible to interference under certain conditions. For example, overmodulation of the sound signal can cause the appearance of a herringbone pattern in the reproduced picture and a buzz in the reproduced sound.

The split-carrier sound method eliminates the potential for beat interference in the reproduced picture and sound. As Figure 9-1B shows, instead of deriving a difference frequency from the video and sound I-F carriers, the split-carrier sound method takes the 41.25-MHz sound I-F carrier from the output of the mixer. Bandpass circuits couple that signal to a set of sound I-F amplifiers, while a 41.25-MHz trap attenuates any frequency-modulated signals within the video I-F circuit. Although the split-sound method eliminates intercarrier interference, it remains open to another type of interference, called incidental-carrier phase modulation, originating in the mixer stage.

The **quasi-parallel sound method** represents an effort to eliminate both sources of interference by using the best portions from the intercarrier and split-sound carrier methods. As Figure 9-1C shows, it utilizes parallel paths from the tuner for the video and sound carriers. Within the sound path, a sample of the video and sound carriers passes through a double-humped bandpass filter that contains a 42.17-MHz rejection trap. The combination of the filter and trap attenuates

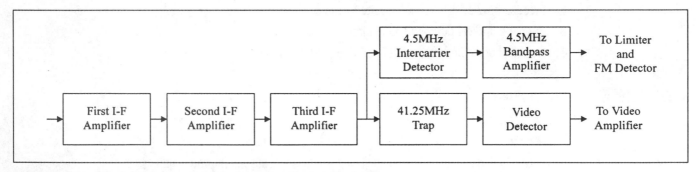

Figure 9-1A   Block Diagram of the Intercarrier Sound Method

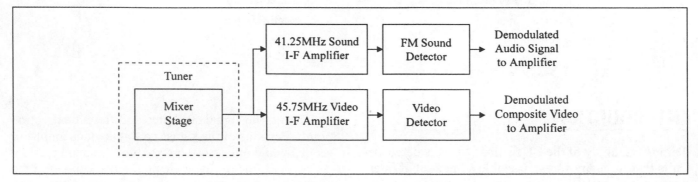

Figure 9-1B   Block Diagram of the Split-Sound Method

the color subcarrier frequency and establishes a flat response curve for both the video and sound carriers. Because of this, the system eliminates the problem of incidental-carrier phase modulation. Following the filter and trap, a diode detector creates the 4.5-MHz sound I-F carrier and passes that signal to the limiting and demodulation circuits. The video I-F path contains traps that attenuate the 41.25-MHz sound signal and eliminate any chance for beat interference.

Figure 9-1C illustrates the differences between the quasi-split-sound and the older intercarrier system. Moving from left to right, the 47.25-MHz trap attenuates any adjacent channel sound signals out of the 45.75-MHz I-F signal received from the tuner. Then Q1, a SAW filter preamplifier, feeds the amplified signal into a specially-designed SAW filter that has two output terminals. While one feeds the video I-F signal into the video I-F integrated circuit, the other terminal ties an inverse I-F waveform into the sound I-F integrated circuit. IC1, the video I-F circuit, processes only the video portion of the I-F signal and passes the 4.5-MHz sound signal.

You will encounter all three methods when you begin to work with video products. For example, while many color television receivers manufactured during the 1980s and early 1990s use the intercarrier method, many monochrome receivers from the same time period may use the split-sound method. Newer receivers that rely on SAW filters and large-scale integration often feature the quasi-parallel sound method.

Regardless of the approach used by the manufacturer, all three methods have common characteristics of basic circuits that, when recognized, can ease your troubleshooting approach. As an example, each method uses some of type of trap for the attenuation of undesired signals. While bandpass filters provide the selectivity required for good audio reproduction, amplifiers boost the gain of weak signals. In every case, you can use your knowledge about basic circuits to determine the nature of a problem and find a solution.

9-1   PRODUCING AND SHAPING THE SOUND SIGNALS

Throughout our discussions of I-F amplifiers, AFT circuits, and AGC circuits, we concentrated on the progression of video I-F signals through the different

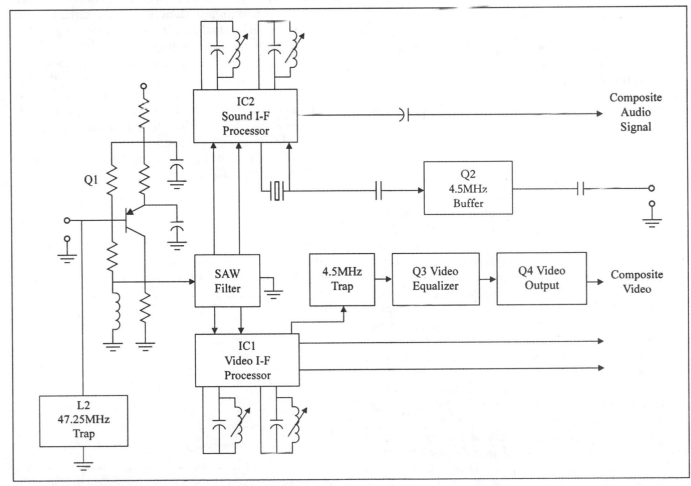

Figure 9-1C   Block Diagram of the Quasi-Parallel Sound Method

amplifier and control stages and into the demodulation stage. Although present in the I-F stages, the sound I-F signal is extracted and attenuated before the I-F signal travels to the video detector. Without the attenuation of the sound carrier, the reproduced picture would contain beat interference. The extraction of the sound carrier allows the signal to travel into processing circuits designed to handle frequency-modulated signals.

Returning to the I-F section for a moment, the sound I-F signal has a low level when compared with the video carrier. Indeed, the video I-F signal has a magnitude approximately ten times larger than sound I-F signal. Without this difference, amplitude modulation of the sound carrier would occur within the last I-F stages, and the video and sound carriers would heterodyne. This is prevented by 4.5-MHz traps that precede the video detector. When the complete I-F signal leaves the mixer, the 41.25-MHz sound I-F signal and the 45.75-MHz video I-F signal have a 4.5-MHz difference. After extraction, the 41.25-MHz sound I-F carrier feeds into the sound section for additional amplification and demodulation.

I-F amplifiers in the I-F section have a broad response that accommodates the video I-F information. A very slight drift in the oscillator frequency does not have a tremendous effect on the picture quality. In comparison, the sound I-F amplifier at the front of the sound section has a narrow response. Any frequency drift within the high-frequency local oscillator distorts the reproduced sound. If the local oscillator drifts from its reference frequency, the sound I-F signal shifts either higher or lower than its assigned frequency.

Pre-emphasis

At the point of transmission, a sound signal is a frequency-modulated signal having maximum deviation of ±25 kHz and capable of providing an audio bandwidth of 50 to 15,000 Hz. The sound carrier is transmitted at a frequency 4.5-MHz above the R-F picture carrier. When considering speech, music, or any reproduced sound, the higher frequency components have less amplitude than the lower frequency components. As a result, the signal-to-noise ratio for any frequency is lower for a high-frequency audio voltage and higher for a low-frequency audio voltage if the noise level remains constant. If the noise level increases at higher frequencies, then an even lower signal-to-ratio results. Given this characteristic, the quality of any high-frequency audio reproduction suffers unless the signal-to-noise ratio improves.

A **pre-emphasis** network such as the one shown in Figure 9-2 improves the signal-to-noise ratio of transmitted high frequencies by boosting high frequencies and attenuating low frequencies. In other words, the values of resistance and inductance in a pre-emphasis

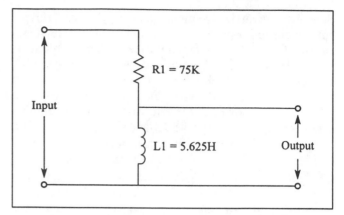

Figure 9-2 Schematic Diagram of a Pre-emphasis Network

network must provide a 75-microsecond time constant. The time constant affects how the circuit responds to certain frequencies.

Our analysis of the pre-emphasis circuit begins with the impressing of an audio voltage across terminals 1 and 2 of the basic RL circuit. In the figure, the RL circuit, consisting of a 75,000-ohm resistor in series with a 5.625-H coil provides a time constant of .000075 seconds. The pre-emphasis function of the circuit occurs because the individual components operate as a voltage divider. Increasing the voltage across the inductor at higher frequencies for a given input voltage causes the attenuation of lower frequencies. As a result, the high frequencies receive emphasis, or a boost.

In all cases, the value of the inductor depends on the need to attenuate specific low frequencies of the audio-modulating voltage found at terminals 3 and 4. We can find the value of the inductance through the use of the time constant equation:

$$\text{Time Constant} = L/R = .000075 \text{ second}$$

The graph shown in Figure 9-3 shows how the output voltage of the circuit—the transmitted sound signal—varies non-linearly with changes in frequency. At the low frequency of 50 Hz, the output voltage equals

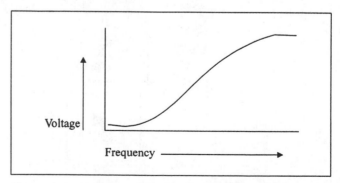

Figure 9-3 Graph of Pre-emphasis Network Output Voltage Varying with Frequency

only 2.5 percent of the input voltage. At 200 Hz, the output voltage rises to almost 10 percent of the input voltage. As the frequency increases to 15,000 Hz, the output voltage rises to nearly 100 percent of the input voltage.

Processing the Sound I-F Signal

Chapter six showed that the first and second I-F amplifiers amplify both the 41.25-MHz sound I-F carrier and the 45.75-MHz video I-F carrier. Two different extraction points for the sound and video carriers in a color television receiver separate the sound signal from the composite video signal. The extraction point for the sound I-F signal precedes the video I-F extraction point because of the possibility that the detected 4.5-MHz sound I-F signal could heterodyne with the 3.58-MHz color oscillator and produce 920-Hz beat interference.

After extraction from the I-F strip at the sound extraction point, the 41.25-MHz sound I-F and 45.75-MHz video I-F signals travel to a sound detector designed to convert the signals into an intercarrier 4.5-MHz sound I-F signal. The mixing of the two I-F signals at the detector produces the 4.5-MHz sound I-F carrier. Although the circuit is technically called a sound detector, a frequency-modulated sound I-F signal continues to reside at the detector output. No demodulation occurs at this point.

Certainly, the conversion of one sound I-F frequency into another may seem confusing. However, the system converts the split-sound I-F to an intercarrier sound I-F for several reasons. Most importantly, the use of a difference frequency for the sound I-F carrier removes the link between the sound carrier and the local oscillator frequency. With this, the fine-tuning of the sound signal is accomplished easily for any picture. Because the

4.5-MHz difference depends on the transmitted carrier frequencies rather than the local oscillator frequency, a problem with tuning the picture will not affect the tuning of the sound.

Working with only one, lower frequency, traps and tuned circuits can effectively eliminate any amplitude-modulated signals. The stages that follow the sound detector amplify the weak 4.5-MHz sound I-F signal and reject video or sync signals. In addition, the limiting circuits also prevent any noise from reaching the next amplifier stages. Finally, the frequency decrease down to 4.5-MHz eases the problem of changing the sound I-F signal into an audio signal.

Figure 9-4 uses a schematic diagram to show how the sound I-F detection, amplification, and limiting circuit precede the sound I-F demodulator. As shown in the circuit, the extracted 41.25-MHz sound I-F signal follows a path from capacitor C3 and transformer T1 to Q1, the sound I-F amplifier. Diode D1 detects the sound carrier and converts the signal to a frequency-modulated 4.5-MHz sound I-F carrier.

De-emphasis

At the receiver end, a de-emphasis circuit compensates for the pre-emphasis applied to the transmitted sound signal. The removal of the non-linear component from the signal allows the audio output section to reproduce a natural sound. **De-emphasis** occurs after the modulation of the sound I-F signal, reduces the higher-frequency noise voltages, and establishes a desired signal-to-noise ratio. With all this, de-emphasis restores a balance between the higher and lower frequencies contained in any sound transmission. As a combination, pre-emphasis and de-emphasis improve the signal-to-noise ratio of the reproduced sound.

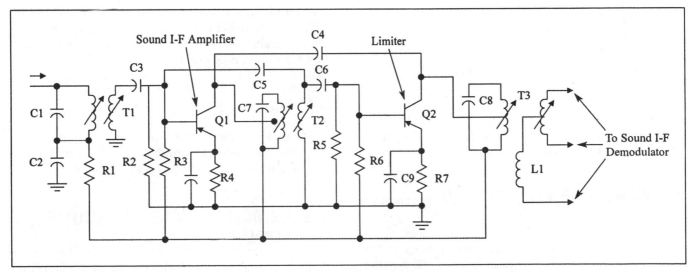

Figure 9-4 Schematic Diagram of the Sound I-F Detection, Amplification, and Limiting Circuit

Figure 9-5 shows a simple de-emphasis filter circuit consisting of a RC series network. Because the reactance of the capacitor decreases with an increase in frequency and the resistance remains constant, an increase in the input frequency and corresponding input voltage causes a smaller percentage of the input voltage amplitude to appear across the capacitor.

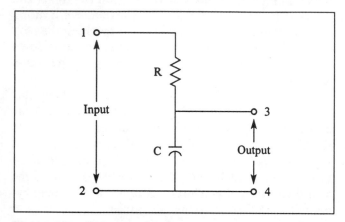

Figure 9-5   Schematic Diagram of a De-emphasis Circuit

In the figure, the values of the resistor and capacitor establish the same time constant seen in the pre-emphasis network. Using the equation for an RC constant, the values equal:

$$\text{Time Constant} = .000075 \text{ second} = RC = 75,000 \times C$$

Figure 9-6 plots the output voltage against the frequency for a constant amplitude input voltage. In this case, the voltage decreases as the frequency increases.

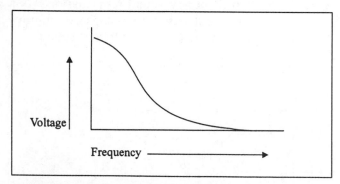

Figure 9-6   Graph of De-emphasis Network Output Voltage Varying with Frequency

↝ FUNDAMENTAL REVIEW

Diodes

A diode is a two-electrode semiconductor device that conducts in only one direction. With a PN-junction diode, conduction occurs when the diode is forward-biased. When reverse-biased, the diode will not conduct.

Forward bias occurs through the either the application of a potential to an n-type material that is more positive than the p-type material, or the application of a potential to p-type material that is more negative than the n-type material. The anode of a diode represents the p-type material, while the cathode represents the n-type material.

PN-junction diodes are used to smooth, or rectify, an ac voltage into a pulsating dc voltage. Common rectifier circuits used in power supplies are the half-wave rectifier, full-wave rectifier, and bridge rectifier circuits. PN-junction diodes are rated by characteristics such as V_{RPM}, or the maximum reverse voltage allowable for a diode; I_O, or the maximum amount of dc forward current allowable for a diode; and $P_{D(MAX)}$, or the maximum amount of power dissipation for a forward-biased diode.

In comparison to PN-junction diodes, zener diodes operate in the reverse breakdown region and conduct in the reverse direction. Zener diodes are used as voltage regulators, or to maintain a relatively constant voltage regardless of any variations of current flowing through the load.

Operating characteristics for zener diodes include: I_R, or the amount of current flowing through a reverse-biased diode; V_{BR}, or the amount of reverse voltage that causes a diode to operate in the reverse direction; V_Z, or the amount of voltage across a zener diode when the diode operates in reverse breakdown; I_{ZK}, or the minimum amount of diode current needed to maintain voltage regulation; I_{ZM}, or the maximum allowable value of diode current; $P_{D(MAX)}$, or the maximum amount of dissipated power for a zener diode operating in the reverse region; and V_Z, or the rated voltage value of the zener diode.

Light-emitting diodes, or LEDs, provide a limited amount of light when biased in the correct direction. LEDs are available in colors such as red, green, white, yellow, orange, or multicolors, and are used as indicators for a wide range of applications.

☑ PROGRESS CHECK

You have completed Objective 1. Discuss pre-emphasis, de-emphasis, and sound I-F detection, amplification, and limiting.

9-2 DEMODULATING THE SOUND SIGNAL

After further amplification by the sound I-F amplifier, the sound carrier travels into a demodulation stage, where an audio detector extracts the audio frequency information from the frequency-modulated 4.5-MHz sound I-F carrier. At this point, the demodulated audio frequency signal matches the audio signal originally seen at the transmission location. The detector stage accomplishes this transition by developing an ac voltage at its output. Variations in the ac voltage correspond with the frequency variations of the I-F carrier found at the detector input.

Detector Circuits

Video receivers use one of several different detector circuits. Of the six circuits described in this section, one—the ratio detector—uses discrete components, while the others reside on an integrated circuit. Before we begin our analysis of the circuits, we should examine the different approaches to FM signal demodulation.

Both the ratio detector and a Foster-Seeley discriminator utilize the phase relationships from input signals to control the amplitude of output voltages. Quadrature detectors measure the phase shift across a reactive circuit as the carrier frequency varies. In contrast, differential peak detectors compare the peak voltages of two reference signals and use the difference between them to produce a demodulated FM signal. Finally, a phase-locked loop detector compares the phase of a reference signal and the incoming sound I-F carrier signal and produces a dc correction voltage. The low-frequency error voltage serves as the demodulated output.

Ratio Detectors   A **ratio detector** uses diodes to compare the phase relationship between two applied signals. Considering Figure 9-7, input signals to the

diodes that make up the detector may consist of either 180 degree out-of-phase signals taken from the transformer terminals, or an in-phase signal found at the junction of two capacitors at the input to the circuit. While the two diodes alternately turn on and off because of the application of a signal at either diode at any given time, the voltage at each diode corresponds with the phase relationship of the two input signals.

L1, a transformer primary, couples the I-F signal into the transformer secondary, L2, and two halves of a single winding, L3 and L4. With the application of the ac signal across the winding halves, an out-of-phase relationship exists between the voltages seen at diodes D1 and D2. The voltage applied by L3 to D1 is 180 degrees out-of-phase with the voltage applied by L4 to D2. Along with those voltages, the secondary winding also applies a voltage across both diodes, yielding a combination of $E_{L2} + E_{L3}$ for D1 and $E_{L2} + E_{L4}$ for D2.

The phase difference between the voltages at L2 and L3, and L2 and L4, varies with frequency. When the sound I-F carrier has a 4.5-MHz center frequency, the L2 voltage is 90 degrees out-of-phase with the voltages across the winding halves. This occurs because capacitor C2 tunes the winding halves to 4.5-MHz; L2 is an untuned inductance. Double-tuning the transformer causes the 90-degree phase shift at the resonant frequency.

Each of the phase relationships also establishes a vector sum relationship for the voltages. The voltage applied to D1 equals the vector sum of the voltages applied to D2. Thus, the diodes produce equal currents. In addition to the inductances and the diodes, the values of several other components also affect the circuit operation. Capacitor C5 along with resistors R1 and R2 provide a long time constant, add limiting to the circuit, and form a center-tapped voltage divider.

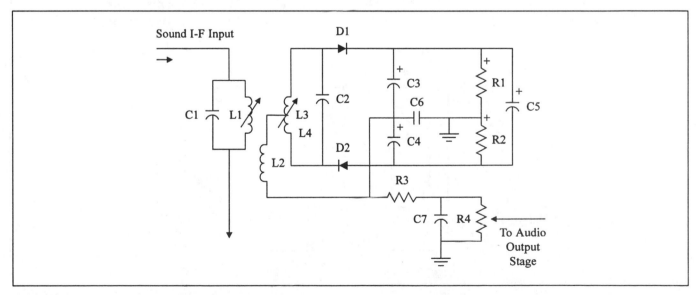

Figure 9-7   Schematic of a Balanced Ratio Detector

An unmodulated I-F signal at the input terminals, the combination of the phase relationships, the long RC time constant, and the voltage-dividing action allow the circuit to absorb any rapid changes in amplitude. Limiting occurs because the absorption does not allow the amplitude changes to appear in the audio output signal. During operation, the divider network consisting of R1 and R2 applies a positive-bias voltage to the cathode of D1 and an equal negative-bias voltage to the anode of D2. With C5 discharging through R1 and R2, the voltage at C5 cannot change suddenly.

Detection of the audio signal occurs because of the vector sum relationships. With an unmodulated signal at the circuit inputs, current flows from the cathode to anode of D1, through L3 and L2, to the negative side of C2, out of the positive side of C3, and to the D1 cathode. Every pulse of current from the diode charges C3. Current also flows from the cathode to the anode of D2, to the negative side of C4, out of the positive side of C4, through L2 and L4, and to the cathode of D2.

The same I-F signal voltage that charges C3 and C4 also charges C5. Because C3, C4, R1, and R2 form a balanced bridge circuit, equal voltages are found across those components. No voltage is present across C6 because of the connection of the capacitor across points of equal potential.

Modulation of the I-F signal allows the signal to shift above and below the carrier and causes the voltage phases at L3 and L4 to shift in reference to the voltage found across L2. As a result, the vector sum of the L3 and L2 voltages alternates between a greater and a lesser value than the vector sum of the voltages across L4 and L2. In terms of conduction, the D1 current alternates between a value greater and lesser than the value of the current through D2. Consequently, the voltage across C3 also shifts between a value greater and lesser than the voltage found across C4, and the voltage across C6 shifts between a positive and negative value.

From our earlier discussion, we know that the rate of voltage variation, or frequency, seen across C6 varies proportionately with the rate of I-F carrier frequency swing. Moreover, the amount of carrier swing determines the amplitude of the C6 voltage. The voltage across C6—or the audio signal—corresponds to the original audio modulation. De-emphasis then restores the balance needed for the low and high frequencies.

Foster-Seeley Discriminator   The **Foster-Seeley discriminator** circuit operates similarly to the ratio detector circuit, in that frequency variations in the I-F carrier cause corresponding phase shifts. In the discrete circuit shown in Figure 9-8, the anodes of both diodes connect to a transformer secondary. Two capacitors connect in series with the cathodes of each diode, while coil L2 connects to the center of the transformer secondary.

When operating, the circuit applies the E_{L2} and E_{L3} voltages in series to D1, and the E_{L2} and E_{L4} voltages to D2. With the carrier variations causing phase shifts in the E_{L3} and E_{L4} voltages, D1 alternately conducts either more or less than D2. When the diodes conduct, the D1 current charges C3 while the D2 pulses charge C4. Because the two diode currents flow in opposite directions, the capacitor plates have opposite polarities.

The difference between the two diode conduction values is proportional to the frequency deviation of the sound I-F carrier. As a result, the voltages produced at the C3 and C4 have an algebraic sum that may be positive, negative, or zero, and that also varies at the same rate as the carrier-frequency changes. The total varying voltage is the audio signal.

Quadrature Detector   A **quadrature detector**, or gated coincidence detector, compares the phase of two input signals applied across a reactive circuit and outputs a signal when the two signals have the same

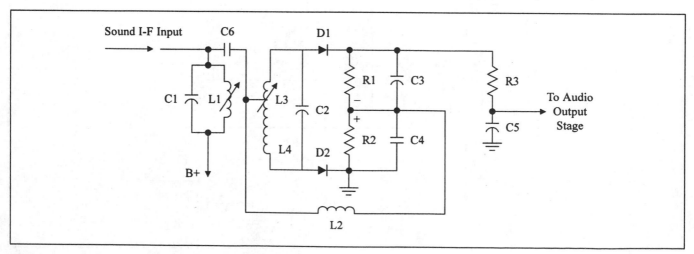

Figure 9-8   Schematic Diagram of a Foster-Seeley Discriminator

polarity. Used in many sound channel designs, quadrature detectors offer:

- the application of a low-cost, easily tuned LC network as a phase-shifting element
- the transformation of the phase shift into amplitude with no linkage to the amplitude characteristics of the phase-shift network
- complete linearity between the phase deviation and the output signal

Figure 9-9 shows how a logic gate compares the direct-coupled sound I-F-output from the 4.5-MHz amplifier with the 90-degree out-of-phase 4.5-MHz output signal taken from a quadrature coil circuit that works as a phase-shift network. As the carrier deviates from the center frequency, the resulting phase shift varies proportionately with amount of deviation and direction. The gated detector compares a square wave signal produced by the amplifier/limiter stage with a sine-wave signal provided by the LC network and generates output pulses. Every output pulse serves as the demodulated signal and has a width equivalent to the angle difference between the zero crossings of the waveforms.

Differential Peak Detector   The **differential peak detector** is another popularly used sound I-F detector. As Figure 9-10 shows, the 4.5-MHz output signal from one differential amplifier feeds into a combination of a coil and capacitors C1 and C2. Each LC combination forms a tuned resonant circuit with L1C1 having a resonant frequency slightly above 4.5 MHz, and L1C2 having a resonant frequency slightly below 4.5 MHz. The differential amplifier compares the peak voltages of each signal and then sends the demodulated FM audio signal to the attenuation/volume-control circuit.

Demodulation occurs because of the amplitude difference between the signals at the differential amplifier input terminals. High-frequency sound signals cause a high-amplitude signal to appear at input 1, while low-frequency sound signals cause a high-amplitude signal to appear at input 2. The deviations from the 4.5-MHz sound I-F correspond with variation in the original audio signal.

Synchronous Detectors and Phase-Locked Loop Detectors   In previous chapters, you had the opportunity to study synchronous detectors and phase-locked loops. Sound channels utilize both types of circuits for the demodulation of the sound I-F signal. Figure 9-11 shows two differential amplifiers simultaneously applying the 4.5-MHz sound I-F signal and the station signal as two out-of-phase input signals to switching transistors. The output of the transistors serves as the demodulated audio signal. The complete synchronous demodulator circuit is enclosed within a single integrated circuit package.

Moving to Figure 9-12, the PLL consists of a phase detector, a local oscillator, a low-pass filter, and a limiter circuit. As with the previously-studied PLL, the phase detector compares the phase of an input signal with the phase of a local oscillator signal produced by a voltage-controlled oscillator. In this situation, the input signal is the 4.5-MHz sound I-F carrier. Any phase difference between the two signals causes the comparator to generate a corresponding dc voltage. Phase differences result from changes in the sound I-F signal modulation. When the dc control voltage shifts the frequency of the VCO closer to the reference frequency, and when the circuit locks onto the reference, the VCO begins to track any FM signal variations. Because

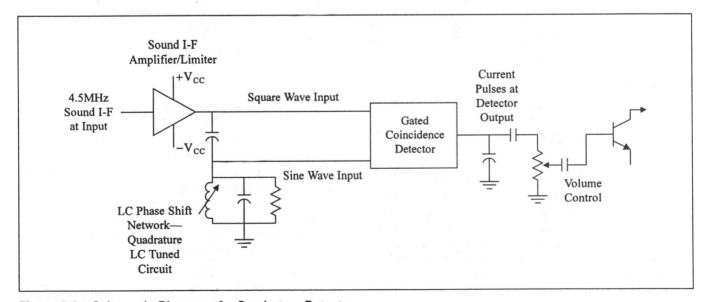

Figure 9-9   Schematic Diagram of a Quadrature Detector

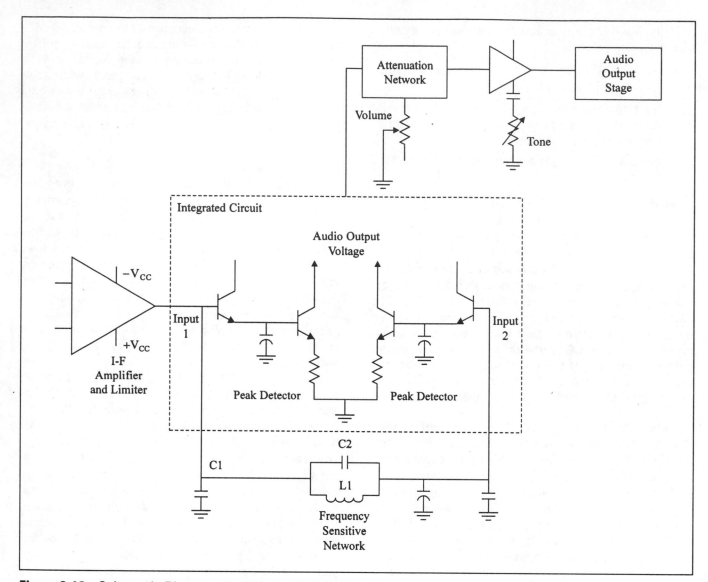

Figure 9-10   Schematic Diagram of a Differential Peak Detector

every change in the dc control voltage corresponds with a change in the original audio signal, the dc voltage serves as the demodulated output signal.

↻ FUNDAMENTAL REVIEW

Differential Amplifiers

A differential amplifier is a circuit that accepts two input signals and produces a signal that is proportional to the difference of the input signals. The input terminals for a differential amplifier are identified as the non-inverting input and the inverting input. The output from the amplifier for a given pair of input voltages depends on the gain of the amplifier, the polarity relationship between the input voltages, and the values of the supply voltage.

☑ PROGRESS CHECK

You have completed Objective 2. Discuss sound I-F demodulation.

9-3 AUDIO AMPLIFIERS

Our discussions of tuner and I-F circuits up to now have covered amplifiers that amplify R-F and I-F signals. An audio-frequency, or A-F, amplifier is the last stage of the video receiver sound circuit. Unlike the amplifier circuits seen in the preceding chapters, audio amplifiers amplify only the narrow spectrum of audio frequencies which ranges from 20 Hz to 20 kHz. Audio-output amplifiers should have the following characteristics:

• High gain

• Very little distortion within the audio frequency range

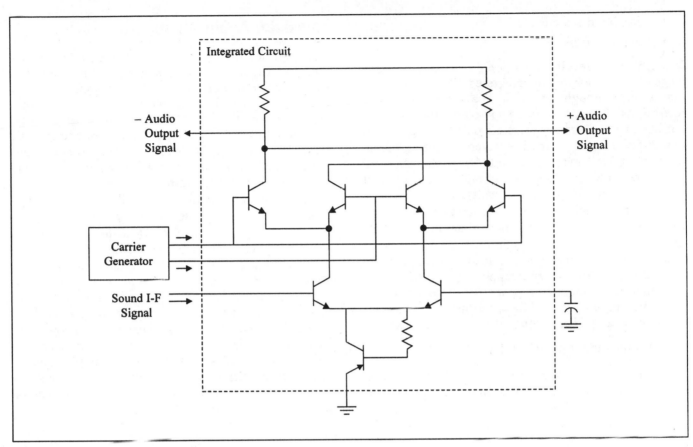

Figure 9-11   Schematic Diagram of a Synchronous Demodulator

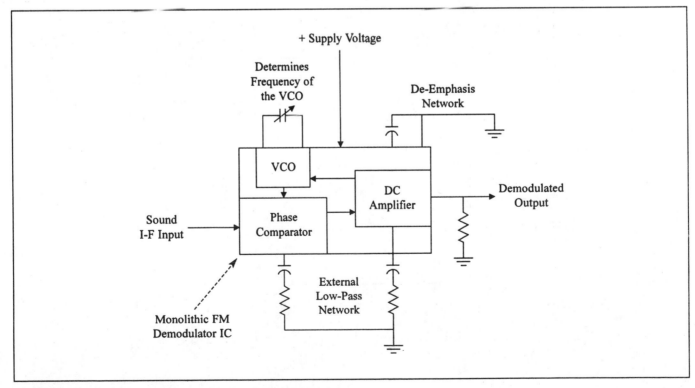

Figure 9-12   PLL Circuit Operating as a Sound Demodulator

- High input impedance
- Low output impedance

Modern video receivers feature in the audio-output stage either transistors or integrated circuit amplifiers that provide enough gain to drive a 30-percent modulated signal. Transistor-based audio amplifiers usually consist of a two-transistor, push-pull Class B or AB output stage and—in small receivers—have an output range of 100 milliwatts to 1 watt. Audio amplifiers incorporated into an integrated circuit have an output power of approximately 2 to 5 watts.

Audio amplifiers are also classified as power amplifiers. In an audio circuit, the power amplifier drives a high amount of power through a low resistance load, with the speaker serving as the load. Although resistances dissipate power, efficient power amplifiers drive the maximum amount of power possible through a load. We can measure the **figure of merit**, or the efficiency of an amplifier, through the following equation:

$$n = (ac\ output\ power/dc\ input\ power) \times 100\%$$

where n represents the figure of merit. Circuit designers rely on the figure of merit when matching an amplifier with an application.

Transistor Audio Amplifiers

The amplifier circuit shown in Figure 9-13 is configured as a Class AB complementary-symmetry amplifier. Note that one of the output transistors is an NPN transistor and the other is a PNP transistor. The two output transistors thus form a complementary pair and work like two variable resistors that are controlled by the audio signal amplitude. The NPN transistor draws current only during the positive half-wave of the audio signal, while the PNP transistor draws current only during the negative half-wave. While complementary-symmetry circuits may use different components for biasing, the use complementary transistors in the output stage will remain constant.

Transistors Q3 and Q4 provide biasing for the power amplifiers and act as diodes. With the biasing transistors configured in this way, each has a shorted collector-base junction and uses only the emitter-base junction in the circuit. Using transistors in this way ensures that the biasing transistors match perfectly with the amplifier transistors. Because a complementary-symmetry amplifier does not require the use of a transformer, the circuit offers a low-cost, low-loss method for amplifying audio frequencies.

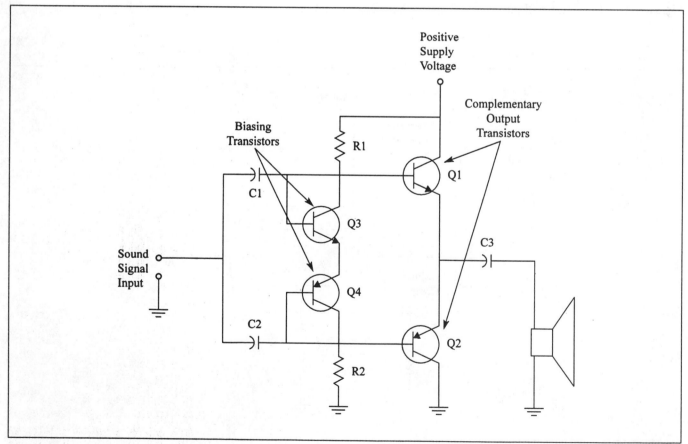

Figure 9-13   Class AB Complementary-Symmetry Amplifier

Because of the potential for **cross-over distortion** when the two transistors go from conduction into cut-off, the biasing circuit prevents both transistors from entering cut-off during the same interval by maintaining a small amount of forward bias for each transistor. Cross-over distortion occurs when neither transistor in a Class B amplifier has any forward bias, and the output voltage of the amplifier circuit equals zero. Figure 9-14 shows type of waveform produced by cross-over distortion.

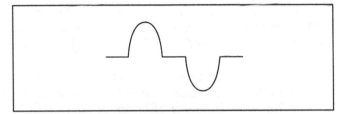

Figure 9-14   Cross-over Distortion Waveform

Applying a small amount of dc bias voltage to the push-pull configuration allows collector current to flow for more than one alternation of the applied signal, but not for the time of one complete cycle. Thus, the circuit operates as a Class AB amplifier and produces a linear output that contains no distortion. Figure 9-15 shows the audio-frequency sine wave produced by the combination of a Class AB amplifier and the application of the dc bias voltage.

The drawback to the need for added dc bias is the potential for **thermal runaway.** Often, power transistors are destroyed when an excessive forward-bias voltage combines with junction leakage to produce progressively higher currents. With a fixed-base current, the collector current increases. This higher than normal internal current causes the transistor to overheat, which in turn breaks down the internal resistance of the device. As the cycle of increased heating and increased current production continues, the transistor eventually destroys itself.

As Figure 9-15 shows, one transistor of the Class AB pair begins to conduct before the other has stopped conducting. For a brief time, a path between power and ground exists through the transistors. To eliminate the

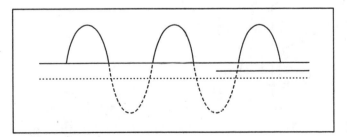

Figure 9-15   Class AB Operating Waveform

thermal runaway problem, Class AB amplifiers utilize a matched pair of transistors that have identical electrical and thermal characteristics. The use of a matched pair allows the same dc current to flow through both transistors and the same collector voltage to split equally between each transistor.

In addition, the biasing circuit establishes a quasi-complementary-symmetry configuration in which the complementary-symmetry section of the amplifier appears before the actual output stage. Rather than use a matched pair of high-cost output transistors, the circuit uses a matched pair of biasing transistors. Going back to Figure 9-13, the circuit consisting of transistors Q3 and Q4 provides the correct amount of forward bias for output transistors Q1 and Q2.

The circuit shown in Figure 9-13 also provides an example of the amount of power dissipated in the resistances of a power amplifier circuit. Current flows through, and voltage is applied across, each resistor. Thus, using the power equation $P = I^2R$, we can find the amount of power dissipated by each resistor. When we compare the sum of the dissipation amounts against the total power drawn by the amplifier, we have the total amount of power going to the load.

Every transistor has a maximum power dissipation rating. When selecting a replacement transistor for a power amplifier application, verify that the power dissipated by the transistor in the circuit does not exceed the rating of the replacement transistor. The $P_{D(MAX)}$, or maximum power dissipation rating of a transistor, indicates the maximum amount of power in milliwatts that a transistor can dissipate. For Class B and AB amplifiers, the maximum power dissipation rating equals:

$$P_{D(MAX)} = V^2_{PP}/40R_L$$

where V_{PP} equals the peak-to-peak load voltage.

Variations in Class AB Circuit Amplifier Designs
Figure 9-13 shows one popular Class AB amplifier design, but others exist. Popular designs include the diode-biased Class AB amplifier, the Darlington complementary-symmetry amplifier, and the split-supply Class AB amplifier. Each alternative offers characteristics that may be more useful for specific amplifier designs. Figures 9-16A, B, and C illustrate the different Class AB amplifier circuits.

The diode-biased amplifier illustrated in Figure 9-16A uses two diodes to match the characteristic base-emitter voltage values of the output transistors. While Class B amplifiers normally operate at cut-off, the addition of the diodes biases the transistors above cut-off. With both amplifiers operating above cut-off and conducting between 180 and 360 degrees of the waveform, the circuit has the response shown in Figure 9-15 and begins to operate in the Class AB range. Thus, because of the matched characteristics,

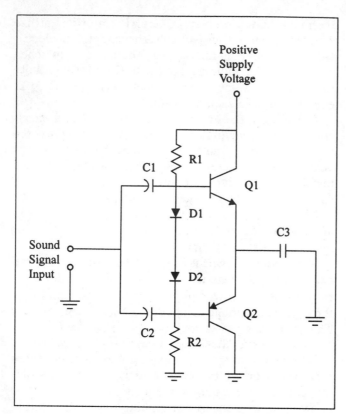

Figure 9-16A   Schematic Drawing of a Class AB Amplifier Using Matching Diodes

the circuit eliminates the chances for thermal runaway or cross-over distortion.

In Figure 9-16B, two Darlington pairs function as the output stage and provide higher overall current gain than the standard complementary-symmetry amplifier. Although a Darlington pair is usually packaged in one case, the term actually describes the configuration of the two output transistors. The first transistor of the pair acts as an input amplifier for the second transistor. Because the dc and ac beta values of the pair equal the product of the individual transistor betas, the Darlington pair has an extremely high beta. With the emitter of the first transistor connected to the base of the second, the base current of the first is multiplied by the beta of the second.

Usually, a Darlington complementary-symmetry amplifier is used for applications that require high load power. Because of the use of an input amplifier, the Darlington configuration provides the transistor pair with better stability, high current gain, and a high input impedance. The four diodes shown in the diagram provide biasing at the bases of Q1 and Q3 and compensate for the base-emitter voltage required for each Darlington pair.

In the **split-supply class AB amplifier** shown in Figure 9-16C, the two power supply connections have equal but opposite polarities. By using matched power

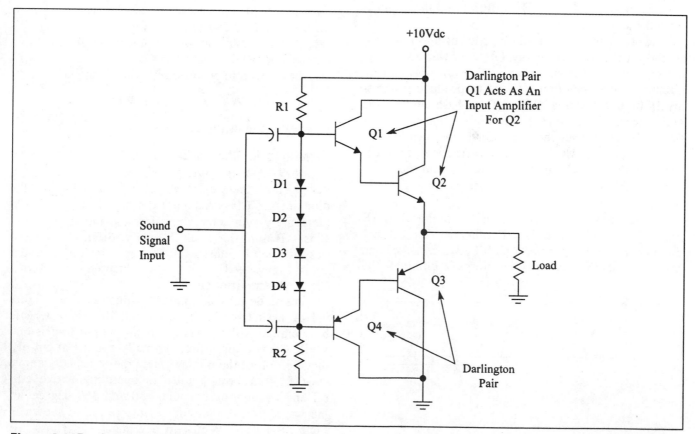

Figure 9-16B   Schematic Drawing of a Complementary-Symmetry Amplifier Using Darlington Pairs

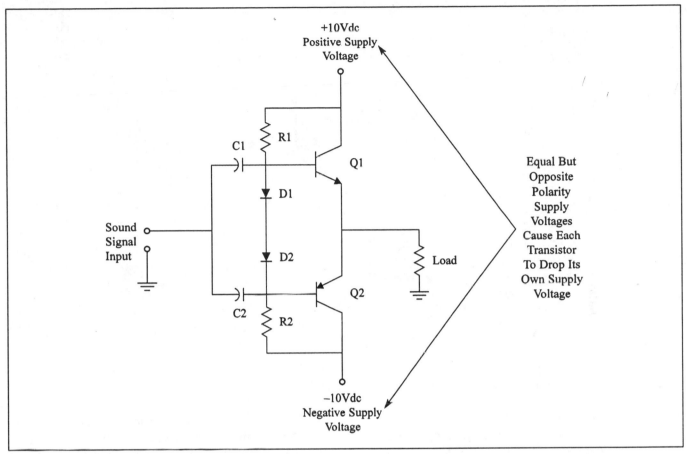

Figure 9-16C   Schematic Drawing of a Split-Supply Class AB Amplifier

supplies, each amplifier drops its own supply voltage. This allows the output of the class AB amplifier to center around zero volts rather than the division of the supply voltage.

↩ FUNDAMENTAL REVIEW

Power Amplifier Classification

Any amplifier that conducts during the entire 360 degrees of the ac input cycle is a Class A amplifier. Class B and Class C amplifiers conduct for less than the entire ac input cycle. With a Class B amplifier, two transistors conduct for 180 degrees of the input cycle. One transistor conducts during the positive alternation of the ac input cycle, and the other conducts during the negative alternation of the cycle. The combination of the conduction cycles yields a 360-degree output waveform.

A Class AB amplifier is a variation of the Class B amplifier. While the Class AB amplifier also uses two transistors, conduction occurs only during the portion of the input cycle between 180 and 360 degrees. Class C amplifiers use one transistor and conduct for less than 180 degrees of the input cycle. Reactive components in the Class C amplifier circuit provide the waveform for the remainder of the cycle.

In terms of efficiency, Class A amplifiers have the lowest efficiency, with a figure of merit ranging from 25 to 50 percent. The range depends on the use of RC or transformer-coupled circuits. Class B and AB amplifiers have an efficiency rating of approximately 78.5 percent, while Class C amplifiers have a maximum theoretical efficiency of 99 percent.

Audio circuits rely on Class B amplifiers because of the good efficiency rating. Although Class C amplifiers have near-perfect efficiency, this type of amplifier works only with a tank circuit tuned to either the same frequency or a harmonic of the input signal.

Integrated Circuit Audio Amplifiers

Figure 9-17 shows the schematic diagram for an audio amplifier found within a single IC package and based on an operational amplifier. The op-amp provides a combination of high input impedance, low output impedance, and low noise distortion. With the low output

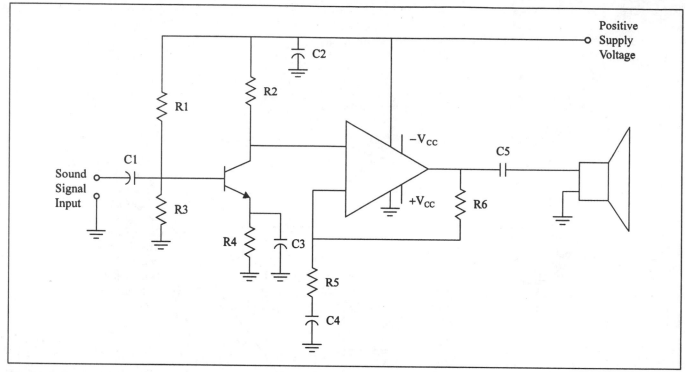

Figure 9-17   Operational Amplifier Working as an Audio Amplifier

impedance and low noise distortion, the amplifier provides optimum coupling to the speaker with minimum distortion in the audio-frequency range.

In the circuit, the grounding of operational amplifier negative-supply voltage input through capacitor C5, a coupling capacitor, limits the amplifier output to a specific range. Grounding the supply in this way places the reference for the speaker closed to ground, and—because of its location in the bias-supply voltage line—protects transistor Q1 from transient current. Without the capacitor in that location, current could feed back from the operational amplifier through the power supply and to the transistor.

FUNDAMENTAL REVIEW

Operational Amplifiers

An operational amplifier is a general-purpose amplifier working as part of a linear integrated circuit. Op-amps offer a very high gain, in the range of 5,000 to 10,000, with no feedback. The bandwidth of an op-amp varies with the amount of gain.

Every operational amplifier has two inputs. In schematic drawings, the non-inverting input is labeled with a plus sign, and the inverting input is labeled with a negative sign. Applying a positive voltage (with respect to ground) to the noninverting input yields a positive output voltage. Applying a more negative voltage (with respect to ground) to the inverting input yields a negative output voltage.

The input stage of an operational amplifier is a differential amplifier. Therefore, an operational amplifier is amplifying the difference between the two inputs. The power supply levels at the operational amplifier inputs establish the limits of the output voltage swing.

☑ PROGRESS CHECK

You have completed Objective 3. Explain discrete and active components in audio amplifier circuits.

9-4 MONAURAL SOUND SYSTEMS

Monaural sound simply describes a sound system that has only one output channel for the audio signal. While the transmitted signal contains the left and right signal components, the reproduced signal does not split the signals. Although every manufacturer offers stereo television receivers, most continue to offer lower-priced units that have only monaural capabilities.

Monaural Sound Circuits Using a Combination of Transistors and an IC

Figure 9-18 shows a monaural sound channel circuit consisting of one I-F processing integrated circuit, an audio driver transistor, an audio amplifier, a transistor, an audio output transistor, and a variety of support

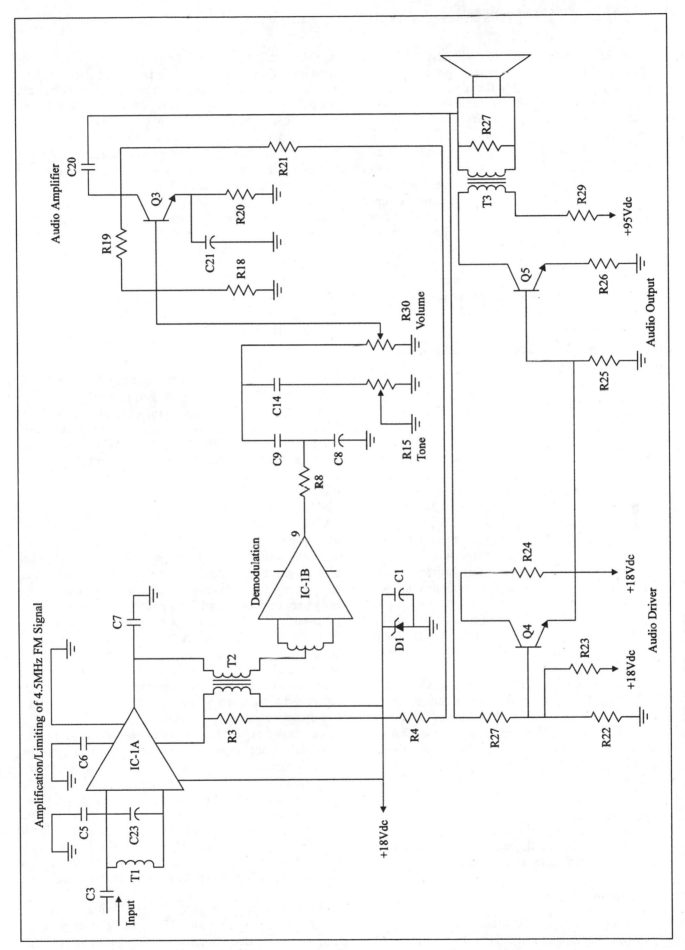

Figure 9-18 Schematic Diagram of a Monaural Sound Channel

components. Moving from left to right, the 4.5-MHz travels from a 4.5-MHz detector, through a coupling capacitor and transformer, and to the input of the sound I-F circuit. The processing IC contains the transistors, diodes, and resistors needed to amplify and the limit the frequency-modulated sound signal.

Note that the drawing splits the integrated circuit into two sections. Although the circuitry is physically contained in one semiconductor package, the schematic illustrates that the circuit offers different and separate functions. The section labeled as IC-1A contains the bulk of the processing circuitry, while IC-1B includes a discriminator circuit used to recover the audio information from the 4.5-MHz sound I-F signal. In addition, the second section of the IC contains an audio preamplifier. At pin 9 of the IC, the preamplifier output couples to the volume and tone controls.

Monaural Sound Circuit Based on Two Integrated Circuits

Newer sound-channel designs enclose the complete audio system in two integrated circuits. Figure 9-19 shows a schematic/block diagram for the sound-processing system of a Sylvania model 19C5 television. IC111 handles the amplification of the 4.5-MHz sound I-F signal, provides limiting, detects the FM audio signal, and controls the level of the signal before coupling it to IC121, the audio output IC. The composite video signal travels from pin 12 of IC51, the video I-F processing IC, through a coupling capacitor, to a 4.5-MHz high-pass filter. From there, the 4.5-MHz sound I-F signal travels to a 4.5-MHz buffer transistor and then to T110, a sound input transformer. After arriving at pin 14 of IC111, the 4.5 MHz is amplified, limited, detected, and coupled to the IC121. The +12-Vdc voltage supply seen at pin 11 of IC111 and the Q36 collector circuit has its source at the integrated flyback transformer.

At pin 5 of the sound I-F IC, a voltage divider controls the volume. R114, the audio preset control, compensates for any variations within the sound I-F IC by controlling the audio level when the volume control is set to its minimum level. The controlled audio signal couples from pin 8 of IC111 through a 1-microfarad capacitor to pin 7 of IC2.

☑ PROGRESS CHECK

You have completed Objective 4. Discuss how monaural sound circuits function.

Stereo Sound Systems

A **stereo sound** system uses separate sound channels to amplify left and right sound signals. Splitting the sound

signals in this way allows the receiver to match sounds with specific portions of the reproduced picture. As a result, when a jet fighter catapults from a carrier deck, we gain the audio sensation that the aircraft is flying from left to right.

As technology has progressed, video receivers have gone from synthesized sound to the true stereo sound of multichannel television. While **synthesized sound** systems work with a monaural sound signal to fool the brain into thinking that left and right channels exist, **multichannel sound** systems rely on the broadcast of a composite audio signal and the splitting of that signal into left and right sound components. Like the monaural system, the stereo system uses an L+R signal that ensures compatibility between the two system types. The stereo system also uses an L−R signal that builds the stereo effect through the modulation of a suppressed subcarrier.

Synthesized Stereo Circuits   During the late 1970s and early 1980s, several manufacturers began to move away from offering only monaural sound and introduced synthesized stereo circuits in which two different audio signals drive a set of two speakers. To accomplish this, a stereo-synthesis circuit divides the sound information into two channels. One channel contains only frequencies that exist in the mid-low, or 60- to 200-Hz range, and the mid-high, or 1,200- to 10,000-Hz ranges. The other contains the remainder of the sound information.

Figure 9-20 shows a schematic diagram of a typical stereo-synthesis system. In the circuit, IC1 includes two matched operational amplifiers. Negative feedback between pins 2 and 7 of op-amp 1, and pins 3 and 8 of op-amp U1-B determines the amount of gain provided by the amplifiers. A filter network for op-amp U1-A determines the frequency response seen at the output of the amplifier, while op-amp U1-B amplifies the remaining frequencies. Consequently, for any original sound signal, the amplifiers send different amplified frequencies to the speakers.

The input signal travels into the synthesis circuit through a filter network consisting of six capacitors and six resistors. The network passes any frequency in the 40-Hz, 640-Hz, and 10-kHz range and blocks any frequency in the 160-Hz and 5,000-Hz range. Once the filter circuit shapes the sound signal, that signal becomes the input for each of the two operational amplifiers. Pin 6 of op-amp U1-A and pin 9 of op-amp U1-B serve as the connecting points for the input signal. Both pins tie to the positive side of the operational amplifiers.

A filter circuit connected to pin 8 shapes the frequency response of op-amp U1-A and passes only the frequencies needed for amplification by op-amp U1-B.

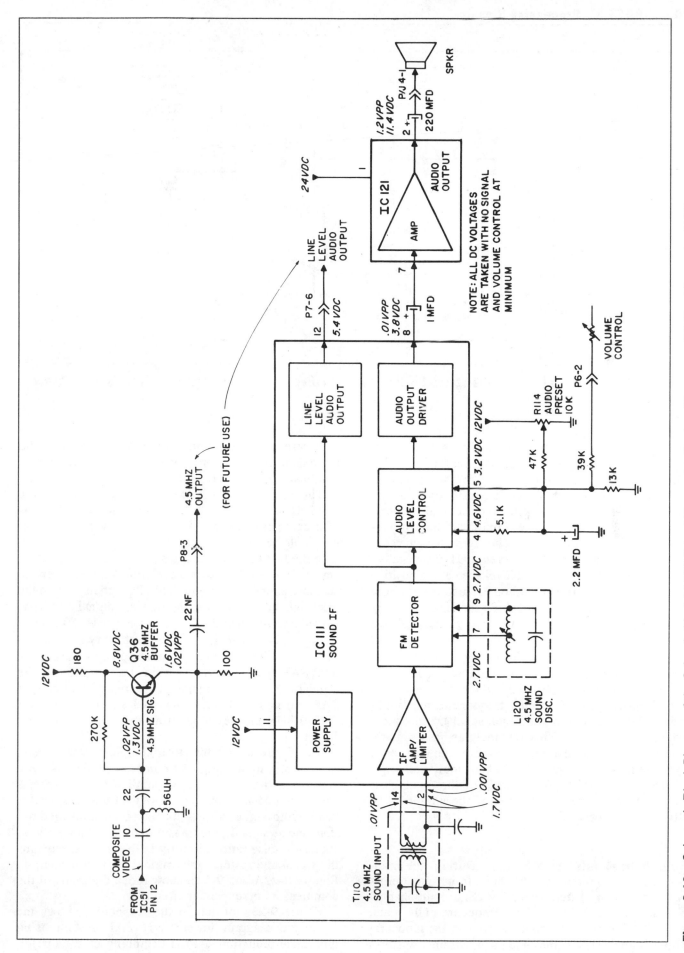

Figure 9-19 Schematic/Block Diagram of the Sylvania Model 19C5 Sound Processing System (*Courtesy of Philips Electronics N.V.*)

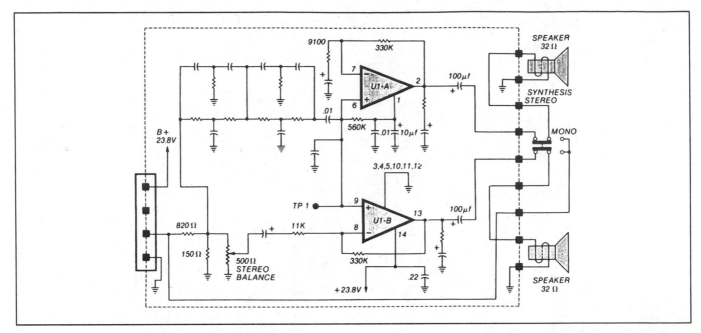

Figure 9-20   Schematic Diagram of a Stereo Synthesis System (*Reprinted with permission of Thomson Customer Electronics, Inc.*)

Feedback defines the amount of gain for each amplifier stage. Again looking at the diagram, the feedback line connects to the negative input side of each amplifier from the amplifier output. When considering op-amp U1-B, also note that a stereo balance control connects between pin 8, or the negative input, and the audio input connection for the circuit. The stereo balance control establishes a signal input level at U1-B's negative input equal to the signal input level seen at the positive input of U1-A. Rotating the stereo balance control from high to low and back to high shifts the amount of signal applied to either the channel-one or channel-two speaker.

Because of the configuration of the amplifiers, the negative and positive inputs of op-amp U1-B cancel any signals already amplified by U1-A. As a result, the second op-amp amplifies only frequencies not amplified by the first op-amp, and each speaker reproduces different frequencies. When channel one has a high response, channel two has a low response, and vice versa. Combined, both channels contain and amplify all original audio frequencies. The output of the op-amps is coupled to the synthesis portion of the circuit by an 820-ohm resistor.

Multichannel Television Sound   During the mid-1980s, several manufacturers began to use an alternative to synthesized stereo called MTS, or multichannel television sound. With MTS, the frequency of the main, or L+ R, channel is identical to that in the monaural sound signal. Like the sound-processing systems

seen earlier in this channel, the system also relies on pre-emphasis and de-emphasis for the accurate reproduction of the original sound.

The change from the standard sound signal to the MTS signal occurs with the introduction of a 15,734-kHz stereo pilot signal. MTS systems use the pilot signal to detect an L–R stereo subchannel—an amplitude-modulated, double-sideband, suppressed carrier signal. Figure 9-21 shows the relationships between the standard sound carrier and the MTS signals. While the pilot signal deviates from the standard signal by 5 kHz, the subchannel deviates from the carrier by 50 kHz.

Figure 9-21 also introduces another type of audio service, called **separate audio programming** or SAP. SAP services provide a bilingual alternative to the standard English-language audio reproduction. The SAP carrier is a 10-kHz deviated signal centered at 78.670 kHz and deviating from the sound carrier by 15 kHz.

The figure also refers to a complementary noise-reduction system using dbx compressed signals. Dbx noise reduction uses a method similar to pre-emphasis and de-emphasis. Compression of the transmitted baseband audio signal allows increased channel modulation, and as a result, the transmitted signal has a higher signal-to-noise ratio. Circuitry in the receiver expands the signal and restores the original audio information. Figures 9-22A and 9-22B show block diagrams of the compressor and expander circuits.

Figure 9-23, located in the Appendix, shows the schematic diagram for an MTS/SAP system using dbx noise reduction. At the beginning of the circuit,

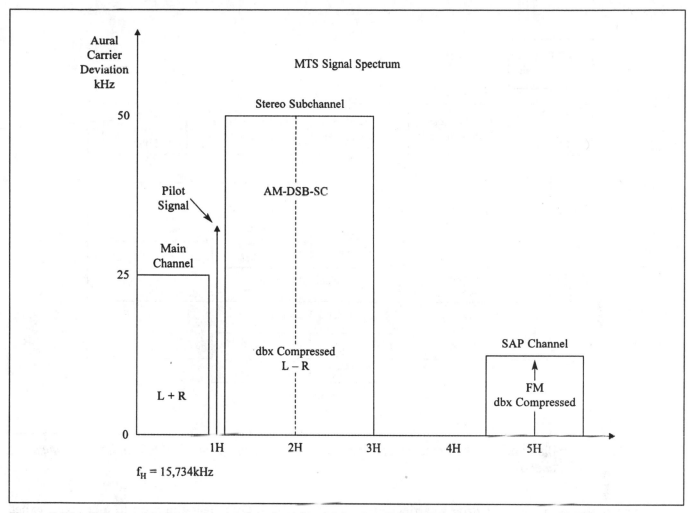

Figure 9-21   Relationship Between the Standard Sound Carrier and MTS Signals

the 45.75-MHz sound I-F signal feeds into IC1601, a combination stereo I-F amplifier/detector that reduces any inherent buzz and noise along with trapping the chroma signal. Preceding the detector, a wide bandpass filter provides the wide response needed to pass the 73-kHz FM-deviated signal. The 4.5-MHz double-tuned quadrature detector has its primary core aligned for maximum output, and its secondary coil aligned for minimum distortion. Because of this, the circuit provides low distortion while passing the composite audio signal to the stereo decoder, SAP detector, dbx expander, and audio controls.

At the input to these circuits, a simple LC resonant circuit attenuates the SAP carrier by approximately 15 to 20 db and reduces the chances for crosstalk between the MTS and SAP signals. The stereo decoder circuit consists of a phase-locked loop, a stereo switch, and a composite stereo decoder. While the PLL locks to the 15,734-kHz pilot signal, the stereo switch provides the capability to force the decoder into a monaural mode. In the stereo mode, the decoder demodulates the L+R and L−R signals.

After the L+R and L−R signals travel through two low-pass filters, each signal takes a separate path. The L+R signal travels directly to a stereo matrix. In contrast, the L−R signal passes through a combination of an audio switch and a level adjustment before passing to the dbx circuitry. The switch allows either SAP or the L−R audio signal to pass to the dbx circuit, while the level adjustment corrects any gain differences within the decoder. A dual emitter-follower switch consisting of transistors Q1617 and Q1618 controls the dbx input signal.

Two variable resistors control the dbx timing and high-band response. Within the dbx circuit, a dual op-amp amplifies the L−R signal before it feeds into the stereo matrix. Because the circuit applies dbx noise control to only the L−R signal, the decoder recovers and outputs both L+R and L−R signals. The matrix adds and subtracts the L−R signal with the amplified L+R signal and produces left and right signals.

As mentioned, the circuit also processes second audio program signals. The SAP subcarrier travels from Q5, an input buffer, into a bandpass filter. Amplification of

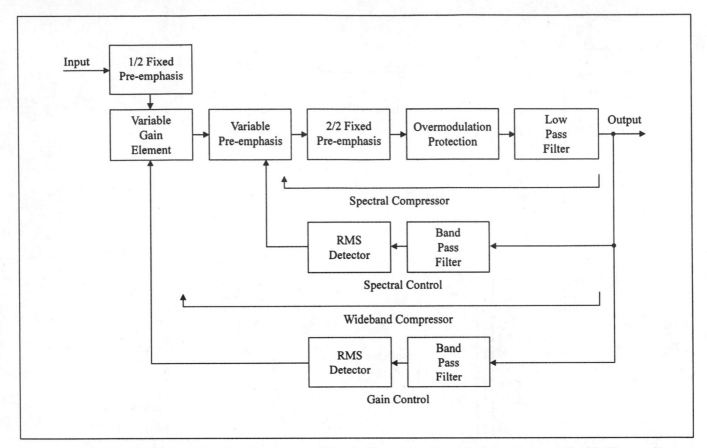

Figure 9-22A   Block Diagram of a dbx Compressor Circuit (*Courtesy of Zenith Data Systems, Zenith Electronics, and Rauland*)

the subcarrier allows the signal to drive a carrier detector and SAP decoder located within IC3. After rectifying and filtering the signal taken from the filter, the detector uses the signal to drive Q1606, the SAP mute transistor, and Q1607, the SAP presence indicator. The phase-locked loop that makes up the SAP decoder locks onto the 78.67-kHz SAP signal.

IC1603 operates as a four-channel stereo switch, allowing the selection of one of four stereo signals by applying a dc voltage to a program-select input. While IC1604 provides dc control of the bass, treble, volume, and balance settings, the IC also enhances the stereo signal for increased separation. A 2-watt stereo amplifier drives the 80 to 15-kHz stereo signals into two 8-ohm speakers.

☑ PROGRESS CHECK

You have completed Objective 5. Discuss how stereo sound circuits function.

Audio Cross-over Networks

A **cross-over network** splits the audio output signal into separate low- and high-frequency signals. It then sends the low-frequency portion to a speaker, called a woofer, specially designed for low-frequency sound reproduction. The high-frequency portion of the signal travels from the cross-over network to a high-frequency speaker, called a tweeter.

Figure 9-24 shows a schematic diagram of a cross-over network based on active filters. In the figure, IC1 and an impedance-matching network labeled as R_{IN} operate as an input buffer. The impedance-matching network matches the output impedance of the amplifier to the input impedance of the cross-over network. While IC2 and its associated circuitry form a low-pass filter, IC3 and its associated circuitry form a high-pass filter. Low-frequency signals pass through the IC2 filter network and travel to the woofer. High-frequency signals pass through the IC3 filter network and travel to the tweeter. The IC2 filter network blocks high-frequency signals, while the IC3 filter network blocks the low-frequency signals.

Audio Switching Circuits

Because many receivers utilize internal and external speakers, audio switching circuits, such as the one featured in Figure 9-25 schematic drawing, are often

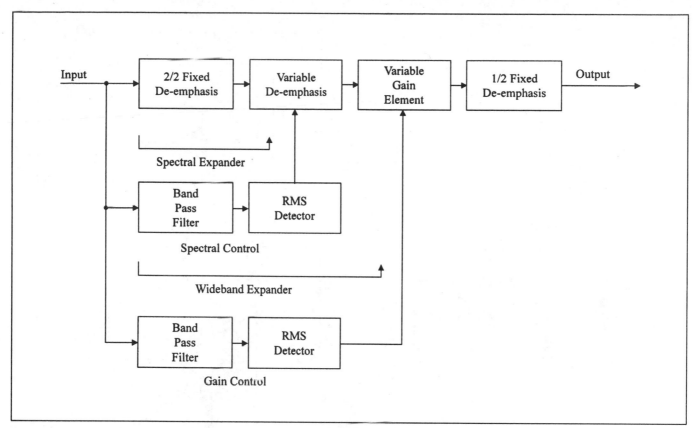

Figure 9-22B   Block Diagram of a dbx Expander Circuit (*Courtesy of Zenith Data Systems, Zenith Electronics, and Rauland*)

used. In the figure, the left- and right-channel audio signals travel into the circuit at points W8 and W12. A quad bilateral switch labeled IC4 controls the switching of the audio signals from the internal/external speaker jacks, the stereo amplifier, and the auxiliary speakers. With this particular television design, switching the TV mode control to its stereo position causes the signals to flow through bilateral switches C and D. Switching the TV mode control to the internal/external position causes the signals to exit through the left and right audio output jacks.

✔ PROGRESS CHECK

You have completed Objective 6. Describe special-purpose sound circuits.

9-5   TROUBLESHOOTING SOUND CIRCUITS

If we consider the audio stage as something that provides the power to mechanically move a speaker cone, then troubleshooting the stage may seem easier. With

audio frequencies, the stage generates power throughout a range from 20 Hz to 20 kHz. Any time that a stage provides power, it also becomes exposed to the possibility of a short or an overload. Either condition can cause symptoms such as blown fuses, open resistors, and shorted power transistors.

Like all troubleshooting efforts, working with the audio section requires a logical approach and basic knowledge about the circuit operation. For example, if the speakers, fuses, and connections appear in working order, measure the power-supply voltage. At the preamplifier stage, filters minimize hum and noise while providing a very stable supply voltage. DC voltages at the driver and power amplifier stages connect directly to capacitors that smooth any ripple.

After verifying the presence of the correct supply voltage, use common signal-injection and tracing techniques to determine where the fault occurs. With the audio stages, you can inject an audio frequency at the beginning of the stage and check for the presence of a signal at the output of the stage. Usually, the schematic will indicate a good point for injecting the signal.

When injecting a signal into an audio channel, use an audio or function generator to supply either a 3-kHz square or sine wave. In addition, connect either an 8- or

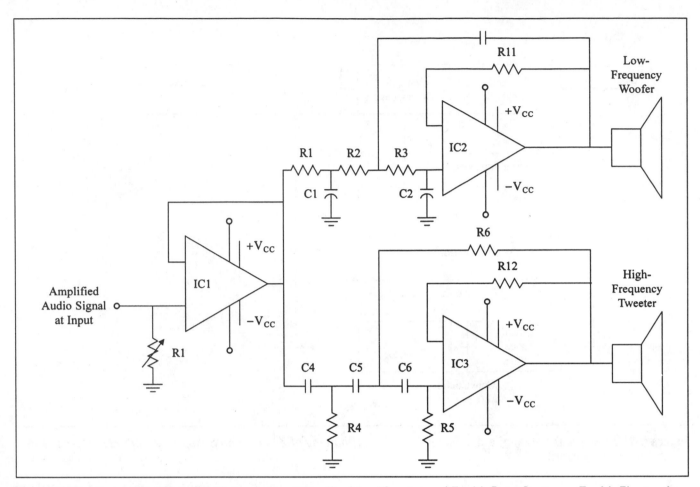

Figure 9-24   Cross-over Network Using Active Filter Devices (*Courtesy of Zenith Data Systems, Zenith Electronics, and Rauland*)

10-ohm resistor across the stereo speaker terminals as a load. The wattage rating of the resistor chosen as a load should match the power output of the amplifier. For transistor amplifiers, connect the audio or function generator to the first audio transistor. Integrated circuit audio channels will require a connection to the pre-amp portion of the circuit.

Then, clip a test lead from the generator probe to the input channel. By using an oscilloscope, you can monitor the performance of the audio channel while injecting the test signal. Set the variable sweep control of the oscilloscope for a steady waveform, and adjust the input signal so that a stable waveform appears. When injecting a sine wave into the circuit, the shape of the sine wave should not change as the signal progresses from the input to the output of the amplifier.

Depending on the shape and amplitude of the waveform at the output of each stage, the injection of a square-wave signal can tell you about the type of problem occurring within the circuit and disclose the source of the problem. For example, a square wave with a rounded leading edge indicates the loss of high-frequency response in the amplifier stage. Using a square wave as an input signal also allows the techni-

cian to test for internal oscillation, or ringing. Any time that a defective component—such as an output transformer—introduces oscillation into an amplifier circuit, a loss of high-frequency response occurs. Figure 9-26 shows several square waves and the corresponding problem as indicated by the shape of the waveform.

Using a square-wave input signal allows a technician to verify that the amplifier passes all the harmonic components of the fundamental frequency and that it has a good bandwidth. When injecting and monitoring a square wave, check the amplifier response at a range of low and high frequencies. As Figure 9-26 shows, the waveform will appear different for poor low-frequency or poor high-frequency response conditions. For example, the waveform at an amplifier output for an injected low-frequency signal may show a perfect square wave, while the waveform for an injected high-frequency signal may show some rounding at the edges.

Some rounding of the square wave is permissible at the higher frequencies. Even when the amplifier is not up to standard, it may continue to produce excellent sound because response is measured relative to a repetitive frequency. An amplifier would need an extremely wide frequency response of −10,000 to +200,000 Hz to

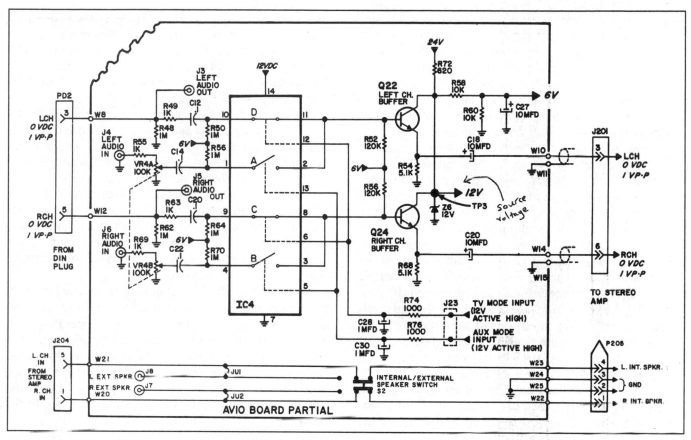

Figure 9-25   Schematic of an Audio Switching Circuit (*Courtesy of Zenith Data Systems, Zenith Electronics, and Rauland*)

produce a perfect square wave. On the other hand, a severely rounded square wave— which almost resembles a sine wave—indicates that the circuit has excessive high-frequency attenuation and that the amplifier has eliminated most of the harmonics associated with the fundamental frequency.

As with many of your checks, you can check the overall response of the circuit by monitoring the amplitude of the injected signal from the output of the circuit back to the first input stage. For example, a circuit that is known to be bad may have a strong output signal with

the injection occurring at the volume control. Moving the injection point further away from the output, however, may disclose that the next stage has a defect. When injecting the signal, remember to adjust the signal-generator input signal to a low level. An overly strong input signal can overdrive the amplifier undergoing the test and distort the waveform found at the output of the amplifier.

Even if no schematic is available, you know that a preamplifier, or low-noise amplifier, increases the signal level between 0.7 to 1 Vdc. You also know that a

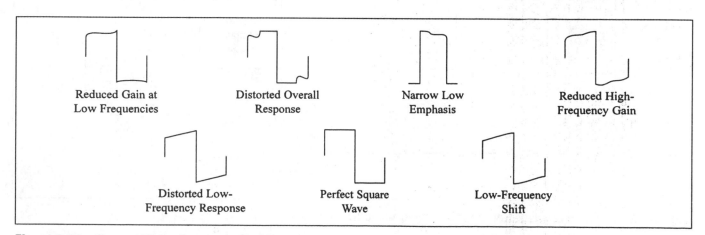

Figure 9-26   Square-Wave Response Problems

driver stage provides enough power to drive the next stage and that the power amplifier drives a speaker. Yet as you move a signal generator toward the speaker terminals, the amplitude of the output signal will decrease, because the number of amplifier stages between the injection point and the speaker has decreased. This knowledge tells you what to expect when you apply test equipment to the troubleshooting process.

Figure 9-27 shows a typical test setup needed for checking the performance of an audio amplifier. Moving from left to right, an audio frequency generator with a frequency range of 10 Hz to 20 kHz works as a signal source, supplying an adjustable output voltage ranging between 0.5 millivolts and 2 Vdc. While the oscilloscope provides a method for monitoring the output signal, you can measure the output voltage at the load resistor with a multimeter. In addition, you can use the oscilloscope to check for noise in the audio channel. Checking the voltages at the preamplifier with an oscilloscope can disclose the source of the noise. In many instances, active components in the signal path can develop characteristics that generate noise. In others, an overloaded amplifier stage can produce distortion.

Whenever an output transistor or integrated circuit becomes leaky, the problem condition will introduce distortion into the signal. Injecting a signal into the audio channel stage-by-stage should allow you to narrow the troubleshooting process. At this point, you can use your knowledge about transistors or integrated circuits to find the faulty stage. Distortion at the output but not the input of either a transistor or integrated circuit audio channel points toward that stage.

If the amplifier stages utilize capacitor coupling, always check the output capacitors. Often a dried electrolytic capacitor in the output stage will either badly distort the signal or cut the amount of power delivered to the speaker. Under normal circumstances, an increase in frequency will cause the capacitive reactance of a capacitor to decrease and the circuit current to increase. When an electrolytic capacitor dries, the internal resistance of the capacitor increases, dissipates additional power, and cuts the ability of the circuit to reproduce high frequencies.

One quick method for checking for a dried electrolytic capacitor involves measuring the voltage drop across the capacitor. In a normally operating circuit, the voltage drop decreases to a negligible value as the frequency increases. A circuit operating with a defective coupling capacitor will have a larger voltage drop. Because of the increased power dissipation within the capacitor, the part may also feel hot to the touch.

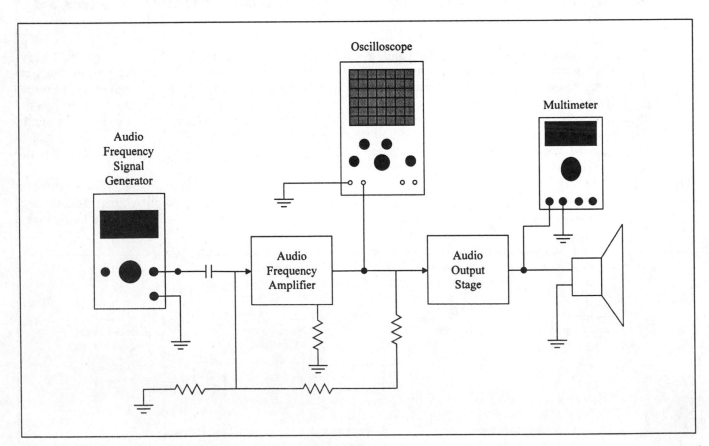

Figure 9-27 Test Equipment Setup for Checking Audio Stages

FUNDAMENTAL REVIEW

Sine and Square Waves

In the discussion of troubleshooting audio channels, you found that you could inject sine- and square-wave signals into an amplifier and then monitor the signal at the amplifier output. A sine wave represents the fundamental frequency and the harmonic components. A square wave represents the fundamental frequency and a large number of odd harmonics of that frequency added in specific phase and amplitude relationships. If an amplifier passes a square-wave signal with little or no distortion, the amplifier has the required bandwidth to pass all the harmonics of the fundamental frequency.

Test Equipment Primer: Radio-Frequency and Audio-Frequency Signal Generators

While a radio-frequency generator produces high frequencies that range from 500 kHz to 30 MHz, audio-frequency signal generators supply a continuous frequency ranging from 0 Hz to 50 kHz. When purchasing a radio-frequency generator, verify that the package includes:

- Output attenuators
- Dial-calibrated accuracy
- A dc-blocking capacitor

The output attenuators prevent the overloading of the R-F/I-F stages, while dial-calibrated accuracy ensures the precise alignment of tuned circuits. A dc-blocking capacitor prevents damage to the test circuit. An audio-frequency generator should have a high output signal ranging 10 volts peak-to-peak, and should provide rectangular, sine, and sawtooth waveforms.

Troubleshooting Transistor Amplifier Circuits

Class B and AB amplifiers pose some interesting troubleshooting problems because of the use of two transistors in the output and biasing stages. Either of the two transistors can develop a shorted or open junction or have intermittent operating characteristics. The most common fault associated with amplifier circuits is "no output signal." This type of problem often leads technicians on a fruitless search for a defective output transistor.

Simple time-tested checks such as signal injection and tracing will usually point toward the problem. More than likely, though, you will need to apply your knowledge of amplifier circuits to your troubleshooting efforts. Class AB amplifiers draw very little idle

current; as a signal is applied, the amount of current increases. When the amplifier operates in Class A mode, the circuit draws a small amount of quiescent current to minimize distortion at low output power. When operating in Class B mode and at a higher output power, the amount of current drawn from the power supply depends on the amount of output power. In addition, each active component of a Class AB amplifier amplifies only a half-wave of the audio signal. The half-waves of the output signal combine at the output stage of the amplifier. Both of these factors affect what you might see if you monitor the audio stages with an oscilloscope or multimeter.

You should also remember that audio output stages usually do not rely on an output transformer such as the one shown in Figure 9-18. Some transformerless power stages will have positive and negative supply voltages, while others will rely on a "center voltage" for both power transistors. The center voltage at each transistor always equals half of the supply voltage. In all situations—because a dc voltage will harm a speaker cone, you should see zero volts (dc) at the output. From a troubleshooting perspective, you can begin to look for a very high dc voltage at the output as one fault indicator, or a very low or negative voltage at the positive supply as another.

Before focusing on only the output stages, check the sound channel for the correct input signal from the previous stages, the correct supply voltages, and good ground connections. After confirming the presence of the signal, voltage, and ground connections, disconnect the load from the circuit and recheck the voltage and signal characteristics. At times, the load can prevent an amplifier from working by adding additional loading. In this case, you would check for a shorted load.

If all these circuit checks point to the amplifier, disconnect the ac signal source—or the output of the previous stage—from the output stage. To do this, either desolder a connecting wire or remove the coupling component that connects the two stages. Either method will isolate the output stage. Then check the voltages at the base, emitter, and collector terminals of each transistor. Also, check for an even distribution of voltages at the voltage dividers. Figure 9-28 illustrates the voltage check-points.

A leaky transistor will have lower than normal voltages at both the collector and base. Along with those checks, verify that the transistor has the correct forward bias between the base and emitter terminals. A leaky transistor will have an improper forward bias. Many times, though, the base or emitter bias resistors can change value and cause the amplifier to appear leaky.

An overload or shorted output will cause a power transistor to short. To find which power transistor has

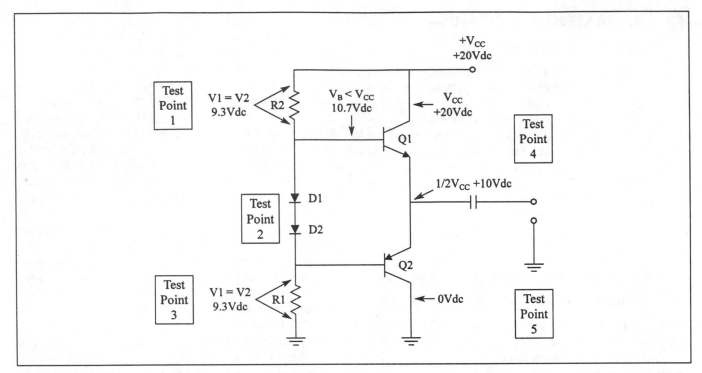

Figure 9-28   Class AB Amplifier Voltage Checkpoints

shorted, check the resistance between the collector and emitter of each transistor. Furthermore, check the corresponding driver transistor. Often a breakdown within the emitter-base or collector-base junctions of the power transistor will cause both the output transistor and the driver transistor to overload. Generally, the driver transistor will have an emitter-base short. As a final rule-of-thumb, if the symptoms direct your efforts toward one amplifier transistor, replace both transistors. More than likely, a defect in one transistor has damaged the other.

Each of the voltage checks tells you about the current operating conditions of the circuit and allows you to pinpoint the problem area. Table 9-1 lists some of the common symptoms and faults found with transistor AB amplifiers. The component numbers listed in the chart correspond with the numbers listed in Figure 9-28. Normal voltage readings for the chart are taken from the schematic drawing for the particular circuit.

Troubleshooting Integrated Circuit Amplifiers

Because of the level of integration seen in IC-based sound circuits, the search for a problem solution is limited to only a few points. The application of basic signal-tracing skills and voltage checks will often disclose the source of the problem. As Figure 9-29 shows, you can apply your knowledge of feedback and gain while checking the circuit from output to input. In the

Table 9-1   Component Level Troubleshooting for a Class AB Amplifier

Symptoms	Check Component
Voltages found at Test Points 1 and 2 will be closer to the source voltage. The voltage at Test Point 3 will be higher than normal.	R2 has opened.
Voltages found at Test Points 1 and 2 are close to zero. The voltage at Test Point 3 is lower than normal.	R1 has opened.
Voltages found at Test Points 4 and 5 are normal. The voltage at Test Point 3 is very low.	Q1 is open.
Voltages found at Test Points 4 and 5 are normal. The voltage at Test Point 3 is very high.	Q2 is open.

figure, R1 is in a negative feedback path. The combination of R1 and R2 determines the gain of the power amplifier, while resistor R3 stabilizes the output voltage.

First, with a signal applied to the receiver, check for an audio signal at the output of the audio amplifier. Next, check for the correct voltage at the source voltage

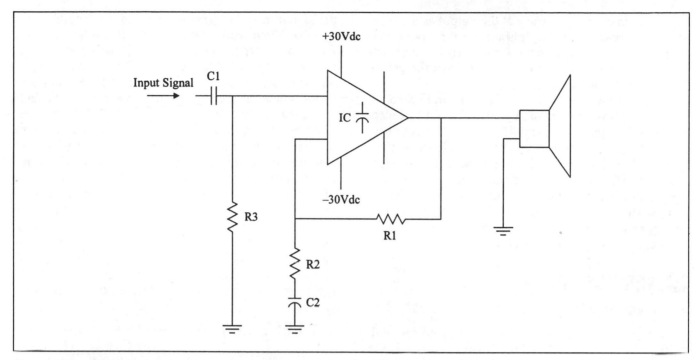

Figure 9-29   Schematic of an IC-based Sound Circuit (*Courtesy of Zenith Data Systems, Zenith Electronics, and Rauland*)

connection of the IC. If those two checks provide no hint of a solution, move your efforts to the IC terminals that connect to the volume control.

Checking Stereo Amplifiers

One of the nice things about troubleshooting stereo amplifiers is that every stage has two identical signal channels. If one channel fails, the other channel works as a reference for verifying the presence of dc voltages and audio-frequency signals. For this reason, you can use a dual-channel oscilloscope to compare the performance of the audio channels with each other. So that you can maintain accurate measurements, adjust the amplifier balance control for even balance between the two channels, and adjust the oscilloscope gain controls for each channel to the same levels. In addition, adjust the variable sweep control of the oscilloscope so that you can see two stable waveforms on the oscilloscope display.

At times, a change in the frequency response of one channel will cause that channel to produce distorted, low-quality sound. Using an oscilloscope, compare the frequency response of both channels with the curve shown in Figure 9-30. A good amplifier will have a flat frequency response for frequencies ranging from 20 Hz to 20 kHz. As with the monaural sound channel, you can also use the oscilloscope and a signal generator to monitor the progress of an injected signal through an audio channel.

When the frequency response of one channel decreases, inject a 400-Hz square wave into the input of the audio stage and use an oscilloscope to check the shape of the signal at the stage output. Checking the frequency response then involves setting the generator to reference value 1000-Hz and the output signal to a value below the maximum output power. At this point, the volume control should be set to the minimum level.

Switching the signal-generator frequency from 20 Hz to 20 kHz should not cause the output voltage to change by more than 1dB. As mentioned, the frequency response should remain flat. The amplifier should begin to symmetrically clip the positive and

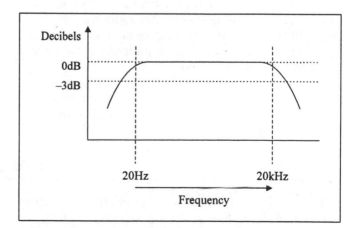

Figure 9-30   Frequency Response Curve for Stereo Sound Channels (*Courtesy of Zenith Data Systems, Zenith Electronics, and Rauland*)

negative peaks of a sine wave at the output as the input voltage increases. The key phrase here is "symmetrically clip." If the oscilloscope shows less than symmetrical clipping, one amplifier does not have the proper frequency response.

After checking the response of the distorted channel, begin to narrow your search by checking for bad electrolytic capacitors in the amplifier line. As mentioned, a bad coupling capacitor will harm the frequency response of an amplifier stage. When using signal injection in an effort to find the bad component, bypass both the preamplifier and the tone control, and consider the point where the signal begins to clip. An amplifier that has poor frequency response will have an output signal that begins to clip at a low output power setting.

➲ SERVICE CALL

Weak or Distorted Sound

One of the most commonly reported problems with audio circuits involves weak or distorted sound. We can use our troubleshooting skills and our knowledge about audio circuits to narrow the search for the problem to a small area. In this case, we will use signal injection and voltage tests to find why the monaural sound circuit of a late-model television produces a somewhat muffled output that the customer describes as "underwater music."

Because this description verifies that both the desired audio and the background noise are weak, we know that the signal path for the audio signals is not completely open. Therefore, we can limit our search to components that may develop a *leaky* condition. An open path—sometimes caused by an open coupling capacitor—in this area would weaken the desired signal but not the background noise.

We start our troubleshooting by injecting a 1-volt audio signal at the input to the audio stages. Given the layout of the circuit board, we can use the volume control, rather than the output of the audio detector, as a convenient test point. Injecting an audio signal at this point and with the control set to midpoint should cause the speaker to produce a loud tone.

With a loud tone produced through injection at this point, we can then move our test back to the sound I-F stage and inject a 4.5-MHz FM signal. We find that

the signal injection produces only a weak tone at the speaker. Even without the benefit of a schematic, we know that a typical sound I-F stage will produce approximately 4 to 5 Vdc at its output. Usually, a transistor in this section will provide a gain of 10. Finding less than 1 volt at the sound I-F output, we know that a fault exists within this circuit.

To narrow the search even further, we can measure the resistance of each transistor base to ground. Lower than normal resistance readings indicate a leaky component in the circuit. Because electrolytic capacitors often become leaky, we can concentrate on a electrolytic capacitor in the emitter circuit of the sound I-F transistor that has become leaky. However, replacing the capacitor does not completely clear the problem. Another check of the transistor shows that the semiconductor has also developed a leaky condition. With that replacement, normal audio is restored to the receiver.

➲ SERVICE CALL

No Sound

Another common audio circuit failure involves a "no sound" condition. Again, we can use signal injection and a few voltage checks to find the source of the problem. For this problem, injecting a 4.5-MHz signal at the sound I-F amplifier and an audio signal at the volume control conveniently divides the circuit into halves. With the circuit passing the 4.5-MHz signal but not the audio signal, we know that the problem lies within the audio amplifier stages.

To narrow the search even further, we can inject a low-level audio signal at the base of each transistor while listening for a tone at the speaker. After using this technique to eliminate most components, we can measure the voltages at the stage most in question. Because checks of the transistor voltages produced normal readings, we can proceed to components in the amplifier circuit. In this instance, an open emitter-bypass capacitor all but eliminated the amplifier stage gain.

☑ PROGRESS CHECK

You have completed Objective 7. Troubleshoot different types of sound-circuit problems.

SUMMARY

In the overview of sound-processing circuits, chapter nine took a close look at circuits and components that affect demodulation, amplification, and switching. The discussion of pre-emphasis and de-emphasis revealed

that the time constant given by a simple RC network can improve the signal-to-noise ratio of audio frequencies and provide the needed balance between low and high frequencies. The section on sound-demodulation circuits such as the quadrature detector, differential peak detector, synchronous detector, and PLL detector examined the use of phase relationships to control signals. As you will see in the next chapters, the concept of using phase to control signals emerges repeatedly.

The overview of audio-frequency amplifiers added to your knowledge of Class B and AB amplifier circuits. This provided an opportunity to discover how transistor and operational amplifiers behave under certain signal conditions. The discussions of cross-over distortion and thermal runaway added to your knowledge of amplifier and biasing circuits. Chapter nine also showed how each fundamental sound circuit fit into a complete monaural or stereo sound system. The discussions of separate audio programming circuits, dbx circuits, cross-over circuits, and switching circuits demonstrated that a modern sound system consists of more than just an amplifier.

The troubleshooting section presented techniques for signal injection and the interpretation of test waveforms. You also learned about the effect of poor amplifier response on the reproduction of low and high frequencies. At the component level, you found that the aging of electrolytic capacitors can dramatically affect the quality of a generated signal, and that a leaky capacitor or transistor can cause as many problems as a shorted or open device.

REVIEW QUESTIONS

1. **True or False**   The sound I-F carrier is extracted before the video detector.

2. **True or False**   The sound signal has ten times the amplitude of the video signal.

3. Define pre-emphasis. Why is there a need for pre-emphasis?

4. After the sound I-F carrier is detected, a signal of which of the following amplitudes is passed to the audio amplifier:
   a. 41.25 MHz
   b. 45.75 MHz
   c. 4.5 MHz
   d. 20 kHz

5. What characteristics allow a de-emphasis circuit to function?

6. De-emphasis restores a balance between the _____ contained in any sound transmission. As a combination, pre-emphasis and de-emphasis improve the _____ of the reproduced sound.

7. The sound I-F section has a _____ response, while the video I-F section has a _____ response. Which section is more sensitive to a change in local oscillator frequency?

8. The following figure shows a pre-emphasis circuit. Find the value of the inductance.

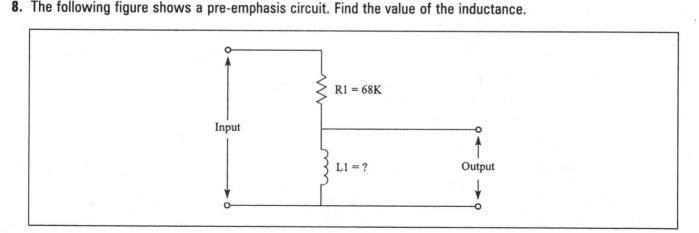

9. Refer to Figure 9-4. **True or False**   The detector shown in this figure demodulates the sound I-F carrier.

10. Why are amplification and limiting necessary after the extraction of the sound I-F carrier?

11. Why does the sound I-F circuit produce a 4.5-MHz carrier? What are the benefits to using a 4.5-MHz carrier?

12. At the audio demodulation point, variations in the _____ correspond with the frequency variations of the _____ found at the detector input.

13. A Foster-Seeley discriminator relies on _____.

14. True or False   A quadrature detector is a NAND gate. This type of logic circuit only produces an output signal if opposite signals are found at both inputs.

15. How many tuned circuits does a quadrature detector have?

16. The following diagram depicts a balanced ratio detector. Diagram the direction of current flow throughout the circuit and describe how the circuit demodulates the sound I-F signal.

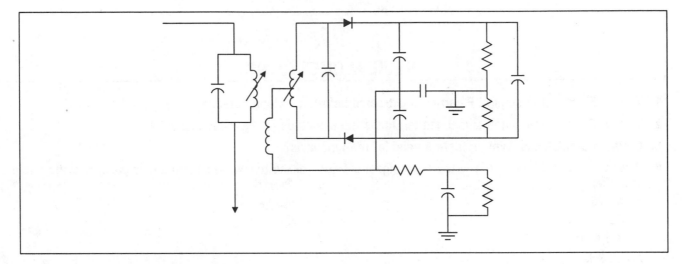

17. The audio output voltage from the differential peak detector results from the comparison of

_____.

18. In the synchronous detector shown in Figure 9-11, the output signal produced by four switching transistors serves as:
    a. the 4.5-MHz sound I-F signal
    b. the demodulated audio output signal
    c. a reference signal
    d. one input for a phase comparator

19. The PLL circuit shown in Figure 9-12 compares the _____ with the output from a _____. The _____ from the phase comparator serves as the _____ signal.

20. Name four characteristics of audio amplifiers.

21. What does the "figure of merit" define?

22. What is a complementary-symmetry amplifier?

23. Cross-over distortion occurs when _____. Show how a waveform with cross-over distortion would appear.

24. What is a common method for preventing cross-over distortion?

25. What is thermal runaway?

26. True or False   A Darlington complementary-symmetry amplifier is used for applications that require high load power.

27. A Darlington pair is defined as _____.

28. In Figure 9-16C, the output of the Class AB amplifier centers around zero volts rather than the division of the supply voltage because _____.

29. True or False   The op-amp provides a combination of low input impedance, high output impedance, and low noise distortion.

30. Why is the output impedance important in the op-amp audio amplifier circuit?

31. In Figure 9-17, what factor(s) limit(s) the amplifier to specific operating range?

32. Monaural sound describes an audio system that has (a. two) (b. one) (c. no) audio channels while stereo sound describes an audio system that has (a. two) (b. one) (c. no) audio channels.

33. What functions are enclosed in the single-IC monaural audio system shown in Figure 9-18?

34. True or False   Monaural sound systems use the L+R audio signal.

35. True or False   Multichannel sound systems rely on the broadcast of a composite audio signal and the splitting of that signal into left and right sound components.

36. New video receivers that have stereo audio systems rely on:
    a. synthesized stereo sound
    b. multichannel sound systems
    c. external amplifiers for all audio reproduction

37. Describe the production of multichannel television sound.

38. What is the frequency of the pilot signal used for multichannel television sound?

39. What is the difference between a SAP decoder and a stereo decoder?

40. An audio cross-over network splits the amplified signal into two parts. The _____ portion travels to a tweeter through a _____ filter. The _____ portion travels to a woofer through a _____ filter.

41. Why are audio switching circuits included in some video receivers?

42. When beginning to check for problems in the audio section, your first step should be:
    a. use a THD meter to check to see if the equipment is operating within specifications
    b. check connections, fuses, and speakers for any visible problems and then check circuit voltages
    c. inject a signal at the input of the circuit with a signal generator and monitor the output with an oscilloscope
    d. replace the coupling capacitors and audio output amplifier with "known good" parts

43. A signal generator used for troubleshooting audio circuits should be able to generate _____ and _____ waveforms.

44. True or False   If an electrolytic capacitor dries internally, the resistance of the capacitor decreases.

45. True or False   A leaky transistor will have lower than normal voltages at both the collector and base.

46. Some transformerless power stages will have _____ and _____ supply voltages, while others will rely on a _____ for both power transistors.

47. Why should you see zero volts (dc) at the output of an audio circuit?

48. An audio amplifier operating under normal conditions will have a flat frequency response from:
    a. 10 Hz to 100 KHz
    b. 20 Hz to 20 MHz
    c. 10 Hz to 20 KHz
    d. 20 Hz to 20 KHz

49. Match the problem with the displayed square wave.

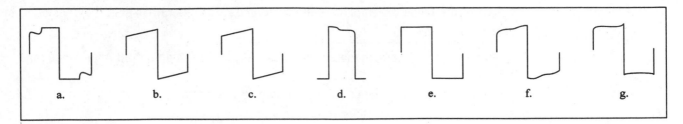

a.   b.   c.   d.   e.   f.   g.

    a. Reduced gain at low frequencies

    b. Narrow low-frequency emphasis

    c. Low-frequency shift

    d. Reduced gain at high frequencies

    e. Poor response at all frequencies

    f. Poor response at low frequencies

    g. Normal square wave

50. Draw a normal sine wave and a clipped sine wave.

51. Distortion in an audio channel can be caused by _____.

52. What size of load resistor would you choose when troubleshooting an audio amplifier circuit?

53. True or False   When injecting a signal and varying the frequency from its minimum level to its maximum level, the output voltage should change dramatically.

54. Class B and AB amplifiers use two transistors in the biasing and amplifier stages. If you find one bad transistor in the amplifier stage, would you replace only that transistor or both amplifier transistors? Explain your answer.

55. Name two routine methods used in the Service Call examples.

56. Why did we suspect a leaky component in the first Service Call example?

57. Select the best answer. The solutions for both Service Call examples tell us that many circuit problems result from defective:

    a. resistors

    b. coils

    c. capacitors

    d. transistors

    e. integrated circuits

CHAPTER 10

Luminance Signals and Video Amplifiers

OBJECTIVES Upon completion of this chapter, you should be able to:

1. Understand how the different components of the composite video signal affect the reproduction of a televised image.
2. Discuss how video amplifier components and circuits react to high, medium, and low frequencies.
3. Explain how video amplifier circuits increase the gain of the luminance signal.
4. Discuss how customer-control and automatic circuits control the amplitude of the luminance signal and the bias of the CRT.
5. Troubleshoot video amplifier and sync-separator circuits.

KEY TERMS

automatic brightness-control circuit	CRT cut-off
automatic sharpness circuit	dc restoration
black level	drive
blanking pulse	luminance signal
brightness control	Miller effect
brightness-limiter circuit	scan-velocity modulation
brightness reference level	sharpness control
clamping circuit	video amplifier
contrast	zero bias
contrast control	

INTRODUCTION

Chapter ten begins a two-chapter series on color television circuits that process the composite video signal after the signal leaves the video detector. Technological innovation has reduced the amount of components and circuitry needed to process the picture signals into one or two packages. We will begin our discussions, however, by considering circuits that use discrete components to handle these chores. This will allow us to firmly define how the circuits operate and how the signals change. From there, we will progress to the integrated circuit packages.

Chapter ten demonstrates how specific circuits—including video amplifiers, customer control circuits, sync-separator circuits, noise-canceling circuits, and output circuits—process the luminance and sync-signal portions of the composite video signal. The chapter uses equivalent circuits to explain how video amplifier components and circuits handle a broad range of frequencies. Given this knowledge, you will find that manufacturers specify certain types of components for specific frequency applications. At the conclusion of the chapter, we will build on those discussions to explore troubleshooting techniques, list potential circuit problems, and study several case histories.

10-1   LUMINANCE SIGNAL BASICS

Figure 10-1 illustrates the different types of signals associated with the complete signal. While the sync pulses have a consistent amplitude and spacing, the changing amplitude and spacing of the luminance and chrominance signals represent the changes occurring in a transmitted picture.

When we look at a picture reproduced on a television screen, we see an orderly arrangement of light and dark areas. As the electron beam deflects across the inside face of the CRT, the beam action produces a tiny spot of light where it strikes the surface. If the beam stayed at one point, it would burn a permanent spot into the CRT. Instead the beam moves from side to side and from top to bottom as it traces successive, closely spaced, horizontal lines. The pattern created by these lines is the raster.

The waveform shown in Figure 10-1 corresponds to the period when the electron beam traces the last four lines at the bottom of the picture, quickly returns to the top of the screen, and then traces the first lines of the next field. When we break the waveform down into its component parts, the eight lines of picture information are represented at each side of the diagram. At the middle of the waveform is the video waveform, when the electron beam returns from the bottom to the top of the picture.

Video Signals   For monochrome receivers, the luminance signal is the video signal. **Luminance signals** represent the amount of light intensity given by a televised object. With no need to produce color, the monochrome receiver uses amplitude differences of the video signal to produce different shades of gray. Because each color shade has a different amount of light intensity, a multicolored object reproduced through a monochrome transmission will appear to have different shades. The luminance signal and sync signals separate at the first video amplifier in monochrome receivers.

In a color television receiver, the different components of the composite video signal contain the

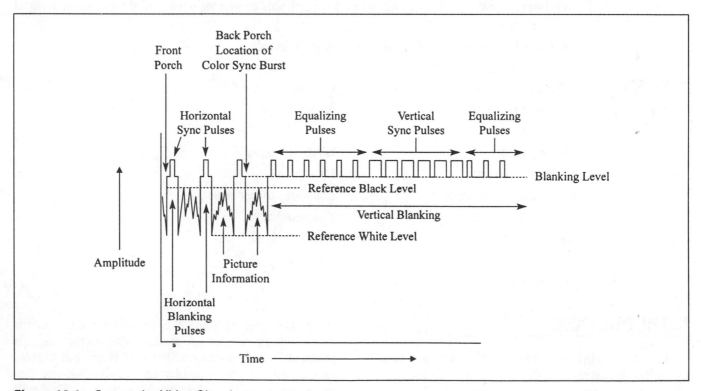

Figure 10-1   Composite Video Signal

information needed to reproduce both a monochrome and a color picture. After the composite video signal leaves the detector, the signal splits into the luminance, chrominance, and sync components. The changing amplitude and spacing of the luminance and chrominance signals represent the changes occurring in the transmitted picture.

Luminance signals consist of proportional units of the red, green, and blue voltages, and contain the brightness information for the picture. In addition, luminance signals cover the full video frequency bandwidth of over 4 MHz and provide the maximum horizontal detail. After the weak luminance signal is amplified by the video amplifier stage, the signal travels a relatively short distance to the CRT. Because of the difference in the paths that the luminance and chrominance signals travel, the luminance circuit always contains a delay that allows the signals to arrive simultaneously.

At the video output section, the luminance and signals combine and proportionately reproduce the complete picture information. When each part of the luminance signal amplitude combines with the chrominance signal, the chrominance signal variations shift to the axis of the luminance signal. To accomplish this,

the luminance information is inserted as the average level of the chrominance signal variations. Depending on the receiver, the injection of the luminance signal into the chroma circuits may occur at the three color demodulators, the video output section, or the CRT. The receiver depicted in Figure 10-2 has the luminance signal injected at the video output stage.

↻ FUNDAMENTAL REVIEW

Measuring Pulses

Any waveform consisting of high and low dc voltages has a pulse width and a space width. The pulse width is a measure of the time spent in the high dc voltage state, while the space width is the measure of the time spent in the low dc voltage state. Adding the pulse and space widths gives the cycle time, or duration of the waveform. While the pulse width and space width for a square wave are equal this is not the case for rectangular pulses, such as the equalizing pulses, horizontal sync pulses, and vertical sync pulses seen in the composite video waveform. The values for the pulse width and space width are always taken at the halfway points of the waveform.

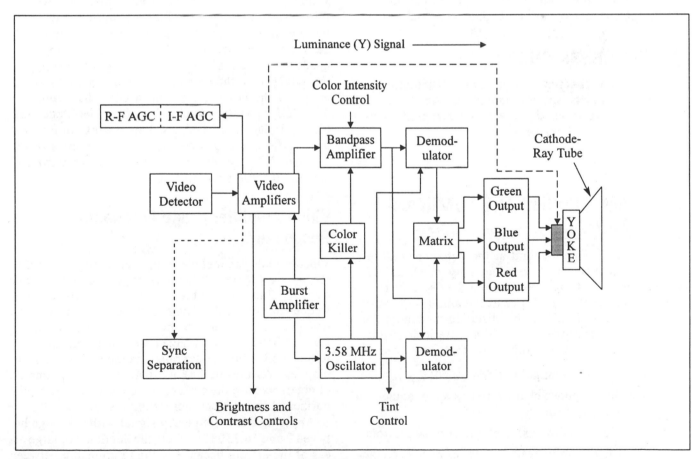

Figure 10-2 Block Diagram of Video, Chroma, and Sync Paths

Processing the Luminance Signal

For a moment, think back to the conclusion of chapter eight and then to the opening chapters of this text. At the end of chapter eight, we found that a detector circuit demodulates the video I-F signal and produces a composite video signal that contains picture, luminance, chrominance, and sync signal information. All the information found within the composite video signal fits within a bandwidth of 4 MHz. This wide band of frequencies is necessary because of the different elements that make up a televised scene.

In any video picture, motion causes a succession of continuously changing voltages to occur. Also, every scene contains different amounts of light and shade that are distributed unevenly across the picture. Each light and shade element corresponds to horizontal lines and vertical fields that, in turn, match with amplitude variations. Because the amplitude variations correspond to the horizontal and vertical scan rates, a horizontal scan contains rapid amplitude changes, while vertical scans contain lower frequency variations.

Large areas of a constant white, gray, or black produce signal variations that occur at low frequencies. With no rapid changes in intensity or shade, the amplitude has fewer variations. If we take the same area and break it down into small areas of light and shade, the amplitude changes occur at a higher frequency.

 PROGRESS CHECK

> **You have completed Objective 1. Understand how the different components of the composite video signal affect the reproduction of a televised image.**

10-2 AMPLIFYING THE LUMINANCE SIGNAL

The luminance signal found at the detector does not have enough amplitude to adequately drive a CRT. With an amplifier circuit built around a CRT, the term **drive** describes the capability of the amplifier to vary the light output of the tube by changing the bias of the tube over a wide range of conditions. Because of this, televisions rely on video amplifiers that:

- provide sufficient gain to drive a CRT
- have the response characteristics to cover the required bandwidth
- have tie points to customer and service controls
- either maintain or restore a dc reference voltage related to the original transmission

- ensure that the proper phase video signal reaches the CRT
- accept horizontal and vertical blanking waveforms that cut off the CRT during blanking intervals.

To accomplish these functions, the **video amplifier** circuitry includes subcircuits that increase gain, attenuate specific signals, shape other signals, and clamp voltages to a specified level. In addition, the video amplifier circuits interface with other sections of a television including the video detector, AGC, sync separation, and video output sections. Most importantly, though, the video amplifier increases the magnitude of the detector output voltage without changing any frequencies.

The type and impedance of the video detector circuit determine the number of video amplifier stages used in a receiver. For example, a diode detector that has a relatively low output amplitude of 1 to 2 Vdc and a load impedance of 3 to 5 kilohms will need a two-stage video amplifier. In contrast, a video detector built within an integrated circuit has a higher output amplitude of 2 to 5 Vdc and a low output impedance that matches with only one video amplifier stage.

The drive requirements of the CRT used in the television govern the output requirements of the video amplifier stage or stages. CRT drive requirements cover the transition from **CRT cut-off**, or a black screen, to **zero bias**, or a white screen, and the voltage swing needed to prevent any of the video stages from clipping the sync tips of the composite video signal. As an example of the last requirement, a CRT that requires a 40-Vdc drive voltage must pair with a video stage capable of producing a 64-Vdc voltage swing. In terms of peak-to-peak voltages, and depending on the size and type of CRT used, the signal swing may vary from 30 Vdc to 200 Vdc peak-to-peak.

Video Amplifier Stage Frequency Response

Video signals can include frequencies ranging from 20 Hz to over 4 MHz. Within that range, high frequencies carry the fine detail of the reproduced picture, and low frequencies cover the detail of large areas. Applying a high-frequency signal to the CRT cathode in a range from 15,750 Hz to 4.2 MHz causes the beam current to turn on and off at a faster rate than the horizontal scanning rate. As a result the CRT displays a higher amount of detail because the current switches on and off as the horizontal line deflects across the raster.

Applying a low-frequency signal within a range between 0 and 15,750 Hz affects the detail in the large areas of the picture because of the relationship between time duration and the scanned horizontal line. At a low frequency, the off and on periods of the beam current will exceed the time needed to deflect one horizontal line across the raster. Consequently, one or more of the scan lines may be black for the turn-off period or white for the turn-on period.

Slow changes in background brightness occur along with changes in fine picture details. Because of this, a video amplifier stage must provide uniform or near-uniform amplification across the frequency band. A typical transistor video amplifier utilizes feedback and peaking to achieve the necessary high-frequency bandwidth and transient response. Acceptable frequency response is also obtained through resistance and direct coupling.

⟳ FUNDAMENTAL REVIEW

Frequency

The term "frequency" describes the number of electromagnetic waves that pass a point each second, and the rate of polarity change seen with the waves.

Low Frequencies and Video Amplifier Stage Response

Without the proper low-frequency response, large areas of the reproduced picture will appear either smeared or washed out. To protect against the attenuation of frequencies ranging as low as 60 Hz, video amplifier circuits include components that will not affect the low-frequency response. Considering Figure 10-3, the coupling capacitor and the load resistor couple the video signal from the last video amplifier to the CRT. To prevent attenuation, the two components must have

values that translate into an RC time constant longer than 0.1 second. For the components in the figure, the RC time constant is:

$$T = C_C R_G = .1 \times 10^{-6} \times .33 \times 10^6 = 0.33 \text{ second}$$

Medium Frequencies and Video Amplifier Stage Response

We can use equivalent circuits to observe how amplifier circuits behave at different frequencies. In contrast to a schematic drawing of an entire circuit, an equivalent circuit shows only the components that either transfer or amplify signals. With Figures 10-4A–E we can compare the schematic diagram for a resistance-capacitance coupled transistor amplifier with equivalent circuits and discover how the circuit functions with the application of medium-level frequencies.

In Figure 10-4A, R_L represents the Q1 collector load while R3 is the Q2 collector load. Capacitor C3 couples the sections of the amplifier together. While the voltage divider consisting of R_A and R_B forward-biases Q2, resistors R1 and R2 prevent thermal runaway from occurring in the circuit. C1 and C2 bypass each emitter resistor and prevent degeneration.

Figure 10-4B uses an equivalent circuit to show how the components in the amplifier circuit respond to medium frequencies. At all but the very low frequencies, C1, C2, and C3 offer no impedance and effectively act as a short-circuit. In addition, the power supply filter capacitors place the voltage supply terminal, or $+V_{CC}$, at ac signal ground.

In the Figure, R_{BE} represents the operation of Q2 and all components associated with the transistor. With the

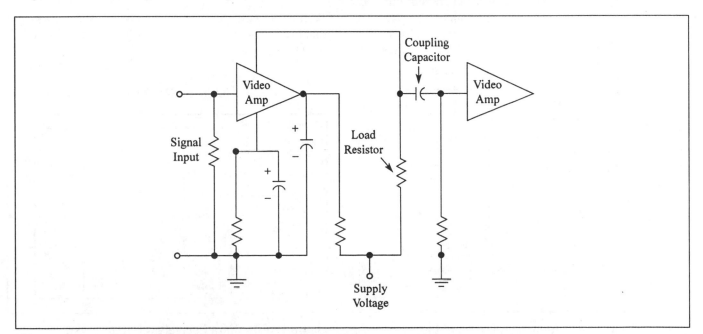

Figure 10-3 Simplified Drawing of a Video Amplifier Circuit

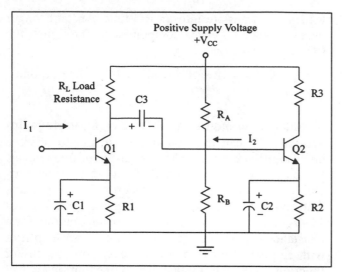

Figure 10-4A   Schematic Diagram of a Resistance-Capacitance RC-Coupled Transistor Amplifier

application of medium frequencies, those components act the same as a resistance. Because Q1 amplifies the input signal, I_1, and causes a larger signal current, I_2, to flow through R_{BE}, the transistor remains as part of the diagram.

Figure 10-4C further modifies the original drawing by considering additional component/signal relationships. Because one end of each resistor attaches to signal ground, we can redraw the resistors as shown in the figure. Also, we can combine R_A and R_B into an equivalent parallel resistance, or R_{EQ}, and show Q1 as a current generator. We can combine this current generator with the input signal and have a constant current source. The current equals the signal current multiplied by the forward current gain of the transistor.

With Figure 10-4D, we have the final version of the equivalent circuit for Figure 10-4A. Here we discount

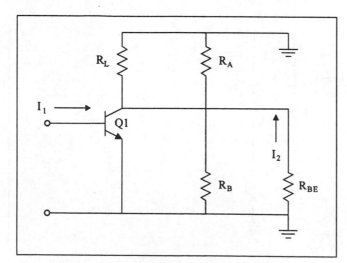

Figure 10-4B   First Equivalent Circuit of the RC-Coupled Transistor Amplifier

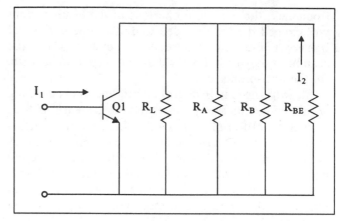

Figure 10-4C   Second Equivalent Circuit of the RC-Coupled Transistor Amplifier

the internal collector resistance of Q1 and begin to assign typical values to the resistances shown in the figure. For the purposes of our study, R_L has a value of 4 kΩ, R_{EQ} equals 4 kΩ, and R_{BE} has a value of 2 kΩ. Then we can apply a standard rule to the circuit by remembering that the total signal divides inversely proportional to the resistance of each branch. Referring to Figure 10-4E, one-fourth of the current flows through R_L, one-fourth through R_{EQ}, and one-half through R_{BE}.

We can apply this rule through the following equations by showing that the input signal current equals 10 μA, the beta of Q1 equals 40, the total signal current equals 400 μA, and that the I_2 current through R_{BE} equals 200 μA. Because we need to use only the current gain through R_{BE} to find the voltage gain, our first equation is:

$$A_I = I_2/I_1 = 200 \ \mu A/10 \ \mu A = 20$$

To find the voltage gain for the circuit, we can assume that the base-emitter resistance of Q1 equals R_{BE}. Thus, the voltage gain equals:

$$A_c = I_2/I_1 \times R_{BE(Q2)}/R_{BE(Q1)}$$
$$= 200 \ \mu A \times 2 \ k\Omega/10 \ \mu A \times 2 \ k\Omega = 20$$

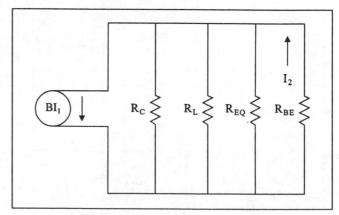

Figure 10-4D   Third Equivalent Circuit of the RC-Coupled Transistor Amplifier

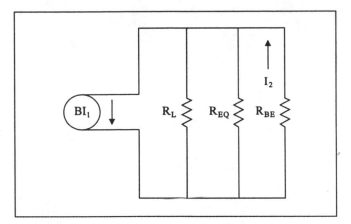

Figure 10-4E Fourth Equivalent Circuit of the RC-Coupled Transistor Amplifier

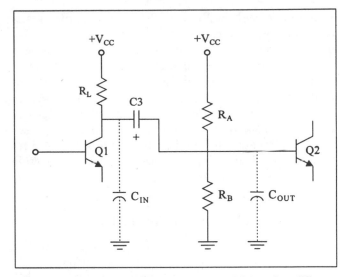

Figure 10-5A Two-Stage Transistor Video Amplifier

The last equation verifies that the voltage gain always equals the current gain multiplied by the ratio of the output resistance to the input resistance. When the two resistances equal one another, the voltage gain equals the current gain. We can then use Ohm's law to calculate the amplitude of the input and output signal voltages:

$$E_{IN} = I_1 R_{BE(Q1)} = 10 \times 10^{-6} \times 2 \times 10^3$$
$$= 20 \text{ millivolts}$$

$$E_{OUT} = I_2 R_{BE(Q2)} = 200 \times 10^{-6} \times 2 \times 10^3$$
$$= 400 \text{ millivolts}$$

High Frequencies and Video Amplifier Response

Any video amplifier faces the task of providing a flat response across the desired bandwidth. At high frequencies, shunt capacitance builds within semiconductor devices and attenuates high frequencies. In short, the elements of a transistor may become the plates for a small input and output capacitor during the application of a higher frequency. Offsetting these capacitive effects requires some type of high-frequency compensation. The capacitive effects that result from the application of high frequencies to an amplifier circuit occur because of the **Miller effect**, a rule that states that the value of an input capacitance will increase with the gain of the stage.

Figures 10-5A and B show a two-stage transistor video amplifier and the equivalent circuit for the circuit operating at high frequencies. The input capacitance and output capacitance generated because of the high frequencies are designated as C_{IN} and C_{OUT}. As Figure 10-5B shows, C_{IN} and C_{OUT} parallel the resistances in the circuit. While C_{IN} consists of the Q2 base-to-emitter capacitance, stray capacitance, and the Miller capacitance, C_{OUT} consists of the Q1 collector-to-emitter capacitance and stray wiring capacitance. The values assigned to the resistors are typical values used for this example.

For the circuit represented by all these figures, the Miller effect capacitance results from the multiplication of the Q2 collector-to-base capacitance by the Q2 forward-current gain. As a result, C_{IN} can be shown as:

$$C_{IN} = C_{BE} + C_S + C_{CB} (B+1)$$

and C_{OUT} can be shown as:

$$C_{OUT} = C_{CE} + C_S$$

If we substitute typical values for each capacitance, we have:

$$C_{CE} = 10 \text{ pf}, C_S = 5 \text{ pf}, C_{BE}$$
$$= 10 \text{ pf, and } C_{CB} = 4 \text{ pf}$$

Assuming that the Q2 has an average beta of 40, the values for C_{IN}, C_{OUT}, and C_T are:

$$C_{IN} = C_{CE} + C_S = 10 \text{ pf} + 5 \text{ pf} = 15 \text{ pf}$$

$$C_{OUT} = C_{BE} + C_S + C_{CB} (B + 1)$$
$$= 10 \text{ pf} + 5 \text{ pf} + 4 \text{ pf}(41) = 179 \text{ pf}$$

$$C_T = C_{IN} + C_{OUT} = 15 + 179 = 194 \text{ pf}$$

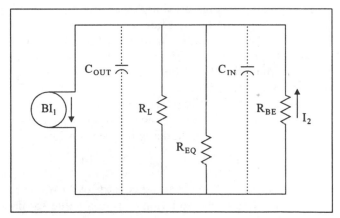

Figure 10-5B Equivalent Circuit for the Two-Stage Video Amplifier

Looking back at Figure 10-5B, the parallel resistance of the three resistors equals 1 kilohm. The frequency response of the amplifier remains flat up to a frequency where the reactance of the total capacitance equals the amount of parallel resistance. To find that frequency, we can substitute the capacitance values into the equation. Therefore, we find that the amplifier circuit has a flat response until the frequency surpasses 820 kHz. Because the amplification of video signals requires a much broader response, we need to provide high-frequency compensation.

High-Frequency Compensation

Decreasing the value of the load resistor in an amplifier circuit is one easy method for extending the upper limit of the frequency response. However, the relationship between a decrease in load resistance and frequency response is not linear. In the circuits that we just studied, decreasing the collector load resistance from 4 kilohms to 2 kilohms only extends the frequency response by 20 percent. This is because any decrease in the load resistance also changes resistance of a parallel resistance combination.

Decreasing the load resistance also decreases the amplifier stage gain by the same percentage that frequency response improves. That is, a 20 percent improvement in frequency response causes a 20 percent drop in gain. This drop in gain limits the amount of possible decrease in load resistance. On the hand, if we increase load resistance, we lose upper frequency response.

Shunt and Series Peaking   Because of all the potentially negative consequences resulting from changing the load resistance, we need another method for extending the upper limit of the frequency response. Adding small inductances, called peaking coils, in parallel with the input and output capacitances of the circuit provides a better method for improving frequency response. Each peaking coil shunts the capacitances and, as a result, eliminates the effect of the capacitance.

Figure 10-6 shows a transistor amplifier circuit with the addition of a peaking coil. Transistor Q1 operates as the last video amplifier. The video signal feeds through a coupling capacitor and travels as an input to the CRT.

Designated as L_{sh}, the peaking coil forms a relationship with the input capacitance, output capacitance, and load resistance that appears in equation form as:

$$L_{sh}/(C_{IN} + C_{OUT})R_L^2 = .414$$

Because we know the capacitance and load resistance values, we change the equation to find a value for the peaking coil:

$$L_{sh} = .414(C_{IN} + C_{OUT})R_L^2$$

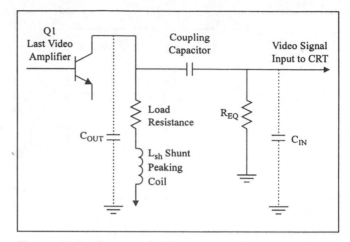

Figure 10-6   Schematic Diagram of a Transistor Video Amplifier Circuit with a Peaking Coil

and then substitute the known values:

$$
\begin{aligned}
L_{sh} &= .414 \, (8.6 \times 10^{-12} + 17 \times 10^{-12})(5.6 \times 10^3)^2 \\
&= .414 \, (25.6 \times 10^{-12})(5.6 \times 10^3)^2 \\
&= .414 \, (25.6)(5.6)^2 \times 10^{-12} \times 10^6 \\
&= 332 \times 10^{-6} \\
&= 332 \ \mu H
\end{aligned}
$$

With the addition of the peaking coil, we improve the frequency response of the amplifier circuit by 1.7 times the uncompensated frequency response. Given the 820-kHz frequency response of the original circuit, our new frequency response equals 820 × 1.7 or 1.394 MHz.

In comparison to shunt peaking, the use of series peaking electrically separates C_{IN} and C_{OUT}. In Figure 10-7, a shunt peaking coil and a series peaking coil form a series-shunt peaking network for the same

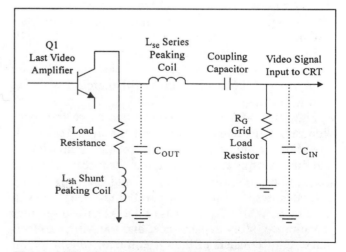

Figure 10-7   Schematic Diagram of a Transistor Amplifier Circuit with a Series-Shunt Peaking Network

transistor video amplifier shown in Figure 10-6. With this configuration, the series peaking coil compensates only for the input capacitance, and the shunt peaking coil compensates only for the output capacitance. The use of the series-shunt peaking network and the electrical separation of C_{IN} and C_{OUT} extends the frequency response by nearly 2.7 times the uncompensated response. As a result, we see the frequency response of the circuit improving from 1.394 MHz to 2.214 Mhz.

Using the Sharpness Control to Enhance Video Peaking
When a consumer purchases a television, one of the basic criteria for judging the quality of the receiver is the ability to reproduce the fine detail of an image. An image contains detail because of the brightness differences between small adjacent areas in monochrome pictures, or the combination of brightness, hue, and saturation differences in color pictures. The amount of detail in a picture determines whether the human eye can discern small or distant objects in the picture and whether the edges of the objects are well defined.

As mentioned earlier, the high-frequency components of the luminance signal contain the detail portion of the picture. A **sharpness control** enhances this detail by adjusting the picture so that a clearly defined line separates light and dark areas. Usually located within the third video amplifier circuit, the sharpness control adjusts the amount of high-frequency degeneration occurring throughout the reproduced picture. Figure 10-8 uses a block diagram to illustrate the location of the sharpness control.

At one end of its rotation, the variable resistor increases the amount of degeneration and reduces the overall sharpness of the picture. At the other end, the control decreases the high-frequency degeneration and increases the amount of detail of the picture. All this occurs because of the increasing and decreasing of the amount of resistance in series with a small capacitor. As the variable resistor rotates from end to end, the resistance either increases and reduces the ability of the capacitor to pass high frequencies, or decreases and allows the high-frequency signal to travel through the capacitor.

Many modern receivers include an **automatic sharpness circuit**. This circuit consists of an adaptive feedback loop that uses a high-pass filter and detector to determine the amount of high-frequency content in the picture. The loop drives the sharpness circuit. With one input of the automatic sharpness circuit connected to the circuit, high-frequency peaking decreases if the receiver tunes to a weaker signal.

Scan-Velocity Modulation

The quest for improved picture detail has also resulted in the implementation of **scan-velocity modulation**. Here the selective modulation of the constant horizontal scanning velocity of the CRT electron beam improves the brightness and definition of fine picture detail. As the electron beam scans back and forth across the 525 horizontal lines, scan-velocity modulation speeds up and then slows down the velocity of the scanning beam during a black-to-white transition. When a white-to-black transition occurs, the electron beam velocity slows down and then speeds up.

Changing the speed of the electron beam allows the transition from a dark scene to a bright scene to occur over a small distance on the screen of the CRT. As a result, the sharpness of the picture improves dramatically. Scan-velocity modulation also improves the contrast of small areas and the overall focus of the picture. Because of the changing electron-beam speed, the amount of white overshoot on bright scenes is reduced. Additionally, the size of the black areas increases slightly.

Figure 10-9 provides a block diagram of a scan-velocity modulation circuit. Scan-velocity modulation uses the same principles as a deflection yoke, and operates through SVM driver and output stages. To the far right of the drawing, the small auxiliary SVM yoke mounts on the neck of the CRT directly behind the main deflection yoke. The strength of the SVM yoke field automatically changes with variations in the picture information. As the figure shows, the SVM circuitry monitors the luminance signal through connections at the AGC stage.

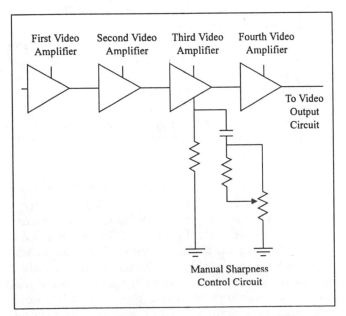

Figure 10-8   Block Diagram of a Sharpness-Control Circuit

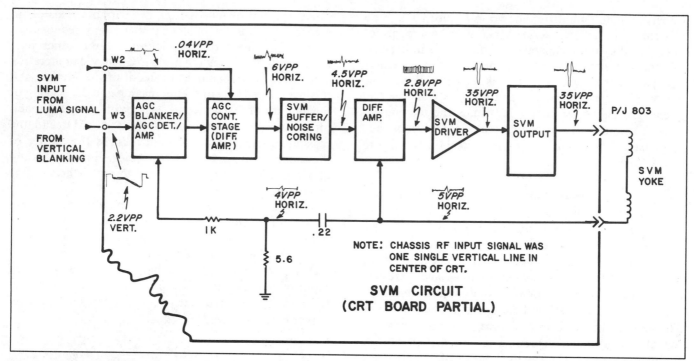

Figure 10-9 Block Diagram of a Scan-Velocity Modulation Circuit

☑ PROGRESS CHECK

You have completed Objective 2. Discuss how video amplifier components and circuits react to high, medium, and low frequencies.

10-3 VIDEO SIGNAL GAIN REQUIREMENTS

Given an approximate 1-Vdc signal from the video detector, and again depending on the type and size of the CRT, the video amplifier stages may need to provide a gain ranging anywhere from 30 to 200. Because of this, the dc supply voltage for the video amplifier circuit must have a value more than the peak-to-peak voltage swing. In many cases, the video amplifier circuit will use a separate, dedicated voltage supply to maintain the proper video output signal.

The phase of the composite video signal found at the output of the video amplifier stage depends on whether the system uses grid or cathode drive. Feeding the signal to the grid of the CRT requires a positive-polarity composite video signal at the amplifier output. Applying the video signal to the cathode of the CRT requires a negative-phase composite video signal with positive-polarity sync pulses and negative-polarity picture information.

The sync and blanking portions of the composite video signal provide the bias needed to drive the CRT into cut-off. As you may recall from our preliminary discussion in chapter three, the blanking portion of the composite video signal is often called the black portion of the signal. During operation with normal bias, the amplitude of the blanking portion equals the amplitude needed for cut-off. The sync pulse level exceeds the blanking amplitude and enters a region called blacker than black. These two signal levels establish a reference for the fluctuating picture signal as its level changes the amount of light given by the CRT over a range from white to black.

In a cathode-ray tube, grid one and the cathode can control the amount of beam current need to drive the CRT. Most systems use the cathode as a driver, because of the ratio of beam-current change to input-voltage change is greater for cathode drive than for grid drive. In addition, cathode drive requires less operating voltage than grid drive to produce the same beam-current range. Both factors ease the requirements at the power supply level.

Changing the bias of the tube so that the cathode is less positive than the control grid causes an increase in the light output of the CRT. With the cathode more positive than the control grid, the CRT brightness decreases. Feeding the composite video signal to the cathode allows the control grid to maintain a constant potential. Variations in the composite video signal cause the cathode to become either more or less positive than the control grid. Any bias setting between saturation and cut-off yields different shades of gray or color brightness.

Controlling CRT Bias and Amplifier Stage Gain

The transmitted video signal contains a **brightness reference level** that provides a basis for adjusting the brightness controls. This level may range from the signal level that corresponds with a maximum white raster, the signal level that produces a black raster, or a level corresponding with a definite shade of gray. Almost all transmission standards, however, use the level that produces a black raster as the brightness reference level. The **black level** remains fixed so that, as the camera moves across a scene, the background brightness varies with respect to black. Every change in the brightness level corresponds with the intensities of the elements that make up the picture.

At the receiver, the luminance signal consists of an ac voltage component that corresponds with the detail in the reproduced picture, and a dc voltage component that corresponds with the average background brightness. Maintaining a picture background that has proper brightness level with respect to black requires that the luminance signal retain the dc component. Using the black level as a standard allows any receiver circuit or CRT type to match the varying degrees of brilliance seen with the original picture.

The composite video signal also includes vertical and horizontal blanking pulses, or retrace blanking pulses, that have an amplitude greater than the black level. **Blanking pulses** have a greater amplitude than any of the picture signals and rise above the designated blanking level. Because of this, a blanking pulse always reduces the scanning spot intensity to zero. Retrace blanking pulses blank the electron beam out during the horizontal and vertical retrace.

Controlling the CRT Bias   Television broadcast stations use a negative transmission system based on the reduction of power. To make this easier to understand, Figure 10-10 again enlarges the composite video signal seen at the beginning of the chapter and places the waveform against the horizontal and vertical retrace periods. At the transmission point, the tips of the sync pulses cause the maximum radiated power; blanking pulses cause a 25 percent reduction in the radiated power; black causes another 5 percent reduction of radiated power; and white causes the radiated power to decrease to approximately 15 percent of maximum.

Every change in the background brightness of a televised scene produces a dc voltage that becomes part of the video signal used to modulate the transmitter. Figure 10-11 compares the composite video signal for one

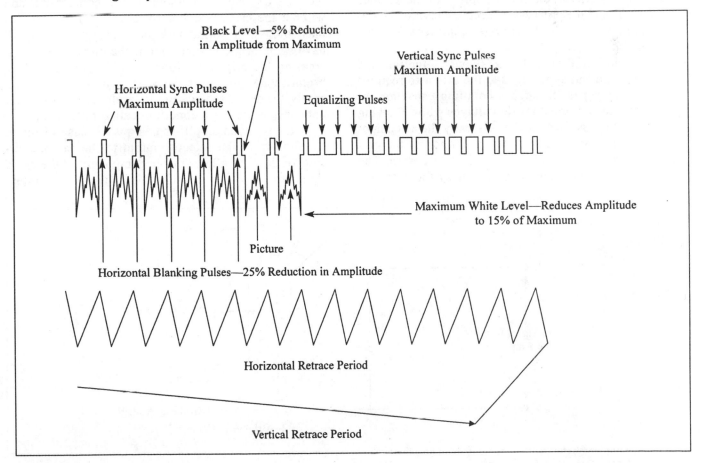

Figure 10-10   Composite Video Signal

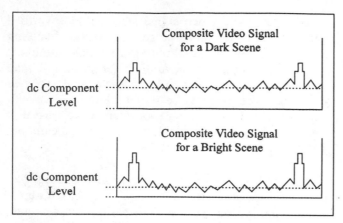

Figure 10-11 Comparison of Composite Video Signals for Dark and Bright Scenes

horizontal line of a dark scene with the composite video signal for one horizontal line of a bright scene. In both figures, the dotted line represents the dc component of the video signal. As you can see, the video signal for the bright scene has a lower dc component than the signal for the dark scene.

The blanking pulses depend on the polarity of the video signal when cutting off the CRT and producing a black screen during retrace. When a signal voltage with a positive polarity is applied to the cathode of the CRT, and the processing circuitry inverts the signal, the negative polarity voltage is impressed on the CRT grid.

However—as Figure 10-12 shows—inverting the signal voltage and applying it to the grid also requires that the signal pass through a coupling capacitor. As you know, capacitors block dc voltages while allowing ac signals to pass. Because of this, only the ac component of the luminance signal appears across the grid resistor, and the picture has a dark appearance. The blocking of the dc signal component by the capacitor

changes the fixed reference for the signal from the blanking level to the average signal level. Consequently, the blanking pulses are less negative than normal, and the image details contained within the signal are too bright. In addition, the blanking pulses cannot drive the CRT into cut-off and cause the tube to produce a black screen.

Reproducing the correct image intensity requires that video amplifier stage designs include a method for allowing the dc voltage to control the grid bias of the CRT. Controlling the grid bias allows the blanking pulses to drive the CRT into cut-off. Without the correct dc voltage level at the CRT, several problems will occur because the sync pulses and blanking level no longer have identical voltage levels. The loss of the dc component can cause the reproduction of incorrect colors or an incorrect balance between light and dark scenes.

Direct-Coupling the DC Voltage The video detector output signal contains the dc component of the video signal because detection of the amplitude-modulated video I-F signal recovers the dc component. The use of coupling capacitors, however, blocks the dc component from traveling to the cathode of the CRT while passing the ac video signal. Without the dc component, and without the brightness control adjusted for increased bias on the CRT during dark scenes, bright scenes will have some brightness but not at the correct level. Increasing the brightness for normal brightness during bright scenes decreases the bias of the CRT. Unfortunately, the decreased bias produces dark scenes that are not dark, as well as a condition called retrace lines.

In many video amplifier designs, the use of direct coupling from the detector output to the cathode circuits of the CRT allows the dc component of the video signal to pass. As a result, the circuitry accurately

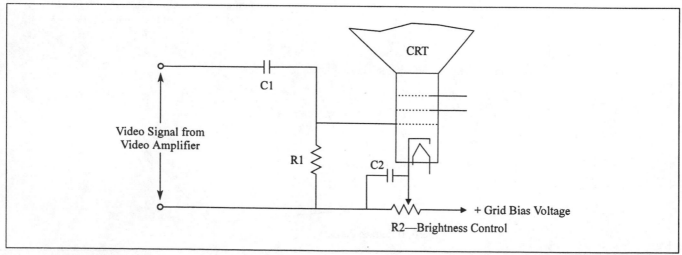

Figure 10-12 Blocking the dc Component of the Video Signal

reproduces the brightness levels of the transmitted picture. Applying separate blanking pulses to the CRT eliminates the retrace lines. Negative pulses from the horizontal output transformer and the vertical oscillator blank the beam out during the horizontal and vertical retrace.

Figure 10-13 uses a schematic diagram to show how direct coupling occurs in a television receiver. D1, the video detector, produces a composite video signal with negative-going sync pulses. Inductors L7—the 4.5-MHz trap—L8, and L9 direct-couple the dc portion of the luminance signal to the base of Q1, a video amplifier transistor. Q1 inverts, amplifies, and couples the dc signal to the base of Q2, the second video amplifier. Q2 amplifies the combined ac and dc portions of the luminance signal, and inverts the signal. From there, L5, delay line DL1, L2, and R19 couple the signal to the video output stage.

The amplifiers can produce a stage gain of 60. Therefore, a 2-volt peak-to-peak signal at the output of the video detector becomes a 120-volt peak-to-peak signal at the output of the video output stage. With a completely black televised scene, the output signal found at the video detector will have a dc component that measures approximately −1.4 Vdc. Amplifying this voltage by the gain figure of 60 produces an 84-Vdc increase at the video output stage and the bias of the CRT. Increasing the bias to this level causes the CRT to enter cut-off.

A scene with normal background brightness conditions causes the video detector output signal to contain only −0.4 Vdc of the dc component. Because the lower-amplitude dc component causes a much smaller increase in the CRT bias, the electron guns conduct heavily. As a result, the reproduced picture has a bright background.

Restoring the DC Voltage   Because of several factors, other designs retain coupling capacitors or networks and rely on dc restoration circuits. Certainly, direct coupling seems to offer a simple method for maintaining the dc component of the luminance signal. Nevertheless, a direct-coupled circuit requires exact parameters for voltage dividers, a highly regulated power supply, and increased thermal protection.

With **dc restoration**, a clamping circuit inserted between the video output transistor and the CRT restores the dc reference voltage. The **clamping circuit**, illustrated in Figure 10-14, holds the sync tips of the composite video signal to a fixed voltage level.

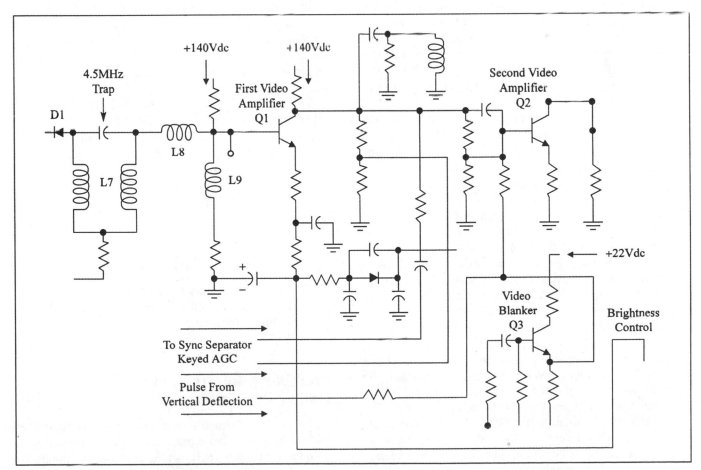

Figure 10-13   Video Amplifier Circuit that Direct-Couples the dc Component of the Luminance Signal

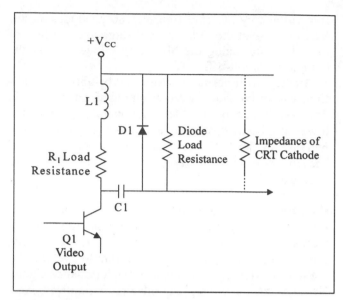

Figure 10-14   dc Restorer Circuit

Clamping the video signal to the fixed level restores the signal to the form originally seen at the video detector output.

In the circuit, the parallel combination of resistor R1 and diode D1 work in series with the coupling capacitor, C1, between the video amplifier output and ground. The combination also is in series with the dc bias voltage. The capacitor blocks the dc component found at the output of the video amplifier and has a charge that varies with the ac signal portion of the luminance signal.

When the capacitor charges and discharges, the R1D1 combination carries the current. Because the resistor and diode are in a parallel configuration, any voltage across the resistor is also found across the diode. When the capacitor charges, the polarity of the voltage across the resistor reverse-biases the diode. When the capacitor discharges, the polarity of the voltage across the resistor changes, forward-biases the diode, and allows the diode to conduct. Thus the capacitor charges through the resistor and discharges through the parallel combination of the resistor and the diode.

Blanking pulses initiate a sudden drop in the video amplifier output voltage and cause C1 to discharge. Discharging the capacitor causes the diode to conduct and—because of the low resistance presented by the R1D1 parallel combination—also causes the voltage across R1 to decrease. With the capacitor discharging and the voltage across the resistor decreasing, the supply voltage becomes the only voltage in the CRT grid-cathode circuit, and the CRT is biased into cut-off.

When diode D1 rectifies the video signal, the rectified signal contains the amount of dc voltage needed to restore the video signal to its original level. The diode conducts only during the sync period and remains off during the scan period. Conduction during the scan period could allow the introduction of noise spikes into the video signal. If those spikes have an amplitude the same as or greater than the sync tips, the bias level for the video amplifier would be incorrect.

The values of C1 and R1 provide an RC time constant equal to the interval used by approximately 10 horizontal lines, and allows the charging current across the resistor to remain constant with each individual scanning line. When the blanking pulse ends, the output voltage at the video amplifier transistor rises to the average signal level and causes C1 to charge. The charging current across the resistor produces a voltage, which becomes applied to the grid-cathode circuit of the CRT. Consequently, the supply voltage is more positive than the voltage across R1; the amount of average CRT beam current increases; and the screen shows a bright picture.

Receiving a dark picture decreases the difference between the signal level and the blanking pulses to a lower level than that seen with bright picture conditions. As a result, a blanking pulse initiates a smaller decrease of the video amplifier output voltage and causes C1 to discharge less. Once the blanking pulse ends, the output voltage rises back to the average signal level, and the capacitor begins to charge. Because the difference between the signal level and blanking pulse amplitudes are lower, the voltage increase caused by the charging capacitor is also smaller. Therefore, the brightness of the reproduced picture decreases to the proper value.

The dc restorer circuit is always located in the last video amplifier stage. Restoration of the dc component occurs because the circuit automatically varies the CRT screen brightness to correspond with the difference between the dark and average levels of the transmitted signal. In the next section, we will discuss another version of the dc restoration circuit, the automatic brightness limiter.

☑ PROGRESS CHECK

You have completed Objective 3. Explain how video amplifier circuits increase the gain of the luminance signal.

Brightness-Control and Brightness-Limiter Circuits

As Figure 10-15 illustrates, the **brightness control** sets within the last video amplifier stage and precedes the video output stage. It allows the customer to change the bias of the CRT and adjust the picture from black to white display. Brightness controls may appear either in

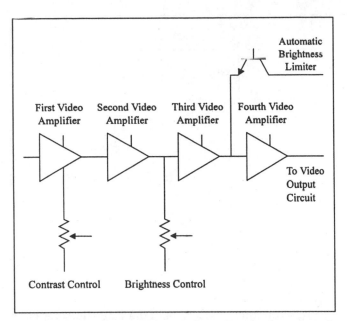

Figure 10-15 Block Diagram Showing the Location of the Brightness-Control, Automatic Brightness-Control, and Automatic Brightness-Limiter Circuits

the cathode or grid circuits of the CRT, or in the base circuit of the video output stage.

Placed in the cathode circuit of the CRT, the brightness control governs the cathode voltage of the CRT and, therefore, controls the CRT bias. With the brightness control set so that the cathode voltage is zero volts, little or no bias exists between the grid and cathode, and the beam current is at the maximum level. Under these conditions, the CRT has its maximum light output. Setting the brightness control at the other extreme applies the maximum voltage to the cathode and causes the CRT to cut off and remain black.

Placing the brightness control in the grid circuit again adjusts the CRT bias. The difference occurs, however, because of a stationary cathode voltage and changing grid voltage. Within the grid circuit, the brightness control becomes part of a voltage divider that controls the boost voltage for the CRT. The brightness control in the grid circuit is part of a voltage-divider circuit within the video output stage; it adjusts the base voltage of the amplifier.

Moving the control in either direction makes the base more positive or more negative with respect to the collector. As a result, the collector voltage increases or decreases with a change of the brightness control. To take advantage of this technique, video amplifier designs direct-couple the amplifier collector to the cathode of the CRT. Thus, changing the brightness control affects not only the bias of the transistor but also the amount of voltage applied to the CRT. Moving the control in either direction causes the cathode voltage to

increase or decrease and changes the brightness of the reproduced picture.

Most modern television receivers supplement the customer control with an **automatic brightness-control circuit** that monitors the existing light in the room housing the receiver. The circuit automatically adjusts both the brightness and the contrast so that the picture will have the correct settings for the current environment.

In addition, most receivers also include **brightness-limiter circuits**. An increase in the average CRT beam current caused by increased white signal levels or brightness can overdrive the picture tube. When this occurs, the entire picture will bloom, that is, become overly bright and appear to lose focus. The brightness-limiter circuit monitors changes in the beam current and changes the bias of the CRT to prevent the change from occurring.

Controlling the Video Amplifier Stage Gain

Video amplifier circuits require some method for controlling the amount of gain and for allowing the adjustment of the picture contrast. We define **contrast** as the difference between the light and dark areas of a scene. The contrast of a picture is determined by the amplitude of the video signal.

Figure 10-16 uses a block diagram to show the relationships between the video stages and gain-control circuits. Adjustment of the **contrast control**, which is located in the video amplifier stage, varies the amplifier gain for the ac video signal without changing the dc operating point of the circuit. If changing the contrast control affected the dc operating point, then any adjustment by the customer would affect the signal found at the AGC and sync circuits.

Contrast Control Many factors, such as unwanted signal pick-up, stray capacitance, and circuit stability, can affect the performance of the contrast control. Consequently, most video amplifier circuit designs place the contrast control in the low impedance emitter circuit portion of the amplifier rather than in either the base or collector circuits. Figure 10-17 shows a typical contrast-control circuit in which a fixed resistor and the adjustable contrast control operate as the emitter-resistor for the video amplifier.

The combination of fixed resistor and adjustable potentiometer provides a variable ac-voltage ratio of 4:1, and yields up to two times the lowest amount of required gain. The capacitors shown in the circuit protect against transient response within the control circuit, since adjusting the control for maximum contrast could introduce transients into the circuit.

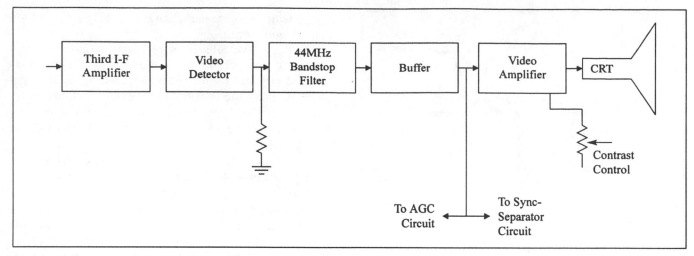

Figure 10-16   Block Diagram of the Video Amplifier Section

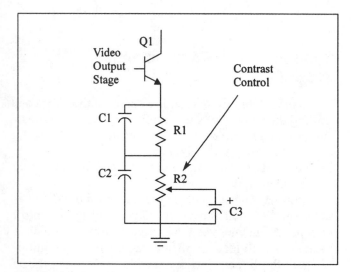

Figure 10-17   Schematic Diagram of a Contrast-Control Circuit

While almost all televisions feature a customer-accessible contrast control, many also utilize an auto-contrast circuit that consists of a second feedback loop or AGC system placed within the luminance channel. Because the auto-contrast circuit compensates for signals that have less than full modulation, the system has a more uniform level of contrast. This in turn reduces the requirements for the high-voltage supply and, as a result, the wear-and-tear on the CRT during the broadcast of scenes that contain large amounts of white.

AGC and Sync Circuit Connections

Referring once again to Figure 10-16, the AGC line ties to the video amplifier circuit as well as the I-F amplifiers and the tuner. A buffer stage placed between the video detector and the video amplifier stages provides impedance matching between the video detector and the AGC circuits. The output of the video detector has a high output impedance, while the AGC circuits have a low input impedance.

The AGC circuit uses a sample of the amplified composite video signal as a reference voltage to control the gain of the R-F and I-F amplifiers. With this system, the gain of those amplifiers remains inversely proportional to the received signal strength. Controlling the amplifier gain in this way guarantees that undesired changes in the received signal strength do not cause changes in the picture contrast.

Building the Blanking Signals into the Process

As mentioned, video amplifier circuits utilize horizontal and vertical retrace waveforms to cut off the CRT during the blanking intervals. In many modern television receivers, the vertical waveform is introduced at the video amplifier interstage, while the horizontal waveform is coupled directly to the CRT grid. As shown in Figure 10-18, the retrace blanking circuit applies a positive pulse from the horizontal deflection section through R7 and C4 and to the base of the video blanking amplifier Q3, which is configured as an emitter follower.

Working from the emitter of Q3, the positive horizontal blanking pulse travels through R13 to the base of Q2. The retrace blanking circuit obtains a positive vertical blanking pulse from the vertical deflection and applies the pulse through R6 and R13 to the base of Q2, the second video amplifier. Both pulses ensure the cut-off of the CRT-beam currents during the retrace intervals.

Amplitude differences between the blanking and picture signals allow the separation of the sync signals from picture signals before the signals travel to the

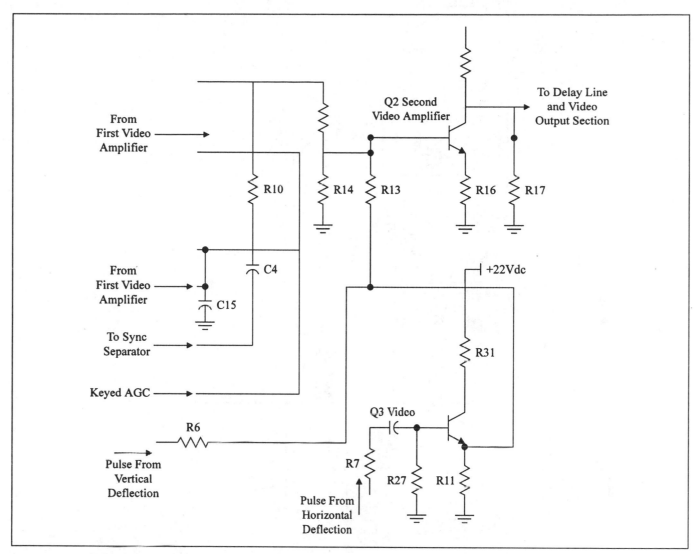

Figure 10-18 Schematic Diagram of the Retrace Blanking Circuit

deflection circuits. Both the vertical and the horizontal blanking pulses reduce the spot illuminated by the electron beam to black. Given the position of the sync signals and their placement within the retrace intervals, the sync pulses do not interfere with the picture.

☑ PROGRESS CHECK

You have completed Objective 4. Discuss how customer-control and automatic circuits control the amplitude of the luminance signal and the bias of the CRT.

10-4 LUMINANCE PROCESSING APPLICATIONS

Figure 10-19 shows the schematic diagram of a luminance circuit from a Sylvania C5 color television

chassis. The circuit fits between the output of a comb filter and the input to a luminance/chrominance processor, and provides a combination of frequency and phase equalization. In chapter eleven, we will see how the comb filter takes advantage of frequency interleaving to separate the luminance and chrominance signals. For now, however, we will look only at the luminance signal arriving at one output of the comb filter.

According to the schematic, the luminance signal arriving from the comb filter travels to the base of Q155, a luminance amplifier, and becomes inverted at the collector of the same transistor. The Q155 emitter circuit provides the initial equalization of the luminance signal. From there, the luminance signal can follow two possible paths. High-frequency components of the signal travel to the base of Q157, the luminance equalizer, while low-frequency components travel to the Q157 emitter. While the base-emitter circuits provide some signal phase equalization, the circuit has the basic task of operating as an all-pass filter.

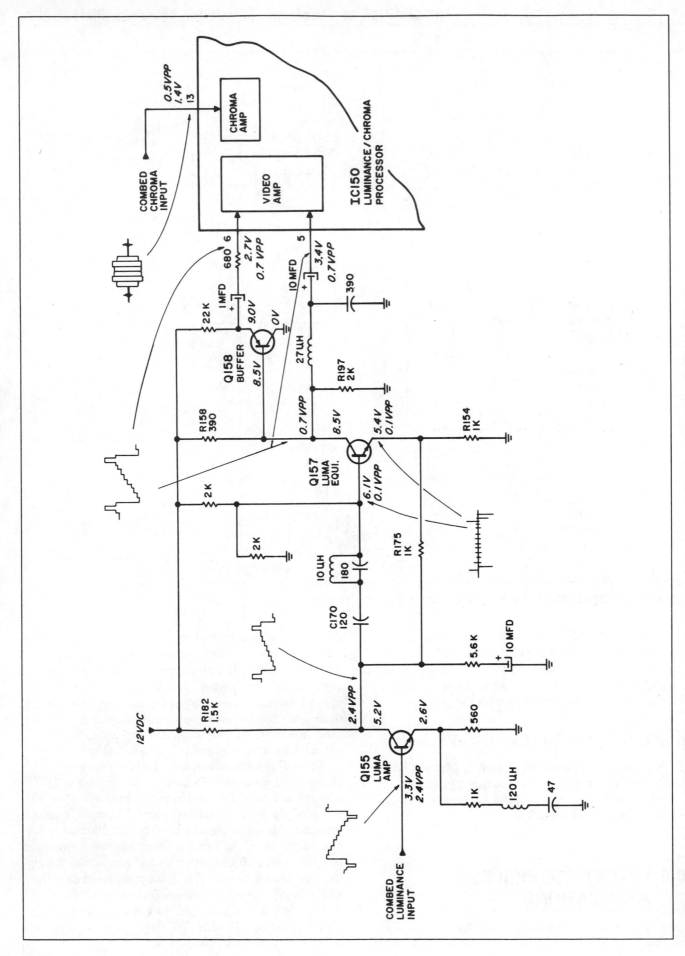

Figure 10-19 Schematic Diagram of a Sylvania C5 Chassis Luminance Circuit (*Courtesy of Philips Electronics N.V.*)

Any increase in the voltage at the collector of Q155 results in an equal increase of voltage at the emitter of Q157. As a result, any increase in voltage reverse-biases Q157 and opposes any current flow through resistor R154. With this opposition to current in place, the increased voltage also cancels any signal that might appear at the Q157 emitter. As the transistor begins to turn off, the collector voltage rises and generates the same in-phase signal seen at the collector of Q155.

A Luminance/Chrominance Integrated Circuit

Most modern televisions enclose the luminance and chrominance signal-processing operation within one integrated circuit. Combining the operations saves cost and space, and allows the integration of color and contrast tracking. Figure 10-20 shows a block diagram of a popular design used by RCA, Sylvania, and Zenith.

During operation, the luminance signal enters the IC and travels into the video amplifier. While the video amplifier provides amplification for the weak signal,

input signals from a peaking circuit and the sharpness control produce a sharper, well-defined picture. Moving to the right upper corner of the diagram, a combination of the brightness control, the brightness limiter, the sub-bright control, and the internal brightness-control pedestal clamp controls the bias of the video amplifier. The sub-bright control is placed within the receiver chassis and is not a customer control.

In addition to the brightness control circuitry, a dc voltage applied to pin 26 not only establishes contrast, or picture, control but also provides additional control over the video amplifier. The contrast control governs the video amplifier by sending a level-controlling signal to the brightness-control pedestal clamp. With the output of the picture-control amplifier applied to both the luminance and chrominance amplifiers, the circuit achieves a combination of contrast and color tracking.

Toward the lower left corner of the drawing, a high-voltage resupply line combines with the brightness control to control the beam-limiting circuit. In turn, a comparator compares the amplitude of the brightness-control voltage with the level of the blue blanking output signal during the retrace interval. The dc voltage

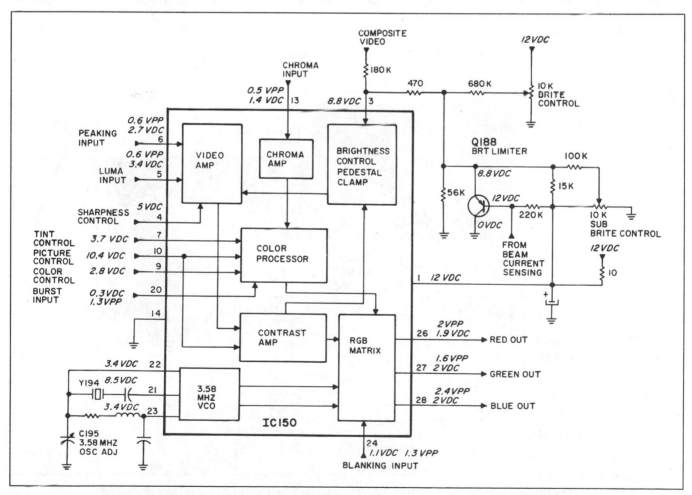

Figure 10-20   Block Diagram of a Luminance/Chrominance Processing IC (*Courtesy of Philips Electronics N.V.*)

developed by the comparator controls the brightness level found at the luminance amplifier. Additional contrast tracking occurs through the modulation of the picture-control amplifier by the beam-limiting circuit.

Figure 10-21 shows an enlarged schematic diagram of the luminance/chrominance IC, emphasizing the luminance-processing operation. The connection of a resistor between the high-voltage resupply line and pin 26 of the IC allows the sampling of the dc beam current. With a normal picture, the voltage at pin 20 of the IC remains clamped at 12 Vdc. An increase of the beam current causes the voltage at pin 28 to decrease and the beam limiter to conduct. Conduction of the beam limiter only occurs when the pin 28 voltage drops below 12 Vdc, and reducing beam current by changing the luminance-amplifier bias.

Automatic brightness limiting occurs through the sampling of the dc current produced at the windings of the high-voltage transformer through the resistor connected in the high-voltage resupply line. A picture with normal background brightness causes the voltage at the output of the luminance/chrominance matrix to remain clamped at 12 Vdc. Any increase in beam current causes the voltage seen across the high-voltage

resupply line to drop, which in turn causes the clamped dc voltage to fall below 12 volts. With this, the internal beam-limiter circuitry conducts and reduces the beam current by changing the bias of the luminance amplifier.

The luminance and chrominance signals matrix together within a matrixing amplifier. (Chapter eleven will take a closer look at the matrixing amplifier and the chroma circuits that complete this particular luminance/chrominance processor design.) As Figure 10-21 indicates, a large part of the processor is dedicated to the reproduction of an image that has the proper hue and color.

10-5 TROUBLESHOOTING VIDEO AMPLIFIER CIRCUITS

Troubleshooting video amplifier circuits involves the use of almost all the techniques that we have discussed. While voltage checks are a common part of the routine, efficient troubleshooting also requires the use of a signal generator and an oscilloscope. However, as with all troubleshooting approaches, a common-sense

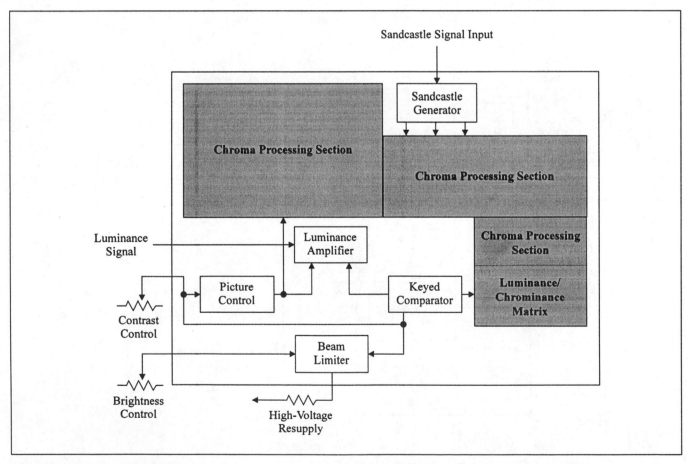

Figure 10-21 Schematic Diagram of the Luminance/Chrominance Processing IC (*Courtesy of Philips Electronics N.V.*)

approach that includes cause-and-effect logic dramatically speeds the process.

Test Equipment Primer: Oscilloscopes

The oscilloscope offers versatility as a troubleshooting tool through functions such as frequency, time, phase, and peak-to-peak voltage measurements. Before considering those applications, it may be useful to review a few fundamentals. Alternating currents energize the deflection coils of the yoke. As the alternating current changes polarity at specific intervals, it moves from positive to negative and back to positive. During these alternations, the current crosses the zero axis. The distance from the point where the current leaves zero to go positive, then negative, and then back to zero is a wavelength. Every wavelength has a time duration, or a period, which is measured in seconds.

⮌ FUNDAMENTAL REVIEW

Hertz

Any number of wavelengths per second equals the frequency of the alternating current and is measured in hertz.

Figure 10-22A marks the wavelength, period, and frequency of a sample waveform. In the figure, the wavelength appears identical to the period. You can measure wavelength from successive positive or negative peaks of the waveform, or choose two accepted points of measurement on the waveform. In addition, the units of measurement makes it clear measuring length is measured against time. Therefore, the period is the amount of time required for a wave to travel the

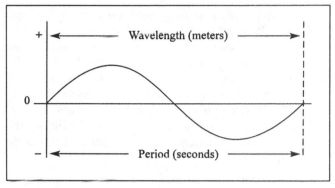

Figure 10-22A Wavelength and Period of an Alternating Current

distance equal to one wavelength. We can use the following equations to determine values for wavelength, period, and frequency and show the relationships between the three characteristics:

$$\text{Frequency or F (kilohertz)} = 3 \times 10^5/\text{wavelength or } \lambda \text{ (meters)}$$

$$\text{Wavelength or } \lambda \text{ (meters)} = 3 \times 10^5/\text{frequency or F (kilohertz)}$$

$$\text{Time or T (seconds)} = 1/\text{frequency or F (hertz)}$$

Time Measurements with the Oscilloscope Using the characteristics and the capabilities of an oscilloscope, you can measure both the rise time and the period of a waveform. Rise time is defined as the amount of time required for a waveform to increase from 10 to 90 percent of its total amplitude. Referring to Figure 10-22B, the square wave has expanded enough in both the horizontal and vertical directions for measurement of one rise-time interval. If you use the horizontal sweep rate control to set the oscilloscope sweep speed to 0.05 microsecond per centimeter, the rise time or RT equals 0.1 microsecond.

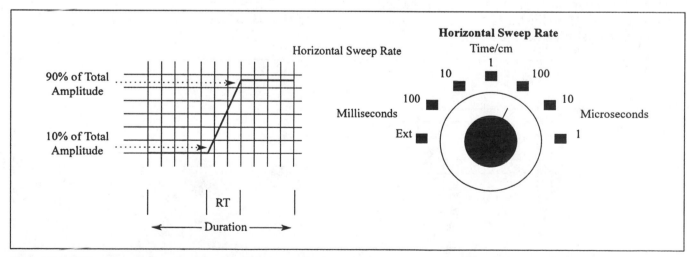

Figure 10-22B Measuring the Rise Time and the Period of a Sample Square Wave

Given the control setting, each horizontal division on the oscilloscope screen has a relationship with time. That is, if each square centimeter shown on the face of the oscilloscope equals 0.05 microseconds, then you can measure the period of a waveform as it is displayed on the screen. As shown in the figure, approximately one-half of the wavelength crosses 10 centimeters. To measure the time duration of this portion of the waveform, multiply the distance by the setting of the horizontal sweep control:

T (time duration) = 10 cm × 100 ms = 0.001 seconds

Frequency Measurements with the Oscilloscope

Measuring the frequency of the waveform calls for almost the same procedure used for time measurement. Again looking at Figure 10-22B, we apply the time duration of the waveform to the following equation. The frequency of the waveform is:

frequency = 1/0.001 seconds = 1000 Hz

Peak-to-Peak Voltage Measurements with the Oscilloscope

All waveforms have accompanying peak-to-peak voltage measurements. To illustrate such measurements, we will use the sample sine wave shown in Figure 10-22C. The vertical step attenuator of the oscilloscope is set to 20 volts p-p/cm. Each setting of the volts/cm switch establishes the number of volts required to produce a pattern 1 centimeter high on the oscilloscope screen graticule. Because one cycle of the sine wave covers 4 vertical centimeters, it has an amplitude of 4 centimeters multiplied by 20 volts p-p, or 80 volts peak-to-peak.

Phase Measurements with the Oscilloscope

Technicians can measure phase with either a single-trace or a dual-trace oscilloscope, but the latter offers greater accuracy, as well as the capability to measure the phase differences between complex signals that have different amplitude, frequency, and waveshape characteristics. Figure 10-22D illustrates the measurement of phase using a dual-trace oscilloscope.

To measure the phase difference between two signals, measure the distance covered by one complete cycle of the reference signal. In the figure, one cycle covers 5 centimeters. To find the phase factor, divide one cycle by the number of centimeters covered. If one 360 degree cycle equals 5 centimeters, then 1 centimeter equals 360/5 or 72 degrees. The waveform thus has a phase factor of 72 degrees.

To find the phase difference of the two signals, measure the horizontal distance between two points on the waveform. For the reference signal and signal 1, the distance is approximately 0.33 centimeters. Then multiply the distance by the phase factor to find the phase

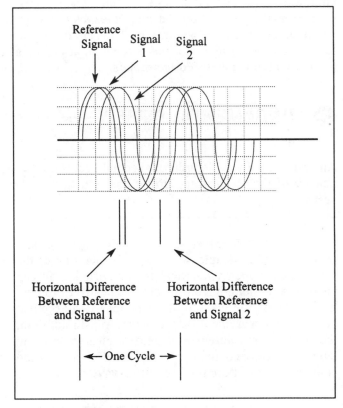

Figure 10-22D   Measurement of Phase Using a Dual-Trace Oscilloscope

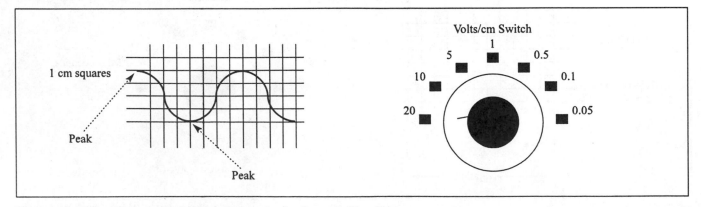

Figure 10-22C   Peak-to-Peak Measurement of a Sample Sine Wave

difference. In this case, the phase difference between the two signals equals:

$$0.33 \times 72 \text{ degrees} = 23.76 \text{ degrees}$$

Waveform Analysis with the Oscilloscope   The display characteristics of an oscilloscope allow us to see the effects of circuit defects on signals. Often a technician can connect a square-wave generator to a circuit such as an amplifier and check for different types of distortion. For example, introduction of low-frequency hum combined with a phase shift will distort the square wave so that it resembles a trapezoid.

Video Amplifier Circuit Troubleshooting

Obviously, audio amplifiers handle sound signals and video amplifiers handle video signals. However, many of the same methods used for troubleshooting the sound circuit problems seen in chapter seven also apply to the video amplifiers studied in this chapter. In quick review, those checks are dc voltage measurements, signal tracing, and the monitoring of the amplifier response. With video amplifiers, the problem symptoms center around:

- the loss of the raster
- a weak or distorted raster
- a weak or distorted picture
- no contrast or brightness control
- problems with the AGC

An oscilloscope works well for signal tracing, measuring of gain, square-wave response checks, and frequency-response checks of video amplifier circuits. When utilizing an oscilloscope for these tests, attach a low-capacitance probe so that the test equipment will not load the amplifier. Signal tracing with an oscilloscope can be easily accomplished with a regular on-air television signal and the oscilloscope set to a deflection rate of 7875 Hz.

As you learned earlier, the polarity of the composite video signal depends on the polarity of the video detector and the point where the oscilloscope probe connects to the circuit. To illustrate this last point, you may find a positive pulse at the base of a transistor configured as a common-emitter video amplifier. Because of the amplifier configuration, the waveform at the collector of amplifier will have a negative polarity.

From earlier discussions you know that you can measure gain by comparing the amplitude of waveforms found at the input and output of an amplifier stage. An increase in signal amplitude by a factor of 10 translates into a gain of 10. When attempting to measure gain, rely on a generated signal rather than the

constantly changing on-air signal that you used for signal tracing.

The troubleshooting methods for audio amplifiers in chapter seven emphasized the importance of square-wave response tests and showed how the appearance of the square wave could point to a problem area. With only a few changes, you can apply the same methods to video amplifier circuits. When performing the square-wave tests, set the oscilloscope repetition rate to 100 kHz, use the low-capacitance probe, and disconnect the video detector from the circuit.

The need for a stable oscilloscope display dictates that you set the square-wave generator rise time to a faster rate than the video amplifier rise time. Because video amplifiers usually have a rise time of 0.1 microseconds, a 0.08-microsecond rise time will produce a stable waveform display. If the test equipment does not have the correct rise time, the displayed waveform will exhibit the conditions called overshoot and undershoot, as shown in Figure 10-23.

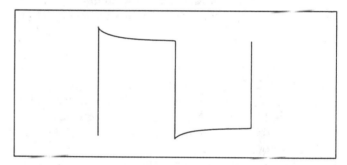

Figure 10-23   Square Wave Overshoot and Undershoot

The rise time of a square wave is important because of the relationship between rise time and the bandwidth of the amplifier: the rise time equals one-third of the period for the high cut-off frequency of the amplifier. For example, a typical video amplifier has a bandwidth extending from 30 Hz to 4 MHz and has a rise time of 0.083 microseconds. The period corresponding to the 4-MHz bandwidth equals 0.25 microseconds; one-third of that period equals 0.083 microseconds.

Several different component failures can cause poor square-wave frequency response. An open decoupling capacitor or a load resistor that has increased in value can increase the square-wave rise time so that the waveform appears similar to Figure 10-24. Because the values of the peaking coils are critical, any change in value due to a partial short or completely shorted inductance will cause the distortion shown in the figure.

When viewing the square-wave response of a video amplifier, note that the tilt of the waveform describes the difference between the amplitudes of the trailing and leading edges. From this, we see that a relationship

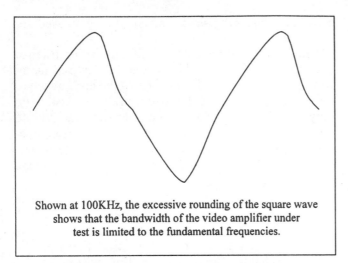

Shown at 100KHz, the excessive rounding of the square wave shows that the bandwidth of the video amplifier under test is limited to the fundamental frequencies.

Figure 10-24 A Square-Wave Response Caused by an Open Decoupling Capacitor or Load Resistor

exists between the tilt and the low-frequency cut-off point of the amplifier. Mathematically, this relationship is represented as:

$$f_c = 2f (E_2 - E_1)/3(E_2 + E_1)$$

where E_2 is the amplitude of the leading edge, E_1 is the amplitude of the trailing edge, f is the square-wave frequency, and f_c equals the low cut-off frequency.

To check the frequency response of a video amplifier, connect a demodulator probe from the oscilloscope to the output of the video amplifier. Although schematic drawings will not show a frequency-response curve for a video amplifier, we can compare the frequency response of an amplifier suspected to be bad with the response of an amplifier known to be good. In most cases, components such as a decoupling capacitor, a load resistor, peaking coils, or the amplifier itself can harm the frequency response.

Troubleshooting Luminance/Chrominance Processors

Problems with the luminance portion of the combined processor can cause the following symptoms:

- Receiver has raster but no picture (no luminance information)
- There is no control of the brightness or contrast
- The picture is weakly reproduced and lacks contrast

As with all other circuits, check the dc voltages for the circuit before proceeding to other checks. For this circuit, the loss of the +10.9-Vdc supply for the circuit can cause the loss of the picture information or the loss of

control over either brightness or contrast. In addition, a voltage of less than +11.5 Vdc at the beam-current limiter circuit with a normal low-brightness picture can cause either the loss of the picture or a washed-out picture.

Several other conditions may also cause the loss of the picture information. Although the loss of the red, blue, and green signals at key test points indicates an internal circuit problem, check also for the proper I-F output signal and the proper video input signals.

➔ SERVICE CALL ─────────────

Washed-Out Picture

While carrying the portable television into the service shop, the customer told the technician that the picture lacked detail and seemed "washed out." She also commented that the picture would completely fade away at times. Powering up the receiver disclosed that normal sound was present and pointed towards the video circuits. Because the picture blacked out after a period of time, logic dictated that a heat-sensitive component, such as a video amplifier, was causing the problem. The use of cooling sprays, however, did not cause the problem source to surface.

Checks of the video amplifier supply voltages during normal operation and during the time that the problem occurred showed that the supply voltage dropped to nearly zero when the picture blacked out. A further check in the power supply section showed that the operation of a regulator transistor had become intermittent. As the receiver operated, the transistor opened and cut off the supply voltages.

Using a signal generator and an oscilloscope, the technician traced the video signal from the receiver antenna to the video amplifier circuits. A cold solder joint on a delay line in the video amplifier circuit blocked the video signal as the receiver operated for a short period of time. Careful resoldering of the delay line restored the normal operation of the receiver.

➔ SERVICE CALL ─────────────

Sylvania Television with No Video

The late-model Sylvania television arrived with a symptom of no video. Because the receiver relied on a luminance/chrominance processing circuit, troubleshooting centered around measuring the +12 Vdc at the brightness control circuit and checking for the luminance signal at the input of the circuit. The voltage check disclosed that the +12 Vdc had decreased to almost zero, and that the problem existed in the switched-mode power supply of the receiver. Replacing a rectifier diode in that section restored the +12 Vdc and the normal operation of the luminance/chrominance amplifier.

 **PROGRESS CHECK**

You have completed Objective 5. Troubleshoot video amplifier and sync-separator circuits.

SUMMARY

Luminance signals represent the amount of light intensity given by a televised object. While the luminance signal is the video signal for a monochrome receiver, the complexity of a color television requires a different approach for luminance signals. The luminance signals consist of proportional units of the red, green, and blue voltages, and contain the brightness for the picture. The luminance and chrominance signals add together and proportionately reproduce the complete picture information.

In the discussion of the processing and amplification of video signals, you learned that motion in a video picture produces a succession of continuously changing voltages. In addition, every televised scene contains varying amounts of light and shade that distribute unevenly across the reproduced picture. The amplitude variations caused by this uneven distribution correspond to the horizontal and vertical scan rates. Areas of the picture that have many variations in white and black intensity have higher frequency-amplitude changes.

Television receivers must amplify the luminance signals, which have a very low amplitude after leaving the video detector. The amplification provides sufficient amplitude to drive or vary the light output of the CRT by changing the bias of the tube over a wide range of conditions. As the chapter demonstrated, the video amplifier section must have the capability to either maintain or restore a dc reference voltage related to the original transmission, ensure that the proper phase video signal reaches the CRT, and accept horizontal and vertical blanking waveforms that cut off the CRT. Further, the phase of the composite video signal found at the output of the video amplifier stage depends on whether the system uses grid or cathode drive. In turn, the sync and blanking portions of the composite video signal provide the bias needed to drive the CRT into cut-off.

Chapter ten also revealed that video amplifier circuits may use either direct coupling or dc restoration as a method of providing the amount of dc voltage needed to restore the video signal to the level seen during transmission. The brightness control allows the customer to change the bias of the CRT, and an automatic brightness-control circuit monitors the existing light in the room housing the receiver. Brightness-limiter circuits prevent increased white signal levels or brightness from overdriving the CRT. The chapter concluded by providing examples of different luminance processing circuits and by illustrating how those circuits interface with the AGC and sync sections of a television receiver.

Troubleshooting techniques for video amplifier circuits included a combination of voltage checks, the use of a signal generator, and the use of an oscilloscope.

REVIEW QUESTIONS

1. The composite video signal includes _____,
   _____, and _____ signals.
2. The horizontal scanning frequency is:
   a. 15,734 Hz
   b. 60 Hz
   c. 3.58 MHz

   The vertical scanning frequency is:
   a. 15,734 Hz
   b. 60 Hz
   c. 3.58 MHz

The color subcarrier frequency is:

a. 15,734 Hz

b. 60 Hz

c. 3.58 MHz

3. True or False   The luminance channel uses the entire 4 MHz bandwidth.

4. Does the picture information included in a composite video signal have as consistent spacing as the sync signals?

5. Luminance signals represent the _____ given by a televised object and provide the _____ detail.

6. What would occur if the electron beam was not deflected vertically and horizontally?

7. In a positive-phase composite video signal, the white portions of the signal point:

a. upward

b. downward

c. there are no white portions

In a negative-phase composite video signal, the sync tips point:

a. upward

b. downward

8. What is the purpose of the blanking and blacker-than-black levels of the composite video signal?

9. Because the amplitude variations correspond to the horizontal and vertical scan rates, a horizontal scan contains _____ changes while vertical scans contain _____.

10. True or False   Large areas of a constant white, gray, or black produce signal variations that occur at low frequencies.

11. Video signals can include frequencies ranging from _____.

12. Name five functions of the video amplifier stage.

13. The _____ and _____ of the video detector circuit determine the _____ of video amplifier stages used in a receiver.

14. Why do higher-frequency video signals carry fine details?

15. Why do low-frequency video signals carry detail for large areas?

16. True or False   A video amplifier stage must have an even response across the entire frequency band.

17. Explain how you would draw equivalent circuits for the circuit shown in Figure 10-5A.

18. Again, refer to Figure 10-5A. True or False   At all but the low frequencies, capacitors C_{IN} and C_{OUT} act as an open circuit.

19. The voltage gain always _____ the current gain multiplied by the ratio of the output resistance to the input resistance.

20. Name the effects that high frequencies can have on a video amplifier circuit.

21. What is the Miller effect?

22. Describe how series, shunt, and series-shunt peaking circuits offset the effects of high frequencies.

23. What is a symptom of poor low-frequency response?

24. The sharpness control varies the amount of _____ peaking.

25. What is the function of a scan-velocity module circuit?

26. Every change in the background brightness of a televised scene produces a dc voltage that becomes part of the video signal used to modulate the transmitter. A very bright background produces a

    _____ dc voltage, while a black background produces a

    _____ dc voltage.

27. Without the correct dc voltage, the video signal will _____.

28. A clamping circuit _____.

29. What would occur if diode D1 opened in the dc restoration circuit shown in Figure 10-14?

30. If a CRT is cut off, it has a completely (black) (white) screen. A CRT with zero bias has a completely (black) (white) screen.

31. The dc supply voltage for the video amplifier circuit must have a value more than the

    _____. Why?

32. Name two methods for controlling the bias of a CRT.

33. Define the terms "brightness" and "contrast."

34. How do circuit designs control the gain of a video amplifier stage?

35. If a completely white screen is displayed, the amplitude of the luminance signal is:
    a. maximum
    b. average
    c. minimum

    What causes this?

36. The _____ of the CRT used in the television governs the

    _____ of the video amplifier stage or stages.

37. In a CRT, _____ and the _____ can control the amount of beam current need to drive the CRT. Most systems use the

    _____ as a driver for the CRT.

38. Why do most systems use the _____ as a driver for the CRT?

39. True or False   Changing the bias of the tube so that the cathode is less positive than the control grid causes a decrease in the light output of the CRT.

40. Why does it matter if the composite video signal has a negative or positive polarity?

41. True or False   The AGC line connects to the video amplifier stage, the I-F section, and the tuner.

42. In the circuit shown in Figure 10-19, high-frequency components of the luminance signal take one path while the low-frequency components of the signal follow another. Given what you know about these parts of the luminance signal, discuss why this separation occurs.

43. In the circuit shown in Figure 10-20, the video amplifier is controlled by _____

    and _____. Describe the process used to control the video amplifier.

44. True or False   The sub-bright control is a customer control mounted next to the picture and brightness controls.

45. Another description for the picture control is _____.

46. How is the beam current limited in the circuit shown in Figure 10-21?

47. In Figure 10-21, the clamped voltage in the automatic brightness-control circuit is _____.

48. Refer to Figure 10-21. Why is a resistor connected between the luminance/chrominance processor and the high-voltage supply line?

49. True or False   Most luminance circuits are contained within a single IC package that combines the sync-separation functions with other functions.

50. What are four symptoms of a problem in a video amplifier circuit?

51. Why is the injection of a square-wave signal important during the troubleshooting of a video amplifier circuit?

52. When checking the gain of an amplifier with an oscilloscope, you would use a _____

    probe and look for changes in _____.

53. To check the frequency response of a video amplifier, connect a _____
    from the oscilloscope to the output of the video amplifier.

54. When injecting a square-wave signal into a video amplifier circuit, you should look for

    _____.

55. Name three component failures that can affect the response of an injected square wave.

56. In the first Service Call, why did the video amplifier circuit lose its supply voltages?

57. Choose the best answer. In the second Service Call, the source of the problem was disclosed by a check of the:
    a. supply voltages
    b. signal waveforms
    c. circuit connections
    d. components

58. Why did the technician use a cooling spray and hot air during the checks of the circuits?

59. Choose the best answer. In the second Service Call, the source of the problem was disclosed by a check of the:
    a. supply voltages
    b. signal waveforms
    c. circuit connections
    d. components

60. _____, _____, and

    _____ are interrelated characteristics of a waveform.

61. What oscilloscope control is used for time measurements?

62. Define rise time.

63. What oscilloscope control is used for peak-to-peak voltage measurements?

64. Refer to Figure 10-22A. What is the frequency of the waveform if the time duration is doubled? What is the frequency of the waveform if the time duration is cut by 55 percent?

65. Wavelength is measured in _____. Time is measured in

    _____. Frequency is measured in _____.

66. Refer to Figure 10-22D. What is the phase difference between the reference and signal 2?

67. What are the advantages of using a dual-trace oscilloscope for phase measurements?

68. Refer to Figure 10-22C. What is the peak-to-peak value of the sinewave if the control is set to 5 and the screen is divided into 0.5-centimeter squares?

69. Draw a waveform that is distorted by a damped oscillator.

CHAPTER 11

Chrominance Signals and Chroma Amplifiers

OBJECTIVES Upon completion of this chapter, you should be able to:

1. Describe the characteristics of color and hue signals.
2. Discuss vector addition and subtraction.
3. Define frequency interleaving.
4. Explain how comb filters work.
5. Discuss the operation of the bandpass amplifier, color killer, burst amplifier, AFPC, color demodulator, and color matrix circuits.
6. Explain how various controls and signals affect chroma signal processing.
7. Discuss the operation of video output circuits.
8. Troubleshoot chroma-processing circuits.

KEY TERMS

automatic chroma control

automatic frequency and
    phase control

automatic tint control

burst amplifier

burst gate

burst separator

charge-coupled device

chroma phase detector

chrominance

color bandpass amplifier

color-bar/cross-hatch
    generator

color burst signal

color control

color-difference amplifier

color-difference drive matrixing

color killer

color oscillator

comb filter

complementary color

frequency interleaving

hue

hue control

in-phase signal

keyed amplifier

keyed rainbow generator

low-level RGB matrixing and
    CRT drive

matrix circuit

primary color

quadrature signal

saturation

subcarrier reference system

tint

tint control

vector

INTRODUCTION

Now that we have established that the luminance signal travels from the video amplifier to the CRT, we move back to the point where the chrominance signal separates from the composite video signal. At this point, a chroma bandpass amplifier stage prepares the chroma signal for demodulation by attenuating the high and low frequencies found in the 4-MHz band and by shaping the frequency response curve. Demodulators compare the phase of a reference signal with the modulated sync signals, and output the color and hue information to additional amplifiers, which pass the signals to the CRT.

In this chapter, you will learn more about the individual components—such as the color burst subcarrier and comb filters—that make up the signal. The chapter describes how proportionate amounts of the chrominance signal combine to produce a luminance signal, and how the video output stage connects the chroma-processing circuitry with the CRT.

Chapter eleven spends considerable time describing the fundamental circuits that combine to process the chrominance signals. The chapter also revisits the luminance/chrominance processor first seen in chapter ten. As always, the chapter concludes with a look at troubleshooting techniques for chrominance circuits and reviews several case histories.

11-1   WHERE DO CHROMINANCE SIGNALS COME FROM?

Primary colors are colors that do not result from the mixing of other colors. Video systems use red, green, and blue as primary colors because combining the three colors in various ways can produce a large number of color mixtures. The complete transmission/reception system begins with red, green, and blue signals at the camera and finishes with red, green, and blue signals at the receiver. Most systems use mixtures of primary colors during encoding and decoding, because two color-mixed signals can contain all the color information found in the three primary colors.

Previously, we identified the different parts of the composite video signal as the luminance, sync, and chrominance components. Within the **chrominance** components, a 3.58-MHz subcarrier signal modulates the video carrier and produces a side frequency. For example, the color subcarrier modulates the video carrier for channel three and produces a 3.58-MHz + 61.25-MHz, or 64.83-MHz, modulated picture-carrier signal. The 3.58-MHz frequency is constant for all channels as a modulating signal.

Using the 3.58-MHz frequency for the color subcarrier separates the chrominance signal from any portion of the lower video signals found within the luminance signal. It establishes enough distance between the subcarrier and the sound signal to lessen the chances of chroma interference from showing in the luminance signal, and to prevent the 3.58-MHz color subcarrier from interfering with the 4.5-MHz sound signal.

At the transmission point, the subcarrier is modulated through the use of 90 degree phase separated **in-phase** or **quadrature signals**. Modulating the subcarrier with the in-phase, or I, signal does not cause a phase shift to occur. Modulating the subcarrier with the quadrature, or Q, signal shifts the subcarrier by 90 degrees. Consequently, as part of the entire 3.58-MHz modulated chrominance signal, the two color-information-carrying signals remain separated by a maximum of 90 degrees. The transmission of the I and Q sidebands without the carrier within the composite video signal is called suppressed-carrier transmission, and it eliminates any potential 3.58-MHz interference.

As the following equations show, the I and Q signals contain proportional amounts of the red, green, and blue signals:

$$I = -0.60R + 0.28G + 0.32B$$
$$Q = 0.21R + 0.31B - 0.52G$$

Combined, the I and Q signals must contain all the red, blue, and green color information. Thus, the two signals consist of mixtures of the colors so that the combination provides the correct information.

Hue, Saturation, and Chrominance

Although the equations show the I and Q signals with positive polarities, both signals may also have negative polarities. The polarities of the voltages that correspond with the primary colors will have the opposite polarities from those seen in the equations. Given opposite polarities, the new colors have a hue that is the direct opposite of the primary colors. Therefore, with the positive-polarity I and Q signals representing mixtures of the primary colors, the negative-polarity I and Q signals represent mixtures of complementary colors. A **complementary color** is a color that produces white light when added to a primary color.

With a positive polarity, the I signal appears orange, while the negative-polarity I signal appears cyan. As a result, the I signal contains all color mixtures between orange and cyan. Every color within these mixtures is a derivative of the red, green, and blue primary colors. Given these colors, the I signal shows small details of color. A positive-polarity Q signal appears purple, while a negative-polarity Q signal appears yellow-green. Consequently, the Q signal contains mixtures of every color between purple and yellow-green, or a set of red, blue, and green mixtures.

Applying in-phase and quadrature-phase modulation to the chrominance signal concentrates almost all the color information into one of the chrominance signals. Thus, the color corresponding to a complete chrominance signal resulting from the mixing of a strong I signal and a weak Q signal will have an orangish tint. A weak I signal and strong Q signal will produce a purplish tint. Equal applications of the I and Q signals will yield a color somewhere between orange and purple.

Every color signal represents hue, saturation, and luminance. Essentially, the **hue**, or **tint**, is the color and is represented as an angular measurement. For example, a blue sweater has a blue hue. When the human eye senses different wavelengths of light for an object, it sees different tints. **Saturation** represents the degree of white in a color and is represented through amplitude measurements. A highly saturated color is intense and vivid, while a weak color has little saturation. For example, a red signal that has a low amplitude will have less saturation and will appear as pink.

Within the chrominance circuitry, separate red, blue, and green video channel voltages equal proportionate combinations of the luminance signal and the two subcarrier frequency signals:

R = Y + 0.96(I) + 0.62(Q)
G = Y − 0.28(I) + 0.65(Q)
B = Y − 1.10(I) + 1.70(Q)

Amplitude variations of the complete chrominance signal indicate the amount of saturation within the color information. Therefore, when we measure the amplitude of the chrominance signal, we are measuring the saturation of the signal.

If we subtract luminance from hue and saturation, we have chrominance. In terms of signals, the chrominance signal is the 3.58-MHz-modulated subcarrier signal contained within the composite video signal. Because the human eye cannot detect small color details, the full bandwidth of the chrominance signal ranges from 0 Hz to 1.5 MHz, and the used bandwidth ranges from 0 Hz to only 0.5 MHz. At the transmission point, color information modulates the subcarrier and produces sidebands that extend from 0.5 MHz and 1.5 MHz away from the subcarrier frequency.

☑ PROGRESS CHECK

You have completed Objective 1. Describe the characteristics of color and hue signals.

Vectors and Phase Angles

When we move to the study of chrominance signals, you will find that each hue represents signals that have an equal frequency but different phase angles. The chrominance-processing circuits either add or subtract the phase angles to produce a particular color signal. Vector addition and subtraction provide additional insight into this process.

A **vector** defines both quantity and direction. When working with electrical quantities, phase is the angular measure, while amplitude corresponds with the size of the vector. As you know, the phase of an electrical sine-wave signal continuously moves through a complete circle called a cycle. Because of this, the vector also continuously rotates through a sine-wave voltage cycle. As this rotation occurs, we can define the vector through its angular position (phase) and its length (size), or by showing the vertical and horizontal components of the vector. Figures 11-1A and B illustrate both methods for measuring the vector. Looking at the figures, we can describe the vector as 10 volts at 45 degrees, or as 7.07 horizontal plus 7.07 vertical. When studying the transmission and reproduction of chrominance signals, the method shown in Figure 11-1B is more widely used.

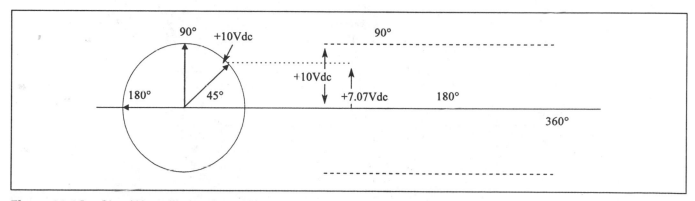

Figure 11-1A Sine-Wave Illustration of Vector Angles

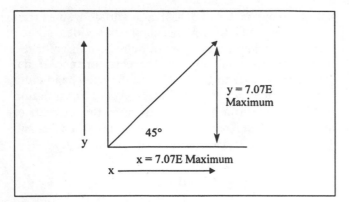

Figure 11-1B   Phase-Angle Measurement of a Vector

Figure 11-1B illustrates the use of several mathematical functions—the geometry of a right triangle, the sine of an angle, and the cosine of an angle—to find the vector relationship. While it is highly unlikely that you will perform algebraic equations during your work on video receivers, knowing how these mathematical functions are applied to color signals will help you understand why the chrominance signals have a specific form and how the chrominance-processing circuits operate. Every chrominance-processing process seen later in this chapter will involve vector addition and subtraction.

As mentioned, vectors include phase and size measurements. Because of this, we cannot add vectors with two different phases. Instead, as Figure 11-2 shows, we need to add those vectors graphically. The adding of the two different phase vectors produces a new vector with a new angle. Although we cannot add the phase angles or length of the vectors together, we can combine the x and y coordinates of each vector to find the quantity of the vector. With this, the new vector equals:

$$\text{Angle A} + \text{Angle B} = (x_1 + x_2) + (y_1 + y_2)$$

To subtract vectors, we can simply add a positive quantity or polarity to a negative quantity or polarity.

If we use a graphical representation of vector subtraction, the negative vector has the opposite or reverse direction of the positive vector, as shown in Figure 11-3. As with the adding operation, we can also subtract the vectors algebraically by adding the difference of the two x-coordinates with the difference of the two y-coordinates:

$$\text{Angle A} - \text{Angle B} = (x_1 - x_2) + (y_1 - y_2)$$

Vector addition and subtraction in terms of electrical signals becomes easier to understand when we consider the sine waves shown in Figures 11-4 and 11-5. In Figure 11-4, sine waves A and B have a 45-degree phase difference. Adding the two sine waves together produces the third sine wave, C. At each point of sine wave C, the total amplitude equals the sum of the amplitudes of sine wave A and sine wave B. When the sine wave dips below the zero axis and takes on a negative value, the amplitude is subtracted from the amplitude of the portion above the axis.

The vector representation of the three sine waves depicts the relationship of the three signals at every instant and as the three signals rotate together at the same frequency. For the purposes of this example, both sine waves A and B have an amplitude of 10 volts. In actual practice, color signals will have different phase angles and different amplitudes and produce a very irregular waveform. At the receiver, one voltage will

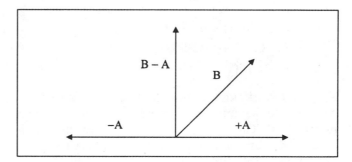

Figure 11-3   Vector Subtraction

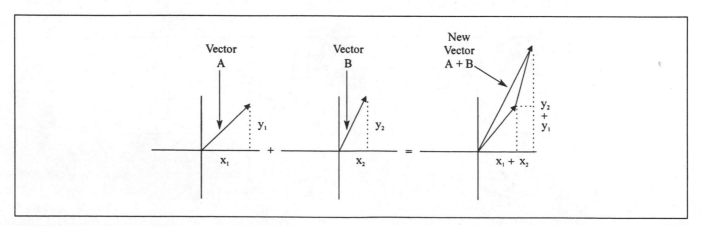

Figure 11-2   Vector Addition

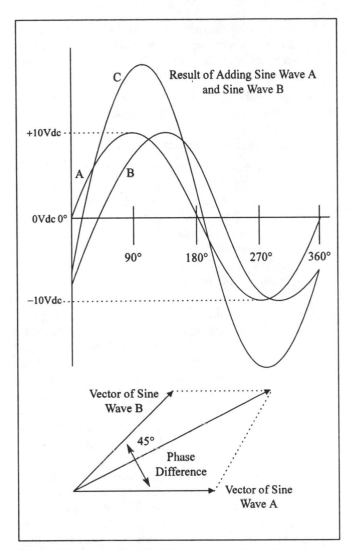

Figure 11-4 Vector Addition of Two Sine Waves

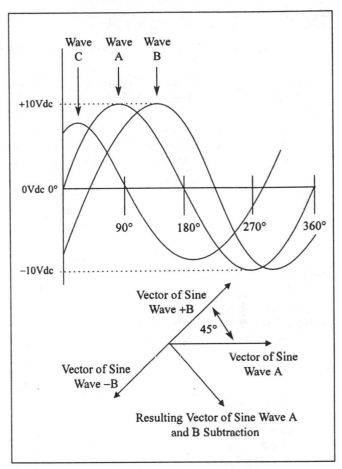

Figure 11-5 Vector Subtraction of Two Sine Waves

have a constant phase and amplitude while the other will have a varying phase and amplitude. The resulting waveform represents the demodulated video information contained within the varying signal.

In Figure 11-5, we subtract sine wave B from sine wave A and produce sine wave C. Again, the sine waves have a phase-angle separation but have the same amplitude. However, the difference between the two sine waves produces a wave with an amplitude that is the difference between the amplitudes of waves A and B. Like Figure 11-4, Figure 11-5 shows the vector representation of the sine wave subtraction.

Both Figures 11-4 and 11-5 illustrate one other factor that we should consider. Although the original sine waves in each figure have the same amplitude and a constant phase difference of 45 degrees, the resulting waveforms appear different than they should if they were the result of only arithmetic addition or subtraction. In both cases, the introduction of harmonics has altered the appearance of the waveform.

⟳ FUNDAMENTAL REVIEW

Harmonics

The basic or lowest frequency of a tone is called the fundamental frequency. Harmonics have frequencies that are multiples of the fundamental frequency.

Phase Angles and Chrominance Signals

The modulation of the chrominance signals occurs because of shifting of the phase of an R-F signal with respect to a specific reference signal that has the same frequency. The two 90-degree out-of-phase I and Q signals are transmitted on the 3.58-MHz color subcarrier. Demodulation occurs because of the sensitivity of a detector to the phase difference between the two signals. The detector recovers the original video signal. When we consider the modulation of the I and Q signals, we should remember that both the I signal and the Q signal are amplitude-modulated.

The Color Sync Burst Signal

The phase reference signal—called the **color burst signal** and consisting of 8 to 11 cycles of the subcarrier frequency—rides on the back porch of each horizontal blanking interval. Although the unmodulated color burst signal is included in each horizontal interval, it does not appear on the CRT screen as any type of visible signal. As shown in Figure 11-6, the peak-to-peak amplitude of the burst signal equals the amplitude of the sync signal. In terms of timing, the color burst occurs at the same time as the blanking pulse—which corresponds with a zero value for the sync pulses. Because of this timing, the color burst signal does not interfere with the sync signals or, as a result, the timing of the deflection oscillators. Later in this chapter, we will examine how the color burst signal becomes a reference for control of the hue.

The 3.58-MHz color subcarrier frequency results from a relationship with the 455th harmonic of one-half of the horizontal scanning frequency:

$$455 \times (15{,}750 \text{ Hz}/2) = 3.58 \text{ MHz}$$

Receivers rely on 3.58 MHz as a color signal frequency for several reasons:

- Monochrome receivers use only the luminance signal. With the chrominance signal placed at 3.58 MHz, the high-frequency chrominance signal has no effect in monochrome receivers.

- Video receivers rely on heterodyning between the picture- and sound-carrier frequencies to produce the 4.5-MHz beat frequency.

- Placing the chrominance signal at 3.58 MHz minimizes interfering beat frequencies caused by heterodyning between the chrominance and sound-carrier frequencies.

- Placing the chrominance signal at 3.58 MHz minimizes interfering beat frequencies caused by heterodyning between the chrominance and medium video frequencies.

During modulation of the 3.58-MHz subcarrier for broadcast, there are two key phase differences: a 57-degree phase angle between the I signal axis and the zero phase signal, and a 90-degree phase angle between the I and Q signal carriers. Figure 11-7 illustrates the differences between the I and the Q signals. Each combination of the I and Q signals produces color difference signals at the receiver which, in turn, cause a matrix in the receiver to produce specific voltages for each reproduced color.

Figure 11-8 uses a simple color wheel to illustrate that the hue of the chrominance signal is based on phase angles. When we place the colors produced by the I and Q signals as coordinates on the color wheel, we find that any color in either the I or Q signal range matches a particular angle. Shifting the chrominance signal toward either the I or the Q signal causes the phase angle of the chrominance signal to change, and affects the tint of the reproduced picture.

During our study of the amplification of video signals in chapter ten, we found that frequencies ranging from 100 Hz to 4 MHz can carry varying amounts of detail. Higher frequencies correspond with fine lines, edges, and other details, while lower frequencies correspond with larger areas. Like the luminance signals, the bandwidth of color signals also has a relationship with the sensitivity of the human eye. Our eyes are sensitive to the amount of white in a color and less sensitive to differences in hue. Given our ability to perceive certain differences in saturation and lesser differences

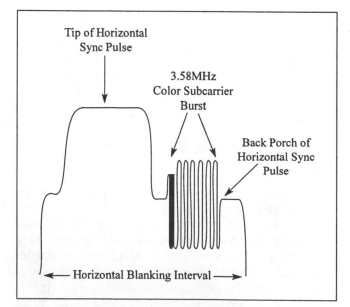

Figure 11-6   Comparing the Amplitude of the Horizontal Sync Pulse and the Color Burst

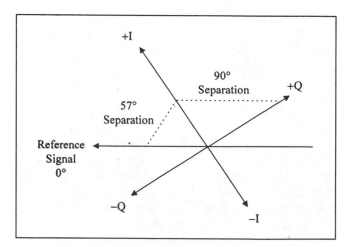

Figure 11-7   Comparison of I- and Q-Signal Vectors

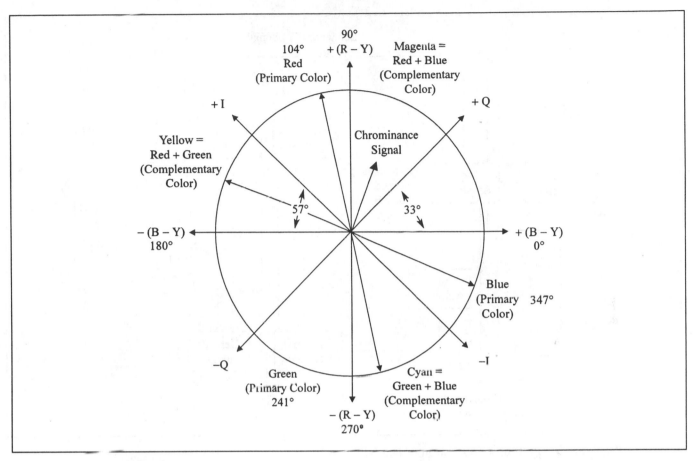

Figure 11-8   Color Wheel

in hue, the bandwidths of the transmitted chrominance and luminance signals have specific values. Those values are:

Luminance	Y signal	4 MHz
Saturation	I signal	1.3 MHz
Hue	Q signal	0.4 MHz

When looking at these values, it is easy to see that the I signal has a much wider bandwidth than the Q signal.

The I signal represents orange-yellow, but the human eye cannot discern the extra detail in the orange region. Consequently, television manufacturers do not use the full 1.3-MHz bandwidth of the I signal during demodulation. Instead, receiver demodulation systems use either narrower-bandwidth color difference signals for the vector addition and subtraction, or—because of the low costs associated with IC technologies—equal-bandwidth processing.

☑ **PROGRESS CHECK**

> **You have completed Objective 2. Discuss vector addition and subtraction.**

11-2   SEPARATING THE CHROMINANCE AND COMPOSITE VIDEO SIGNALS

A color television receiver contains circuitry that separates the chrominance signals from the remainder of the composite video signal, and processes the chrominance signals so that a synchronized color image appears on the CRT screen. The chrominance signals separate from the luminance signals either after the composite video signal exits the detector or after the first video amplifier stage. From there, the chrominance signal travels through a group of circuits that:

- remove and amplify the color burst signal
- reinsert the suppressed color subcarrier
- recover the original color difference signals
- control the hue and saturation of the reproduced colors
- disable the color signal during a monochrome broadcast

Moving to the partial block diagram of a color television shown in Figure 11-9, the chrominance and sync signals enter the video amplifier stage as part of the

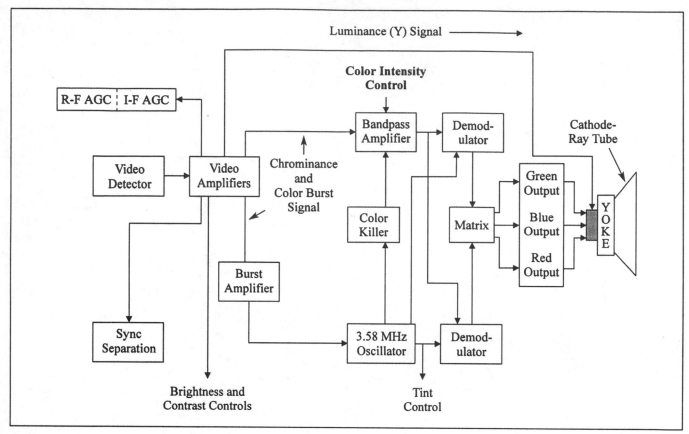

Figure 11-9   Block Diagram of Video, Chroma, and Sync Paths

composite video signal, but then separate to follow different paths. Complete passage of the chrominance signal requires a uniform amplifier response of at least 3.6 MHz due to the presence of the color subcarrier and sidebands. Because of that factor and because it passes both the luminance and chrominance signals, the first video amplifier must have a response that extends to 3.6 MHz.

After the first video amplifier, the luminance signal takes a relatively direct path to the CRT, while the chrominance signal is delayed because of its passage through several layers of amplification. A delay line built into the luminance path slows the luminance signal and allows the luminance and chrominance signals to arrive at the same time.

⟳ FUNDAMENTAL REVIEW

Delay Lines

Delay lines consist of an inner conductor, which provides an inductance in series with the signal path, and a distributed capacitance, which shunts the signal path. Capacitance occurs because of the spacing between the inner and outer conductors of the delay line. Delay lines have a frequency response that remains flat to at least 4 MHz.

A color-separator system separates the chroma signals from the luminance signals and passes the signals onto the remainder of the chroma-processing circuitry. First, circuits must demodulate the subcarrier and produce the chrominance sidebands. Then, processing circuits use the three primary colors to reproduce exact color and monochrome images of the transmitted scene. Exact processing of these images requires that the circuits monitor and mathematically mix different quantities of the primary colors to produce different color shades. In addition, the circuits must regulate the saturation of all reproduced colors.

Frequency Interleaving

When the original standards for the transmission of color television signals evolved, the best method for transmitting those signals was to share the bandwidth with the monochrome signal transmission. However, luminance signals in a color transmission may have a bandwidth anywhere between zero and 4.2 MHz. To provide a transmitted signal with color information, a chroma subcarrier is inserted at 3.58 MHz. Color information modulates the suppressed subcarrier and creates sidebands that extend 0.5 MHz above and 1.5 MHz below the subcarrier frequency. The sound carrier is located 4.5 MHz from the video carrier and has at least a 0.1-MHz bandwidth.

All this information fits within the 6-MHz channel because of the characteristics of the luminance signal. From 60 Hz to 4 MHz, the luminance signal does not consist of a continuous band of energy, but rather varies as bursts of energy spaced in 30-, 60-, and 15,534-Hz intervals. The spacing conforms to the frame scanning rate (30 Hz), the field scanning rate (60 Hz), and the horizontal scanning frequency (15,750 Hz). In Figure 11-10, the frequency components of the luminance signal appear as harmonics clustered around each frequency.

Frequency interleaving—or the interlacing of odd and even harmonic components of two different signals to minimize interference between the signals—allows the transmission of the chrominance signal within the same 6-MHz channel as the luminance and sound signals. Interleaving of the signals occurs because of the time and phase relationships seen with the luminance and chrominance signals. With the luminance signal, a concentration of energy occurs at multiples of the horizontal scanning frequency. Chrominance information peaks at a frequency of approximately 3.58 MHz.

Luminance signal information appears as an energy burst between every horizontal sync pulse. Chrominance information also appears as an energy burst at the horizontal rate but, because of the introduction of a 3.58-MHz chroma subcarrier, is offset by one-half the horizontal rate. With this offset, the chroma information is displaced by 15 cycles and falls between the luminance signal harmonics.

Referring to the figure, luminance frequency pairs fit above and below the horizontal scanning frequency and are separated by 60-Hz gaps. As a result, a repro-

duced scene that has 40 pairs of luminance frequencies requires a bandwidth of ± 40 × 60 Hz or 2400 Hz. The 60-Hz separation creates open spaces between each luminance signal cluster. Each of those spaces occurs at an odd multiple of one-half of the horizontal line-scanning frequency. Because of the relationship between the chrominance signal and the horizontal scan frequency, the chrominance signal frequencies fit, or interleave, between the clusters of luminance signal frequencies. This occurs because the 3.579545-MHz color subcarrier frequency is the 455th harmonic of half the line-scanning frequency.

Limiting the bandwidth of the luminance signal to approximately 3 MHz prevents the subcarrier sideband information from mixing with the 4.5-MHz sound carrier and appearing as interference in the luminance portion of the video signal, which would show as patterns in the reproduced picture. Bandwidth limiting prevents the interference from occurring between the low-frequency sidebands and the high-frequency luminance signals. Unfortunately, bandwidth limiting also results in some loss of picture detail.

☑ PROGRESS CHECK

You have completed Objective 3. Define frequency interleaving.

Comb Filter Circuits

Given the appearance of the spectrum, chrominance/luminance-processing circuits incorporate a filter that has a comb-like amplitude response, which allows the total transmission of desired signals and the total attenuation of any undesired signals. Thus, a **comb filter**

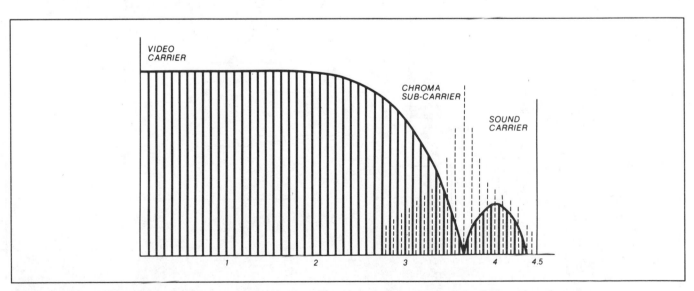

Figure 11-10 Illustration of Frequency Interleaving (*Reprinted with permission of Thomson Customer Electronics, Inc.*)

can separate the chroma and luminance components from the composite video signal. As a result, comb filtering recaptures video information residing between 3 and 4.2 MHz and restores the lost picture detail.

The comb filter accomplishes this separation by delaying the composite video signal one horizontal scan period, or approximately 64 microseconds, and then either adding or subtracting the undelayed composite video signal. Along with separating the two signals, the comb filter also rejects the chroma subcarrier from the luminance channel. By improving the signal-to-noise ratio of both the luminance and chrominance channels, the comb filter enhances the rejection of crosstalk between the two channels.

The luminance and chrominance channels result from the subtracting and adding of the signals in a comb filter circuit. Delaying the composite video signal by one horizontal scan period produces a difference channel, or luminance path, that consists of only the in-phase luminance signals that occur during that interval. All out-of-phase signals are rejected. The sum channel, or chrominance path, contains the interleaved chrominance signals and rejects the in-phase luminance signals.

Using a Delay Line in a Comb Filter Circuit

Figure 11-11 shows the schematic diagram of a comb circuit that relies on a delay line to slow the luminance signal. While the chrominance signals go through a multistage process that includes amplification, demod-

ulation, and inversion, the luminance signal travels through only one stage of amplification. Because of this, the chrominance signals require more time to reach the output stages than the luminance signals. The delay line ensures that the two different signals will arrive at the matrix stage at exactly the same time. Although this comb filter design includes a delay line, many manufacturers will use a separate delay line in their circuit designs.

In the figure, the composite video signal feeds into a single input, while the luminance and chrominance signals feed out of separate output terminals. A delay line at the center of the figure introduces a time delay of 63.5 microseconds, or the equivalent of one complete horizontal cycle. Arriving simultaneously at an adder, the delay line output signal and the inverted delay line output signal have a phase difference. Yet, because the second signal is inverted, the two signals add together and double the amplitude of the chrominance signal. With the delay line slowing the composite video signal by one complete cycle, the output of the delay line is in-phase with its input.

A composite video signal with a peak-to-peak amplitude of 2.5 Vdc is applied to Q154, a delay line driver. After the amplifier inverts and amplifies the signal, it travels to the delay line, DL152. Amplification of the signal overcomes any possible loss that occurs during the insertion of the composite video signal into the circuit. With the ultrasonic glass delay line operating as a piezoelectric bandpass device, external

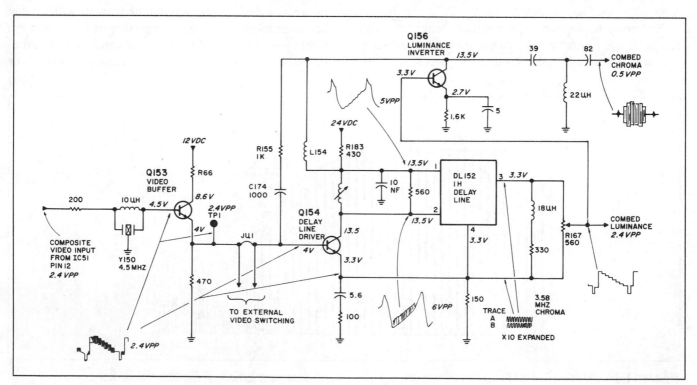

Figure 11-11   Schematic Diagram of a Comb Circuit with a Delay Line (*Courtesy of Philips Electronics N.V.*)

resistors and inductors tune with the capacitance of the delay line. Because the delay line must pass the range of the chrominance sidebands, it has a bandpass characteristic of 3 to 4 MHz.

The delay line driver also emitter-couples the composite video signal to R167, an adjustable resistor that nulls, or eliminates, any undesired chroma signals at this point. A luminance signal found at the R167 wiper consists of the sum of the delay line output and the original composite video signal. Luminance signals feed from the R167 wiper to a luminance equalizer for further amplification, and from the delay line to a luminance phase splitter. After comparison of the two luminance signals and correction of any phase differences, the signals pass to a luminance/chrominance processor IC.

The luminance signal inverter adds the output of the delay line to the inverted video signal, cancels the luminance signal through inversion, and provides a combed chroma signal. Inverting the combined luminance signal and then adding the luminance signal to the original composite video signal cancels the luminance signal. Any remnants of the luminance signal are eliminated because the signal falls outside the passband. If the amplitude of each of the chrominance signals decreases to zero, the red, blue, and green signals equal one another and produce white. From another perspective, white contains no chrominance information. Consequently, we can deduce that the luminance signal provides all the information needed to produce white and all shades of white.

Charge-Coupled Devices in Comb Filter Circuits

Although the frequency spectrum establishes an ideal model for comb filtering, actual television pictures may not have an exact fit within the spectrum. The delay line cannot offer the bandwidth needed to filter frequencies ranging below 3.58 MHz. Therefore, the use of a delay line will not provide the length of delay needed to prevent chroma subcarrier information from blending with the luminance signal and degrading the reproduced picture. To counter this problem, some comb circuits may use a **charge-coupled device**, or CCD, to achieve the necessary signal delay.

In Figure 11-12, one integrated circuit contains the charge-coupled device and summing circuits, provides comb filtering, and amplifies the luminance and chrominance signals. At the upper left of the diagram, the composite video signal couples into the comb filter circuit and divides along three paths. While one signal path travels through the charge-coupled device and is delayed by 63.5 microseconds, another path goes through the amplified luminance-processing channel. The third path goes through the amplified chroma-processing channel, where it is amplified and applied to the input of a summing circuit. The portion that travels through the charge-coupled device also reaches the summing circuit but, because of the delay, arrives one horizontal line later. With the luminance information appearing the same to the summing circuit, the circuit doubles only the in-phase components of the input signals. Because the input signal information contains both in-phase luminance and 180-degree

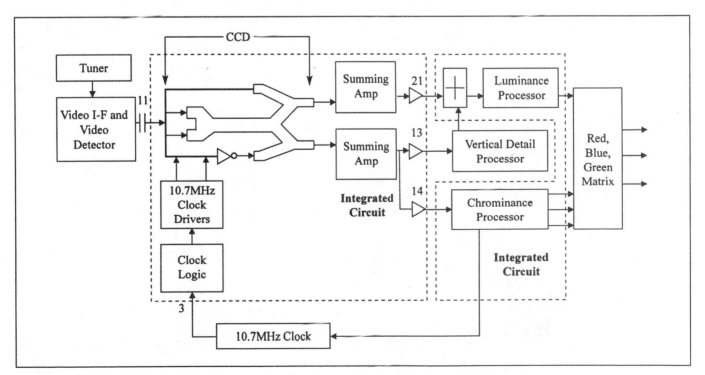

Figure 11-12 Schematic Diagram of a Comb Circuit and Charge-Coupled Device

out-of-phase chrominance information, the circuit cancels the phase-shifted chroma signal. As a result, only luminance information travels to one output of the comb filter circuit.

At the chroma-processing channel, the circuit inverts the incoming composite video signal and then applies the signal to an amplifier. As with the luminance channel, the signal then travels to one input of an adder circuit. Because of the inversion at the beginning of the chroma-processing channel, the signal would appear to be 180 degrees out-of-phase with the signal at the output of the CCD. Yet, because the line-by-line delay provides another 180-degree phase shift, the inverted signal and the delayed signal remain in phase.

The chrominance signal passes because the frequency of the input signal deviates from whole multiples of 15,750 Hz. With the output reaching its maximum amplitude when the input signal is an odd multiple of 15,750 Hz divided by 2, the amplified chrominance signal passes. Chroma bandpass filtering eliminates any remaining undesired luminance signals.

Circuit operation begins with the application of +16 Vdc, +9.1 Vdc, and −5 Vdc at power supply pins on the circuit. While the composite video signal enters the circuit at pin 11, a 10.7-MHz clock pulse is coupled to the circuit at pin 3. The clock pulse results from a 3.58-MHz tripler circuit. Combed luminance output, vertical detail information, and combed chroma information output signals appear at pins 21, 13, and 14. The circuit requires the addition of the vertical detail information at pin 13 because of the combining of the composite video spectrum. Taking the low-frequency output from the difference channel and adding that information to the combined luminance signal yields the vertical detail information signal. Adding a greater amount of low-frequency signal than the amount of low-frequency signal originally lost enhances the vertical detail.

The charge-coupled device shown in Figure 11-12 provides a full bandwidth and a time delay independent of frequency changes. With the CCD, a fixed number of elements depend on a clock frequency locked to a harmonic of the bandpass characteristics of the comb filter. A charge-coupled device is a semiconductor, but operates much like a capacitor, storing a given amount of charge. In this particular circuit, the CCD stores video information in 682.5 storage elements. During operation, a 10.7-MHz clock pulse supplies the energy needed to shift the charge in one CCD element to the next element in the series. The necessary delay of the luminance signals occurs through the element-to-element shifting of information. Because each shift represents a delay of 93 nanoseconds, and the CCD has 682.5 elements, the total delay provided by the CCD equals 63.5 microseconds. With this system, the luminance information is delayed without sacrificing signal quality.

The 10.7-MHz clock pulse develops through the tripling of the 3.58-MHz chroma oscillator signal. With each pulse, the comb circuit samples the video information and applies a charge equal to the video information to a CCD element. Moving to Figure 11-13, the

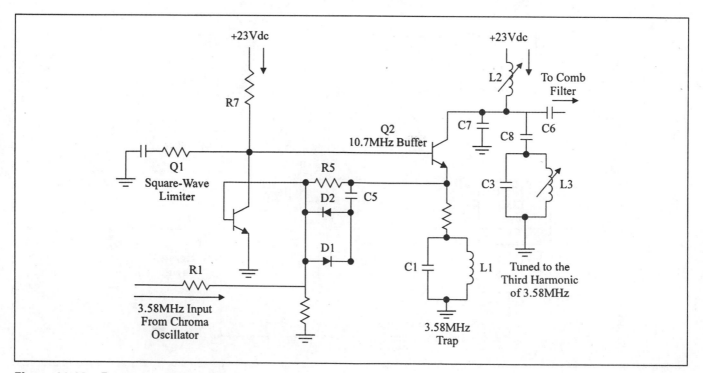

Figure 11-13 Frequency-Tripler Circuit

frequency-tripler circuit includes Q1, a square-wave limiter, Q2, a buffer, diodes D1 and D2, and several LC networks. With the 3.58-MHz chroma oscillator signal applied to the base of Q1, the transistor amplifies and applies the signal to the base of Q2. Diodes D1 and D2 provide a feedback path back to the base of Q1 and clip the waveform. The clipping establishes the steep rise time seen in the 3.58-MHz square wave at the emitter of Q2. Frequency tripling occurs because of the action within the tank circuit consisting of L2, C8, L3, C3, and

C6. The LC circuits resonate at the third harmonic of the 3.58 MHz and generate the 10.7-MHz clock pulse.

Additional circuits reestablish the vertical detail initially lost during the comb filter operation and combine that information with the luminance signal. The circuit depicted in Figure 11-14 extracts vertical detail by bandpassing the chroma information, and then adds the vertical detail output signal to the combed luminance information. In this way, the circuits peak the vertical pulse and provide a whiter-than-white leading edge

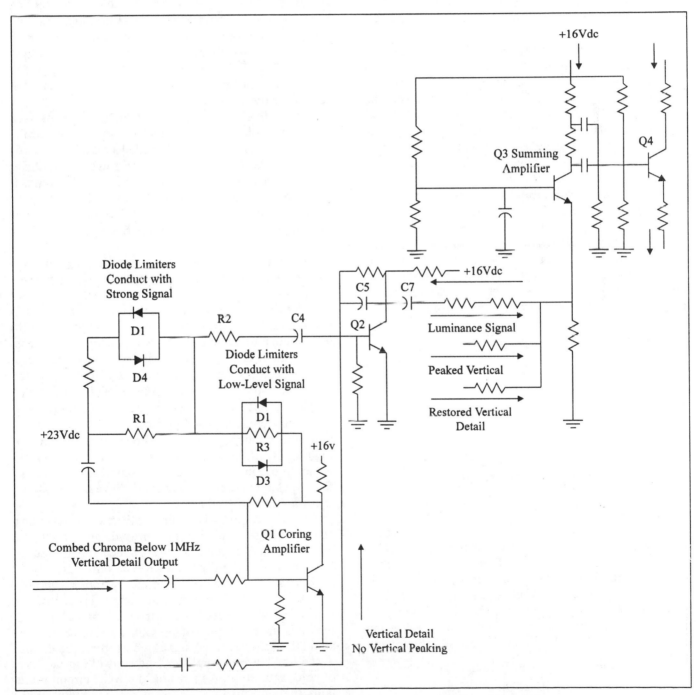

Figure 11-14   Vertical Detail and Summing Amplifier Circuit (*Courtesy of Philips Electronics N.V.*)

and a blacker-than-black trailing edge. Because the whiter-than-white and blacker-than-black transitions can either push the CRT into saturation or cause video smearing, the circuit varies the amount of peaking with modulation levels.

In Figure 11-14, the ac modulation level of the input signal and the feedback paths established through diodes D1, D2, D3, and D4 control the gain of Q1, a nonlinear amplifier. With a low-level input signal, diodes D1 and D3 do not conduct, the signal flows through R3, and the circuit has a low output. As the input signal level increases, the two diodes conduct and apply the signal to the base of Q1 through R2 and C4 and to the output coupling capacitor, C5.

Further increases in the input signal level cause diodes D2 and D4 to conduct. With this, R2 parallels R1 and reduces both the impedance of the feedback path and the gain of the amplifier. Consequently, the nonlinear processed signal flows into the base of Q2, the peaking amplifier, through coupling capacitor C5. Here the processed signal adds to the nonpeaked vertical detail signal and produces a difference signal that contains only the vertical peaking information.

This peaking information travels to the emitter of Q3, a summing amplifier, through capacitor C7. The summing amplifier adds and amplifies the combed luminance information, the vertical detail output information, and the peaked vertical detail signal. After the combined signal is applied to the base of Q4, the video buffer, it serves as the input signal for the luminance portion of the luminance/chrominance processor.

✓ PROGRESS CHECK

You have completed Objective 4. Explain how comb filters work.

11-3 REVISITING THE LUMINANCE/ CHROMINANCE INTEGRATED CIRCUIT

Chapter ten introduced the combination luminance/ chrominance processing circuit by emphasizing the luminance portion of the circuit. Figure 11-15 is a block diagram of the luminance/chrominance processor. It illustrates that a large part of the integrated circuit package is dedicated to chrominance signal processing. In this section, we will use the chrominance-processing portion as a platform for our study of the following:

- the chroma bandpass amplifier stage
- the burst separation stage
- the color killer circuit

- the color sync reference system
- the demodulators
- the hue and color controls
- the matrix

The Chroma Bandpass Amplifier

The **color bandpass amplifier** stage, or BPA, separates the color subcarrier and burst signals from the video signal, and amplifies the 3.58-MHz subcarrier sidebands. At the same time, the color sync section generates the color reference signal that locks in with the color sync burst. The amplified color component drives the demodulators and burst amplifier. Looking back at Figure 11-15, the color bandpass amplifier stage consists of the three chroma amplifiers found at the left-center of the diagram.

Although early color television designs utilized broadband color bandpass amplifiers that covered the entire I and Q sidebands and needed a total bandpass of 2 MHz, modern designs rely on a narrowband amplifier stage that does not pass signals above 0.5 MHz and that has a total bandpass of 1 MHz. Typically, though, a chroma bandpass amplifier stage will consist of two or three chroma bandpass amplifiers that operate as Class A tuned devices.

Referring to the diagram, the combed chrominance signal feeds into pin 3 of the integrated circuit, while a dc voltage is applied to the color-level adjustment through pin 2. Tuned circuits in the input path allow the chrominance portion of the composite video signal to pass while attenuating any low-frequency components of the signal. The chroma amplifiers provide the necessary bandpass characteristics by eliminating all the higher and lower frequency components of the video spectrum. The attenuation of these frequencies prevents interference caused by stray I-F signals and provides the frequency response shown at the bottom of Figure 11-16. Because of the tuned circuit component values, the signal traveling into the first chroma amplifier has a flattened frequency band with a range extending from 3.08 to 4.08 MHz.

As a whole, the two to four amplifiers that make up a color bandpass amplifier stage provide an overall gain of 50, resonate at the 3.58-MHz subcarrier frequency for any channel, and produce the chrominance signal at the amplifier output. While the customer can control the gain of the amplifier through the use of the color control, the **automatic chroma control**, or ACC, provides an even control of gain by monitoring the strength of the received signal, controlling the dc bias of the first chroma amplifier, and controlling the level of color saturation. As a result, the ACC circuit maintains the chrominance signal at the bandpass amplifier at a constant level under changing input signal

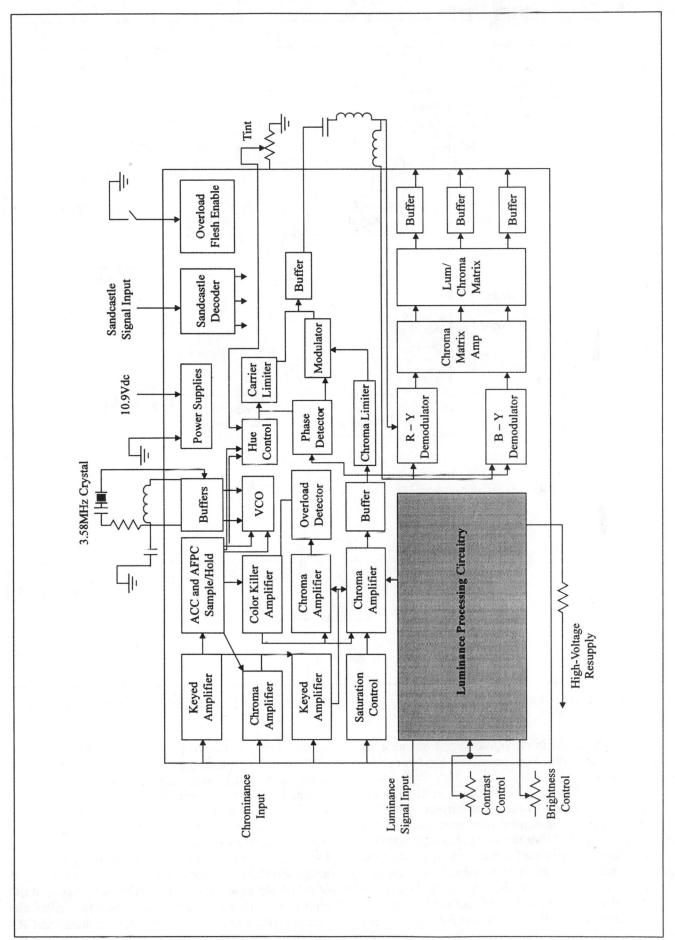

Figure 11-15 Block Diagram of a Luminance/Chrominance Processor IC (*Courtesy of Philips Electronics N.V.*)

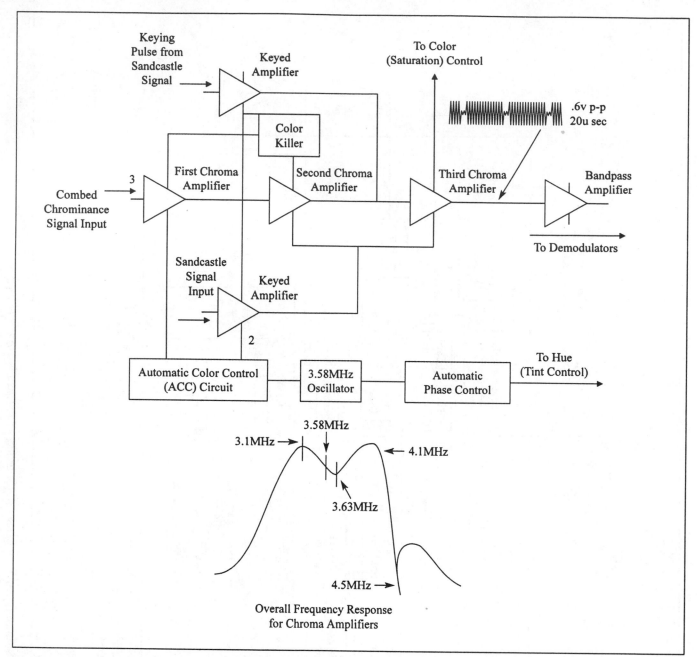

Figure 11-16   Chroma Bandpass Amplifier Circuit and Frequency Response (*Courtesy of Philips Electronics N.V.*)

conditions. The color burst signal found on the back porch of the horizontal sync pulse is the reference for the ACC circuit.

As the circuit shows, the **color** (saturation) **control** potentiometer attaches to the third chroma amplifier. The color control establishes the amplitude of the chrominance signal. With the control at one extreme, the reproduced picture will have saturated color; the other extreme should produce a monochrome picture. As you may suspect, moving the color control setting to its midpoint allows part of the current to flow through the amplifier and establishes a normal level of color for the reproduced picture. By sending the amplifier into

cut-off during the color burst interval, the application of the blanking pulse prevents any color burst signal from reaching the video output circuit.

Another control circuit, called the **color killer**, also interfaces with the last bandpass amplifier and checks for the presence of the color burst signal. If the color burst is not present, the color killer circuit increases the reverse bias on the bandpass amplifier and turns the amplifier off. The color killer circuit controls the gain of the bandpass amplifier in the same way that an AGC circuit controls gain. As a result, the color killer circuit prevents any color signals from reaching the CRT during a monochrome-only broadcast or during the blanking interval. Many modern television circuits eliminate the color killer.

The amplified chrominance signal produced at the output of the bandpass amplifier contains the chrominance modulation and color sync information for the color-processing section of the receiver. While the modulated signal contains the color information for the original picture, the color sync information controls the phase of a 3.58-MHz local oscillator, which restores the suppressed subcarrier signal. By controlling the phase of the local oscillator signal, the color sync information also controls the hue of the reproduced colors.

Because of this factor, the **tint control** or **hue control** connects to the first chroma amplifier through the AFPC network and the 3.58-MHz oscillator and adjusts the phase of the chrominance signal. Moving the tint control from one extreme to another changes the phase of the reference signal presented by the local oscillator. With this adjustment, the phase relationship between the chrominance signal and the reference signal also changes, which, in turn, causes the hue of the reproduced picture to change. The dc voltage for the tint adjustment is applied through pin 14 of the integrated circuit.

Most new television systems also include an **automatic tint control** that corrects any flesh-tone problems that may occur. The automatic tint control works by phase shifting the incoming chroma signal by a small percentage. While the phase shifting affects only the hues that make up the flesh tone, it does not produce an amplitude change. Because of this, the automatic tint-control circuit shuts down if the saturation level increases past the 70 percent region.

The Burst Separator or Burst Amplifier Stage

Suppressing the subcarrier leaves the processing circuits without any type of reference signal for detection or for matching the phase of the original transmitted color signal. The burst signal locks the frequency and phase of the 3.58-MHz chroma subcarrier oscillator to the frequency and phase of the oscillator found in the transmitting equipment. With the burst signal in place as a reference, the chroma-processing circuits can accurately measure the phase of the chrominance signal. The measurement of the chrominance signal phase translates into the correct reproduction of the hue information.

The second color processing stage—known as the **burst separator, burst amplifier, keyed amplifier,** or **burst gate**—selects and amplifies the color burst signal from the chroma subcarrier signal. The burst

separator also phase locks the local oscillator to the burst frequency used as a reference for processing. In practice, the burst separator consists of a gated amplifier "keyed" into conduction by high-amplitude horizontal retrace pulses taken from the horizontal output section.

Keying pulses turn the amplifier on during the time that a color burst signal appears at its input, and allow the circuit to remove the color sync signal from the back porch of the horizontal sync pulse. During monochrome transmissions when no color burst signal should appear at the amplifier input, keying pulses turn the amplifier off. When we move to the next section, we will find that a signal called the sandcastle signal provides the burst keying pulses for the circuit. Without those pulses and the horizontal and vertical blanking pulses obtained from the sandcastle signal, the chrominance/luminance processor will not have an output.

Conduction of the burst amplifier occurs during the burst period and provides a sample of the burst signal for the reference circuit. Thus, the output of the burst amplifier consists only of the eight cycles of the color burst signal. Using the transmitted burst signal to control the frequency and phase of a carrier signal, the reference portion of the **chroma phase detector**—a phase-locked loop—controls the frequency and phase of a locally generated reference signal so that it matches the frequency and phase of the subcarrier.

The Sandcastle Signal   Each of the two prior illustrations have shown another type of signal, called a sandcastle signal, as an input. With a circuit that handles a variety of functions, the sandcastle signal provides an efficient method for transferring information. As Figure 11-17 shows, a sandcastle signal is applied to pin 7 of the processing IC, U701. Moving to Figure 11-18, the sandcastle signal contains a matrixed combination of vertical and horizontal blanking information and a color burst keying signal. The use of one signal to carry the information allows the application of those signals to one pin of the IC. Within the IC, a decoder network decodes the three signals and applies each signal to its proper circuitry.

Along with associated circuitry, the burst keying amplifier transistor, Q801, and the vertical blanking transistor, Q702, generate the sandcastle waveform. During the operation of the receiver, vertical blanking information couples to the base of Q702, causing the transistor to conduct during the vertical blanking interval and increasing the voltage at the anode of diode D703. At the same time, the positive horizontal blanking pulse is coupled through R712 to the anode of D703. the sandcastle signal contains a matrixed combination of vertical and horizontal blanking information and a color burst keying signal.

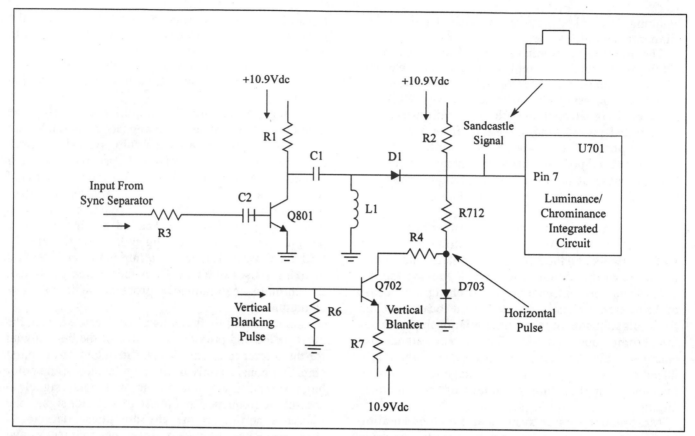

Figure 11-17   Sandcastle Signal and Circuit (*Courtesy of Philips Electronics N.V.*)

Providing a Reference for the Subcarrier

During operation, a **subcarrier reference system** within the luminance/chrominance processor converts the synchronizing bursts obtained from the burst gate into a continuous carrier with a stable frequency and phase. The **automatic frequency and phase control** or AFPC, loop shown at the top center of the processor in Figure 11-18 is a phase-actuated feedback system that consists of a chroma phase detector, a low-pass filter, and a voltage-controlled oscillator. Figure 11-19 shows a block diagram of the AFPC loop. The AFPC loop operates like the phase-locked loops seen in earlier chapters. The phase detector compares the phase of the burst

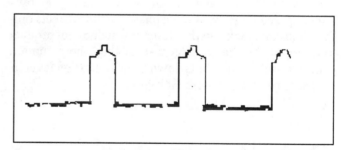

Figure 11-18   Sandcastle Signal (*Reprinted with permission of Thomson Customer Electronics, Inc.*)

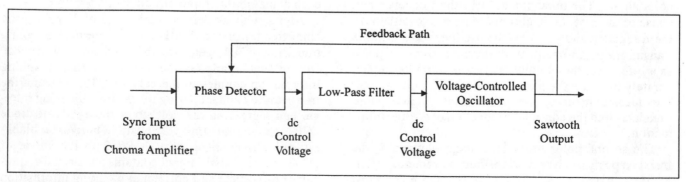

Figure 11-19   Block Diagram of the AFPC Loop

signal with the signal provided by the local oscillator. Depending on the difference in phase, the detector generates an error voltage that controls the resonant frequency of the oscillator. When the oscillator is off-frequency, the detector generates the error voltage until the oscillator has reached the correct frequency. Thus, the AFPC system supplies the proper phase reference signal to the demodulators and determines the phase of the oscillator signal.

To accomplish these two tasks, the AFPC circuitry must reproduce a sine wave at 3.579 MHz that is exactly the same frequency and phase as the burst signal transmitted at the end of each horizontal line. In addition, the phase of the locally generated signal must remain within approximately 10 electrical degrees of the burst phase with a constant amplitude. During operation, two samples of the burst signal—each with the sample amplitude but 90 degrees out-of-phase—are applied to the detector circuit through a buffer. At the same time, the 3.58-MHz oscillator signal is applied to the detector circuit. When the reference signal matches the phase of the burst signal, the circuit does not pass current.

While the gain of the loop represents the maximum possible frequency difference handled by the system, the magnitude of the gain determines the amount of allowable phase error. The low-pass filter controls the amount of loop gain. Because an external crystal and phase-shift network can cause the output of the VCO to shift by as much as 45 degrees, the AFPC circuit provides a symmetrical frequency pull-in range.

As the major processor diagram in Figure 11-15 showed, the AFPC loop also contains the automatic color-control circuitry. As a result, the loop supplies the control voltage for the color killer and the first chroma amplifier. When the ACC circuitry detects a signal with no color burst, it uses the color killer to cut off the second chroma amplifier. The reception of a color signal allows the second amplifier to supply the 3.58-MHz subcarrier to the demodulators.

Again looking at Figure 11-15, the hue control determines the phase of the color burst. The AFPC loop uses the control voltage to control the color oscillator phase and frequency. A **color oscillator** consists of a crystal oscillator tuned to 3.579545 MHz; it supplies the reference signal for demodulating the color signal. If we traced the signal from the demodulators back to the color oscillator, we would find a reconstituted color synchronizing signal. This signal is supplied to the demodulators at the proper phase angle.

Signals from pins 11, 12, and 13 of the integrated circuit control the 3.58-MHz local oscillator. Looking back at Figure 11-17, the circuit demodulates the I and Q signals described earlier in this chapter. The phase-shifted 3.58-MHz reference oscillator signal required for demodulation flows through pins 18 and 19 to the I and Q signal demodulators. The demodulated signal then travels from the I and Q demodulators, through a luminance/chrominance amplifier, to the luminance/chrominance matrix.

From all this, the frequency and phase of the restored subcarrier match the frequency and phase of the color carrier from the original transmission. The local oscillator portion of the chroma phase detector supplies the subcarrier signal to the next stage—a set of synchronous detectors. Along with the subcarrier, the chrominance signal makes up the detector input signal.

Processing the Chroma Signal

Any type of receiver that reproduces color images on a display screen uses demodulation to recover the color and hues from the chroma signal. Demodulation of the color signal involves reinserting the 3.58-MHz color subcarrier signal at the proper phase angles. Reproduction of color and hue is accomplished through a blue-minus-luminance, or B − Y signal, and a red-minus-luminance, or R − Y signal. The B − Y signal lies 180 degrees out-of-phase from the burst phase, and the R − Y signal lies in quadrature with the burst.

Every possible color reproduced by a video receiver has a phase-angle relationship with the 3.58-MHz color burst signal. For example, yellow has a 12.5-degree angle with reference to the zero-degree phase angle of the burst signal. Chroma demodulation involves decoding the chroma signal encoded at the transmitter and recovering the specific phasing and amplitude of each chroma signal.

Referring to the color wheel in Figure 11-8, different methods may be used to measure phase angles between the colors represented by the R − Y, B − Y, and G − Y signals. While the method shown in that figure uses the burst phase as a reference, another method counts the angles clockwise in the positive direction from zero. For that method, the B − Y signal and the color sync burst signal would rest at zero degrees. With burst phase as a reference, the B − Y signal is at 180 degrees.

Obviously, the receiver requires the reproduction of the color green to accurately reproduce any televised image. Manufacturers may use one of four methods to produce the green-minus-luminance, or G − Y signal. The demodulators may take the form of an I-Q signal-demodulator system; a two-demodulator system that uses the R − Y and B − Y signals to produce the G − Y signal; a three-demodulator system that directly produces the R − Y, B − Y, and G − Y signals; or X and Z demodulators. Regardless of the configuration, each demodulator requires a chroma signal input and the 3.58-MHz signal from the color oscillator.

R − Y, B − Y, and G − Y Signal Demodulation Referring to Figure 11-20A, a voltage divider consisting

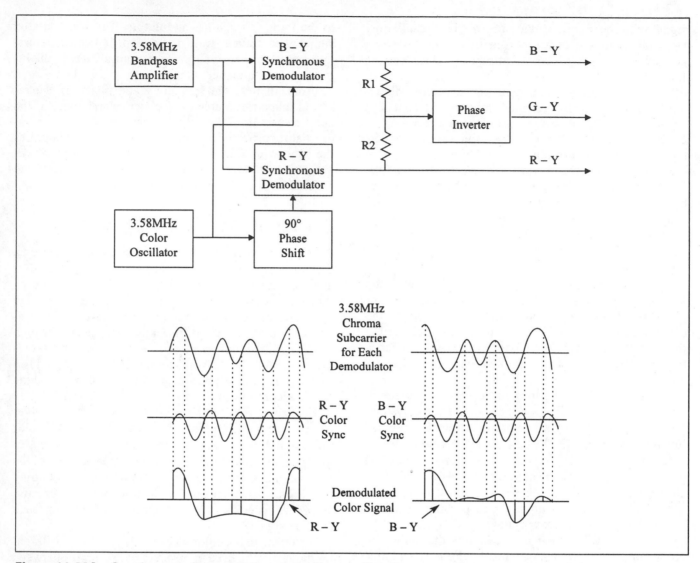

Figure 11-20A Synchronous Demodulation of the Chroma Signals

of R1 and R2 develops the G – Y signal from the signals produced by the R – Y and B – Y demodulators, while a phase inverter develops the proper G – Y phase. Two synchronous detectors demodulate the sidebands of the color subcarrier separately and reproduce the I and Q signals as the chroma signals needed by the receiver. The demodulator circuits compare the phase of the phase- and amplitude-modulated color subcarrier signal with the fixed phase of the color-synchronizing signals. The output signal taken from the demodulator circuits differs according to the phase of the color sync signal applied to the particular demodulator circuit. For R – Y and B – Y color sync signals, a 90-degree phase shift occurs between the sync signals and the subcarrier reference. In addition, the two signals are separated by a 90-degree phase difference.

The synchronous detectors take advantage of a quadrature relationship between the subcarriers of the two chroma signals. Because of the synchronous char-

acteristics of the demodulator circuit, one detector demodulates the I, or B – Y, signal, while the other detects the Q, or R – Y, signal. The complete demodulation circuit employs phasing networks between the subcarrier oscillator and the detector so that the two recovered signals will have the proper phase relationship. During operation, the color subcarrier and a reference signal taken from the oscillator drive the detector circuits and establish accurate phase control of the color demodulation.

In the block diagram shown in Figure 11-20B, three demodulators directly produce the R – Y, B – Y, and G – Y signals, while two phase-shifting networks develop the proper phase angles for the signals. The R – Y demodulator uses a 90-degree phase shift, while the G – Y demodulator relies on an inverter and an additional phase-shift network for a 236.6-degree phase shift. The B – Y demodulator has no phase shift. With this type of demodulator system, the luminance signal

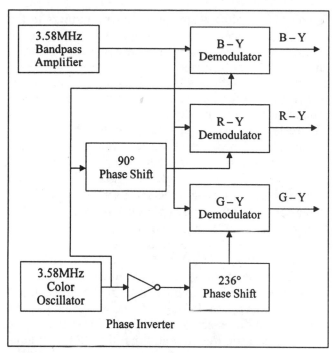

Figure 11-20B Block Diagram of Chroma Signal Demodulation Using Three Demodulators

adds directly to the demodulator output. The complete video signal then travels to the CRT.

↻ FUNDAMENTAL REVIEW

Synchronous Demodulation

Synchronous demodulation is the process of demodulating a modulated signal with a suppressed carrier. The term "synchronous" tells us that the circuit detects a signal or portion of a signal that is in sync with some quantity. In this instance, the detector has a maximum

output only when it demodulates information that exists in sync with the restored subcarrier.

Other Chroma Demodulation Systems Although the R – Y/B – Y and R – Y/B – Y/G – Y demodulator systems dominate the market, manufacturers have also used the I-Q signal-demodulation system and the X/Y demodulator system. These systems rely on different methods to control the phase of the chroma signals seen at the output. Because of the phase differences, both require additional circuitry for phase and polarity conversion.

The I-Q Signal Demodulation System One method for ensuring that the demodulated signal has the same phase as the encoded signal is to use demodulators that operate on the same phase axes as the transmission system. The I and Q demodulators seen in Figure 11-21 use 90-degree out-of-phase oscillator signals, as input signals, as well as the chroma signals from two chroma amplifiers. Because the oscillator signal for the I demodulator travels through an inverter, the I oscillator signal has a phase at 270 degrees with respect to the Q signal. As a result, demodulation occurs on the positive I axis. The low-pass filters following the demodulators restrict the demodulated I signal bandwidth to 1.3 MHz, and the demodulated Q signal bandwidth to 600 kHz. Phase splitters in each channel develop the positive and negative I and Q signal polarities for the output stages.

Past designs rarely used the I-Q demodulator system because of the complexity of reproducing the wide-bandwidth I signal. As the block diagram shows, the I-Q demodulation scheme requires an additional chroma amplifier and an additional delay line in the I channel. Benefits of the I-Q system involve producing the highest possible color resolution. As we learned

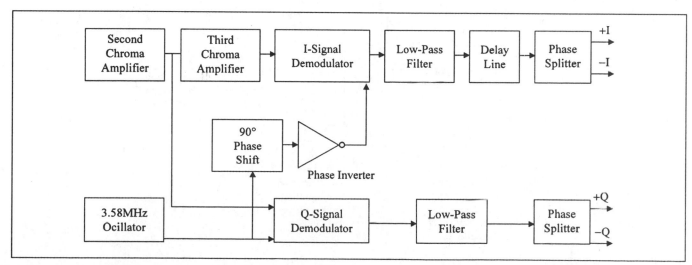

Figure 11-21 Block Diagram of I-Q Signal Demodulation

earlier, the orange-cyan I signal has a maximum bandwidth of 1.3 MHz, which compares with the 0.5 MHz of other color video signals.

X and Z Demodulators   Referring to Figure 11-22, voltages from the reference oscillator are applied to phase-shift networks before reaching the X and Z demodulators. The phase-shift networks shift the phase for the X demodulator by 102 degrees and the Z demodulator by 166 degrees, yielding a phase difference of 64 degrees. The color-difference amplifiers form a matrix that converts the negative-polarity X and Z signals into the R – Y, G – Y, and B – Y chroma signals. From there, the signals feed into the CRT grids.

Matrix Circuits

Matrix circuits combine the demodulated color-difference signals with the luminance signal by adding proportional amounts of input voltages to form new combinations of an output voltage. Returning for a moment to Figure 11-15, at the lower right-hand corner the luminance and chrominance signals matrix together in a matrixing amplifier. Within the matrix, processed color signals mix with the 3.58-MHz oscillator signal and the luminance signals. As a result, the output of the matrixing amplifier consists of the red, blue, and green video signals. Those signals couple through buffers to the output stage of the receiver.

We can use a simple resistive network like that shown in Figure 11-23 to illustrate how the matrixing amplifier combines the signals to produce the video

signals. Two voltages appear as separate inputs to the network and become a single composite signal at the network output. Because of the network configuration and the values of the resistors, the voltages combine instantaneously, with positive values adding and negative values subtracting. Two negative input signals will produce a negative output signal. Looking at the circuit, resistor R3 has lower value than the additive resistors R1 and R2 because of the need for isolation at the inputs. The additive resistors have equal values.

The particular design seen in Figure 11-15 uses a technique called **low-level RGB matrixing and CRT drive** because of the type of guns used in the CRT. With this design, the RGB signals matrix at a level of only a few volts. The CRT drive circuits then amplify the signals to the 100- to 200-Vdc level needed to drive the CRT. Because of the specific demands of a color receiver, the matrix circuit provides direct current stability, the appropriate frequency response, and signal linearity. All this ensures that the receiver will have the proper gray-scale tracking and that the CRT will have the proper bias.

Other receivers may use another matrixing technique called **color-difference drive matrixing**. With this technique, the R – Y, B – Y, and G – Y signals travel to the CRT control grids, while the luminance signal travels to the red, blue, and green CRT cathodes. Matrixing of the red, blue, and green colors occurs within the CRT. The advantage of this design lies within the tracking of the gray scale. Because the gray scale does not rely on matched red, blue, or green color signals, the color-difference stages remain at the same

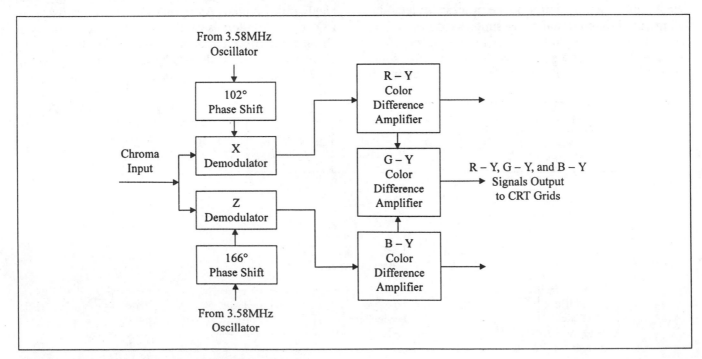

Figure 11-22   Block Diagram of X-Z Demodulation

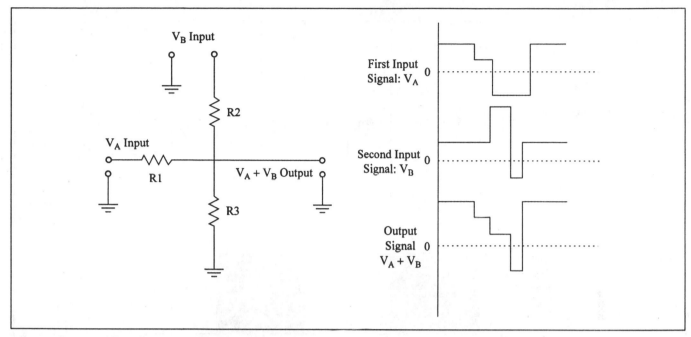

Figure 11-23   Simple Resistive Network Matrix and Resulting Signals

dc level. The color-difference drive matrixing design has the disadvantage of requiring higher CRT drive voltages.

Given this need, receivers may employ **color-difference amplifiers** to amplify the R − Y, G − Y, and B − Y signals to the level needed to drive the CRT. Color-difference amplifiers are only used in receivers that rely on a separate video amplifier for the luminance signal, and usually fit between the processing and the output stages. Generally, receivers will feature separate red, blue, and green video amplifiers that amplify already mixed luminance/chrominance signals.

☑ PROGRESS CHECK

You have completed Objective 5. Discuss the operation of the bandpass amplifier, color killer, burst amplifier, AFPC, color demodulator, and color matrix circuits.

☑ PROGRESS CHECK

You have completed Objective 6. Explain how various controls and signals affect chroma signal processing.

11-4   RGB DRIVER AND VIDEO OUTPUT STAGES

The luminance, or Y, signal contains information about brightness variations that occur in a televised picture. At the transmission point, the luminance signal equals the sum of proportional amounts of the primary red, green, and blue video signals. In equation form, the proportional amounts for a white screen appear as:

$$30\%\text{RED} + 59\%\text{GREEN} + 11\%\text{BLUE}$$
$$= Y \text{ or Luminance (WHITE)}$$

with the percentages corresponding to the brightness of each of the three colors. At its maximum relative amplitude of 100 percent, the luminance signal contains all three color signals and reproduces as a completely white raster. Changing the values of any of the colors will cause the reproduced raster to have a tinted appearance. As an example, we can drop the value of the red video signal to zero while leaving the other values unchanged. In equation form, this change appears as:

$$0\%\text{RED} + 59\%\text{GREEN} + 11\%\text{BLUE}$$
$$= Y \text{ (CYAN)}$$

Eliminating the red video signal allowed the green and blue signals to mix and tint the raster cyan.

With this example in mind, we can find that the proportional values of the red, blue, and green video signals combine as voltages to form the luminance signal voltage. To make this point even clearer, Figure 11-24 compares the waveforms of the red, green, blue, and luminance video signals with a color bar pattern. Each bar in the pattern corresponds with a different signal combination. As the figure illustrates, the amplitude of the luminance signal varies with the color displayed on the television screen. While a white color bar contains all three primary color signals and has a maximum luminance value, a blue color bar contains only the blue video signal and has a low luminance value.

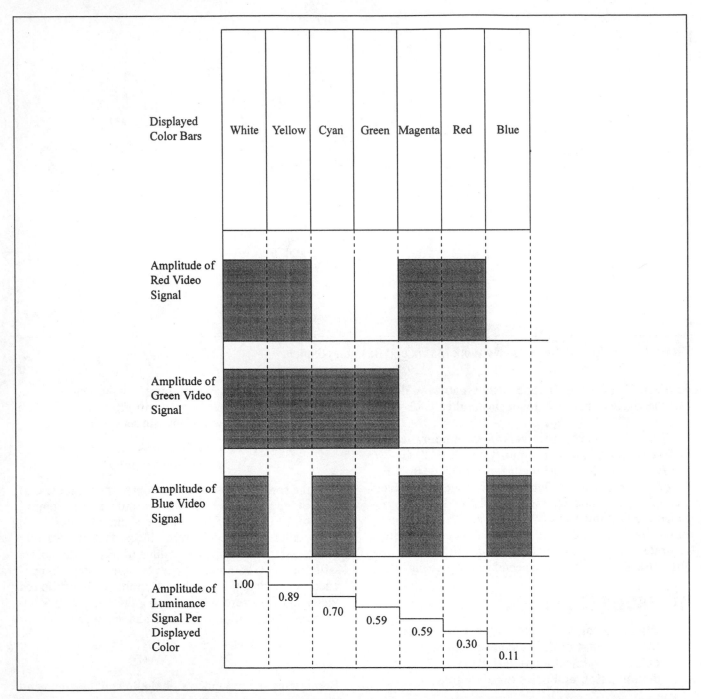

Figure 11-24 Comparing the Waveforms of the Red, Blue, and Green Luminance Signals with a Color Bar Pattern

Video Output Amplifier Circuits

Most video systems rely on the low-level RGB matrixing and CRT system because of the use of CRTs that have unitized guns. The common G1 and control-grid connections in this type of CRT require differential cathode bias adjustments and drive adjustments for the proper gray-scale tracking. With the low-level system, the output from the RGB matrix has a low level of only a few volts. Amplifiers at the CRT amplify the voltages to the 100 to 200 Vdc needed to drive the CRT.

Figure 11-25 shows an example of a video output circuit commonly used by many manufacturers. When the amplifier black bias voltage equals the black level of the signal taken from the matrix, any drive adjustment will not affect the amplifier black-level output voltage level or change the cut-off point of the CRT. The figure shows one of three identical video output circuits used for the amplification of the red, blue, and green signals.

We can expand our analysis of the video output subcircuit shown in Figure 11-25 by considering the

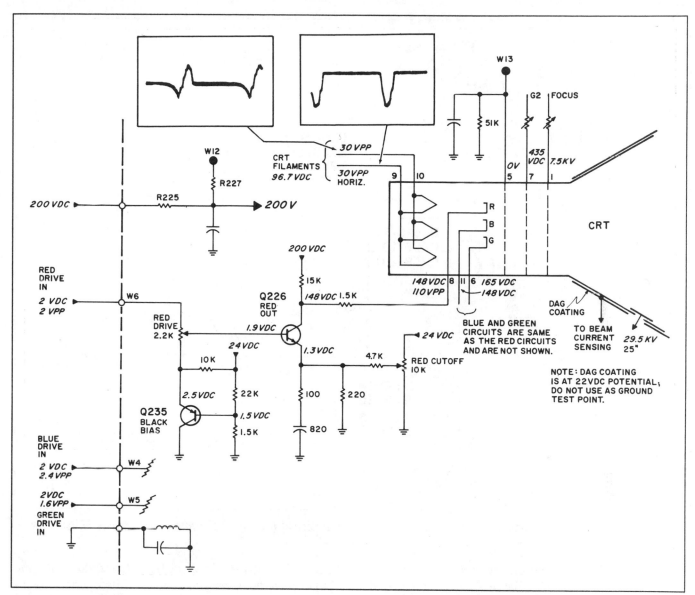

Figure 11-25   Schematic Diagram of a Common Video Output Circuit (*Courtesy of Philips Electronics N.V.*)

complete circuit of Figure 11-26. The amplifier system illustrated by the figure consists of a common-emitter circuit with the NPN stages driving the CRT cathodes through the PNP transistors. The capacitors and coils L2584, L2585, L2586, and L2587 provide the peaking needed to establish an adequate frequency response. Each of the inductors self-resonate at the second harmonic of the chroma signal and prevent picture interference by trapping out the 7.2-MHz frequency.

With each of the output amplifier transistor emitters connected together, a 5.6-Vdc zener diode sets up the amplifier bias. When considering the cathodes of the CRT, current flows through each of the PNP transistors to all three cathodes. During positive output transitions of the video signals, the conduction of diodes establishes the correct video levels at the cathodes. The automatic CRT tracking system contained within IC2501

uses a sample of the video signal to determine the value of the currents without blocking the passage of normal video signals.

Automatic CRT Tracking   As mentioned, IC2501 makes up a large part of the automatic CRT tracking system found in the video output circuit. The system automatically adjusts the cut-off points of the three CRT guns. Thus, the circuit eliminates many manual controls and gray-scale tracking procedures formerly set and monitored by a technician.

Within the integrated circuit, each separate color-signal channel features a gate-controllable amplifier. Rather than rely on a manual control to establish the bias point and amplifier gain, the circuit uses a dc control voltage to electronically adjust the background. The circuit also compares the already-mentioned

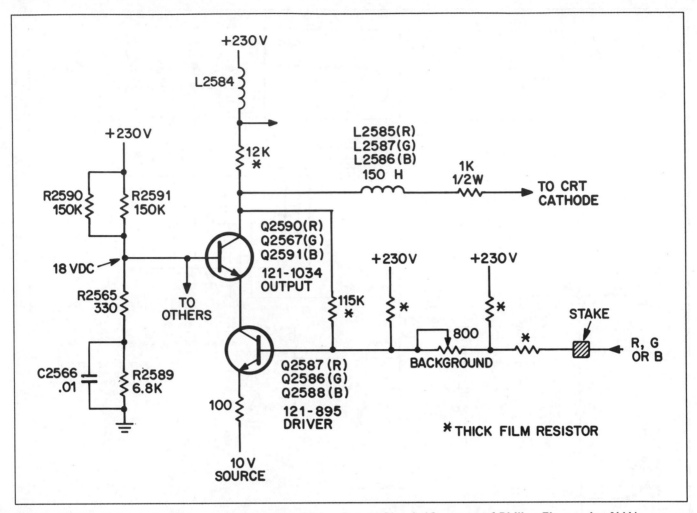

Figure 11-26 Schematic Diagram of a Complete Video Output Circuit (*Courtesy of Philips Electronics N.V.*)

sample of the CRT cathode current with an internal reference and—if the current changes—automatically adjusts the gray scale by varying the bias and gain of the amplifier.

Another circuit within the IC uses an input from the vertical amplifier to set the CRT at the black level, and a generated sample pulse to monitor the current at the cathode. The circuit also relies on a sample-and-hold subcircuit to check the level of CRT against a predetermined reference, and adjusts the output voltage so that the current conforms to the reference. As you will discover in a later chapter, the CRT has specific temperatures for the red, blue, and green channels. The references for the cathode current are derived from three currents within the IC that have values corresponding to the color temperatures.

During the period that the circuit generates the sampling pulse, the IC checks CRT cathode currents. The sampling pulse, which has a short interval, occurs before the normal video signal and at the end of the vertical blanking interval. If one of the cathode currents ranges higher or lower than normal, then the capacitors

will either charge or discharge. Once the sampling pulse ends, the capacitor returns to its operating voltage and allows the normal video signals to pass.

An internal sampling pulse that corresponds with the trailing edge of the vertical retrace pulse is derived from the capacitors and resistors that connect in series to the vertical deflection circuit. With the sampling pulse matching the edge of the vertical retrace pulse, it also matches the end of the vertical retrace period. A second internal circuit prevents a mistimed retrace pulse from forcing the sampling pulse to occur during the video signal. A pulse during this time would turn the automatic tracking circuit off, because the tracking system relies on a black-level signal from the IC as a reference. With the retrace occurring at the correct time, the chroma image remains blanked, and the sample arrives at the blank level.

Cascode Video Output Amplifiers

The cascode circuit shown in Figure 11-27 offers an improvement in frequency response and dc voltage

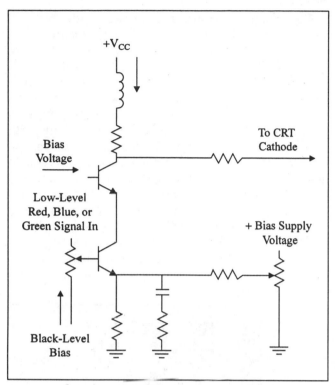

+V_CC

Bias Voltage

Low-Level Red, Blue, or Green Signal In

Black-Level Bias

To CRT Cathode

+ Bias Supply Voltage

Figure 11-27 Schematic of a Cascode Amplifier Video Output Circuit (*Courtesy of Philips Electronics N.V.*)

stability over the characteristics given by the previous circuit. The shunt peaking coil is common to all three channels because of the relatively narrow bandwidth of the color-difference signals found at the input of the circuit. Cascode video output circuits are often used in computer and video monitors.

☑ PROGRESS CHECK

You have completed Objective 7. Discuss the operation of video output circuits.

11-5 TROUBLESHOOTING COLOR-PROCESSING CIRCUITS

Because the correct reproduction of color depends on both the proper phase angle and amplitude, any troubleshooting effort should involve careful monitoring of those characteristics. For example, if a chroma waveform has an incorrect phase angle, a demodulator phasing error occurs. In addition to monitoring those characteristics, chroma circuit troubleshooting also involves many of the same techniques that we have seen in the past: 1) checks of the supply and output voltages, 2) signal injection and signal tracing, and 3) checks of input and output waveforms.

Test Equipment Primer: Color-Bar Generators

Used for television receiver set-up adjustments, a **color-bar/cross-hatch generator** should supply color bars for the primary and complementary colors. Each of the twelve gated-rainbow color bars should have a consistent amplitude and luminance value. Most color-bar/cross-hatch generators also supply the luminance signal, a video sweep signal for testing video circuits, and a black-and-white cross-hatch pattern for adjusting convergence and checking linearity. Other units may provide variations of the cross-hatch pattern for close-tolerance checks of the dynamic convergence, screen blanking for purity checks, and adapters for testing computer monitors.

As with many of the circuits that you have studied, signal tracing provides a good technique for troubleshooting the chroma-processing circuits. In this case, though, the characteristics of a color television allow the technician to make several easy checks of the receiver performance. To make the task of checking for the proper luminance, saturation, and hue easier, technicians can use a color-bar generator as a useful troubleshooting device.

We can divide color-bar generators into two basic categories. The keyed rainbow generator provides the capability to perform basic tests and accurately align the color demodulators. More specific tasks may require the use of an NTSC, or National Television Systems Committee, generator that produces the same signals found at the transmission site.

Keyed rainbow generators contain an oscillator tuned to 3,563,795 MHz (3.58 MHz minus 15,750 Hz), or one scanning line below the color-carrier frequency. With the color modulator inside the generator scanning one line lower in frequency than the carrier, the generator loses one complete cycle for each horizontal line in the picture. As a result, the generator starts out-of-phase for each horizontal line, produces colors that are phase-modulated on the carrier, and supplies all colors.

Because the phase-modulation of the color subcarrier produces various colors, the red color signal appears within a range of 90 to 105 degrees away from the burst signal, and the blue signal appears at approximately 180 degrees away from the burst reference. The name "rainbow generator" is derived from early color generators that featured a display in which the test colors had no definite dividing line and blended into a rainbow. One color after another appeared as the CRT beam moved across the screen.

Keyed rainbow generators use the phase modulation seen with the original rainbow generators, but also utilize a crystal-controlled circuit that cuts off twelve times during each horizontal line. The cut-off divides the display into twelve evenly spaced, sharply defined bars. Because the sync-retrace-color-burst portion of the cycle causes two bars to disappear, we see only the ten bars represented in Figure 11-28. Because the color bars are not accompanied by the luminance signal, each color displayed has its maximum saturation.

As shown in Figure 11-29, color-bar generators may attach to either the R-F antenna terminals or the I-F input of the receiver. A chroma control located on the front of the generator allows the adjustment of the amplitude of each 3.58-MHz burst associated with the displayed color bars. Each burst has a slightly different phase when compared to the others. As you know, increasing the amplitude also increases the saturation of the color and the intensity of the bar. In addition, increasing the amplitude of the color burst also allows a technician to see how the color-sync portion of the receiver functions under varying signal conditions.

Connecting the color-bar generator to the receiver antenna terminals and then injecting a standard signal into the receiver provide an easy method for checking the alignment of the tuner, I-F amplifiers, video detector, video amplifiers, and bandpass amplifier. Technicians can either monitor the reproduced pattern on the receiver CRT or, as the block diagram shows, attach an oscilloscope to convenient test points. Along with checking the alignment of those stages, a technician can also use the signal provided by the generator to quickly check the adjustment of the color and tint controls. Checking for the proper reproduction of the primary and complementary colors becomes easier through the use of the receiver display as a monitor.

While most technicians use a keyed rainbow generator to check and align the color-processing circuits, another signal generator provides the same signal as a

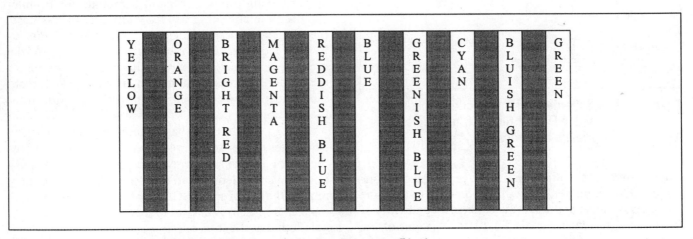

Figure 11-28   Illustration of a Keyed Rainbow Color-Bar Generator Display

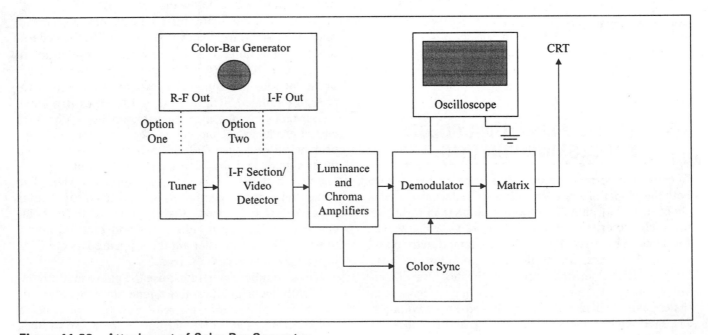

Figure 11-29   Attachment of Color-Bar Generator

television station. NTSC color-signal generators have an output signal consisting of the 30-percent red signal, 59-percent green signal, and 11-percent blue signal needed to establish the correct balance of color, light, and hue. Because of this, a technician using an NTSC generator can adjust both the demodulator phase and the amplitude of the color signals. Generally, NTSC signal generators are used by manufacturers and broadcast stations.

General Troubleshooting Tips for Chroma Problems

In many cases, the aging of low-cost components such as capacitors or resistors affects the performance of a chroma circuit. For example, leaky capacitors between the 3.58-MHz reference circuitry and the subcarrier oscillator often affect the chroma demodulation. Zero output from the demodulator may result from a shorted capacitor in the oscillator circuit. Amplitude problems usually originate within the demodulator circuit. Using another example, if a demodulator diode opens, the output waveform will have a lower than normal amplitude, and a cross-over error will result.

Problems in the bandpass amplifier stage almost always cause symptoms such as no color, weak color, or excessive color. A failure of components in the bandpass amplifier stage can block the flow of the chroma signal to the demodulator stage and prevent the production of color-difference signals. Without those signals, the color matrix receives only the luminance signal, and a monochrome picture results.

A "no color" symptom may also surface if the output from the color oscillator is blocked. The lack of the oscillator signal prevents the color demodulators from operating and producing an output. In addition, a defect in the color-killer circuit may also prevent the receiver from displaying anything but monochrome images.

An "incorrect color" symptom may occur if the color oscillator operates at the wrong phase, if one demodulator fails, or if the matrix output signal is missing a color-difference signal. With the color oscillator operating at the wrong phase, the receiver will produce all colors, but place the colors in the wrong location. The sky may take on a reddish appearance, for example, and grass may appear as magenta rather than green. A failed demodulator will cause particular colors to disappear.

Reduced gain in the bandpass amplifier stage causes the colors to lose saturation. Failures in the automatic chroma-control section can push the bandpass amplifier to operate at maximum gain and cause excessive saturation to occur. Along with extremely vivid colors, the colored confetti-like interference will also float through the picture.

Other symptoms occur because a defective AFPC stage will not lock onto the subcarrier signal. With no locking, the phase of the color oscillator drifts continuously with respect to the chroma-signal reference provided by the bandpass amplifier. As each frequency shift occurs, the hue of objects shown in the picture changes with the rate of the frequency error. At a frequency error of 60 Hz, the hue changes 360 degrees for the entire spectrum from the top to the bottom of the picture. Higher-frequency error rates cause the colors to break into diagonal bands, with rainbow colors floating through each band. The loss of color locking also points to problems within the phase-locked loop portion of the AFPC system. Several symptoms—no burst input, no feedback from the color oscillator, or a phase-comparator failure—may become apparent with the PLL failure.

Problems with the color demodulator stage often involve the failure of one demodulator. If one demodulator fails, the demodulator circuit becomes unbalanced, producing a steady dc voltage at the output, which will change the gray-scale balance for the CRT. With a monochrome-only picture, this type of failure becomes apparent through the tinting of the picture.

The loss of color sync will cause color bars to drift slowly through the picture or cause the hues to shift. Receivers will lose the color sync if the burst separator fails or if the color oscillator runs at the incorrect frequency. More than likely, the loss of color sync will occur because of a failure in the bandpass separator, burst amplifier, or phase detector.

Because of the number of possible symptoms produced by faults within the chrominance circuitry, Tables 11-1 to 11-6 summarize these faults, their general symptoms, and the most likely solutions. First, however, we should examine methods for troubleshooting the chrominance/luminance integrated circuit that has been covered throughout this chapter.

Troubleshooting the Chrominance/ Luminance IC

Many of the problems illustrated in the last section involve receivers that use separate, discrete components, such as transistors, integrated circuits, and diodes, as bandpass amplifiers, oscillator, burst amplifiers, and demodulators. Despite the obvious difference between those receivers and receivers that combine color circuits within one integrated circuit device, the symptoms remain the same. Many of the problems, however, result from components that couple signals and supply voltages to the IC.

As you begin troubleshooting suspected chroma problems in a single IC receiver, always check the supply voltages for the integrated circuit. A lower than normal reading or the lack of a reading at any one of these checkpoints often points to a defective power supply source. At times, though, a leaky condition within the chrominance/luminance processor can also drive the supply voltages lower. If you suspect the latter, disconnect the supply from the integrated circuit and re-measure the supply voltage. The combination of a now-correct voltage and a low-resistance measurement from the IC pin to ground almost always points toward a leaky condition.

Along with checking the supply voltages, also check the voltages at the color output terminals of the integrated circuit. Correct voltages illustrate that the demodulators are operating normally. One method for verifying the operation of the color demodulators involves comparing the voltage readings with the demodulator waveform amplitudes. Figure 11-30 shows the waveform found at the R – Y demodulator output of the processor. The absence of the keyed-burst portion of the waveform would indicate that the processor is not providing a color output signal.

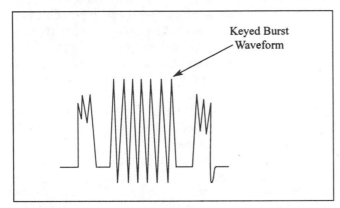

Figure 11-30 Waveform at the R – Y Demodulator or Red Output of the Chrominance/ Luminance Processor (*Courtesy of Philips Electronics N.V.*)

As with the discrete-based circuits, the chrominance/luminance processor requires an output signal from the color oscillator before the processor can operate normally. With no oscillator signal, the chrominance circuit will not produce a color picture. When we discussed the sandcastle waveform in this chapter, we found that it provided the keyed pulse for the processor. The loss of the sandcastle signal affects not only the chrominance-signal processing, but also causes the loss of the luminance signal. In addition, we discovered that the comb filter supplies the chroma signal for the circuit. All these points—oscillator signal, sandcastle waveform, and comb filter output—tell you that you should use the oscilloscope to carefully monitor the input and output waveforms for the integrated circuit. Connect a keyed rainbow generator to the VHF antenna

terminals of the receiver as an input signal and measure both the input and output waveforms.

When checking for proper oscillator operation, connect the oscilloscope to the 3.58-MHz crystal and check the waveforms on both sides of the crystal terminals. Figure 11-31 depicts the correct waveform shape; a color oscillator waveform should have an amplitude of 1 to 2 volts peak-to-peak. Often, waveform checks will show that bypass or coupling capacitors have developed problems. As mentioned, a leaky component can draw supply voltages down to almost zero. Waveform checks at the output terminals of the chrominance/luminance processor can point to defective components tied to the processor pins.

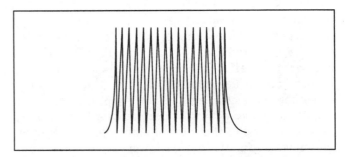

Figure 11-31 Waveform at the 3.58-MHz Crystal (*Courtesy of Philips Electronics N.V.*)

Although this last point may seem obvious, it is worth repeating. When working with the luminance/chrominance processor, never jump to the conclusion that the processor or the comb filter circuit has developed internal problems. Instead, carefully use the voltage and waveform checks to confirm that components such as capacitors, resistors, rectifiers, and regulators are functioning correctly. Nearly eight of ten color-signal defects associated with the chrominance/luminance processor occur because of shorted filter capacitors, leaky coupling and bypass capacitors, off-tolerance resistors, or leaky zener diodes.

Study the following tables (Table 11-1 through Table 11-6) carefully. Most of the problems you will encounter can be solved by using these problem/solution tables to help diagnose and correct color-signal defects.

➔ SERVICE CALL

Sylvania Model CLB536AR03 Has Normal Sound, No Video, and a Narrow Dim Raster

The Sylvania Model CLB536AR03 (Chassis E51-46) came into the shop with normal sound, no video, and a narrow dim raster that illuminated only the right side of the screen. Following good troubleshootingprocedures, the technician checked the regulated

Table 11-1 Comb Filter Problems

Problem	Symptom	Solution
Defective comb filter: Incorrect frequency at the 10.7-MHz input to the comb filter	No color—missing chroma signal at comb filter output	Check B+ supply voltages, check for presence of 3.58-MHz signal at comb filter input, replace comb filter
Defective comb filter: Incorrect reference signal at comb filter input—comb filter should have 3.58-MHz signal to develop delay-line pulses	Receiver has sound and raster but no video or color	Check B+ supply voltages, check for presence of 3.58-MHz signal at comb filter input, replace comb filter
Defective comb filter: Incorrect reference signal at comb filter input—comb filter should have 3.58-MHz signal to develop delay-line pulses	No video—missing luminance signal at comb filter output, chroma signal is present	Check B+ supply voltages, check for presence of 3.58-MHz signal at comb filter input, replace comb filter
Incorrect supply voltages at comb filter	Smeared or noisy picture	Check voltage regulator and bias circuits
Defective comb filter	Extremely vivid colors—no luminance	Replace comb filter

Table 11-2 Color Bandpass Amplifier Circuit Problems

Problem	Symptom	Solution
Defective 3.58-MHz crystal	No or weak color, no 3.58-MHz reference signal	Replace crystal
Bad solder joint	No color—no chroma signal at amplifier input	Resolder connections
Defective amplifier	No color—no signal at amplifier output	Replace amplifier
Defective amplifier	Weak color	Replace amplifier
Incorrect voltages at chroma/luminance processor, defective bypass capacitor defective chroma/luminance IC	Improper tint	Check supply voltages, replace capacitor, replace IC
Incorrect voltages at chroma/luminance processor, defective bypass capacitor defective chroma/luminance IC	Improper color	Check supply voltages, replace capacitor, replace IC

Table 11-3 Color Killer Circuit Problems

Problem	Symptom	Solution
Lower-than-normal color killer supply voltage	Monochrome-only picture—no color	Replace filter capacitor in color killer supply voltage line
Misadjusted color killer threshold control	Monochrome-only picture—no color	Adjust color killer threshold control

Table 11-4   Color Burst Amplifier Circuit Problems

Problem	Symptom	Solution
AFPC needs realignment (older-model designs)	Weak color	Realign AFPC using oscilloscope, color signal, manufacturers instructions
No horizontal keying pulse	No or poor color sync	Check for keying pulse at input to circuit, check components in horizontal deflection system, check coupling components (resistors, capacitors in bandpass circuit)
No horizontal keying pulse, no sandcastle input waveform	No chroma or video signals	Check sandcastle generator

Table 11-5   Color Demodulator Circuit Problems

Problem	Symptom	Solution
Defective demodulator	Raster shows one excessive color	Replace demodulator that corresponds with excessive color
Oscillator voltage does not reach one demodulator	Pictures lack one of the three primary colors	Check coupling components
Open coupling capacitor, open demodulator diode	Excessive color (unbalanced demodulation in a dc-coupled circuit)	Check coupling capacitor, check demodulator diode
Open component in demodulator input circuit	Loss of one color, no oscillator voltage to demodulator	Check components in demodulator input circuit
Demodulators do not produce an output voltage	No raster	Check supply voltages to demodulators

Table 11-6   Video Output Circuit Problems

Problem	Symptom	Solution
Open video output transistor collector load resistor, defective video output transistor	One primary color in all images is missing	Replace red video output transistor (leaky), open load resistor in video output transistor collector circuit
Defective video output transistor (corresponds with excessive color condition)	Raster shows one excessive color	Incorrect CRT bias, replace video output transistor (shorted)
Defective video output amplifier	Raster blooms (goes out of focus, picture expands and grows darker as brightness control is adjusted)	Replace video amplifier
Video output amplifiers are cut off	No raster—incorrect bias at CRT	Check video output amplifier supply voltages

power supply and the high voltage. Both gave normal readings. But a check of the 220-Vdc supply from the horizontal output transformer showed a much lower than normal reading of 39 Vdc. Such a low voltage at this point pushes the red, blue, and green amplifiers and each accompanying regulator into cut-off.

An investigation of the schematic showed that the horizontal output transformer develops a 250-Vdc peak, 5-microsecond pulse that passes through diode SC566. After capacitor C564 filters the pulse, it drops to 1.2-Vdc. Looking at the schematic, the technician determined that three possible problems could cause the 220-Vdc voltage:

1. a shorted SC566
2. a leaky or open C564
3. a short-circuit in any one of the three color amplifiers

Standard continuity tests of the diode and the three color amplifiers showed that the devices had no problems. This narrowed the tests to the filter capacitor, C564. An out-of-circuit test of the capacitor with a capacitor checker revealed the open condition of the component. Replacement of the capacitor restored the normal picture for the receiver. Figure 11-32 shows a partial schematic of the receiver and the problem area.

 SERVICE CALL ────────

RCA CTC 145 Chassis Has Intermittent Color Problems

The RCA receiver (chassis CTC 145) arrived with an intermittent color problem. At times, the color portions of the picture would disappear and leave a normal monochrome picture. At other times, the picture would change from color to monochrome and back to color.

The CTC 145 relies on one integrated circuit package for video I-F, sound I-F, chroma, vertical, horizontal, and AFT signal processing. Therefore, troubleshooting chroma problems quickly becomes a matter of checking for the correct signals and voltages at key IC pins. While checks at all other pins showed normal readings, the voltage at pin 5 of the IC had dropped from the normal 5.3 Vdc to 2 Vdc. A further check of the diagram showed that a small, 0.012-microfarad capacitor is placed in the line leading to pin 5 of the IC. A check of the capacitor disclosed a leaky condition. Replacement of the component allowed to the receiver to operate normally.

☑ PROGRESS CHECK

You have completed Objective 8. Troubleshoot chroma-processing circuits.

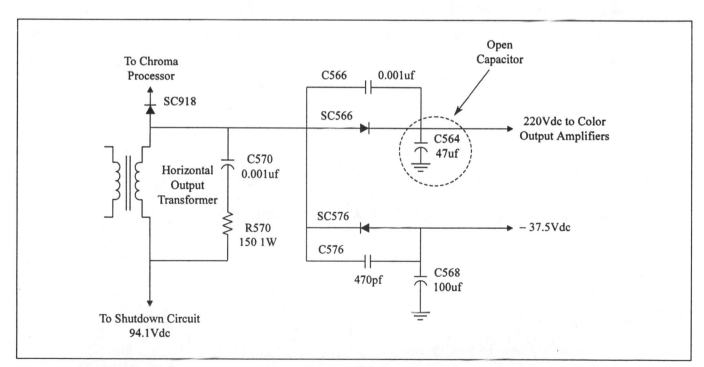

Figure 11-32   Schematic Diagram of a Sylvania CLB536AR03 Chroma Circuit (*Courtesy of Philips Electronics N.V.*)

SUMMARY

Chapter eleven introduced you to concepts of chrominance signals and their circuit requirements. Primary and complementary colors were covered, and you found that the hue of a color corresponds with a specific phase angle, while saturation corresponds with a specific amplitude. The chapter discussed the use of I (in-phase) and Q (quadrature) signals as carriers for the transmitted color information, and the use of the 3.58-MHz color-burst sync signal as a reference.

In considering the separation of the chrominance signal from the remainder of the composite video signal, you learned that comb filters take advantage of the process of frequency interleaving to separate the luminance and chrominance signals. Other circuit functions included removing and amplifying the color burst signal; reinserting the suppressed color subcarrier; recovering the original color-difference signals; controlling the hue and saturation of the reproduced colors; and

disabling the color signal during a monochrome broadcast.

The chapter also examined how the different controls and control circuits—such as the hue and color controls and the automatic chroma control and color-killer circuits—interact with the chroma-processing circuits. The operation of the AFPC circuit and the process of color demodulation demonstrated another design application for phase-locked loops. The topics concluded with a discussion of color matrix and video output circuits.

The section on troubleshooting color-processing circuits introduced another piece of test equipment, the color-bar generator, which allows technicians to inject chrominance signals into the video receiver and then trace the progression of those signals through the receiver. The different problems that can occur with chrominance-processing circuits, their symptoms, and solutions were examined in detail.

REVIEW QUESTIONS

1. True or False   The 3.58-MHz subcarrier signal modulates the video carrier and produces a side frequency. Regardless of the selected channel, the subcarrier signal is always a modulating frequency.

2. Separating the chrominance signal from any portion of the lower video signals found in the luminance signal lessens the chances of chroma interference from showing in the _____ signal, and prevents the 3.58-MHz color subcarrier from interfering with the _____ signal.

3. Modulating the subcarrier with the quadrature signal shifts the subcarrier by:
   a. 0 degrees
   b. 45 degrees
   c. 90 degrees
   d. 120 degrees
   e. 180 degrees
   f. 360 degrees

   Modulating the subcarrier with an in-phase signal shifts the subcarrier by:
   a. 0 degrees
   b. 45 degrees
   c. 90 degrees
   d. 120 degrees
   e. 180 degrees
   f. 360 degrees

4. With a positive polarity, the I signal appears as _____; the negative-polarity I signal shows as _____. A positive-polarity Q signal appears as _____, while a negative-polarity Q signal appears as _____.

5. True or False   When we place the colors produced by the I and Q signals as coordinates on the color wheel, we find that few colors in either the I or Q signal range match a particular angle.

6. List the four reasons why the 3.58-MHz frequency is used as a color subcarrier frequency.

7. The _____ consists of eight to eleven cycles of the subcarrier frequency, and rides on the back porch of each horizontal blanking interval. The signal has a

   _____ amplitude that equals the amplitude of the sync signal.

8. Does the chrominance channel use the entire bandwidth? Why or why not?

9. The chrominance sidebands contain:
   a. no information
   b. luminance and color information
   c. color information

   They extend out from the carrier:
   a. 0.5 MHz
   b. 1 MHz
   c. 1.5 MHz
   d. 0.5 MHz and 1.5 MHz
   e. 0.5 MHz and 1 MHz

10. Define hue, saturation, and chrominance.

11. The primary colors are:
    a. cyan, orange, and blue
    b. red, blue, and white
    c. red, blue, and green
    d. red, purple, and green

12. Explain why frequency interleaving is used.

13. With frequency interleaving, _____ fit above and below the horizontal scanning frequency and are separated by 60-Hz gaps.

14. Because of the relationship between the _____ and the horizontal scan

    frequency, the _____ interleave between the clusters of luminance-signal frequencies.

15. A filter that has a comb-like _____ response will allow the total

    transmission of desired signals and the total attenuation of any undesired signals. With this

    type of _____ response, a comb filter can separate the

    _____ and _____ components from the composite video signal.

16. True or False   The comb filter separates signals by delaying the composite video signal one horizontal scan period, or approximately 64 microseconds, and then either adding or subtracting the undelayed composite video signal.

17. Why is a delay line incorporated into the luminance-signal channel?

18. How does a comb filter built around a charge-coupled device operate?

19. A comb filter circuit using a CCD has better _____ and

    _____ than a comb filter using a delay line. Why?

20. What is the purpose of the 10.7-MHz clock pulse in the CCD comb filter circuit?

21. The circuit shown in Figure 11-14 improves the appearance of the picture by _____
    _____.

22. The color bandpass amplifier stage separates the _____ and
    _____ signals from the video signal, and amplifies the
    _____.

23. The color control adjusts the following in the chrominance signal:
    a. phase
    b. amplitude
    c. frequency

24. True or False   The color killer interfaces with the last bandpass amplifier and checks for the presence of
    the color burst signal.

25. Refer to Question 3 above. What happens if no color burst signal is present?

26. The tint control adjusts the following in the chrominance signal:
    a. phase
    b. amplitude
    c. frequency

27. The burst separator selects and amplifies the _____ from the chroma
    subcarrier signal.

28. _____ turn the burst separator amplifier on during the time that a color
    burst signal appears at its input, and allow the circuit to remove the color-sync signal from the back porch of
    the horizontal sync pulse.

29. A color oscillator consists of a crystal oscillator tuned to _____ MHz and
    supplies the _____ signal for demodulating the color signal.

30. What is the purpose of the AFPC loop? Describe how the AFPC loop operates.

31. What is the difference between an automatic chroma-control circuit and an automatic tint-control circuit? In
    your answer, describe the purpose of each circuit and how the circuits operate.

32. Chroma demodulation involves _____ encoded at the transmitter and
    recovering the _____ signal.

33. Describe how synchronous demodulators operate.

34. Most manufacturers use:
    a. RGB demodulator systems
    b. I and Q demodulator systems
    c. X-Y demodulator systems
    Why?

35. Matrix circuits combine the _____ signals with the
    _____ signal by adding proportional amounts of input voltages to form
    new combinations of an output voltage.

36. True or False   With color-difference drive matrixing, the RGB signals matrix at a level of only a few volts.
    The CRT drive circuits then amplify the signals to the 100- to 200-Vdc level needed to drive
    the CRT.

37. True or False   With low-level RGB matrixing, the R – Y, B – Y, and G – Y signals travel to the CRT control
    grids, while the luminance signal travels to the red, blue, and green CRT cathodes.

38. Color difference amplifiers are used only in receivers that rely on a _____
_____ for the luminance signal. They usually fit between the
_____ stage and the _____ stage.

39. What are two key characteristics of chroma signals that should be monitored during the troubleshooting process?

40. Standard troubleshooting techniques used for solving chroma circuit problems are

_____, _____, and

_____.

41. When troubleshooting video receiver problems, technicians generally use a:
    a. standard rainbow color-signal generator
    b. keyed rainbow color-bar generator
    c. NTSC color-bar generator

42. The color-bar display provided by a keyed rainbow color-bar generator has _____
bars. Each bar is a different _____ and a slightly different
_____ than the other color bars.

43. A technician using an NTSC generator can adjust both the _____
_____ and the _____ of the color
signals.

44. Name three service applications for a color-bar generator.

45. _____ between the 3.58-MHz reference circuitry and the subcarrier oscillator often affect the chroma demodulation.

46. Amplitude problems usually originate within the _____.

47. A failure of components in the bandpass amplifier stage can block the flow of the chroma signal to the demodulator stage and prevent the production of color-difference signals. This problem causes:
    a. excessive color throughout the picture
    b. confetti-like colored snow throughout the picture
    c. a blooming raster
    d. a monochrome-only picture
    e. a raster with no video signal

48. With the color oscillator operating at the wrong phase, the receiver will produce all colors but

_____.

49. Choose all the answers that are correct. A failure of components in the automatic chroma-control stage can cause:
    a. excessive color throughout the picture
    b. confetti-like colored snow throughout the picture
    c. a blooming raster
    d. a monochrome-only picture
    e. a raster with no video signal

50. A failure in the AFPC circuit will cause _____.

51. True or False   If only one demodulator fails, the receiver will operate normally.

52. True or False   The loss of color sync will cause color bars to drift slowly through the picture or cause the hues to shift.

53. True or False   When troubleshooting a chroma problem in a receiver that utilizes a luminance/chrominance processor, immediately assume that the processor is defective.

54. If a receiver has a smeared or noisy picture, the starting point for troubleshooting efforts would be the

_____ circuit. What specific portion of that circuit should be checked?

55. What are the symptoms of a defective 3.58-MHz crystal?

56. What is the sandcastle signal? Why is the sandcastle waveform a factor in troubleshooting chroma-processing circuits?

57. Refer to the first Service Call. What discovery led the technician to check and find the leaky capacitor?

58. Refer to the second Service Call. Outline the procedure that was followed during the troubleshooting process.

59. True or False   The luminance signal equals the sum of proportional amounts of the primary red, green, and blue video signals.

60. At its maximum relative amplitude of 100 percent, the luminance signal reproduces as a completely white raster and contains:
   a. only the luminance portion of the composite video signal
   b. all three color signals
   c. only the red and blue color signals—green is added at a later point

61. True or False   Changing the values of any of the colors at the CRT will cause the reproduced raster to have a tinted appearance.

62. With a yellow raster, the luminance signal has an amplitude of _____,

while a magenta raster produces a luminance-signal amplitude of _____.
Use a diagram to illustrate why the difference in signal amplitudes occurs.

63. Most video systems rely on the low-level RGB matrixing and CRT system because of the use of CRTs that have the following guns:
   a. unitized
   b. in-line
   c. four

64. True or False   In Figure 11-26, zener diode sets up the amplifier bias.

65. Why does the circuit shown in Figure 11-26 rely on peaking components?

66. What is the purpose of an automatic tracking system? What benefits are given through the use of an automatic tracking system?

67. The automatic CRT tracking system contained within IC1 uses a sample of the video signal to determine the

value of _____.

68. What are the advantages of a cascode video amplifier circuit? Name a common application for the cascode video amplifier.

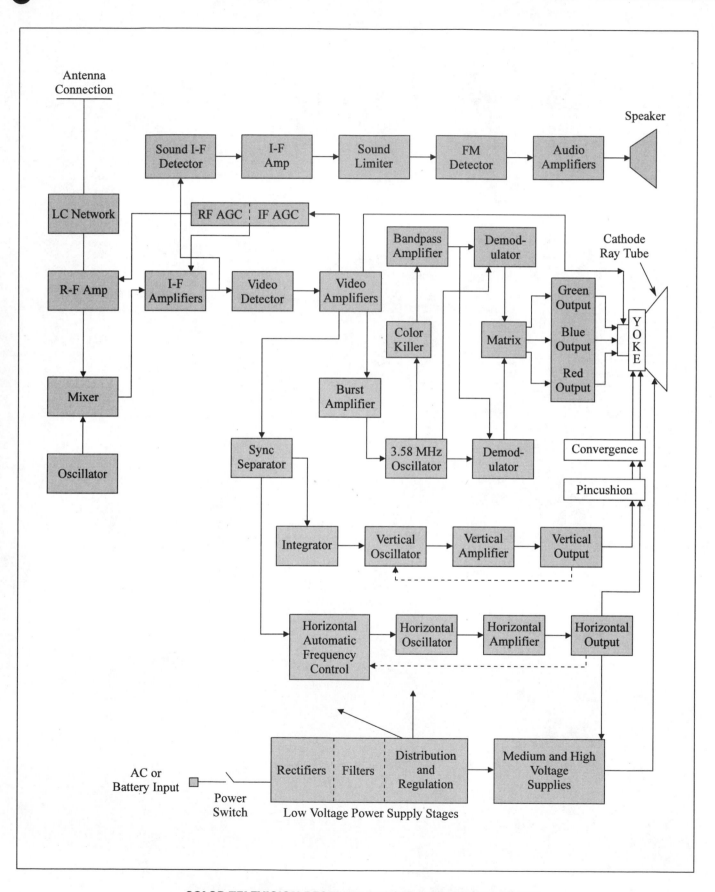

COLOR TELEVISION RECEIVER STAGES AND SIGNAL PATHS

PART 4

Reproducing the Image

Chapters twelve and thirteen show how the circuitry illustrated in the previous chapters can recreate an image on a display screen. While chapter twelve covers cathode-ray tubes and projection television designs, chapter thirteen introduces the newest method for transmitting and receiving a televised image, HDTV, and provides an analysis of computer monitor operation. Each chapter will present the processes used to develop a properly shaped and detailed picture on a viewing screen. The concepts described in both chapters rely on the knowledge that you have gained about power supplies, deflection circuits, and signal-processing circuits.

CHAPTER
12

CRTs and Projection Television Systems

OBJECTIVES   Upon completion of this chapter, you should be able to:

1. Discuss the basics of cathode-ray tube displays.
2. Explain adjustments and circuits that affect the quality of the displayed picture.
3. Discuss projection television displays.
4. Troubleshoot CRT-related problems.

KEY TERMS

aluminum screen	focus adjustment	Novabeam
aperture	focus grid	phosphorescence
Aquadag	front-projection system	pincushion
automatic degaussing circuit	glass faceplate	pincushion circuit
base	gray-scale adjustment	projection television
cathode	grid four	purity magnet
control grid	grid one	rear-projection system
corrector lens	grid three	refractive system
deflection angle	grid two	Schmidt optical system
degauss	in-line configuration	screen grid
delta-gun configuration	light valve	shadow mask
dynamic convergence	low electron gun emission	static convergence
electron gun	low-voltage focus	Trinitron
electrostatic focus	luminescence	ultor
envelope	magnetic focus	unipotential focus
evacuation		

INTRODUCTION

Throughout this text, we followed signals from the transmission point to the their functions within the receiver. Each chapter pointed towards reproducing an exact duplicate of an original, transmitted image onto a CRT display. Chapter twelve takes you inside the display technology currently used in today's video industry, shows different variations of that technology, and considers how to troubleshoot CRT-related problems.

After defining the fundamentals of picture-tube technologies, the chapter shows how different circuits and adjustments—such as purity, convergence, and focus adjustments—affect the quality of the reproduced picture. This is followed by an examination of projection television technologies, ranging from CRTs to laser beams and liquid-crystal displays. As before, the chapter closes by defining methods for troubleshooting CRT problems.

12-1   CATHODE-RAY TUBE BASICS

The picture tube is the result of studies about luminescent materials, magnetic qualities, electric fields, and vacuum tube characteristics. During the operation of a television receiver, electrons bombard a thin coating of phosphors on the inside of the picture tube face, producing varying intensities of light. A group of electrodes, called an **electron gun**, forms the thin beam of electrons and directs the beam at the screen. Magnetic and electric field action provide the electron beam and a means for the beam to sweep back and forth across the entire screen.

Construction of the large tube that contains the phosphor screen at one end and the electron guns at the other relies on vacuum characteristics, the even distribution of phosphors, and the proper alignment of each set of electron-gun electrodes. Monochrome CRTs with only one electron gun, are usually found in smaller displays. Color picture tubes have three electron guns, corresponding to the primary colors. Each gun controls the electrons that strike only the red, green, or blue phosphors.

A thin sheet of metal called a **shadow mask** mounted inside the tube maintains the separation of the electron beams and allows the electrons to converge at the precise angle. While the correct convergence of the electron beams ensures that the correct beam will strike the correct phosphor target, black coatings prevent the reflection of the beams from the shadow mask and enhance the contrast. Like the circuits seen in the previous chapters, the CRT operates as a system. The dynamics of that system produce a finely detailed, stable picture for the viewer.

Luminescence

Any CRT uses luminescence to recreate an image on a screen. **Luminescence** is the absorption of energy and the emission of light without the absorbing material becoming hot enough for incandescence. In the CRT screen, the thin coating of phosphor crystals converts electron energy into light. When a television uses electrical signals to control the electron energy, it converts those signals into variations of light, which become a picture.

The phosphors used for picture tube screens are classified as sulfides and silicates. Adding a small quantity of an impurity, or activator, to those base ingredients causes crystallization. If we add different quantities of the activator to the base ingredients, different phosphor types result. Each phosphor has a unique set of characteristics that involve the color and the period of **phosphorescence**, or persistence. For example, the P1 green phosphor is often used for oscilloscope displays, the P4 white phosphor works well for monochrome displays, and color picture tubes utilize the P22 phosphor. Given the need for producing different color shades, the P22 phosphor contains zinc sulfide for blue, zinc silicate for green, and rare earth elements such as europium and yttrium for red. The manufacturing process used to coat the screen of a color picture tube deposits the phosphor in dots or vertical lines for each color.

Manufacturers use several methods to improve the contrast given by the phosphor screen and, as a result, enhance the quality of the reproduced picture. While one method involves a slight increase in high and screen-grid voltages, another involves the reduction of the spot made by the electron beam when it strikes the shadow mask. The reduction of the spot size creates a higher-density beam where the electrons strike the screen and a whiter white. Other techniques increase the blackness of any black area by reducing the reflectivity of the phosphor screen. Newer screens include a black mask, which covers the area between dots and prevents the reflection of any light.

Parts of the CRT

Picture tubes used in televisions consist of four major parts:

- the glass envelope
- the electron guns
- the screen
- the base

In Figure 12-1, the **envelope** is a large, bell-shaped, glass bulb with a cylindrical extension called the neck. The envelope houses and supports the screen at the large end and the electron gun assembly in the neck.

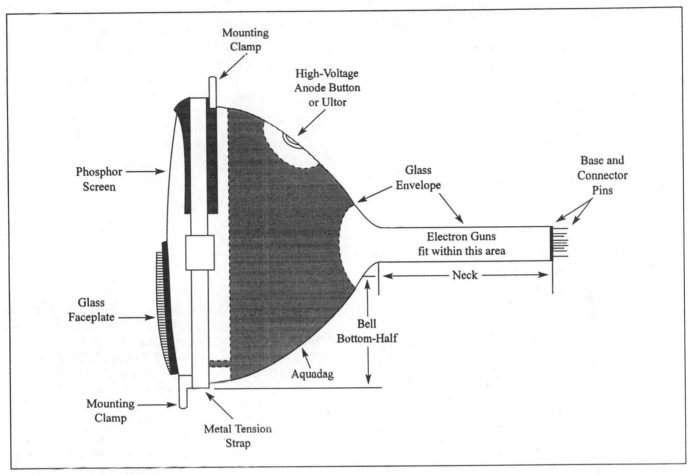

Figure 12-1   Parts of the CRT

The **base** contains connectors that allow the attachment of external electronics to the electrodes.

We can consider the picture tube as a series circuit that utilizes a high-voltage supply. The electron beam forms a path from the cathode to the phosphor screen of the tube. In turn, the conduction between the cathode and the high-voltage anode creates a complete circuit for the electron beam. The load current, or beam current, for the high-voltage supply ranges from 100 microamps for monochrome to 1000 microamps for color electron guns.

The Electron Gun   Each electron gun in a tricolor CRT provides a narrow beam of electrons, directs the electrons towards the screen, and accelerates them to the high speeds needed to cause light emission on the screen phosphors. When the electron beam strikes the phosphor, it produces a tiny spot of light. The intensity of the electron beam varies with the video signal applied to the electron gun. As discussed in earlier chapters, the vertical and horizontal coils in the deflection yoke control the up-and-down, and back-and-forth movements of the electron beam.

An electron gun is a group of closely spaced electrodes that produces a small-diameter electron beam.

As shown in Figure 12-2, the essential parts of each of the three electron guns are the:

- Cathode
- Anode
- Grids

Starting at the left of the figure, the indirectly-heated **cathode** consists of a closed-end cylinder surrounded by a larger-diameter tube, or grid one (G1). Another tube-shaped electrode serves as grid two (G2), while the fourth tube is the anode.

In order to maintain the electrons in a tight beam, each of the grids and the anode contain a metal disk with a small hole, or aperture, in the center. The apertures provide the only path for the electrons to reach the screen. Depending on the CRT size and type, the anode has a potential ranging from +8000 to 30,000 Vdc with respect to the cathode. This difference in potential creates a strong electric field between the two electrodes and causes the electrons in the beam to accelerate to a high velocity.

Electron guns appear in the three different configurations—called the delta, the in-line, and the Trinitron—shown in Figure 12-3. With the **delta-gun**

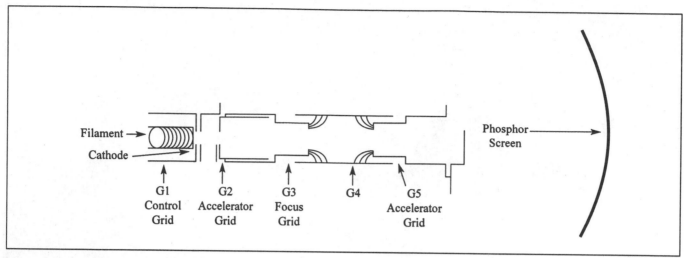

Figure 12-2   Parts of the Electron Gun

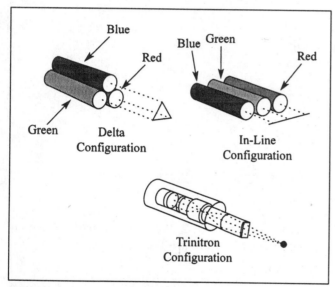

Figure 12-3   Electron Gun Configurations

configuration, the three electron guns form an equilateral triangle, or delta. This configuration allows the use of a larger-focus electrode in each of the guns. In the **in-line configuration**, the guns lie along one horizontal plane of a width equal to the diameter of the tube neck. The in-line configuration yields a smaller spot size, excellent focus, and high resolution. **Trinitron** tubes, produced by the Sony Corporation, feature three cathodes and one electron gun that contains all three electrodes. The metal disks that make up each grid assembly have three holes that match with the three electron beams from the single gun. When the three beams exit from grid one, the beams travel first to a cross-over point and then through a large Einzel lens. The lens uses a common field to focus the beams.

Electron-Gun Grids   For each of the electron guns in a tricolor CRT, **grid one**, or G1, works as the **control**

grid. Given a negative potential with respect to the cathode, the control grid controls the number of electrons in the beam. The voltage applied between the G1 grid and the cathode of the CRT controls the beam current and the resulting light output of the tube. With the control-grid voltage varying from zero to −80 Vdc, the beam current can increase to a maximum of 1.5 milliamps. At −80 Vdc, the control-grid voltage cuts off the beam current and causes the picture to black out. Reducing the control grid voltage to −60 Vdc and then to −40 Vdc causes the beam current to increase and produce more light for the white picture information.

At every point of the control-grid voltage measurements, the control grid remains negative with respect to the cathode. Any voltage between the G1 grid and the cathode determines the amount of beam current. Even though the voltage measured from the control grid to the chassis ground may stand at +100 Vdc, the grid must have a negative potential when measured against the cathode. Thus, with +100 Vdc at the grid and +140 Vdc at the cathode, the negative difference in potential remains within the zero to −80 Vdc range at −40 Vdc.

Grid two (G2), called the accelerating or the **screen grid**, has a positive potential with respect to the cathode, and prevents any interaction between the control grid and anode electric fields. Because of the electric fields existing between the adjacent electrodes, the beam remains confined to an extremely small diameter. Therefore, the electron beam passes through the apertures without striking the disks.

We can consider grid two as the first anode as it accelerates electrons towards the screen. Within the grid two cylinder, small internal baffles maintain the narrow path for the beam. A **focus grid** or cylinder, designated as **grid three (G3)**, combines with grid two to form an electrostatic lens. With a voltage of several kilovolts at the focus grid, the lens combination forces electrons to stay on paths that come to a point at the phosphor-coated screen. **Grid four (G4)** is part of the high-voltage connection.

The Glass Envelope

Looking back at Figure 12-1, we can begin to see the relationships between the components that make up the electron guns, the screen, and the coatings layered onto the inside of the glass envelope. During the manufacturing process, the inside surfaces of the glass envelope receive a coating of conductive material. The anode, or grids three and five, make electrical contact with that coating through curved spring supports found at the forward end of the electron gun.

Picture tubes undergo a process called **evacuation**, which establishes a vacuum inside the glass envelope. The phosphor-coated **glass faceplate** must have the proper thickness to withstand the air pressure exerting a force against the vacuum, and to protect the viewer from possible X-ray radiation. To further minimize the chances of an implosion—a violent collapse of the faceplate that scatters glass, the glass faceplate either consists of a laminate of thick glass layers, or has a prestressed steel tension band mounted around it. The tension band forces the faceplate to hold its shape even if cracked. Most television picture tube manufacturing processes apply the second method.

Aluminizing the Screen   In addition to the phosphor coating on the inside of the screen, a very thin layer of aluminum is placed on the inside surface of the phosphor screen facing toward the electron gun. This **aluminum screen** is transparent to the electron gun but continues to reflect light from the screen. Thus, the light emitted by the phosphors reflects toward the viewer rather than back to the gun, and the picture tube achieves a higher level of brightness.

The aluminized layer also blocks any heavy-ion charges that could burn a brown spot into the center of the screen. Electron emission at the cathode of the CRT generates the ion charges, while the phosphors react to the charges by burning. Blocking the ion charges is necessary because the ions do not deflect as easily as electrons.

Along with reflection and blocking, the aluminum layer collects any secondary electrons emitted by the phosphor screen during contact by the electron beam. Collecting the secondary electrons allows the screen to charge to the same potential as the high voltage entering at the secondary anode. With this positive potential in place, the phosphor screen can attract the electron beam.

The Aquadag   As we discussed in chapter three, the high-voltage power supply applies the low-current, high-dc voltage to the CRT anode. This occurs through the bulb cap, known as the **ultor** or the anode button, located at the top center of the conductive coating, or **Aquadag**. The application of the high anode voltage to the internal conductive coating establishes a uniform electric field for the moving electrons. The Aquadag contacts the screen and provides a path for the electrons to follow from the screen back to the power supply.

The Aquadag also coats the outside surface of the glass envelope, and connects to chassis ground through either a series of clips or a wire harness. The area around the high-voltage anode, or ultor, receives no Aquadag coating. Using the Aquadag as a grounded outer coating lessens the radiation of electrical interference running at the vertical and horizontal frequencies. Electrical interference of 60 Hz entering the receiver at this point would cause a buzzing sound in the audio, because the audio circuits pick up the vertical scanning output signal.

As you may suspect, covering the internal and external surfaces of the tube with a conductive coating forms a large capacitor. The glass envelope serves as the dielectric. Given the large surface area of the tube and the thickness of the glass dielectric, the anode capacitance can have values as high as 2000 picofarads and can hold a charge for an extremely long period. In effect, this anode capacitance becomes a filter for the high voltage induced at the horizontal frequency rate. With the scanning frequency at 15,750 Hz, the anode capacitance filters a ripple frequency of 15,750 Hz. When we discuss safety issues in the troubleshooting section of this chapter, we will review safe methods for handling picture tubes.

CRT Deflection Angles   Every CRT is measured by a standard called the **deflection angle**, in which the size of the raster produced on the face of the tube is proportional to the angle of the electron beam deflection and the distance between the tube face and the point of beam deflection. Older picture tubes had a small deflection angle ranging from 52 to 72 degrees, which resulted in the lengthening of the glass envelope. New picture tubes have deflection angles of 90, 100, and 110 degrees and a much shorter length.

To compensate for the decreased length and larger deflection angle, the newer picture tubes rely on increased magnetic field strength. To accomplish this, the CRT manufacturers reduced the diameter of the electron gun and the diameter of the tube neck. Reducing the CRT neck diameter also allowed a reduction in the diameter of the deflection yoke, which, in turn, brought the windings of the yoke closer to the electron beam. Given these changes, the decrease in the inside yoke diameter produced a stronger magnetic field. A greater deflection angle requires more deflection power and additional load current, so manufacturers keep the deflection angle within the 90-degree range. The deflection angle of the yoke matches the deflection angle of the CRT.

All modern television picture tubes have rectangular faceplates that have the same shape as the raster produced by the scanning beam. Picture tubes are measured on the diagonal, rather than by length or width. That is, a 19FLP4 picture tube has a diagonal measure of 19 inches. However, the *viewable* diagonal measure and usable screen area of the tube shrink to 17.56 inches and 172 square inches.

The CRT Base Except for the grids that make up the anode, each electrode of an electron gun has a wire lead that extends through the glass end of the neck. Each wire lead has a soldered connection to external base pins that connect the electron gun to the video output circuitry. Looking at the schematic symbols shown in Figures 12-4A and 12-4B, the grids are placed above the cathode and are marked in the order that the electron beam passes through each grid. Therefore, we can refer to the control grid as G1 and the accelerating grid as G2.

Figure 12-4A illustrates the base layout for a monochrome picture tube. Figure 12-4B uses the base diagram of a color CRT to illustrate the differences between monochrome and color tubes. Comparing the drawings, we can see that the color CRT has more connections because of the number of grids and cathodes per electron gun. The capacitor symbol inside each base diagram represents the glass dielectric for the anode capacitance.

The Shadow Mask In Figure 12-5, two light sources suspended above a perforated plate at different angles illuminate specific points on a metal plate. If we assume that the points on the plate are painted red and blue, light from source A would strike only the red spots, while light from source B would strike only the blue spots. The drawing illustrates the principle of using separate beams for the primary colors in a tricolor CRT, and the purpose of the shadow mask in keeping the beams separate before they contact the phosphor screen.

Properly aimed electron beams strike specific perforations on the shadow mask. Rather than have the source emit a given color, as we saw in the prior illustration, the electron beam emitted from the CRT gun is directed to a phosphor screen that emits the color. A shadow mask may have round, rectangular, lozenge-shaped, or continuous perforations. While monochrome CRTs and projection tubes do not contain shadow masks, all direct-view color picture tubes include a shadow mask.

Figure 12-6 depicts the shadow masks used for the delta and in-line CRT gun configurations. An actual round-hole shadow mask has approximately 400,000 holes, or **apertures**, placed at regular hexagons along a vertical axis. Each aperture is tapered to present a sharply defined target for the electron beam and to prevent the scattering of the beam. Because the slot mask aligns the vertical perforations with the vertical axis, we could consider each slot as a series of vertically merged, oblong dots. The vertical alignment of the perforations in the slot mask eliminates the need for color purity and gun registration adjustments in the vertical direction, and simplifies the design of the CRT.

Phosphor dots on the screen contain the chemicals needed to produce the primary colors and match with the holes in the shadow mask. As Figure 12-7A shows, during operation the deflection center of the electron beam places the end of the beam in the precise location that matches phosphor dot to shadow mask hole.

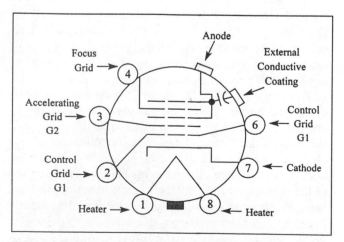

Figure 12-4A Schematic Diagram of the Base Layout of a Monochrome Picture Tube

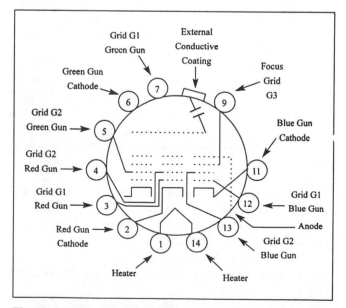

Figure 12-4B Schematic Diagram of the Base Layout of a Color Picture Tube

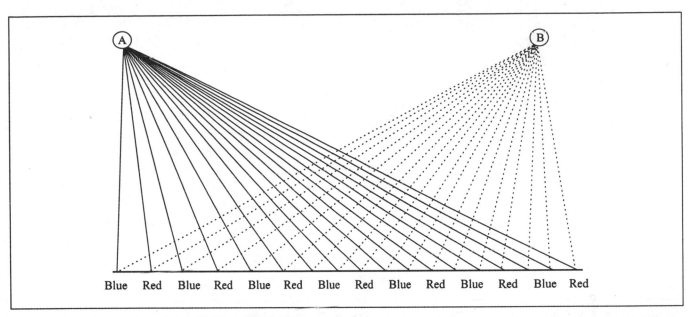

Figure 12-5   Illustration of the Shadow Mask Effect

Although most picture tube manufacturers use shadow masks that have acid-etched, conical holes, the Sony Trinitron shadow mask relies on continuous vertical slots instead of holes. The vertical slots match with vertical phosphor stripes painted onto the screen.

In each case, the shadow mask serves as a pattern for printing the phosphor onto the screen. Using photoresist techniques, manufacturers sequentially print color phosphor segments onto the screen and use a different light exposure for each primary color. During operation, the specific electron beam must have a precise alignment with the phosphor dot representing the designated color of the beam. Purity, or beam landing, adjustments refer to the adjustments needed to ensure that the beam strikes the correct area.

Figures 12-7A and 12-7B both refer to a black material surrounding the phosphor on the screen. Not surprisingly, the entire phosphor-coated surface of the screen reflects light in all directions. To improve contrast and to ensure that the electron beams strike only the correct phosphor materials, manufacturers coat the screen between the phosphor areas with a black mask. The mask eliminates the problems of light reflection within the tube and causes black areas in the picture to appear blacker.

☑ PROGRESS CHECK

You have completed Objective 1. Discuss the basics of cathode-ray tube displays.

12-2   DISPLAY ADJUSTMENTS

This section of the chapter leads us away from the actual CRT to circuits and adjustments that direct the electron beams to the correct positions and refine the reproduction of the picture. Because of ever-present magnetic fields, every receiver includes an automatic degaussing circuit that removes the residual effects of a magnetic field on the CRT. Those effects appear as dull blotches around the corners of the picture. Along the degaussing circuit, small magnetic rings mounted on the neck of the CRT also establish the purity of the reproduced picture.

Other circuits, called static and dynamic convergence circuits, correct the trajectory of the electron beams so that the beams strike the correct phosphor dots on the screen. The static convergence adjustments

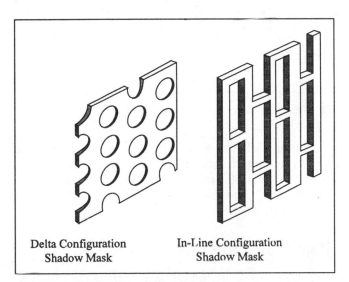

Figure 12-6   Shadow Masks Used in Delta and In-Line Configurations

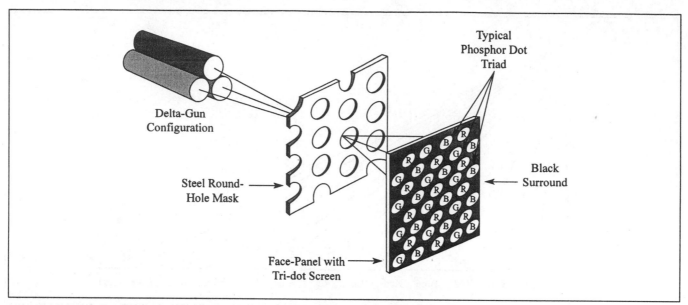

Figure 12-7A   Matching the Screen Phosphor Dots with the Shadow Mask

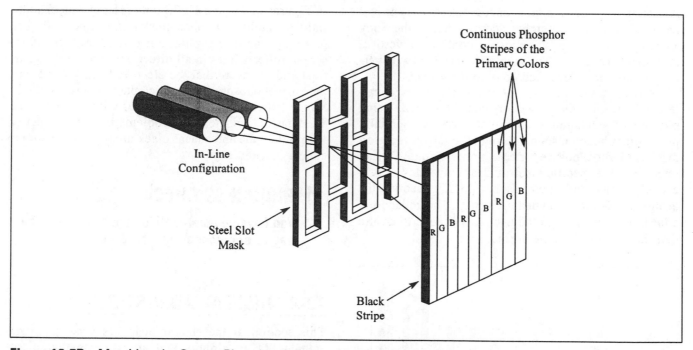

Figure 12-7B   Matching the Screen Phosphor Stripes with the Shadow Mask

adjust the beams for the center of the screen, while the dynamic convergence adjustments control the convergence for the outer edges. Additional picture refinement is produced by pincushion circuits, which establish the correct shape of the rectangular picture; focus adjustments, which ensure that the electron beam will retain a narrowly-defined shape rather than spreading as it leaves the electron gun; and gray-scale adjustments, accomplished through red, blue, and green screen controls, which provide a correctly balanced picture for the viewer.

Figure 12-8 shows a CRT with the purity rings and convergence and deflection yokes, as well as the degaussing coils that wrap around the edges of the bell of the CRT. The focus and gray-scale controls are mounted in the chassis.

Degaussing the CRT

Any magnetic field—ranging from the field presented by a loudspeaker to the magnetic field surrounding the earth—can adversely affect the quality of a color

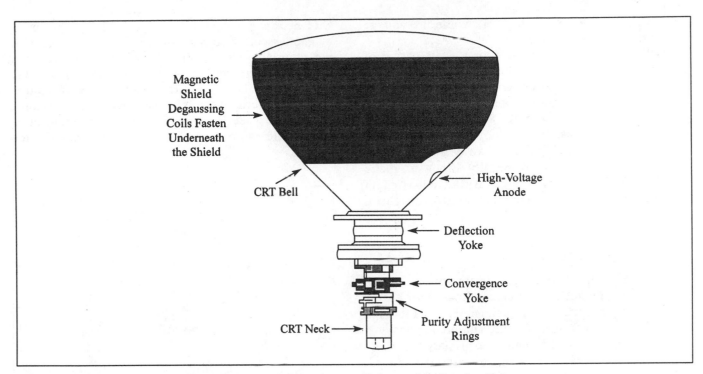

Figure 12-8   Location of the Purity Magnets, Convergence Yoke, and Deflection Yoke

television picture by changing the path of the electrons from the electron gun to the phosphor screen. As a result, the magnetic fields can downgrade the purity of the raster and cause a misconvergence of the three electron beams.

Degaussing removes the induced magnetic field and builds a local magnetic field to offset the effects of the earth's magnetic field. While a technician can use an external, ac line-connected degaussing coil to demagnetize problem CRTs, all picture tube assemblies include an **automatic degaussing circuit** to remove the residual effects of a magnetic field. Turning on the receiver causes a strong current from the ac input to pass through a set of degaussing coils attached around the CRT support harness. As a result, the CRT is demagnetized during every powering on of the receiver.

Figure 12-9 shows the schematic diagram for a typical automatic degaussing circuit. The thermistor has a large resistance when cold, and a very small resistance when hot. The voltage-dependent resistor, or VDR, has a large resistance when a large voltage is connected across the resistor, and a small resistance with a small voltage. When the receiver powers on, the large resistance of the cold thermistor causes almost all of the 60-Hz current to flow through the VDR and the degaussing coil. As the thermistor heats, less current

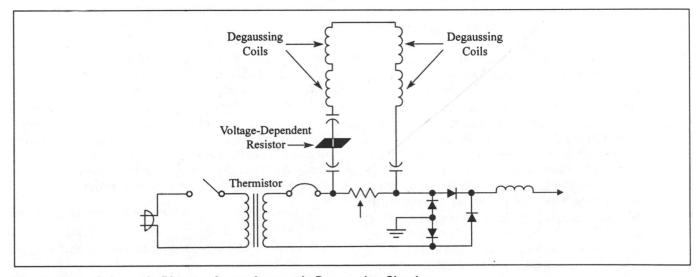

Figure 12-9   Schematic Diagram for an Automatic Degaussing Circuit

flows through the degaussing coil, while the voltage across the VDR increases.

With the receiver fully on after a short period of time, the resistance of the VDR increases, and most of the 60-Hz current flows through the low resistance presented by the thermistor. Almost no current flows through the degaussing coils at this point. Consequently, the circuit has the effect of applying current to the coil and then moving the coil away from the CRT. This occurs because of an initially strong magnetic field that gradually grows weaker. As the field strength decreases, it demagnetizes any nearby stray magnetic object.

Purity Adjustments

At the time of manufacture, the electron beams are adjusted so that the deflection centers for all three beams are at precise points. This adjustment ensures that the electrons emitted from red, blue, and green guns will strike only phosphors that match the primary color designated for the specific gun, with the result that the raster will display pure colors. For example, if we biased the blue and green guns off, the raster would display a pure red color. An impure raster with the same conditions would show a mostly red, somewhat out-of-focus raster with blotches of blue and green at the corners.

Correct color purity adjustments produce pure red, green, and blue rasters. The correct balance of the beam currents will produce a uniform, neutral white raster with all three electron guns biased on. Achieving the correct purity and beam-current balance requires the movement of the deflection yoke, adjustments of a color **purity magnet** mounted on the CRT neck, the correct convergence settings, and the demagnetization of the CRT.

While picture tubes with a delta arrangement usually use a completely red raster for the purity adjustments, the in-line configuration purity adjustment often works best with a green raster. The use of the red raster has become popular because of the higher beam current needed to excite the red phosphors on the screen. Any purity problems will show quickly with the higher beam current.

If the deflection yoke has an adjustable mount, the first step for making a purity adjustment involves moving the yoke either all the way forward against the bell of the picture tube, or all the way back toward the socket. Moving the yoke to either extreme creates large errors in the beam landing positions at the edges of the screen, isolates the center of the screen for the purity check, and causes a medium-sized red spot to appear at the center of the screen. Adjustment of the purity magnet moves the red spot to the center of the screen.

The purity magnet is used to adjust all three beams to obtain pure primary colors. It is mounted on the CRT neck and is depicted in Figure 12-10. The purity magnet assembly consists of two magnetic rings with indicator tabs. Moving the tabs in opposite directions produces a stronger magnetic field and affects the centering of all three electron beams. Once the adjustment of the purity magnet has centered all three beams, moving the yoke back to the normal operating position should cause the red cloud to uniformly fill the screen.

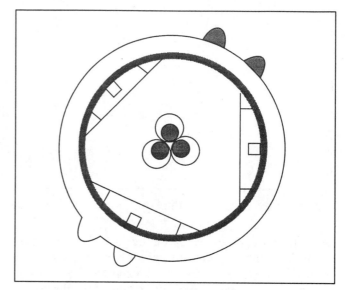

Figure 12-10   Purity Magnet Assembly

Convergence Adjustments

As mentioned, the purity magnet provides a method for adjusting the beams for the purpose of obtaining pure primary colors. Nevertheless, most receivers require a more comprehensive method for achieving the correct registration of the three colors. Another adjustment, called convergence, allows the individual adjustment of the electron beams and produces white. The lack of convergence becomes evident with the display of red, blue, green, cyan, or magenta fringes to the left, right, or bottom of objects reproduced on the screen.

When comparing the purity and convergence adjustments, the color purity adjustment applies to the solid color of the raster and the background of the picture, while the convergence adjustment applies only to the picture and not the raster. Convergence adjustments are completed through the use of a color-bar-pattern generator. At the minimum, the generator provides:

- Small, white dots in horizontal rows and vertical columns
- A cross-hatch pattern of horizontal and vertical white lines
- Ten vertical color bars

Good convergence becomes evident through the appearance of the distinct white dots and cross-hatch pattern; neither should have color fringing. Convergence occurs through static and dynamic adjustments.

Static Convergence   Static convergence involves the adjustment of three permanent magnets mounted around the CRT neck in order to converge the electron beams to a target at the center of the screen. Pictured in Figure 12-11, the entire convergence assembly is referred to as the external convergence yoke. Adjusting the magnets on the convergence yoke affects the polarity of magnetic flux lines.

Because the electron beams move at right angles to the field lines, adjusting a magnet moves the beam back and forth or up and down. In tubes using the delta configuration, an adjustment of the blue magnet mounted on top of the assembly moves the beam either up or down. Adjusting the red or green magnets—located below and at 30-degree angles from the CRT neck—causes the beams to move in a line 30 degrees from the horizontal. In-line configurations rely on a slightly different method for achieving static convergence. With the beams placed along the same horizontal plane, adjusting the magnets has a minimum effect on the green gun. Adjustments of the red and blue beams cause the beams to bend and, as a result, to converge with the green beam. Another set of magnets moves the beams slightly up or down to compensate for any possible horizontal alignment problems with the guns.

Dynamic Convergence   After completing the static convergence of the electron beams, the next step is **dynamic convergence** which controls the convergence of the beams to the outer edges of the screen. Because the screen has a slight curvature and because the electron beams static-converge short of the screen edges, dynamic convergence is necessary to spread the beams and cause convergence at a longer distance. When the beams converge short of the screen edges, the viewer can see color fringes surrounding any image in that area. Dynamic convergence increases the deflection of the beams and causes convergence to occur at the edges of the screen.

This type of correction occurs through the modulation of static convergence magnetic fields with an ac component from both the horizontal and vertical deflection signals. Adjusting the controls mounted on the convergence panel establishes current in the convergence winding and generates a parabolic convergence-correction current. With up to twelve available controls and the use of a cross-hatch pattern, a technician can adjust the amplitude, tilt, and vertical and horizontal directions of each of the electron beams for the top, bottom, left, and right edges of the screen.

Vertical and Horizontal Dynamic Convergence Circuits

Figures 12-12 and 12-13 show the circuitry used for the vertical and horizontal dynamic convergence of a color

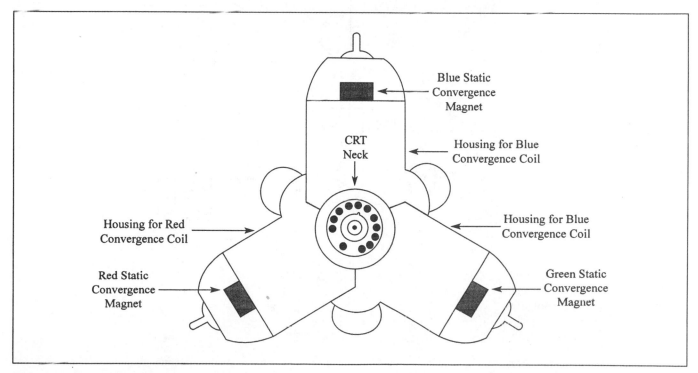

Figure 12-11   Convergence Yoke Assembly

television receiver. In the first figure, the convergence for the blue gun occurs through the adjustment of the blue convergence magnet and the combined effect of resistors R1 and R5, capacitor C2, diode D1, and the blue vertical convergence coils. While the blue convergence magnet receives vertical-correction current from the vertical output amplifier through capacitor C1, the resistor-capacitor-diode combination produces a parabolic waveshape.

Potentiometer R4 controls the amplitude of the parabolic current waveform, while the addition of a small sawtooth current obtained from the vertical output circuit controls the waveshape. Potentiometer R3 provides additional control over the waveshape. Setting R3 in either direction adds a small sawtooth signal to the parabolic waveform, which results in adding tilt to the parabola.

Potentiometer R1 controls the amplitude of the parabolic waveform for the red and green convergence magnets. R7, connected in series with R3, and R8, connected across the vertical output circuit, adjust the tilt of the red and green parabolic waveforms. Potentiometers R1 and R7 establish the total convergence of the red and green electron beams. In combination, controls

R1 through R7 adjust the vertical convergence of the three electron beams.

Moving to Figure 12-13, a special winding on the flyback transformer provides a signal that travels through C1 and T1 to the blue convergence magnet. The blue horizontal convergence coil, along with resistor R1 and diode D1, establishes a parabolic waveshape for the blue convergence control. Because the horizontal convergence circuit operates with higher frequencies, a tuned circuit consisting of R2, C2, and L1 controls the amplitude of the waveform. Tuning L1 to its resonant frequency establishes a high impedance and allows more current to flow through the convergence coil. In addition, potentiometer R3 controls the time constant of the C3/R1 circuit and, as a result, affects the phase of the parabolic current.

The same flyback winding supplies a correction current through C4 and L2 for the red and green horizontal convergence coils. At coil L3, the signal divides into two opposite-polarity signals. Those signals flow into the convergence coils and, because of the opposite polarities, reverse the polarity of one coil. As before, a series of resistor and diode combinations operates with the convergence coils and the center-tapped

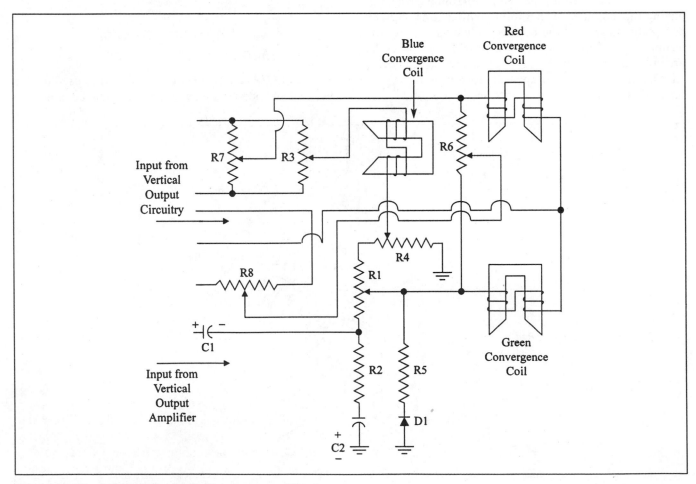

Figure 12-12   **Vertical Dynamic Convergence Circuitry**

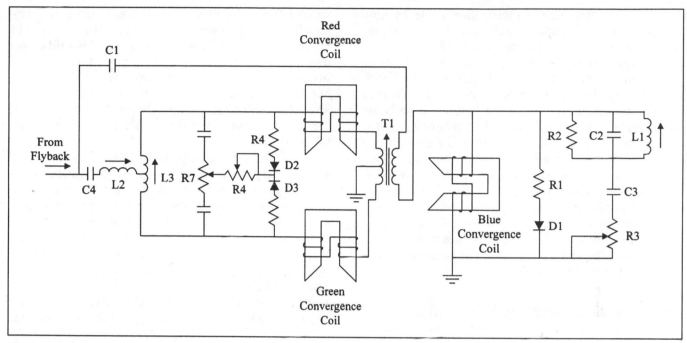

Figure 12-13 Horizontal Dynamic Convergence Circuitry

winding of transformer T1 to produce the parabolic waveform needed for red and green convergence. While potentiometer R7 adjusts the amount of current flowing through the coils, potentiometer R6 adjusts the phase of the parabolic waveform. With this, R7 controls the convergence of the red and green electron beams to the left, while R6 adjusts the convergence of the red, blue, and green electron beams on the left side. Adjustable coil L3 controls the convergence of the beams on the right side.

Pincushion Correction

Pincushion describes a condition where the raster has the distorted shape shown in Figure 12-14. This condition occurs because the deflection center for the electron beams is farther from the corners of the screen than from the middle. During the scanning process, the beams move at extreme angles horizontally and verti-

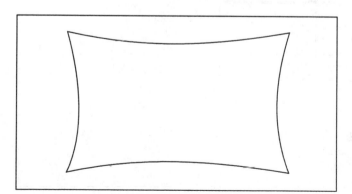

Figure 12-14 Illustration of Pincushion

cally when reaching the corners of the screen. Within the receiver, correction signals from pincushion circuits travel to the deflection coils of the yoke and straighten the top, bottom, and sides of the raster. In modern televisions, the improvements in deflection yokes and picture tubes have reduced the chances for pincushion distortion.

Focusing the Electron Beam

In chapter three, we briefly discussed the reasons for using a focus-control voltage to control the electron beams. As the electrons leave the electron gun, the physical laws of attraction and repulsion cause them to repel one another and diverge. If this were not corrected, the reproduced picture would consist of a mix of badly blurred images. Instead, **focus adjustments** force the electrons to converge back into a narrow beam and produce a crisp, clear picture. These adjustments are accomplished through the application of either an electric or a magnetic field. Video camera picture tubes use the magnetic deflection provided by the yoke horizontal and vertical deflection coils, while direct-view and projection CRTs rely on electrostatic focus.

Focusing with Magnetic Fields Magnetic deflection of the electron beam is less complicated than the electrostatic focus method because of the difference in obtaining a usable deflection angle. With electrostatic focusing, the deflection angle in magnetic deflection is inversely proportional to the square root of the high voltage. For example, a nine-fold increase in anode voltage would decrease the deflection angle by only one-third. In this example, the amount of deflection angle would require an immensely long CRT.

The older **magnetic focus** method utilizes the magnetic fields developed through the placement of horizontal and vertical yoke deflection coils placed around the picture tube neck. Since the horizontal coils are placed on the top and bottom of the beam axis, and the vertical coils are placed to the left and right, the two magnetic fields produced by the sawtooth current flowing through the coils react with the magnetic field that accompanies the electron beam. As a result, the electrons deflect at right angles to both the beam axis and the deflection field. The deflection of the electron beam caused by the yoke coils combines with the control given by the low-voltage focus to both narrow and direct the beam to its target.

Focusing with Electrostatic Fields With **electrostatic focus**, the narrow beam of electrons travels through two electron lenses. The first lens consists of the electrostatic field produced by the difference in potential between the cathode and control grid. Because of that voltage, the electron beam converges to a spot just beyond the control grid called the cross-over point. In effect, the difference of potential between the positive G2 voltage and the negative anode voltage establishes a forward-accelerating force that causes the electrons to travel through a grid aperture. The cross-over point is the point where the electron beam narrows the most and has a much smaller area than the cathode area supplying the beam.

The second electron lens evolves from either the high-voltage or the low-voltage area. With high-voltage focus, the focus voltage has a value approximately one-fifth of the high-voltage value. Pictured in Figure 12-15, a gap between the G3 and G4 grid cylinders causes an electric field to form and forces diverging electrons toward a center axis. While the G3 grid has the fixed focus-voltage value, the G4 grid includes a variable control that permits the precise adjustment of the G3 voltage for a picture that has the sharpest possible details.

Low-voltage focus or **unipotential focus** uses a low voltage, ranging from 0 to 400 Vdc, from the low-voltage power supply to control the electron beam. In this system, grid three or the focus grid, which is a large cylinder, fits between two smaller grids that have the same potential as the high-voltage anode. With the low-potential focus grid sandwiched between two high-potential grids, the electron beam enters a decelerating field and the electrons in the beam converge. Many monochrome picture tubes and some Sony Trinitron CRTs utilize the low-voltage focus technique.

Gray-Scale Adjustments

When we first discussed the use of red, blue, and green as primary colors, we found that merging the three primaries produced white. **Gray-scale adjustments** for a color television receiver provide the balance between the three primary colors and produce a neutral white screen. During the normal operation of the receiver, the balance must remain steady from low to high brightness values. Without the correct balance, the screen color will change slightly with any changes in the picture brightness. From an overall perspective,

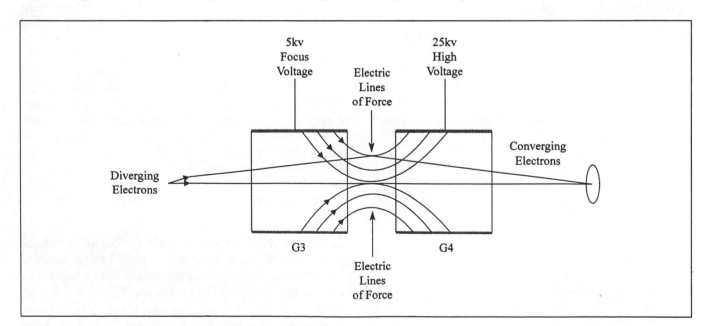

Figure 12-15 An Electric Field Forms During High-Voltage Focus

low-lighted whites in a picture may have a different tint than high-lighted whites.

The procedure for adjusting the gray scale varies with the age of the television. For older receivers, adjustment of separate red, blue, and green drive and screen-grid controls changes the proportions of ac video signals flowing to the electron guns, and establishes gray-scale tracking. Generally, those controls are located at the rear of the receiver chassis.

Adjustment of the gray scale for the older receivers also involves setting the normal setup switch to the setup position. This position disables the vertical deflection and removes the video signal, and consequently, the receiver displays a stable, horizontal line. From there, the slight adjustment of three screen controls should produce a faint white line and the proper cut-off characteristics. Returning the switch to the normal setting and the brightness and contrast controls set to normal operating positions, adjustment of the three drive controls for a neutral white establishes the correct balance between the low- and high-lighted areas of the picture.

Newer receivers combine the screen and drive controls into one set of red, blue, and green controls. Because of this, the receiver lacks the independent setting of cut-off for each electron gun. However, the adjustment of the background controls at the minimum setting of the master brightness control will establish the proper grid-cathode bias for the electron guns. Setting the gray-scale balance involves turning the color control and the brightness and contrast controls to minimum settings, and setting the master screen control for a dim monochrome picture. Adjustment of the background controls should produce a neutral gray screen. With the brightness and contrast controls adjusted to the normal settings, the video drive control should produce neutral white in bright areas of the picture.

☑ PROGRESS CHECK

You have completed Objective 2. Explain adjustments and circuits that affect the quality of the displayed picture.

12-3   PROJECTION TELEVISION DISPLAY METHODS

Although large-screen, direct-view CRTs have become more common, **projection television** systems still command a large market. This type of system involves projecting separate red, blue, and green images onto a common reflecting screen. The resulting light mixtures produce a color picture. Projection television provides the advantage of a much larger picture than that of direct-view picture tubes. That advantage, however, is accompanied by two disadvantages.

The first disadvantage involves the need to produce sufficient brightness for a picture that may cover ten times the area of a standard 25-inch diagonal direct-view picture. Every increase in picture size must be accompanied by a proportional increase in brightness. Recent technology counters this potential problem with projection CRTs that produce an extremely intense light output. Such CRTs' usually have a screen that ranges in size from 1 to 5 inches, and utilize an anode high voltage that ranges from 30 to 80 kilovolts.

The second disadvantage involves the maximum number of scanning lines and 4-MHz resolution presented by current transmission and reception technologies. Given this current technology, every horizontal scanning line becomes visible. In addition, the reproduced picture loses some of the sharpness seen with a smaller, direct-view picture. All this occurs because the projection television operates with the same amount of details and contrast as the direct-view television. As the next chapter will show, the advent of high-definition television will eliminate this disadvantage.

Projection Television Systems

As shown in Figure 12-16, projection televisions require four basic elements:

- a viewing screen
- an optical assembly
- an imaging source
- an electronics system

While projection televisions may follow different design models—such as front or rear projection—the basic concept for moving a reproduced picture from a CRT to a large screen remains the same.

The type of viewing screen used in the system controls the amount of illumination, picture resolution, and the capability to transmit light back to the viewer. All this affects whether the viewer can enjoy a quality video reproduction while sitting at various angles from the viewing screen. Optical assemblies involve the type of mirrors and lenses that combine with the CRT

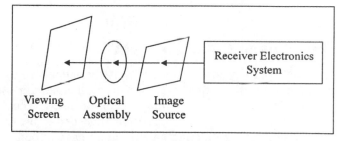

Figure 12-16   A Basic Projection Television System

mounting scheme to project a picture on the viewing screen. The quality of the optical assembly depends on image geometry and resolution. Without the correct geometry, the projection of three images on the viewing screen could result in a keystone effect, pincushioning, or a barrel effect. The quality of resolution depends on the capability of the lens to reproduce a defect-free, magnified image.

While image devices for projection televisions include high-intensity CRTs and light valves, a projection system uses the same electronics design as in direct-view televisions. However, the unique characteristics presented by projection televisions require several modifications. These include the use of several different types of heavy-duty, low- and high-voltage power supplies; added overvoltage and overcurrent protection; three focus coils; extra controls for the linearity and height of each CRT raster; and the addition of parabolic and keystone generators. The parabolic and keystone generators ensure that the image projected on the viewing screen will retain a rectangular shape. Most projection television systems also include enhanced tuning and sound systems.

Projection television systems vary with the type of video display system used to place an image on the large screen. While one manufacturer continues to use the Novabeam CRT and the Schmidt optical system, developed by Henry Kloss, most use the refractive system. Both the Novabeam and the Schmidt optical systems rely on the conventional set of three high-intensity picture tubes. Other manufacturers have begun to move to a system that discards the three-CRT model in favor of a high-intensity light and three liquid-crystal screens. The last system does not use any type of cathode-ray tube.

Figures 12-17 and 12-18, depict the two basic categories of projection televisions. The **front-projection system** depicted in Figure 12-17 allows the image to be viewed from the same side of the screen as that used for the projection. Front-projection systems depend on reflectivity for brightness. Illustrated in Figure 12-18, the **rear-projection system** has the image viewed from the opposite side as that used for the projection. Rear-projection televisions use a higher transmission scheme, diffusers, and corrector lenses to achieve acceptable brightness levels.

The Schmidt Optical System

Adapted from a system designed for astronomical cameras, the **Schmidt optical system** uses a spherical mirror to fold the reproduced image back on itself. The spherical mirror decreases the distance needed to project the image by reflecting an enlarged image in both front- and rear-view projection systems. A **corrector**

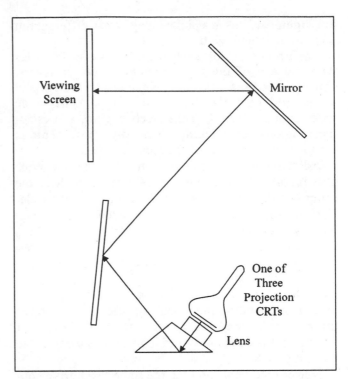

Figure 12-17   Diagram of a Front-Projection System

lens mounted around the neck of each CRT reduces possible optical irregularities around the edges of the projected image.

The Schmidt optical system relies on a CRT design, called the **Novabeam**, that has a large aperture and low surface reflectance. With this design, a relatively

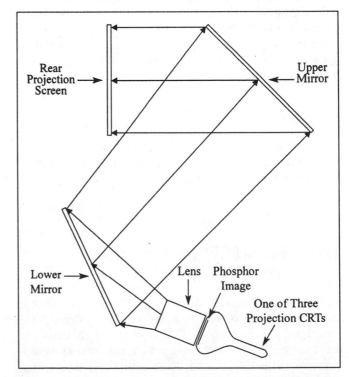

Figure 12-18   Diagram of a Rear-Projection System

small, metal-backed target screen and magnetic focusing provide the small-dot resolution needed to produce a viewable picture. Shown in Figure 12-19, an optical mirror provides the reflective surface within each of the three CRTs used in the system.

Refractive Systems

The **refractive system** takes advantage of the magnification produced by convex lenses to reproduce a picture on the viewing screen. Within the projection television console, three CRTs—such as the one pictured in Figure 12-20—project either a red, blue, or green image onto the viewing screen. As with the Schmidt system, the combination of the three guns produces the correct gray scale and colors. In addition to the CRT, a convex mirror mounted approximately 1 foot in front of the picture tube screen refracts the light coming off the tube. Because the convex lens provides a mirrored, inverse image of the actual televised object, each CRT has a specially designed yoke with reversed windings. While the reversal of the vertical yoke coils permits the vertical scanning to invert the

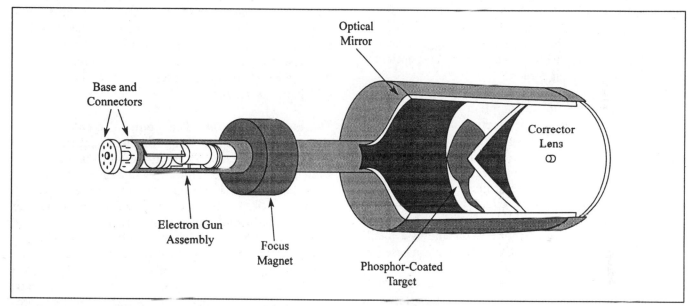

Figure 12-19   Kloss Novabeam CRT Used in the Schmidt Optical System

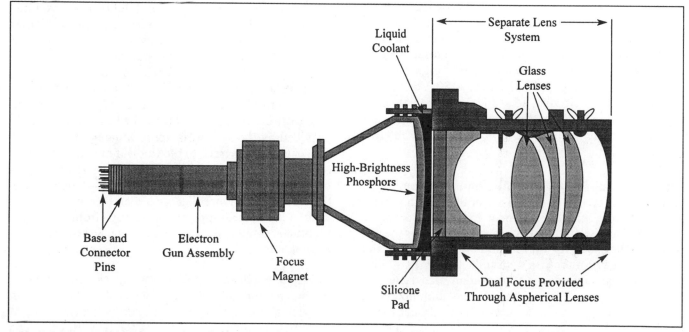

Figure 12-20   Refractive System CRT

picture, the reversal of the horizontal yoke coils reverses the picture.

A Zenith PV4541P Projection Television System

The Zenith Model PV4541P projection television relies on a three-tube, in-line refractive system and features the rear-projection design seen in Figure 12-18. Although the system includes the term "in-line," you should not confuse the projection design with the direct-view, in-line CRT gun design discussed in section 12-1. Instead, this in-line system places the red, blue, and green projection tubes on a horizontal axis. Images from those CRTs are magnified approximately ten times, reflected through two mirrors, and optically projected to a viewing screen.

The system uses a combination of electronic and optical subsystems to accomplish this projection. Looking again at the rear-projection drawing in Figure 12-18, the upper light path for the system consists of the projection screen and an upper mirror mounted with the cabinet. The lower light path consists of the lower mirror, the three projection picture tubes, and the lenses. As we saw earlier, each liquid-cooled projection tube mounts within an individual pod and corresponds with a lens. Figure 12-21 shows a block diagram of the circuits and the interconnections used to achieve the projected picture.

The projection television system uses the same type of switched-mode power supply that we explored in chapter four to supply the auxiliary dc power supply voltages. However, the use of liquid-cooled projection tubes and higher drive levels also causes increases in the B+ voltages for the video output drivers. Since the B+ voltages are scan-derived voltages, the horizontal output transformer also requires a different set of windings. Shown in the schematic diagram of Figure 12-22, the switched-mode supply provides +132 Vdc for the horizontal sweep circuits, +12 Vdc for the remote control circuits, +35 Vdc for the audio circuits, and +62 Vdc for the vertical circuits, as well as the CRT heater voltages.

Projection Tubes Used in the Zenith System Although the projection tubes used in the Zenith system share the same function, the need to project a properly converged image on the viewing screen requires the use of slant-face tubes for the red and blue projection; the green projection tube has a straight face. Along with the tube face differences, the system also relies on self-convergence. That is, the system minimizes the number of procedures needed to achieve proper registration. The raster registration circuit board shown in the middle-right portion of Figure 12-21 includes the

controls needed to achieve the proper alignment of the three images.

Each of the three tube assemblies includes a metal jacket, which includes a small glass window. Because of the need to increase picture brightness, the design uses liquid cooling to limit faceplate temperatures as CRT drive levels increase under high power conditions. The optically clear liquid coolant fills the space between the clear glass window and the faceplate, and allows the doubling of the actual safe power driving level over tubes not using liquid cooling.

Application of the liquid-cooling design also permits the use of optical coupling to reduce multiple light reflections along the lower light path. A pliable, optically clear silicone pad fits between the glass window in the metal jacket assembly and the rear element of the lens assembly. The silicone pad makes contact with the two light-path interconnecting surfaces and reduces the chance reflections.

The Viewing Screen The Zenith projection television system uses a two-sided screen to capture the projected image and display the image for the viewer. As shown in Figure 12-23, the front side of the screen uses a vertical lenticular design to provide the necessary brightness and contrast. Black striping on the viewer screen adds to the contrast and enhances the picture brightness under normal room lighting conditions.

Figure 12-24 shows the other side of the projection screen, which consists of a vertically off-centered Fresnel section. The off-centering of the Fresnel lens changes the vertical angle of the maximum light projection and, as a result, directs the maximum amount of picture brightness at the eye level of a seated viewer. As we discussed earlier, projection televisions have limitations regarding viewing angles. In this particular case, the Zenith projection television has a horizontal viewing angle of plus or minus 35 degrees.

The projection screen offers a focal length of 20 feet, which establishes an optimum viewing distance of 10 to 20 feet from the receiver. Because of the effect given by the off-centered Fresnel lens, the best viewing angle provided by the system is approximately $3\frac{1}{2}$ feet from the center of the screen to the floor. If the viewer moves outside of the recommended viewing angles, the picture will have decreased brightness and—depending on the side—will have either a reddish or bluish appearance. The altered appearance results from the fixed placement of the red, blue, and green CRTs along a horizontal axis.

Safety and Shutdown Circuits In the introduction to this section, we briefly discussed that projection televisions require additional safety and shutdown circuits because of the higher CRT drive levels. The Zenith system applies a vertical-sweep shutdown

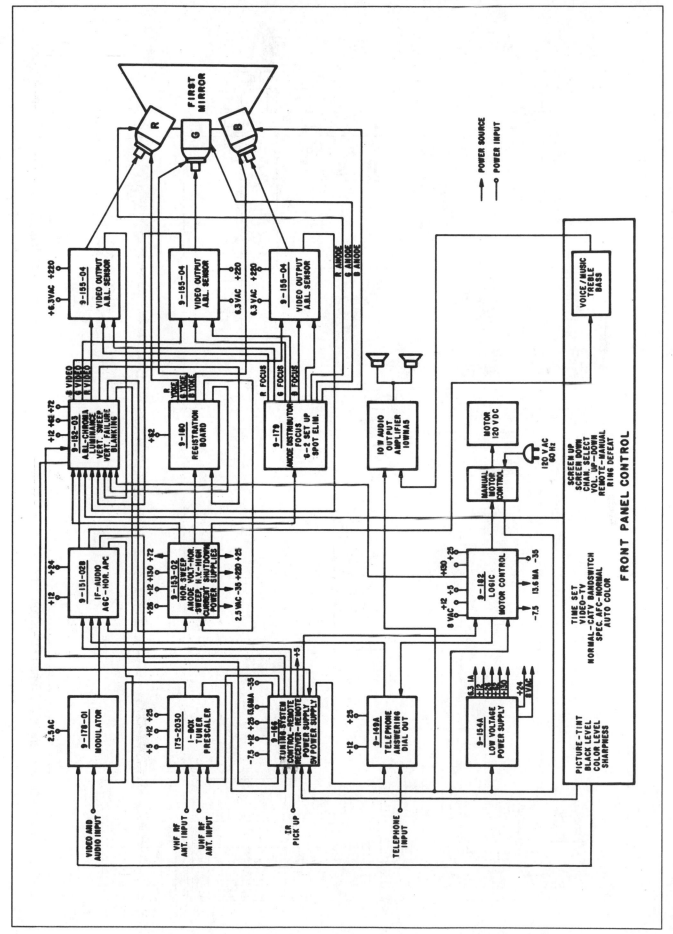

Figure 12-21 Block Diagram of the Zenith PV4541P Projection Television System (*Courtesy of Zenith Data Systems, Zenith Electronics, and Rauland*)

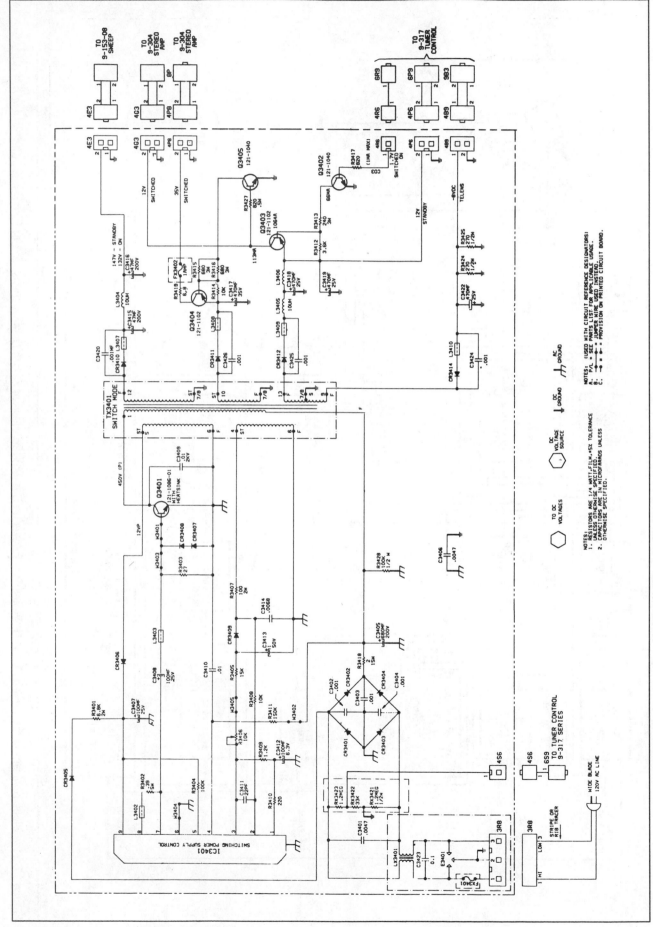

Figure 12-22 Schematic Diagram of the Zenith Projection Television Switched-Mode Power Supply *(Courtesy of Zenith Data Systems, Zenith Electronics, and Rauland)*

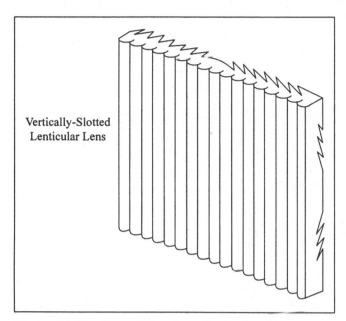

Figure 12-23   Front Side of the Projection Screen

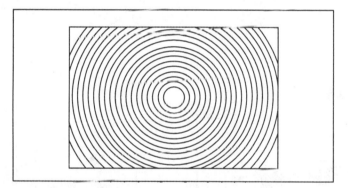

Figure 12-24   Fresnel Lens

circuit, a beam-current shutdown circuit, a high-voltage shutdown circuit, and horizontal yoke open/short-circuit protection.

Looking first at the vertical-sweep shutdown circuit, a transformer located on the registration module has a primary for each of the three vertical yoke windings. Under normal operating conditions, no secondary voltages exist. However, when a vertical circuit fault occurs and attempts to reduce the raster to a single horizontal line, the secondary voltage suddenly increases, produces a dc voltage, and causes a raster blanker transistor to conduct. The conduction of the blanker transistor cuts off the picture tubes and prevents the combination of a vertical fault and extremely high-beam current from rapidly burning a line into the CRT.

Like the vertical-sweep shutdown circuit, the beam shutdown circuit also prevents CRT burn. If the circuit senses an excessive beam current, the beam-current shutdown transistor cuts off. Then a crowbar circuit increases the load on the 5-Vdc power supply line and decreases the 5-Vdc power supply. Because the voltage

line connects directly to the system microprocessor control unit, the decrease in supply voltage turns the receiver off.

While any increase in the high-voltage output to the projection tubes causes two latch transistors to clamp the horizontal drive and disable the high-voltage circuits, an open or short in the horizontal yoke windings will also cause shutdown. Like the vertical shutdown circuit, the secondary windings of an transformer have no output voltage until a fault occurs. Any type of open or short in the yoke circuits causes an imbalance and the production of a secondary voltage. Because the secondary windings connect directly to the high-voltage shutdown latch circuit, the fault triggers the latch, clamps the horizontal drive, and disables the high-voltage circuit.

Light Valves, LCD Displays, and Laser Projection Systems

Used in many large commercial applications such as theaters, the **light valve** system works much like a film projector system. That is, rather than utilize a phosphor screen, an optical valve modulates a high-intensity, fixed light source. Shown in Figure 12-25, the light valve produces extremely high luminance values. The electron gun located to the left of the diagram scans a spherical mirror coated with a thin layer of oil. As the electron beam charges the oil, the action distorts the oil into grooves proportional with video information. Because of the costs associated with light valve technologies, the technology is limited to large-scale applications.

Projection television systems based on liquid-crystal display technologies use three-dimensional displays divided into sections to generate an image. Each display panel operates as a switch and represents one small portion of the entire image. Used primarily in very large-screen business applications, the luminescent-based projection television uses the plasma excitation of phosphors overlying each display cell to produce the sections of the televised image.

Finally, laser-beam technologies, which are largely in the prototypical stages, have provided another method for displaying high-definition images. Laser-screen projection CRTs such as the one shown in Figure 12-26, use an electron gun aimed at a crystalline screen to produce a 1-inch square raster. Conventional optics then magnify the image and project it onto the viewing screen.

☑ PROGRESS CHECK

You have completed Objective 3. Discuss projection television displays.

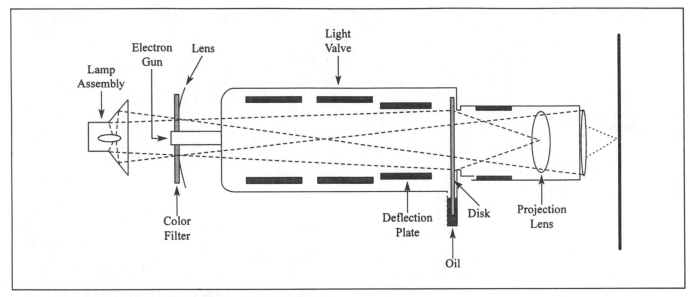

Figure 12-25   Cross-Section of a Light-Valve Assembly

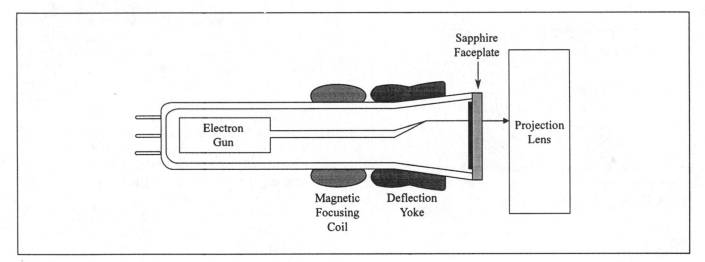

Figure 12-26   Cross-Section of a Laser-Screen Projection CRT

12-4   TROUBLESHOOTING PICTURE-TUBE FAULTS

When troubleshooting CRT problems, part of the problem-solving method involves recognizing whether the fault has occurred within the CRT, in the circuitry associated with the CRT, or within the receiver signal-processing circuitry. Although you could easily conclude that a receiver with no picture or raster has a bad picture tube, common sense and—in the long term—experience reveal that many different types of faults can cause the "no picture or no raster" symptom. As you have already learned, problems with the horizontal drive signal, the high-voltage power supply, and the low-voltage supply can cause the loss of the raster and picture.

Logical troubleshooting evolves from having the patience to check all the probable causes for a problem symptom. When suspecting a CRT problem, always check for the presence of the correct low and high voltages. In addition, always verify that the CRT socket is connected to the CRT base. Those basic checkpoints will allow you to decide which troubleshooting path to follow. Because common problems seen with CRTs often result in the replacement of the tube, the time spent verifying that a CRT defect actually exists is worth the cost for both you and your customer.

Safety First!

During our study of high-voltage power supplies and CRTs, we found that the capacitance that is formed between the outer and inner conductive coatings of the CRT filters the high-voltage power supply. With this capacitance and the high voltage in mind, always follow exact safety precautions when working with an operating CRT or when discharging the high-voltage anode. Although the internal resistance of the high-voltage supply may protect you from a fatal shock as it drops the high voltage to a safe level, accidentally touching the anode still provides a nasty shock.

Along with following safety procedures around the CRT anode, also observe caution when working with the voltages at the CRT base. Both the screen voltages and the focus voltage are at high potentials. Always remain aware of the grounding surface used when checking voltages, and always check the test probe for any breaks in the insulation. The following safety tips emphasize procedures to follow when working with CRTs.

Discharging the Tube   When handling any type of picture tube, always discharge the tube by connecting a lead between the anode and chassis ground. The best method for grounding the anode is by fastening a clip lead to a screwdriver blade. Then, holding only the rubber handle of the screwdriver, carefully insert the tip under the rubber cap that covers the anode connector. You should hear an audible snap when the tube discharges.

A CRT can hold a capacitive charge for a long time. Although the high voltage held in the tube has a low current and probably will not give a dangerous shock, your reaction to the discharge and the shock may cause you to either bump against and break the thin CRT neck, brush against other parts of the receiver that have a high potential, or drop the CRT.

Handling the Tube   When handling a CRT, remember that the tube is essentially a large, evacuated glass bulb. Any crack or break in the glass envelope will cause air to rush into the tube and could cause a violent implosion. Never handle a CRT by its neck, and always wear eye protection.

Measuring High Voltage   When measuring the high voltage at the CRT anode, always use a special high-voltage probe. Check the probe for dirty connections, broken wires, and loose connections. Always follow the probe manufacturer's and receiver manufacturer's recommendations when measuring the high voltage.

X-Ray Protection   Every CRT and receiver contains built-in protection against the accidental emission of X-rays, even though the glass envelope that makes up the largest part of the CRT is thick enough to block most X-rays. Always use exact replacement tubes. In addition, all receivers include overcurrent and over-voltage protection circuits that shut the receiver down in case the high voltage increases past safe parameters.

Common CRT Problems

One of the most common picture-tube faults, **low electron gun emission,** results from the aging of the CRT and causes the display of a dim raster or a raster that lacks one of the primary colors. The condition of the CRT at power-up provides an initial alert to possible low-emission problems. A CRT with a low-emission problem may take longer to establish a clear, bright picture, or may display a gray-scale tinted to one of the primary or secondary colors. Gray-scale adjustments of the red, blue, and green drive controls should provide an even balance between black and white scenes.

Another clue pointing to possible low emission involves the quality of white areas and the amount of detail produced in a picture. A picture tube with low-emission problems will have white areas that appear washed out. Advancing the brightness and contrast controls to enhance the white areas will cause the picture to lose focus.

Other common problems include grid-to-cathode short-circuits, cathode-to-heater short-circuits, and internal arcing. A short between the grid and cathode or between the cathode and heater reduces the CRT bias to zero and causes the tube to have uncontrollable brightness and retrace lines. Internal arcing may take the form of high-voltage arcs around the anode or arcing within the neck of the CRT. With the first problem, a breakdown between the external conductive coating of the tube and chassis ground can be remedied with an insulating chemical. The second problem results from arcing between the cathode and heater, and may cause additional problems with the video output transistors.

➲ SERVICE CALL ─────────────

Color Fades In and Out 10 to 15 Seconds After Turn-on

On arriving at the customer's house, the technician found that the receiver had a normal raster and a normal sound. Although normal at start-up, all the color in the picture faded away 10 to 15 seconds after turning the receiver on. At first, the technician concentrated on the chrominance signal-processing circuits, but found only normal operating characteristics throughout those areas.

Next, the technician began to study the CRT and found intermittent voltage levels at the CRT filaments. A quick check of components located on the video output module showed no problems; all the video output voltages remained normal. On further examination of the CRT, the technician noticed that the CRT filaments went dark momentarily and then came back on.

Flexing the module, the technician noticed that a slight exertion on the module caused the intermittent conditions to appear and then disappear. While looking for bad connections on the module, the technician found several cold solder joints around the connections

for the filaments. After resoldering the joints, the technician found that the problem had disappeared.

 PROGRESS CHECK

You have completed Objective 4. Troubleshoot CRT-related problems.

SUMMARY

Chapter twelve reviewed the basics of cathode-ray tubes, and went on to cover video display adjustments and projection television systems. You learned of the functions of the electron gun, the reasons for constructing the glass envelope and display screen of a CRT in a specific manner, and the need for the external and internal conductive layers of a CRT and the shadow mask.

The chapter examined how a variety of adjustments affect the performance of the picture tube. Those adjustments include degaussing, purity adjustments, static and dynamic convergence, pincushion correction, focus adjustments, and gray-scale adjustments.

The chapter also provided an overview of projection television technologies. After briefly examining the

Schmidt optical system, the discussion concentrated on the refraction systems used in most projection television systems, and on the differences between front and rear projection televisions. The chapter linked theory with practice by describing how a Zenith PV4541 projection television operates. Other projection television technologies covered included light valves, LCD displays, and laser projection systems.

Finally, the chapter reviewed problems associated with CRTs, such as low electron gun emission and internal arcing. CRT troubleshooting methods emphasized that safety precautions should always be observed when working with picture tubes because of the high voltages involved.

REVIEW QUESTIONS

1. Luminescence is the _____.

2. Why is the P22 phosphor used for color picture tubes?

3. Manufacturers use several different methods for improving the contrast given by a shadow mask display. Name two of those methods.

4. Draw a simple sketch of a CRT and label the basic parts.

5. An electron gun is _____, _____, and consists of _____.

6. Define the purpose of the following CRT grids: G1, G2, G3, and G4.

7. An Aquadag is _____ while an ultor is _____.

8. True or False   The shadow mask is a thin, perforated sheet of metal that blocks any part of the electron beams from striking the phosphor screen.

9. What is the difference between a delta configuration and an in-line configuration? When answering the question, consider the electron beam, the shadow mask, and the phosphor screen.

10. Why are CRT screens aluminized?

11. What is the purpose of the black mask seen on most CRT screens?

12. True or False   The electron guns in a tricolor CRT share a control grid and a cathode.

13. What are the advantages and disadvantages of projection television?

14. The two basic types of projection television systems are _____ and
_____.

15. The major types of projection CRTs are the Novabeam and the refractive system. Describe how the refractive system works.

16. What is the advantage of liquid-cooling of a projection tube?

17. Describe the front and back of the viewing screen used in the Zenith system.

18. The Zenith projection system described in this chapter has limitations regarding viewing distance and horizontal viewing angle. What happens if the viewer watches the picture from outside the recommended viewing angles?

19. To protect against and overcurrent conditions, the Zenith projection system relies on a
_____ shutdown circuit, a _____ shutdown
circuit, a _____ shutdown circuit, and _____
protection.

20. With the Zenith PV4541P projection television, any _____ in the high-voltage output to the projection tubes causes two latch transistors to clamp the horizontal drive and disable the high-voltage circuits.

21. True or False   The light-valve projection television system is typically used for applications requiring a very large viewing screen.

22. True or False   Laser-beam technologies are commonly used in projection televisions.

23. Describe CRT purity.

24. Most CRTs require both _____ and _____
convergence.

25. Convergence establishes the _____.

26. Convergence applies to the _____, while purity applies to the
_____.

27. Without some type of focus, the electrons leaving the CRT gun would:
    a. converge
    b. diverge

28. Electrostatic focus relies on a field between grids:
    a. G1 and G2
    b. G2 and G3
    c. G3 and G4
    d. G1 and G3

29. Draw a raster affected by pincushion distortion. What do the pincushion correction signals do?

30. All direct-view picture tubes rely on _____ focus. Why?

31. True or False   Magnetic focus relies on a sawtooth voltage.

32. True or False   Electrostatic focus utilizes the fields developed through the placement of horizontal and vertical yoke deflection coils placed around the picture tube neck.

33. Gray-scale tracking is _____.

34. The parabolic waveform found in dynamic convergence circuits is derived from signals taken from the
_____ circuits.

35. True or False   One of most common CRT problems is low emission.

36. Name three safety precautions that you should follow when working with a CRT.

37. Symptoms of a low-emission problem include _____ ,
    _____ , and _____ .

38. If an arc occurs between the cathode and heater of a picture tube, additional problems may result with the
    _____ .

39. True or False   When handling a picture tube, always pick the tube up by the neck.

40. True or False   A picture tube discharges when the receiver powers down.

41. Potential shock areas around a CRT are the _____ .

42. True or False   Many symptoms that point to a defective CRT may be caused by other circuit faults.

CHAPTER

13

HDTV and Computer Monitors

OBJECTIVES   Upon completion of this chapter, you should be able to:

1. Define video resolution in terms of pixels and dot pitch.
2. Discuss interlaced and noninterlaced scanning.
3. Explain the fundamentals of HDTV.
4. Discuss computer monitor display standards.
5. Discuss computer monitor display adapters.
6. Discuss computer monitor circuits.
7. Troubleshoot computer monitors.

KEY TERMS

1125/60 standard	MPEG-2
aspect ratio	multisync
color graphics adapter (CGA)	National Television Systems Committee (NTSC)
Digital HDTV Grand Alliance	noninterlaced scanning
Dolby AC-3	pixel
dot pitch	progressive scanning
extended graphics adapter (EGA)	refresh rate
Hercules graphics adapter (HGA)	RGB video
high-definition television (HDTV)	super VGA (SVGA)
interlaced scanning	transistor-transistor logic (TTL) video
monochrome display adapter (MDA)	video graphics array (VGA)

INTRODUCTION

Chapter thirteen introduces the concept of high-definition television and explains how HDTV differs from the NTSC standard. The chapter also illustrates the differences between interlaced and noninterlaced scanning, and defines basic terms such as resolution and refresh rate. The discussion of the HDTV standard also illustrates the growing convergence between television and computer systems.

Given that convergence, the chapter provides a thorough explanation of video adapter cards and monitors. With the computer industry experiencing tremendous growth, the explanation of terminology combined with an overview of circuit operation establishes an entry point into a new service arena. As with the previous chapters, chapter 13 concludes with a detailed look at troubleshooting procedures and common faults that appear in computer monitors.

13-1 INTERLACED AND NONINTERLACED SCANNING

When we discuss video displays, the ability of the display to show a clear image is defined through a constant called dot pitch. A video display, is made up of **pixels**, with each pixel consisting of three dots: one each of red, blue, and green. **Dot pitch** is the distance between the center points of adjacent horizontal pixels on the CRT screen. Most advertisements for video display monitors will list the dot pitch measurement in millimeters. Thus, the pixels on a monitor with a .28 dot pitch are closer together than those on a .31-dot pitch monitor. The smaller the distance between pixels the higher the possible resolution.

Each line that results from the vertical and horizontal scanning of the CRT electron beam yields a set number of pixels. The longer horizontal lines will have more pixels than the shorter vertical lines. If the specifications of a monitor list a resolution of 640 × 480 pixels, this means that the horizontal scan lines show 640 pixels while the vertical lines show 480 pixels. Multiplying the two figures gives us the total amount of pixels that the raster will display, in this case, 307,200. Since the number of pixels depends on the deflection signals, varying the horizontal scan rate also varies the number of displayable pixels.

Televisions and older computer monitors have a horizontal frequency of 15,734 kHz. Newer computer monitor standards and the new high-definition television standard use horizontal sync signals of 21.80, 31.50 and 35 kHz. A list of the horizontal scan frequencies as those frequencies correspond with different refresh rates can be found in Table 13-1.

You may wonder why the designs do not change the vertical scan rate. By retaining the 60-Hz vertical scan rate and increasing the horizontal scan rate, more horizontal lines become squeezed into the vertical cycle. An increased number of horizontal lines further improves the clarity produced by the video monitor. Information display monitors also use higher picture bandwidths than television receivers. In other words, the monitor turns its display pixels off and on faster than a television receiver. We know that television receivers have a bandwidth of 4.5 MHz. Information display monitors have a bandwidth of 35 MHz or higher. The higher bandwidth allows the monitor to display more pixels during one horizontal scan. Without the needed bandwidth, a monitor is limited in the resolution that it can provide.

Refresh rate is the rate at which a screen image is redrawn; it shows how many frames are scanned per second; and it is the vertical scanning rate. Because CRTs form images in frames, the refresh rate coincides with the amount of flickering seen on the screen. While a refresh rate between 60 and 75 Hz cuts the flickering, a 75-Hz refresh rate has become commonplace. Table 13-1 shows bandwidth measurements in combination with resolution, number of pixels, and horizontal sync rates at various refresh rates.

☑ PROGRESS CHECK

You have completed Objective 1. Define video resolution in terms of pixels and dot pitch.

Interlaced Scanning

Interlaced scanning is a process in which electron guns draw only half the horizontal lines with each pass. With one pass, the guns draw all odd lines while the next pass draws all even lines. As a result, one complete frame of information is created for every two fields scanned. With a field generated every 1/60th of a second, the human eye cannot discern the scanning motion. The **National Television Systems Committee**, or NTSC, selected interlaced scanning as a standard for broadcast signals because of the limited bandwidth available for delivering picture information. To compensate for any possible flicker, manufacturers of interlaced scanning displays choose phosphors that have a higher decay time.

Because interlaced scanning refreshes only half the lines at a time, it can display twice as many lines per cycle. Thus, this display technique provides an inexpensive method for yielding more resolution. However, interlaced scanning has a relatively slow trace and retrace time, which affects the ability of a display to show animations and video graphics.

Table 13-1 Refresh Rates, Pixels, Bandwidth, and Horizontal Sync Rates

60-Hz Refresh Rate			
Resolution	# of Pixels per Screen	Bandwidth MHz	Horizontal Sync Rate (kHz)
800 × 600	480,000	28.8	36
1024 × 768	786,432	47.2	46.1
1152 × 900	1,036,800	62.2	54
1280 × 1024	1,310,720	78.6	61.4
66-Hz Refresh Rate			
Resolution	# of Pixels per Screen	Bandwidth MHz	Horizontal Sync Rate (kHz)
800 × 600	480,000	31.7	39
1024 × 768	786,432	51.9	50.7
1152 × 900	1,036,800	68.4	59.4
1280 × 1024	1,310,720	86.5	67.6
72-Hz Refresh Rate			
Resolution	# of Pixels per Screen	Bandwidth MHz	Horizontal Sync Rate (kHz)
800 × 600	480,000	34.6	43.2
1024 × 768	786,432	56.6	55.3
1152 × 900	1,036,800	74.7	64.8
1280 × 1024	1,310,720	94.4	73.7

Noninterlaced Scanning

Interlaced scanning has two drawbacks. Because of the higher resolution, any amount of flicker caused by screen phosphor decay would be noticeable and distracting. With all the individual dots displayed, some will dim as others become illuminated. In addition, the scanning lines in an interlaced scanning display are visible. If a person stands too close to a display device, each line of information can be seen as it displays on the screen. For that reason, the optimal viewing distance for an interlaced display is always listed as 4.5 or 6 times the height of the display screen. At this distance, scanning lines seem to merge together and create the illusion that one complete image is displayed. However, with larger display devices, such as projection televisions, the scanning lines are more noticeable.

To counter the flicker and scanning line problems, computer displays and the new HDTV standard use **noninterlaced** refresh or **progressive scanning**, in which every line of information on the display is scanned by the electron gun at each pass across the panel. The technique enhances the vertical resolution of the display while allowing the viewer to sit closer to the display. Viewing distances with noninterlaced scanning shorten to 2.5 times the height of the display.

✓ PROGRESS CHECK

You have completed Objective 2. Discuss interlaced and noninterlaced scanning.

13-2 HIGH-DEFINITION TELEVISION

Throughout this text, we have studied a television broadcasting and reception standard established by the NTSC in 1940. That standard utilizes 525 horizontal scanning lines, interlaced scanning, the transmission of separate luminance and chrominance signals, and relies on a 60-Hz frame rate. Of the 525 scanning lines, only 483 are visible, while the remaining lines are used for interval timing or other functions. The bandwidth for those signals covers 4.2 MHz.

Although many nations use the NTSC standard, most European nations rely on another standard called Phase Alternate Lines, or PAL which relies on a 50-Hz frame rate, uses a color subcarrier frequency of 4.43 MHz, and has 626 scanning lines. The alternate standard surfaced because of detectable shifts in the color subcarrier phase of the NTSC. Still another standard—developed by the French and known as SECAM, or

SEquential Coleur Avec Memoire—is used in the former Eastern Bloc European countries. The introduction of each broadcast standard also introduced incompatibilities between each system. For example, a SECAM system cannot display a PAL broadcast image because of differences in broadcast equipment, and NTSC systems cannot display PAL broadcasts because of the difference in the frame frequencies.

In the early 1980s, the Japan Broadcasting Corporation, or NHK, proposed a system called the MUSE HDTV interlaced system, that would use 1125 scan lines. NHK introduced it as a possible world standard, and established a goal of **high-definition television** playing on a wide-screen format. At the request of broadcasters concerned about America's role in establishing the new technology, the FCC established a rule-making committee called the Advisory Committee on Advanced Television Service, or ATSC. In addition, the FCC decided that new HDTV signals would be broadcast on currently unusable channels, and that broadcasters would be temporarily assigned a second channel for the transition to HDTV.

During the early 1990s, three competing HDTV design teams agreed to combine their efforts and produce a standard, high-quality product. The three design teams—working under the direction of AT&T and Zenith Electronics; the General Instrument Corporation and the Massachusetts Institute of Technology; and Philips Consumer Electronics, Thomson Consumer Electronics, and the David Sarnoff Research Center—formed the **Digital HDTV Grand Alliance**. The high-definition television standard produced by the Grand Alliance establishes a technological framework for the merging of broadcast, cable, telecommunications, and computer technologies. Not surprisingly, the introduction of a high-definition television standard affects both the transmission system and the receiver design for modern video receivers.

With HDTV, the amount of luminance definition doubles both horizontally and vertically, with four times as many pixels as the older NTSC system. In addition, the larger screen format provides more visual information for the viewer. The system provides additional video detail through the application of video bandwidth five times that of the conventional NTSC system.

HDTV offers a larger **aspect ratio**, or the ratio of picture width to picture height. While the NTSC system offers an aspect ratio of 4:3, HDTV establishes an aspect ratio of 16:9, as shown in Figure 13-1. Because of this, the viewer receives almost six times more information. Therefore, high-definition televisions have a place in industrial applications; information capture, storage, and retrieval applications; and educational, medical, and cultural applications. With each of those applications, HDTV provides the picture quality

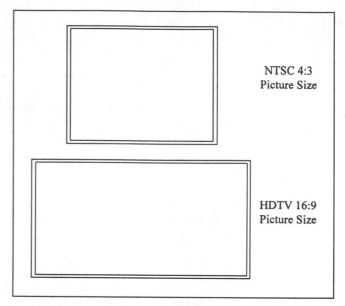

Figure 13-1   Comparison of NTSC Standard Picture Size with HDTV Picture Size

needed for teleconferencing, training, and product promotion.

The HDTV 1125/60 Standard

During the planning and development of the HDTV system, the design team chose to use 1125 scanning lines with a picture refresh rate of 60 Hz. This **1125/60 standard** compares with the type of resolution given by a projecting a 35-millimeter formatted film onto a large screen, and establishes 1035 scanning lines in the active picture display. Also, as an international standard, the 1125/60 system fits within the need to convert from older systems that have 525 or 625 scanning lines. Thus, the 1125/60 standard allows existing television signal distributors to convert from the NTSC 525/59.4 standard through readily available, large-scale integrated circuits, and establishes a format for the global distribution of video information. Currently, the HDTV broadcast system shares television bands with existing services and utilizes unused channels. Television signal broadcasters are temporarily assigned a second channel to accomplish the transition from the NTSC format to the HDTV format.

Other HDTV Standards

Along with the 16:9 aspect ratio and the 1125/60 scanning refresh standard, the HDTV design team also determined that the new system should have:

- 2:1 interlaced scanning combined with noninterlaced scanning
- a luminance bandwidth of 30 Hz

- two color-difference signals with bandwidths of 15 MHz
- an active horizontal picture duration of 29.63 microseconds
- a horizontal blanking duration of 3.77 microseconds
- a new sync waveform

The HDTV standard assembled by the Grand Alliance takes advantage of the interlaced scanning used for television transmission and reception and the noninterlaced scanning commonly seen with computer monitors. With noninterlaced, or progressive scanning, the HDTV system provides a choice of 24-, 30-, and 60-frame-per-second scanning with a 1280 × 720 pixel dot resolution, and a 24- and 30-frame-per-second scan with a 1920 × 1080 pixel dot resolution. As a whole, HDTV supports the following spatial formats:

1280 × 720:	23.976/24 Hz	Progressive
	29.97/30 Hz	Progressive
	59.94/60 Hz	Progressive
1920 × 1080:	23.976/24 Hz	Progressive
	29.97/30 Hz	Progressive
	59.94/60 Hz	Interlaced
	59.94/60 Hz	Interlaced

With this, the HDTV system provides direct compatibility with computing systems. In addition to the noninterlaced scanning formats, the system also offers 60-frame-per-second interlaced scanning at a resolution of 1920 × 1080. The use of interlaced scanning becomes necessary for the two 1920 × 1080 × 60 formats because of the lack of a method for compressing the formats into a 6 MHz channel. Each of the formats features square pixels, a 16:9 aspect ratio, and 4:2:0 chrominance sampling.

The HDTV Sync Waveform The new horizontal blanking interval accommodates the new sync waveform shown in Figure 13-2. By using an improved sync waveform, the design team ensured compatibility across all systems, achieved precise synchronization, and constructed a sync structure that would continue to

have noise immunity in the future. The new HDTV sync signal eliminates jitter by placing the horizontal timing edge at the center of the video signal dynamic range. Moreover, the timing edge has a defined midpoint centered on the video blanking level.

In addition to improving the sync waveform, the design team also improved the capability of the HDTV system to reproduce colors through the use of 4:2:0 chrominance sampling. When compared to the NTSC standard, the HDTV system provides a broader choice of colors which aligns with newer film technologies, computer graphics technologies, and print media. As you may suspect, the capability of the HDTV standard to reproduce a broader spectrum of colors affects both camera and display technology.

The system achieves many of the improvements in resolution and color reproduction through the decision to establish a 30-MHz luminance bandwidth and two color-difference signals with a bandwidth of 15 MHz each. The decision to use the 30-MHz and 15-MHz bandwidths depended on the decision to use 1125 scanning lines, which required a bandwidth of at least 25 MHz. The combining of increased horizontal and vertical resolution with wider luminance and chrominance bandwidths yielded a larger number of viewable pixels. Given 1920 horizontal pixels, the HDTV system becomes a platform for several different applications of computer display technologies ranging from computer-aided design and manufacturing to medical imaging.

The HDTV broadband, 20 megabits-per-second, digital transmission system enables the convergence of the entertainment, industrial, medical, and educational technologies by using a packetized data transport structure based on the MPEG-2 compression format (discussed further in the next section). Each data packet is 188 bytes long, with 4 bytes designated as the header or descriptor and 184 bytes designated as an information payload. With this type of high-compression data transportation, the HDTV system can deliver a wide variety of video, audio, voice, data or multimedia services, and can interoperate with other delivery or imaging systems.

While the digital transmission system may allow the simultaneous transmission and reception of those

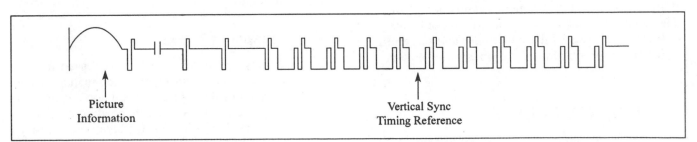

Picture
Information

Vertical Sync
Timing Reference

Figure 13-2 The HDTV Sync Waveform

services, viewers could select services that would substitute for the normal daily programming. For example, a local PBS station could broadcast HDTV programs such as National Geographic specials or ballets during the evening prime-time hours, while also broadcasting data services such as weather forecasts or stock market information. The weather and stock information would be apparent to viewers who had requested the service. During school hours, the same station could deliver five simultaneous educational programs to participating local schools and homes.

Over-the-air broadcasts of HDTV signals will rely on an 8-VSB vestigial sideband broadcast system, while cable transmissions of HDTV signals will use a 16-VSB vestigial sideband system. Thus, the system minimizes any potential interference between the HDTV broadcasts and conventional NTSC transmissions. Each of the standards uses digital technology to provide a high-data-rate transmission and ensure a broad coverage area. The higher-data-rate transmission for the HDTV cable signals allows the transmission of two full HDTV signals in a single 6-MHz cable channel.

The HDTV system uses a video-compression system based on the MPEG-2 video compression standard, while the audio system relies on the Dolby AC-3 five-channel sound system.

MPEG-2 Video Established by the Moving Picture Experts Groups, **MPEG-2** source pictures consist of a luminance matrix and two chrominance matrices, and ensure synchronization between the audio and video playback. In the 4:2:0 format used in HDTV, the chroma matrices are one-half the size of the luminance matrices in both the vertical and horizontal planes of the picture. While the bidirectional-frame motion compensation, or B-Frame, used in MPEG-2 improves picture quality, the MPEG format supports interlaced and progressive scanning.

Dolby AC-3 Sound The **Dolby AC-3** sound format encodes multiple channels as a single channel. As a result, the format can operate at data rates as low as 320 kbps. The Dolby AC-3 algorithm represents five full bandwidth channels representing left, center, right, left-surround. and right-surround, along with a limited-bandwidth, low-frequency subwoofer channel. The audio format is designed to take advantage of the characteristics of the human ear and to permit the noise-free reproduction of the transmitted sound.

✔ PROGRESS CHECK

You have completed Objective 3. Explain the fundamentals of HDTV.

13-3   COMPUTER MONITORS

As technology changes, all the experience and expertise that you have gained helps you to adapt to the changes. In many cases, your ability to adapt makes the difference to surviving in the electronics servicing profession. Given your knowledge of video display technology, and familiarity with terms such as cathode-ray tube, yoke, video signals, and deflection, you can move from servicing televisions to a slightly different product: computer monitors. While moving into this area of electronics servicing may seem challenging, do not let the technology intimidate you. For most service problems, you can apply the same test equipment and professional knowledge to computer monitor servicing that you have acquired in servicing television receivers.

Comparing Display Technologies

Despite their similarities, computer monitors and television receivers have obvious differences. Figures 13-3A and 13-3B compare the block diagrams of a color computer monitor and a color television receiver. Instead of receiving signals through antenna connections, computer monitors receive signals from either a card or set of integrated circuits within a microcomputer. We could think of the monitor and the system graphics adapter card as a video subsystem. When servicing the video display, you will need to provide a signal source such as the microcomputer equipped with the appropriate video card or a generator and the correct video input cable. A variety of cable types are used to connect monitors to microcomputer systems.

Because the video display works off the signals developed within the microcomputer, the monitor does not feature any audio circuitry. All audio for the microcomputer system develops internally through the microprocessor and a small speaker. More recent audio adaptations include the upgrading of the microcomputer system with an add-on sound card. Information video displays show text as well as graphical data at a much higher resolution than television receivers. Indeed, rating video monitors on the number of individual dots displayed and the dot pitch has become common practice.

Video Display Modes

Video displays or monitors have progressed from the fairly simple monochrome displays to much more complex displays. Displays are classified not only as monochrome or color, but also as enhanced color displays, video graphics array (VGA) displays, and multisync displays. Also, the adapter cards that drive the monitors have designations such as Hercules graphics card,

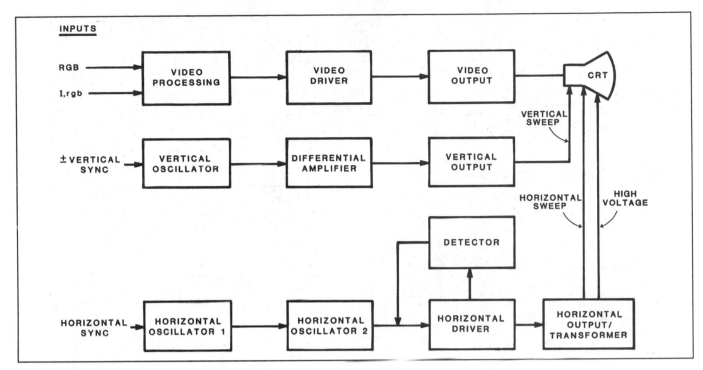

Figure 13-3A Block Diagram of a Computer Monitor

color graphics adapter, monochrome display adapter, extended graphics adapter, super VGA, 8514/A, and 34020.

Computer Video Adapters

Just as processors have gone through a steady evolution and have produced more computing power and speed, the video interface for a computer has also gone through steady improvements. Every personal computer system has some type of adapter for allowing the monitor to interface with the processor. While earlier computers relied on a separate adapter card, all new personal computers directly integrate the interface into the motherboard.

The adapters in early personal computers provided monochrome-only text, a combination of monochrome-only text and graphics, and television-like color. In some cases, schools and other institutions continue to utilize the early color adapters and displays but for the most part, though, those displays and adapters have given way to more advanced technologies. Despite the changes, the principles of the video interface remain relatively constant.

For example, the video interface relies on a CRT controller, video RAM, and a character-generator circuit. The CRT controller is a processor that receives input signals from the computer system main processor through the address bus and data bus. Inputs for the controller include clock signals, data information, and addressing information. In addition to sending and receiving information, the CRT controller also controls the horizontal and vertical synchronization of the monitor through an interface with the system clock.

The CRT controller also works with video RAM that is either placed on the adapter or in separate slots on the motherboard. While the early systems operated with a maximum of 4 kilobytes of video RAM, or VRAM, newer systems have progressed from 1 megabyte of VRAM to a maximum of 8 megabytes. Newer 2-D and 3-D video accelerator cards use the large amount of VRAM to produce studio-quality video effects at the desktop.

Video RAM stores the characters on display. One byte of information contains a code for the character, while a second byte contains the control information so that a specific character will have a certain color or a certain effect. The CRT controller constantly scans the VRAM and, along with associated circuitry, encodes the contents of each byte. With this action, the adapter controls the operation of the electron guns in the monitor. As with television systems, the horizontal and sync signals lock the transmitted signals in sync with the monitor scanning circuits.

Character ROM within the video interface codes each byte into a character set. From there, the video information travels to a shift register, which translates the data into a usable format. Then the information goes through a number of logic circuits and becomes an output signal for the monitor. While this occurs, the CRT controller sends horizontal and vertical information through logic and shift registers, and outputs those

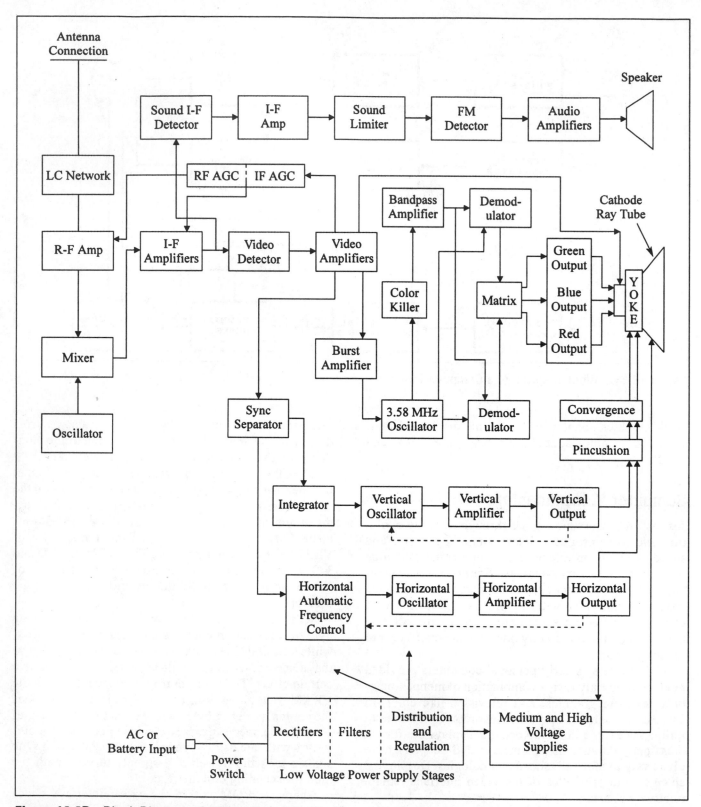

Figure 13-3B   Block Diagram of a Color Television

signals and a brightness signal to the monitor. The three-color output signals, color burst, and clock signals combine within a multiplexer, travel into a buffer, and appear as output signals for the monitor.

Table 13-2 provides a listing of the signals produced by video adapter cards. Only the VGA, SVGA, and multisync modes have any current use, as SVGA and multisync monitors are gradually replacing the older

VGA standards. However, as with all electronic systems, many institutions continue to use the older video standards for many purposes. When considering Table 13-2, pay special attention to the horizontal and vertical scan rates as well as the resolution. As explained in section 13-1, increasing the horizontal frequency produces many more lines in a 60-Hz raster. With more lines, less of the raster line structure is visible.

When we began our study of CRTs and projection televisions, we found that the NTSC system relied on a viewing distance of 8 to 10 feet. With a longer viewing distance, the human eye did not pick up any screen flicker. With computer monitors, the viewing distance obviously reduces to 1 to 3 feet. Because of this, the vertical frequency, or refresh rate, becomes extremely important. If the vertical refresh rate has a high enough frequency, the human eye is fooled into thinking that it has seen a full raster of light. If the number of frames per second drops below 24, the picture will have a disturbing flicker. If the number of frames rises above 24 per second, the flicker becomes less noticeable. Computer monitors use a refresh rate of 60 to 70 frames per second and a high horizontal scan rate to eliminate flicker and to make the human eye and brain think that a steady raster exists.

MDA   Early microcomputers relied on a display that showed only text or pixel-based graphics. These monitors used either a **monochrome display adapter**, or MDA, or a **Hercules graphics adapter**, or HGA, as a source for video signals. As software designers pushed for higher-quality video displays, this type of monitor became inadequate. Often, the adapter and card could not show the graphical images created through the software application.

The earliest microcomputer monitors, which used the MDA video card, could not display colors or graphics. Nevertheless, they worked well to display twenty-five rows of eighty characters each. Figure 13-4 shows a video input connector of a text-only monitor and that of EGA, VGA, and SVGA monitors. While looking at the connector pin-out, note that the MDA adapter has a vertical frequency of 50 Hz and a horizontal frequency of 18,432 Hz. Using these synchronizing frequencies allows the monitor to display the 350 vertical and 720 horizontal raster lines required for good text reproduction.

Since the original monochrome monitors could not support graphics, the Hercules graphics adapter card came into use. This card could attach to the older monochrome monitors as well as to monitors that could

Table 13-2   Adapter Card Specifications

Video System	Horizontal Resolution	Vertical Resolution	# of Colors	Horizontal Scan Rate (Hz)	Vertical Scan Rate (Hz)	Type
MDA	720	350	Mono	18,432	50	Digital
HGA	720	348	Mono	18,432	50	Digital
RGB	640	480	Infinite	15,734	60	Analog
CGA	300 640	200 200	4 of 16	15,750	60	Digital
EGA	640	200 350	16 of 64	21,800	60	Digital
VGA	640	350 400 480	256 of 256,000	31,500	60–70	Analog
SVGA	800 1024 1280	600 768 1024	256 of 256,000	35,200	56–72	Analog
8514/A	1024	768	256			
XGA	640 1024	480 768	65,536 256			
TI 34010	1024	768	256			

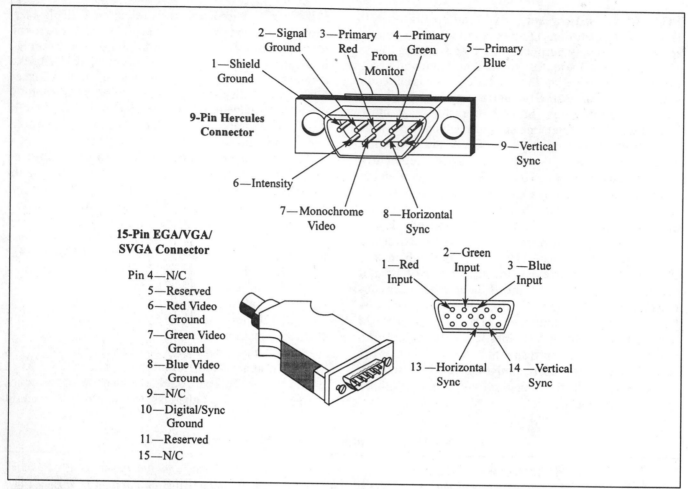

Figure 13-4   Video Input Connectors for Text-Only Monitor and for EGA, VGA, and SVGA Monitors

display monochrome graphics. Since the Hercules adapter card attached to both kinds of monitors and produced 350 vertical and 720 horizontal lines, it retained the unusual vertical and horizontal sync frequencies seen with the MDA video card.

CGA   IBM answered the need for better-quality computer graphics with the **color graphics adapter**, or CGA, and the color display. The early CGA monitors could show sixteen colors of text or graphics. Only a color graphics adapter or extended graphics adapter, described below, could drive the color monitor. Because CGA provided a vertical refresh of 60 Hz and a horizontal scan rate of 18,432 Hz, it had a limited ability to reproduce text.

EGA   As a follow-up to the CGA card, IBM introduced the **extended graphics adapter**, or EGA, and the enhanced color display. With this adapter, users could replace their monochrome, Hercules, or color graphics adapters and monitors with a standardized design. The extended graphics adapter/enhanced color display also

gave the additional bonuses of higher resolution and sixty-four displayable colors.

EGA monitors have eight video input signals, and use a horizontal sweep frequency of 21.50 kHz. Along with the primary red, green, and blue video signals, secondary RGB signals intensify individual color signals. With the analog RGBI monitors, the intensity bit intensifies all three colors at the same time. With the EGA monitor, the secondary video signals intensify only the corresponding primary color. Because the secondary signals will intensify either the individual primary signals or combinations of the signals, the number of possible displayable colors rises from sixteen to sixty-four.

Sync signals also combine with the video signals. The vertical sync signal locks in the vertical oscillator, while the horizontal sync signal locks in the horizontal oscillator. As with television receivers, the sync signals also blank the CRT electron beams between lines, fields, and frames. EGA monitors use a fifteen-pin, D-style connector called a DB-15 connector. Each pin carries an individual video signal. Table 13-3 lists the signals found at the video connector.

Table 13-3 Connector Pin-out for EGA Video Signals

Pin #	Signal
1	Ground
2	Secondary Red
3	Primary Red
4	Primary Green
5	Primary Blue
6	Secondary Green
7	Secondary Blue
8	Horizontal Sync
9	Vertical Sync

VGA Advances in technology along with user demands for better-quality video pushed manufacturers to produce a video system that gave microcomputer owners even more color choices and even higher video resolutions. With the introduction of the **video graphics array**, or VGA, adapter, information-display video technology moved completely away from digital signal transmission to analog signal transmission. Early VGA monitors could display 16 colors at a resolution of 640 × 480, or 256 colors at lower resolutions. Before the advent of VGA technology, only RGB monitors used an analog video signal. While the digital video signals are limited to on or off states, the analog signal causes the electron beam of the CRT to move in smaller increments, resulting in more shades of color. A VGA controller and monitor provide higher pixel addressability and better color graphics than the earlier controller and monitor types. VGA controllers contain the following major sections:

- Graphics controller: Performs logical functions on data written to display memory.

- Display memory: Includes banks of dynamic random-access memory, and stores screen-display data.

- Serializer: Moves data from the display memory and converts it into a serial bitstream.

- Attribute controller: Determines which color will be displayed.

- Sequencer: Controls the timing of circuits found on the video controller.

- CRT controller: Generates sync and blanking signals for the control of the monitor display.

Super VGA Super VGA or SVGA, offers even higher video resolution specifications of 800 × 600 and 1024 × 768. In addition, the combination of monitor and video adapter cards can produce more shades of color at a higher resolution. The increased number of high-resolution colors is a product of the monitor, the video circuitry, and the amount of video RAM on the adapter card. Although four super VGA standards exist, each supports a palette of 16 million colors. The four SVGA resolution standards are:

   800 ×   600 pixels
 1024 ×   768 pixels
 1280 × 1024 pixels
 1600 × 1200 pixels

The VESA SVGA standard produced by the Video Electronics Standards Association covers the capability of an SVGA card to address video RAM as one block. Moreover, the VESA SVGA standard improves the speed and efficiency of transfers between the system RAM and the video RAM. Most new video cards implement the VESA SVGA standard as part of the hardware system.

8514/A and XGA Standards The 8514/A standard produced by IBM extends the resolution of the VGA standard from 640 × 480 pixels to 1024 × 768 pixels. In addition, the 8514/A provides 26,200 colors and 64 shades of gray. When originally produced by IBM, the 8514/A standard relied on interlaced scanning. More recent 8514/A monitors use noninterlaced scanning. The Extended Graphics Array standard provides the same resolution as the 8514/A standard at a higher efficiency.

Multisync Multisync monitors automatically synchronize to a given scan rate. This allows the monitors to be used with any MDA, Hercules, RGB, CGA, EGA, VGA, or SVGA video input signal. Circuits within a multisync monitor scan the incoming signals and set to the received frequency range. Although multisync monitors offer flexibility, fixed-frequency monitors offer lower cost.

Video Accelerators

Although standard video cards feature a CRT controller and digital-to-analog converters, actual processing is always handled by the microprocessor housed in the central processing unit. Because of this, a large amount of data must move from the video adapter card, along the computer bus, into the CPU, and into memory. Video accelerators contain a processor and can perform video operations without the aid of the CPU. As a result, the computer bus and CPU are freed to complete other tasks, and the computer performs faster operations.

Video accelerator cards can provide a large performance gain for common graphics operations, and can provide speed improvements when working with graphics-based software. New video accelerator cards include highly complex 3-D graphics-rendering functions such as polygon shading and texture mapping, and allow on-the-fly magnification of video clips. In addition, the accelerator provides compression and decompression of video images so that video productions shown on a monitor screen will not have a jerky motion.

Most 3-D-accelerated video cards can deliver up to 16.7 million colors at 1280 × 1024, and a still-respectable 64,000 colors (high color) at 1600 × 1200. The latter resolution is useful for computer-aided drafting and manufacturing applications. In addition to increased resolution and millions of colors, video cards contain anywhere from 2 to 8 Mb of video RAM. The large quantity of VRAM speeds the performance of the system and ensures that the computer system will produce a high-quality video image, regardless of the display size.

Although the benefits of 3-D acceleration are not currently utilized in business applications, graphics once seen only in games will migrate into the those applications. Many presentation programs and spreadsheet packages will take advantage of the image quality offered by 3-D acceleration. In addition, individual and business video conferencing has begun to take advantage of the enhanced quality and speed of the new video cards. Moreover, cross-over applications such as digital video disks and image viewers will integrate nicely with the more powerful video adapters.

☑ **PROGRESS CHECK**

You have completed Objective 4. Discuss computer monitor display standards.

TTL and RGB Video Signals

Digital video displays accept **transistor-transistor logic**, or TLL, video signals as an input. Two bipolar transistors work together in a single package and control the logic level of a signal. Four TTL signals—video input, highlight input, vertical sync, and horizontal sync—flow into the monitor through a nine-pin D connector. Unlike composite video signals, which separate at different stages within the monitor, the TTL signals connect directly to their respective stages.

Rather than rely on the same composite video signal found in television signal circuits to carry information, analog computers use **RGB video**. The RGB video signal produces sharper images than the composite video signal. RGB video consists of three separate signals for red, blue, and green. In contrast, composite video mixes the three signals together. While older graphics standards relied on TTL video signals, newer standards ranging from CGA to SVGA, require RGB signals. Some monitors accept digital and analog signals.

☑ **PROGRESS CHECK**

You have completed Objective 5. Discuss computer monitor display adapters.

13-4 COMPUTER MONITOR OPERATION

To illustrate how computer monitors operate, we will use the ZCM-1492 14-inch VGA flat-tension mount monitor as an example. The ZCM-1492 exhibits unique characteristics—a stretched shadow mask, non-glare treatment, a high-contrast and high-resolution CRT, and a completely flat screen. Table 13-4 lists the technical specifications of the ZCM-1492 monitor. As the table shows, the monitor operates at 31.49 kHz and supports most video cards. Support of the video cards includes a text resolution of 720 × 400 pixels, and a graphics resolution of 640 × 480 pixels.

While the smaller, flat-screen monitors support VGA graphics adapters, the larger 17-inch monitors offer multisyncing capabilities and automatically adjust to the frequency of the video adapter. Such a monitor will interface with video technologies ranging through VGA and Super VGA. Despite featuring 100-percent compatibility with existing VGA video sources, Zenith's flat-screen technology monitors differ from other VGA monitor technologies in several ways. Obviously, some of the differences begin with the CRT. Figures 13-5A, B, and C show several views of the Rauland M36AEB22XX03 CRT used in the ZCM-1492.

The CRT utilizes a high bipotential focus, high-resolution electron gun. Moreover, Rauland combines the use of the electron gun with a high-resolution saddle-saddle, toroidal yoke. This combination furnishes a dynamically self-converging deflection system. The phosphor screen for the monitor is a fine-pitch, black-matrix, dot-trio system with an anti-reflective coating. Given the same levels of resolution and contrast, the display is 80 percent brighter than conventional high-resolution CRTs. The CRT also has a 70 percent increase in the contrast ratio, which provides more definite blacks. Along with enhanced brightness and blackness, the design offers a lower trade-off between contrast, brightness, and resolution.

Table 13-4   ZCM-1492 Specifications

CRT	Flat technology, 14", 0.31 mm pitch	**Video Interface Connector Pin Assignment**
Display Area	10.07 × 7.67 inches	1 (Red video)
Number of Displayable Colors	Infinite	2 (Green video)
Characters	80 characters × 25 rows	3 (Blue video)
Character Block	9 × 16 (VGA)	4 (N/C)
	8 × 16 (MCGA)	5 (Self-test)
	8 × 14 (EGA)	6 (Ground for red)
	8 × 16 (CGA, 400-line)	7 (Ground for green)
	9 × 14 (MDA)	8 (Ground for blue)
	9 × 14 (Hercules)	9 (N/C)
Video Input Signal	Analog RGB (0–0.714 v p-p)	10 Digital sync ground
Bandwidth	28 MHz	11 Digital ground (mode)
Horizontal Sync Input	31.49 kHz positive TTL 350-line mode	12 (N/C)
	31.49 kHz negative TTL 400-line mode	13 Horizontal sync
	31.49 kHz negative TTL 480-line mode	14 Vertical sync
Vertical Sync Input	70 Hz negative TTL 350-line mode	15 (N/C)
	70 Hz positive TTL 400-line mode	
	60 Hz negative TTL 480-line mode	
Resolution	640 dots × 480 lines (VGA, MCGA)	
	640 dots × 350 lines (EGA)	
	320 dots × 200 lines (MCGA, CGA)	
	720 dots × 350 lines (MDA, Hercules)	

To further improve the displayed image, Zenith combines the flat-screen technology with a two-stage, anti-glare treatment. Part of the anti-glare treatment extends from a multilayer, anti-reflection coating on the front surface of the viewing screen. An optical anti-reflective coating applied to the inner screen surface of the CRT rounds out the anti-glare treatment, reducing glare problems by nearly 95 percent.

In addition, the ZCM-1492 relies on precision printing of the red, green, and blue phosphor elements and black matrix to its flat faceplate. This replaces the traditional method of using a photolithographic process to apply the phosphors. The traditional process paints the phosphors onto the screen during a multistage process, while the precision-printing process prints the phosphor elements into the respective red, blue, and green apertures within a precise set of parameters. Machining techniques produce the precision glass panels necessary for the screen printing process. Table 13-5 lists the specifications for the CRT.

Placing the Shadow Mask Under Tension

Instead of the traditional shadow mask, which curves to follow the shape of the screen, the ZCM-1492 monitor uses a different technique with the flat display. Zenith's flat-screen technology stretches the shadow mask under tension and places the mask directly behind the flat faceplate. Furthermore, the mask is welded to the faceplate support structure. In contrast, the traditional method supports the shadow mask with a frame and suspends it by springs within the CRT.

The traditional design allows some image distortion because of temperature changes affecting the shape of the shadow mask. With the flat-tension mask design, temperature changes do not affect the shape of the stretched, under-tension shadow mask. This reduces purity loss due to electron-beam heating of the mask. Also, the design prevents shadow mask movement under most display conditions. Zenith's flat-tension monitors offer a display with increased, precise resolution,

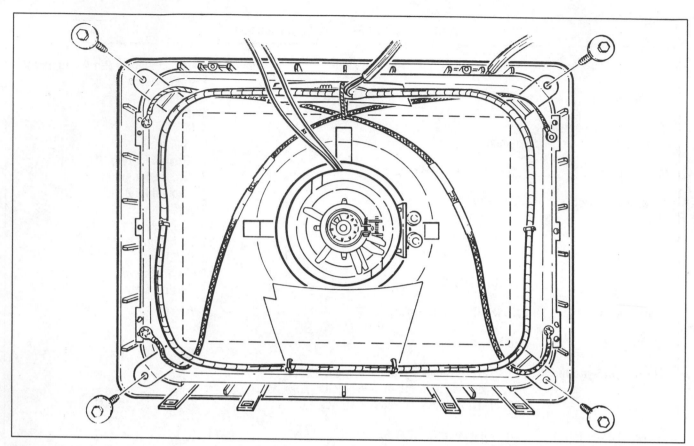

Figure 13-5A Rear View of the CRT (*Courtesy of Zenith Data Systems, Zenith Electronics, and Rauland*)

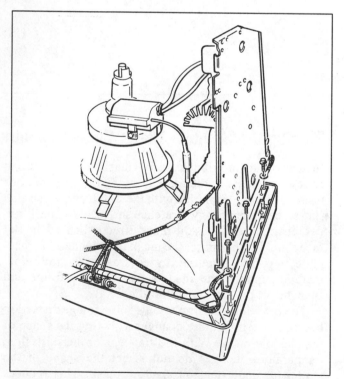

Figure 13-5B Side View of the CRT, Yoke, and Main Circuit Board (*Courtesy of Zenith Data Systems, Zenith Electronics, and Rauland*)

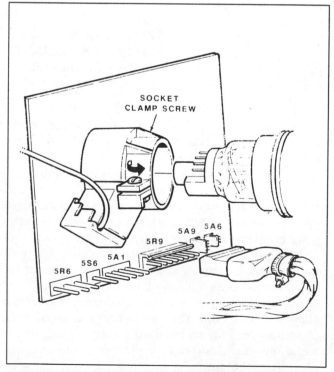

Figure 13-5C CRT Neck and Video Output Board (*Courtesy of Zenith Data Systems, Zenith Electronics, and Rauland*)

Table 13-5   Specifications of the Rauland M36AEB22XX03 CRT

Focusing method	electrostatic
Convergence method	magnetic
Diagonal	90°
Horizontal	77°
Vertical	62°
Spacing between centers of adjacent dot trios	0.028 mm
Screen dimensions	
Diagonal	13.75″
Horizontal axis	11.00″
Vertical axis	8.25″
Anode voltage	
Absolute maximum	27,000 v
Absolute minimum	20,000 v
Cathode voltage	+100 v
G1 Grid voltage	0 v
G2 Grid voltage	+720 v
Focus voltage	7200 v
Dynamic focusing	150 v

along with better color fidelity. Tension on the shadow mask prevents the mask from deforming when bright patches of color display for any length of time, whereas conventional masks deform or "dome" when displaying bright patches of color, and discoloring the images.

13-5   COMPUTER MONITOR CIRCUITS

Figure 13-6 shows a block diagram of the ZCM-1492, and Figure 13-7 shows an exploded drawing of the monitor. As the diagrams show, the monitor consists of four circuit sections and the CRT. The four sections include the video modules, deflection module, and the dynamic focus and pincushion module. As with other monitor designs, the ZCM-1492 video board contains a video amplifier for the analog color signals, video drivers, dc restoration circuits, a beam-current limiter, and video output amplifiers. In the exploded view of the monitor, the deflection module incorporates the sync processing, horizontal deflection, vertical deflection, and high-voltage circuitry.

Video Circuitry in the ZCM-1492

Figure 13-8, located at the end of this book, shows the schematic diagram for the video signal board.

Video input signals enter the board through connector 5R9, while capacitors C5101, C5102, and C5103 couple the red, blue, and green analog color signals to the video inputs of IC5101. The signal inputs at IC5101 are pin 12 for red, pin 15 for green, and pin 18 for blue.

IC5101 operates as a three-channel, variable-gain video amplifier, and controls most of the video signal processes for the monitor. A variable dc voltage at pin 2 of the integrated circuit, and supplied through the contrast control, controls the gain of the three video signals. The voltage varies from +8 Vdc at maximum contrast to zero volts at minimum contrast.

Looking at the diagram, the contrast control operates as a voltage divider, supplying a variable voltage to pin 4 of connector 5A1. R5112 and C5111 make up an integrator that smoothes the action of the control so that the variable voltage stays within 3 percent of its designed range. The gains of the three channels should track to within about 3 percent over the range of the contrast control. Again looking at the schematic diagram, each of the output transistors is configured as an emitter follower.

↻ FUNDAMENTAL REVIEW

Emitter Followers

The emitter follower is a transistor current amplifier that has no voltage gain. The input for the emitter follower is the base, while the output is taken from the emitter. Emitter follower circuits do not have collector resistors or emitter bypass capacitors.

An automatic-brightness limiter circuit within IC5101 also controls the gain of the video signals. In this case, the circuit causes the video signal to decrease if the average anode current goes past 750 microamps. The anode current is sampled at the secondary of the flyback transformer, averaged by R5111 and C5119, and then applied to pin 1 of IC5101. Negative feedback produced by the ABL circuit forces the average anode current back to the 750-microamp level.

After the video amplifier IC processes the input signals, three different stages further amplify and add to the signals. First, three resistors form a resistor attenuator, which attenuates each color signal before it couples to its respective video driver. Using the green video amplifier section as an example, each signal becomes applied to the cascode output amplifier consisting of Q5205 and Q5206. The red and blue video amplifier circuits have identical conditions.

DC Restoration   Because of the original ac-coupling of the video input signals to the video amplifier circuit, the dc portion of the signal is restored by a combination

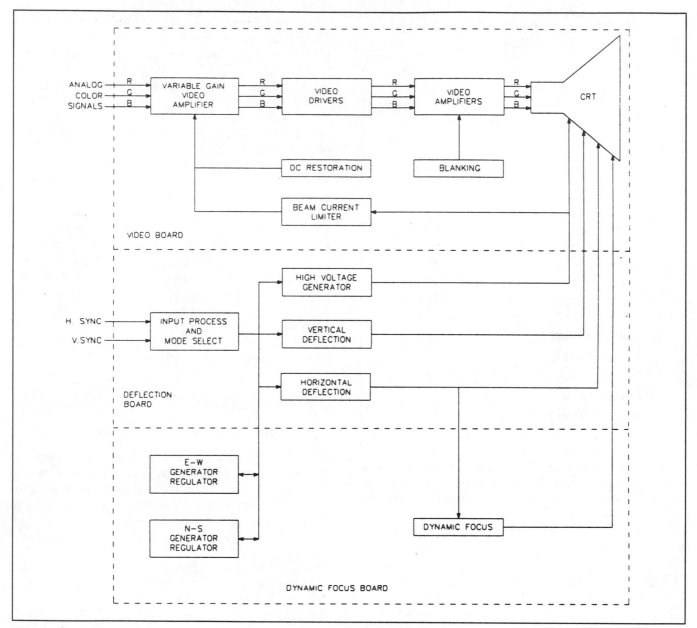

Figure 13-6   Block Diagram of the ZCM-1492 (*Courtesy of Zenith Data Systems, Zenith Electronics, and Rauland*)

Part Identifiers for Figure 13-7
(*see next page*)

1 Tilt-top base	11 CRT and yoke assembly	21 Screw, chassis mounting
2 Tilt-swivel base	12 Control knob assembly	22 Spring, helical
3 Clamp assembly, video output retainer sink assembly	13 Deflection module	23 Swivel retainer disk
	14 Degausser coil	24 Video output module
4 Cabinet front	15 Fastener, PC board retainer	25 Washer, flat
5 Cabinet rear		
	16 Cabinet foot, molded	26 Retainer, spring
6 Interface cable	17 Power switch	27 Brace, upper horizontal
7 Card guide	18 LED and cable assembly	28 Brace, lower front horizontal
8 Chassis plate, left	19 Line cord	29 Brace, lower horizontal
9 Chassis plate, rear	20 Power supply	
10 Chassis plate, right		

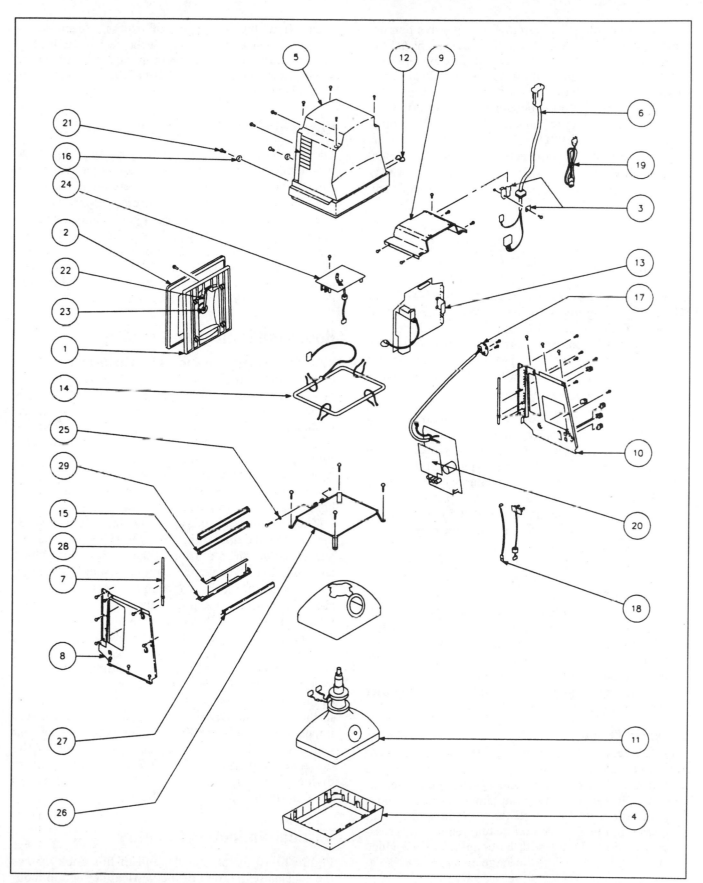

Figure 13-7 Exploded Diagram of the ZCM-1492 (*Courtesy of Zenith Data Systems, Zenith Electronics, and Rauland*)

of transistors, integrated circuits, and passive components. The amplified video signal appears at the Q5206 collector and across load resistors R5212 and R5213.

An RC network formed by R5209 and C5214 samples the voltage at the emitter of Q5205 during the retrace interval. Sampling occurs during the period that the video signal is at the black level, and feeds a sample voltage to pin 14 of IC5101. During this time, the clamp pulse at pin 1 of connector 5R6 is at a logic high. The RC network also combines with C5205 and R5211 to provide frequency compensation for the cascode amplifier.

↻ FUNDAMENTAL REVIEW

Cascode Amplifier

A cascode amplifier consists of either a pair of JFETs or MOSFETs configured so that a common-source amplifier is in series with a common-gate amplifier. The cascode amplifier configuration overcomes the high input capacitances of JFETs and MOSFETs, which could affect the high-frequency operation of the components.

From there, a comparator compares the sample voltage with a reference voltage found at pin 13 of the IC. Depending on the size and polarity of the difference voltage, the comparator uses a push-pull current source at its output to either charge or discharge hold capacitor C5105. The voltage developed across C5105 controls the dc bias of the signal at pin 15 of IC5101. All three video channels rely on the comparison between the sample and reference voltages.

When an active clamp pulse appears, the condition of the dc restoration loop allows the black-level emitter voltage at Q5205 to equal the reference voltage at pin 13 of the IC. When clamping stops and video begins, the dc restoration loop turns off, and the hold capacitor supplies the dc bias. The action of the dc restoration loop and hold capacitor provides a stable dc bias for the cascode output amplifier. As a result, the ac-coupled video signal applied to the input of the video amplifier IC is dc-restored.

The Video Output Stage   Again using the green video amplifier section as an example, we can trace the amplified video signals from the collector of Q5206 to a pair of transistors, Q5207 and Q5208, connected as emitter followers and as a complementary pair. Diodes CR5205 and CR5206 force the complementary pair to operate in the class AB mode and reduce any crossover distortion. Along with ac-coupling of the video output signal to the CRT, the stage also provides isolation and low-impedance drive.

The isolation of the cathode capacitance from the collector of the output transistor protects the output circuit from the possibility of excess capacitive reactance. Because capacitive reactance has an inverse relationship with frequency, a large capacitive reactance would reduce the load impedance of the output stage. In turn, this would reduce the gain of the amplifier.

Horizontal and Vertical Sync Signals

The horizontal and vertical sync circuitry includes the sync input buffering, mode selection, sync waveform adjusting circuits, horizontal and vertical size control, phase control, and the hold controls. Of particular interest, our example monitor can operate in one of three video modes. With mode 1, the monitor uses the EGA mode; mode 2 produces a CGA display; and mode 3 establishes the VGA display. Each of the modes depends on the 31.49-kHz scan frequency and is determined by the polarity of the horizontal and vertical sync signals.

Horizontal Deflection Circuitry

Referring to the schematic of the horizontal deflection circuit shown in Figure 13-9, located at the end of this book, the horizontal oscillator output voltage is applied to the base of Q3403, the horizontal driver. From the collector of Q3403, the output signal from the river goes to an interstage, impedance-matching transformer. TX3401 steps down the B+ voltage, while a waveshaping circuit consisting of C3009, R3007, and R3009 shape the rectangular drive waveform for the horizontal output transistor.

The drive waveform is applied to the base of Q3003, the horizontal output transistor. Q3003 becomes cut off during retrace and a portion of the trace. When Q3003 cuts off after flyback, the damper diode or CR3003 conducts and produces a portion of the trace at the left side of the raster. In addition, the damping diode suppresses oscillations that could produce white vertical bars on the left side of the raster. Retrace capacitor C3008 forward biases the diode.

The action of the transistor and the diode reduces the average amplifier current and increases the efficiency of the amplifier circuit. Looking at other components in the horizontal deflection area, the circuit uses the impedance of coil LX3002 to change the yoke current and improve linearity. Also, the combination of the horizontal centering control or R3001, and a voltage divider consisting of Q3001 and Q3002, provides electrical horizontal centering of the display.

Vertical Deflection Circuitry

Figure 13-10, located at the end of this book, shows the schematic diagram of the vertical deflection module; the module contains circuitry for vertical signal processing and vertical deflection. A differentiator consisting of CR2122, R2101, and C2103 couples the vertical sync signal at the collector of Q2116 to pin 3 of IC2101. Together, IC2101 and Q2101 form the oscillator stage, while C2101 and R2102 determine the time constant for the oscillator.

In the oscillator stage, IC2101 functions as a comparator using positive feedback. The voltage of Q2101, an emitter follower, follows the output of the comparator. When the IC2101 output moves from low to high, Q2101 turns on and charges C2115 through CR2102 and R2102. In addition, the threshold voltage at pin 3 of the comparator rises. C2101 charges until its voltage goes beyond the voltage at pin 3. Then the output of IC2101 goes low, and C2101 discharges. With that action, the oscillator stage develops a free-running frequency.

The output from the driver transistor, Q2106, is applied to the input of a complementary-symmetry amplifier consisting of Q2107 and Q2108. When a positive-going sawtooth appears at the base of Q2106, a negative-going drive is applied to the base of Q2107, increasing the collector current of that transistor. With the same negative drive applied to the base of Q2108, the forward voltage at the Q2108 base and the collector current decreases. If a negative-going sawtooth appears at the base of Q2106, the driver applies a positive-going drive to the base of Q2108. The collector current of Q2108 then increases, while the Q2107 collector current decreases.

Capacitor C2112 increases the load impedance in the collector circuit of Q2106 and keeps Q2108 in a conducting state at all times. With C2112 supplying positive feedback, the Q2106 circuit gain increases. By keeping Q2108 conducting, the capacitor prevents large positive peaks in the signal from placing the base and emitter at the power-supply potential and cutting off the transistor. This occurs because of the voltage stored across the capacitor.

Blanking

Still looking at the vertical deflection schematic, capacitor C2118 ac-couples vertical pulses from the flyback to two transistors. Transistors Q2117 and Q3401 use those 5.6-volt p-p pulses to generate part of the blanking signal. After the pulse is applied to the base of Q2117, its conduction brings the base of Q3401 to ground. Since this action turns Q3401 off, +5 volts appears at the Q3401 base during the vertical blanking interval.

The remainder of the blanking signal results from a similar operation. C3413 ac-couples the −70-Vdc horizontal flyback pulse to the base of Q3401. With the high-amplitude, negative pulse at its base, Q3401 again shuts off. Consequently, a +5-Vdc blanking pulse

appears at the Q3401 collector during the horizontal blanking interval.

Conduction of diode CR3404 during the retrace portion of the horizontal flyback pulse protects Q2117 and Q3401 from any reverse-bias damage. During the trace portion of the horizontal flyback pulse, the conduction of the diode holds the collector of Q3401 to a lower value. Given that condition, a composite blanking pulse appears at the collector of Q3401.

High-Voltage Circuitry

Output from the horizontal oscillator begins to develop the high voltage. From the base of Q3201 and transformer TX3201, the horizontal oscillator output goes to the anode voltage driver transistor, Q3202. The combination of the transistor and the flyback transformer TX3202 produces high voltage. While the focus control determines the amount of voltage applied to the last grid of the CRT, the high-voltage resistor block is the source for the focus and G2 voltages.

IC3201, Q3205, Q3206, Q3203, Q3202, and various passive components make up the anode voltage regulator. An amplifier, labeled IC3201, takes a reference voltage and feedback voltages and uses the difference between them to drive Q3205. Subsequently, Q3205 drives two regulating transistors, Q3204 and Q3206. To insure a stable regulated output voltage from this stage, Q3206 provides additional feedback. Diodes CR3205, CR3206, CR3216, and CR3217 protect IC3201 from transients.

Focus Voltages The Zenith monitor relies on dynamic focus circuitry to vary the focus voltage at the horizontal scan rate. Because of this, the focus voltage at the edges of the raster increases, while the voltage at the center of the raster decreases. With the horizontal sync waveform applied to the base of Q7701, R7703 controls the amplitude of the signal applied to the dynamic focus circuit. Q7701 and Q7703 combine to amplify the incoming waveform, while T7701 steps the resulting waveform up to the generate the 500-Vdc p-p focus voltage.

High-Voltage Shutdown If the high voltage exceeds specified limits, shutdown circuitry consisting of Q3207, R3228, R3232, R3228, Q3208, Q3209, CR3207, and IC5203 shuts the high voltage down. Q3209 senses the anode current at the secondary of the flyback. When excessive beam current occurs, diode CR3207 goes negative and biases Q3207 on.

R3228 holds the collector of Q3207 to a dc voltage, while R3253 and R3256 adjust the shutdown threshold voltage IC5203. After R3232 and R3233 divide the voltage at the Q3207 collector, the voltage divider output turns on Q3208. Therefore, horizontal sync pulses at the collector of Q3208 become shunted to ground. This cuts the base drive of Q3201 and shuts down the high-voltage circuitry.

Pincushion Circuitry for the ZCM-1492

Because the flat-tension mask CRT requires a geometrically perfect display, the ZCM-1492 monitor features a more sophisticated pincushion circuit than most color monitors or televisions. Enhanced pincushion circuitry provides the correction needed to provide the correct display. Pincushion protection breaks down into four basic sections—the east-west waveform generator and regulator, the north-south waveform generator and the north-south output circuit. A perfectly symmetrical display occurs by superimposing the pincushion correction waveforms onto the horizontal and vertical scanning. Figure 13-11, located in Appendix, shows a schematic for the east-west generator/output circuit.

East-West Waveform Generation and Output   The east-west waveform generator affects the left and right sides of the display. It produces three waveforms used for correcting pincushion errors. When combined at a multiplier circuit, the three waveforms at the input of the multiplier combine to form a parabolic correction waveform at its output. The multiplier circuit consists of IC7001, C7001, R7002, and R7014, with the passive components acting as coupling devices for a vertical ramp.

The B+ voltage supplied to the horizontal deflection circuitry modulates the parabolic waveform with a ramp waveform at the vertical scan rate. The modulating waveform corrects any distortion at the left and right sides of the raster. Trap control R7012 allows some adjustment of the display for better symmetry. In effect, the control adjusts the amount of offset so that the vertical scan rate of the ramp waveform has the proper amplitude and polarity.

Another waveform at the vertical scan rate—a sine wave—is also added to the parabolic waveform. From the multiplier, the negative output becomes applied to the base of Q7007. After integration by capacitor C7010, output from the transistor forms the sine wave. The sine wave corrects any phase errors in the pincushion-correction waveform. Q7007 and Q7008 form a differential amplifier. The positive output from the multiplier is applied to the base of Q7008.

While the output of Q7007 forms a sine wave, the output from Q7008 couples to the east-west level control through capacitor C7007. Operation of the differential amplifier controls the production of the proper correction waveform. If the vertical size changes, the vertical ramp compensates for the change. If the horizontal size changes, B+ voltage supplied to the horizontal deflection circuitry is sampled. The circuitry uses the sample to adjust the correction waveform.

Regulator circuitry controls the amount of correction applied against the horizontal scan voltage, and supplies the current to the horizontal deflection circuitry. R7027 controls the amplitude of the parabolic correction waveform. C7008 and R7028 form a network that couples the east-west control output to the noninverting IC1701. Output from the east-west phase control, R7039, goes to both the noninverting and inverting inputs of the IC.

North-South Waveform Generation and Output   The north-south waveform generator corrects the pincushion for the top and bottom of the CRT display by modulating the vertical ramp waveform at the horizontal scan rate. Modulating the ramp waveform increases the vertical deflection at the top center and bottom center of the display. The process begins by coupling a 30-Vdc horizontal retrace pulse found at pin connector 8V6 to a signal-shaping circuit consisting of transistors Q7402 and Q7401. After the two transistors process the pulse, a pulse with a fast rising edge and a delayed falling edge results at the collector of Q7401. Capacitor C7406 and resistor R7409 then differentiate the pulse and apply it to pin 8 of IC7401 as the set input pulse. The integrated circuit is a dual D-type flip-flop, with one-half configured as an astable multivibrator.

⏎ FUNDAMENTAL REVIEW
Astable Multivibrators

An astable, or free-running, multivibrator will lock into a logic state when triggered by an external signal. The device does not require a signal at the input.

As a result of the shaped input pulse at pin 8, the output pulse at pin 13 has a logic high state. An RC network consisting of R7412 and C7409 delays the clock input signal to pin 11 of the IC by approximately 4 microseconds, and sets the duration of the output pulse. When the clock input goes high, the signal at the D-type flip-flop input—pin 9 of the IC—causes the output at pin 13 signal to go low. C7408, R7463, and R7441 differentiate the vertical retrace pulse, and the pulse is applied through diode CR7410 and R7464 to the reset input of a flip-flop contained within IC7401. This action resets the flip-flop at the start of each vertical frame.

During operation, the signal from pin 13 of the IC couples to transistor Q7405. Along with C7414, Q7404, and D7404, Q7405 forms a horizontal ramp generator through charging and discharging C7414. C7414 charges because of a current source formed by Q7404, D7404, and other components. Given a pulse at its base, Q7405 begins to conduct, and discharges C7414. Transistors Q7407 and Q7408 buffer the horizontal ramp and couple the signal to the north and south pincushion level controls.

Q7407, Q7409, and C7417 form a parabola generator that provides a parabolic waveform to correct the upper and lower phase. When C7417 charges, the modulated signal placed at the base of Q7407 causes the transistor to conduct. The conduction of the transistor discharges the capacitor and completes the generation of the waveform. During the same time, Q7410 combines with a resonant LC circuit consisting of C7421 and L7401 to produce a sine wave that has twice the horizontal frequency. Q7411 buffers the sine wave for the north and south pincushion controls.

Processed horizontal rate signals travel through IC7402 from the north-south output circuit, and are applied to the carrier inputs of IC7501, a balanced modulator. The IC and associated circuitry modulate the horizontal rate signal as the carrier with an input signal consisting of a vertical ramp waveform. With this, the amplitude of the signal decreases to zero as the electron beam approaches the center of the screen from the top. When the beam crosses the center of the screen, the amplitude gradually increases. However, the phase reverses while the beam scans the bottom half of the screen.

⏩ FUNDAMENTAL REVIEW

Balanced Modulator

A balanced modulator forms an output voltage that is the product of an input signal and a carrier.

The vertical ramp waveform couples from Q7505 through C7501 to IC7501. A north-south cross-over control adjusts the cross-over point so that the horizontal signal amplitude decreases to zero when the phase reverses. While R7510 and R7545 control the gain for the stage, Q7502 and Q7503 adjust the bias for IC7501. Referring to the schematic shown in Figure 13-12, located in Appendix, the secondary of T7501 is effectively in parallel with the vertical yoke windings and the north-south pincushion-correction waveform. Therefore, the pincushion-correction waveform is imposed on the vertical output. The parallel relationship between the T7501 secondary windings and the yoke windings results from a combination of low impedance presented by the yoke capacitor and resistor along with the ac ground of the vertical scan circuit.

☑ PROGRESS CHECK

You have completed Objective 6. Discuss computer monitor circuits.

13-6   TROUBLESHOOTING COMPUTER MONITORS

Many service organizations shy away from troubleshooting computer monitors because of the lack of schematics and service documentation. When troubleshooting a monitor problem, remember that monitors rely on the same type of circuits seen in a television system. Monitors rely on:

- switched-mode power supplies
- sync signal processing circuits
- video signal processing and output circuits
- chroma signal processing and output circuits
- vertical deflection circuits
- horizontal deflection circuits
- high-voltage circuits
- a CRT

Given the knowledge that you have gained through this text, you should have the background necessary to successfully repair a computer monitor.

By combining that knowledge with experience, you can often find problem sources and solutions without the aid of a schematic. For example, a problem with an RGB drive board becomes easier to define because that type of board relies on three separate drive circuits. Therefore, the measurement of voltages, currents, and resistances in the problem circuit, and the comparison of those measurements with known good values from the other drive circuits, can narrow the search.

You also know that computer monitors rely on a standard electronic parts and circuit designs. If the components have distinct labels, you can use master semiconductor replacement guides to find transistor and IC pin layouts along with voltage and current ratings. From there, you can begin to look for unexpected voltage readings throughout the circuit under test. Because monitors rely on standard designs, it also becomes easier to pinpoint different sections within the entire chassis. For example, the vertical output transistors and deflection ICs are generally located close together on the main board.

When troubleshooting power supply problems in a monitor, use the same techniques described in chapters three and four. That is, use an isolation transformer and a variac to power the set slowly while monitoring the voltages around the switched-mode regulator. In addition, perform out-of-circuit checks of the regulator transistor or MOSFET and the passive components found in the supply. As mentioned in chapter four, always check the filter capacitors, and always observe safety procedures.

While this procedure should allow you to repair most monitor power supplies, always remember that variants of the switched-mode power supply design exist. For example, Sony monitors take advantage of a separated power supply that provides voltage for 1) the flyback transformer and 2) the deflection yoke circuitry. In comparison to the single power that provides voltages to both the flyback and the deflection circuit across a single circuit, the Sony power supply system establishes a more stable power source for both devices and a more stable video image.

Depicted in Figure 13-13, the implementation of the Sony system also allows users to adjust the pincushion balance, keystone, balance, and cancel controls. Given these additional adjustments, consumers can adjust the stability of the picture. As an example, the Sony system will display a spreadsheet application with numerous cells without the pincushioning effect seen with other designs. With separate voltage supplies, the power consumption difference between white and black areas of the spreadsheet highlights balances.

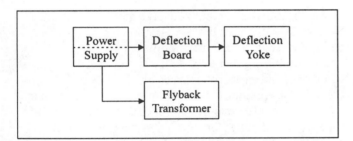

Figure 13-13   Block Diagram of the Sony Separated Power Supply (*Courtesy of Zenith Data Systems, Zenith Electronics, and Rauland*)

Troubleshooting the Video Circuitry

Problems that occur with the video board involve the loss or distortion of the color signals. When checking the video circuits, always ensure that the video input connector and cable are fastened securely and that there are no breaks in the cable. Then use an oscilloscope to verify the presence, amplitude, and shape of the waveforms located at the CRT. Figure 13-14 shows the correct waveforms. If the waveforms are correct, then suspect a problem with the CRT.

Incorrect waveforms lead back to the first set of video amplifiers and a check of the waveforms at the emitters of Q103, Q104, Q203, and Q204. Figure 13-15 illustrates the correct appearance of the waveforms. The lack of a good waveform at any of those points focuses our attention on IC401, the video processor. Good waveforms at those points allow you to move to the collectors of Q102, Q202, and Q302, and expect to find a waveform similar to shown in Figure 13-16. A distorted or zero-amplitude waveform at those locations suggests defects in the processor IC, while good

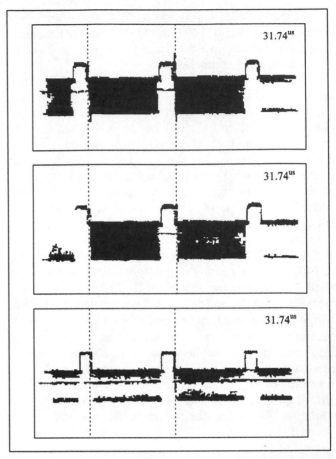

Figure 13-14   Video Waveforms at the ZCM-1492 CRT (*Courtesy of Zenith Data Systems, Zenith Electronics, and Rauland*)

waveforms suggest that defects exist in Q102, Q202, and Q302.

When checking IC401, check the waveforms at pins 12, 15, 18, 5, 8, and 3. A good waveform at pins 12, 15, and 18 indicates that a problem lies within Q101, Q201, and Q301, while the lack of a waveform or a distorted

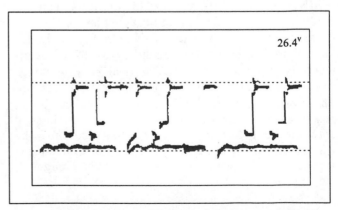

Figure 13-15   Waveforms at the Emitters of Q103, Q104, Q203, and Q204 (*Courtesy of Zenith Data Systems, Zenith Electronics, and Rauland*)

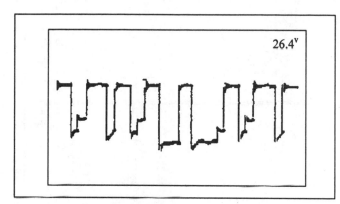

Figure 13-16 Waveforms at the Collector of Q102, Q202, and Q302 (*Courtesy of Zenith Data Systems, Zenith Electronics, and Rauland*)

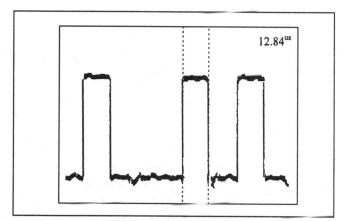

Figure 13-17A Correct Waveform for Pins 12, 15, 18 of IC401 (*Courtesy of Zenith Data Systems, Zenith Electronics, and Rauland*)

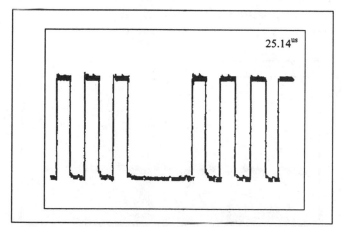

Figure 13-17B Correct Waveform for Pins 5, 8, and 3 of IC401 (*Courtesy of Zenith Data Systems, Zenith Electronics, and Rauland*)

waveform indicates that the IC is defective. Figure 13-17A shows the correct waveform for pins 12, 15, and 18, while Figure 13-17B shows the waveform that should appear at pins 5, 8, and 3. A distorted or low-amplitude waveform at pins 5, 8, and 3 takes you back to the video cable and connection.

Troubleshooting the ZCM-1492 Deflection and Pincushion Circuitry

When attempting to troubleshoot problems with the deflection and pincushion circuitry, always ask these preliminary questions:

- Is vertical deflection present?
- Is horizontal deflection present?
- Does the display have the proper symmetry?

Answering these three basic questions allows you to refine your troubleshooting procedure and narrow the scope of your work. And, as always, verify the quality of the video connectors and cables when performing checks of the vertical and horizontal deflection circuits.

If the monitor lacks vertical deflection, check for the vertical waveform shown in Figure 13-18A at pin 6 of the vertical processor IC, U303. A good waveform tells you to direct your attention toward the vertical deflection amplifier, U302, and transistors Q307, Q308, and Q314.

If there is no waveform or a distorted waveform at pin 6, then check for the waveform shown in Figure 13-18B at pin 11. A good waveform here combined with no vertical deflection suggests a defective U302. A missing or distorted waveform at pin 11 tells you to check for the proper vertical sync signal at connector 5A9.

A symptom of no horizontal deflection leads to an out-of-circuit check of the horizontal output transistor,

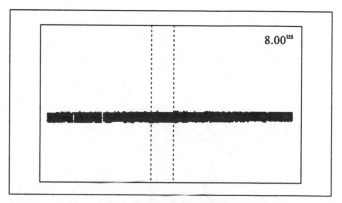

Figure 13-18A Correct Waveform for Pin 6 of the Vertical Processor IC (*Courtesy of Zenith Data Systems, Zenith Electronics, and Rauland*)

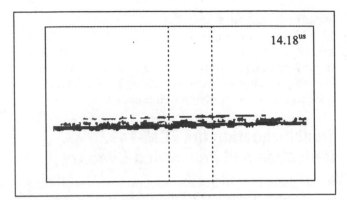

Figure 13-18B Correct Waveform for Pin 11 of the Vertical Processor IC (*Courtesy of Zenith Data Systems, Zenith Electronics, and Rauland*)

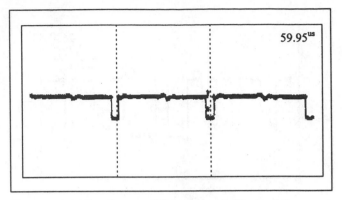

Figure 13-19B Correct Waveform at Pin 3 of the Horizontal Processor IC (*Courtesy of Zenith Data Systems, Zenith Electronics, and Rauland*)

Q404. If the transistor tests good and the symptom persists, check pins 1 and 3 of the horizontal processor IC, U400, for the waveforms shown in Figures 13-19A and 13-19B. If neither of those waveforms are present or correct, check for the proper horizontal sync signal at connector 5A9. If the symptom remains and the waveforms are good, suspect that the horizontal processor has a defect.

The symptom of good horizontal deflection and a nonsymmetrical picture takes you to the pincushion circuitry. Always check the amplitude and phase adjustments in those circuits as well as the size and width adjustments if a problem occurs. As mentioned in the overview of the pincushion circuits, a vertical ramp waveform has a key role in the east-west circuit control of side-to-side symmetry. Using an oscilloscope, check at the base of Q7505 for a good ramp waveform as shown in Figure 13-20. If the ramp is distorted or missing, suspect the C7513 in the east-west pincushion circuit.

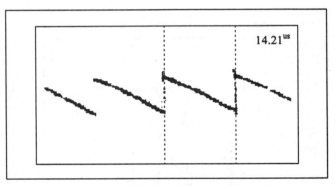

Figure 13-20 Vertical Ramp Waveform for the East-West Pincushion Circuit (*Courtesy of Zenith Data Systems, Zenith Electronics, and Rauland*)

When looking at top and bottom symmetry problems, verify that the horizontal retrace waveform appears as in Figure 13-21, and that it has the proper amplitude and shape. If the waveform is not correct, check capacitor C7405. If the waveform is correct and the symptom continues, suspect the north-south

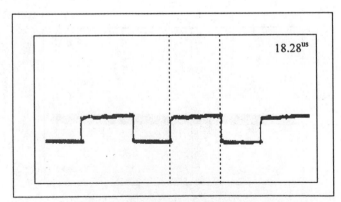

Figure 13-19A Correct Waveform at Pin 1 of the Horizontal Processor IC (*Courtesy of Zenith Data Systems, Zenith Electronics, and Rauland*)

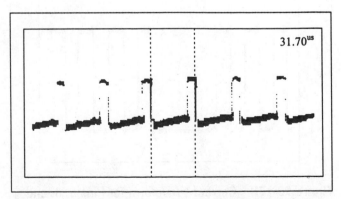

Figure 13-21 Horizontal Retrace Waveform (*Courtesy of Zenith Data Systems, Zenith Electronics, and Rauland*)

pincushion amplifiers—IC7501, IC7402, IC7502, and IC7503. Pins 4 and 8 of IC7501 should have the waveforms shown in Figures 13-22A and 13-22B, while pins 20 and 23 should have the waveforms shown in Figures 13-23A and 13-23B.

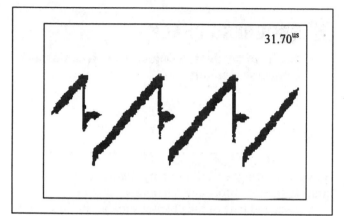

Figure 13-22A   Waveform for Pin 4 of the North-South Pincushion Amplifier IC7501 (*Courtesy of Zenith Data Systems, Zenith Electronics, and Rauland*)

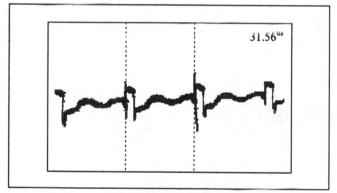

Figure 13-22B   Waveform for Pin 8 of the North-South Pincushion Amplifier IC7501 (*Courtesy of Zenith Data Systems, Zenith Electronics, and Rauland*)

➲ **SERVICE CALL**

Apple Monitor with Bright Red Raster

When the school district technology director brought the Apple monitor in for repair, it displayed a bright red raster. The technician removed the cover from the monitor and began looking for the three main video output transistors on the RGB board. The transistors were labeled as Q6B2, Q6G2, and Q6R2. A quick check of the collector voltages of each transistor showed that the voltage at transistor Q6R2 was zero. However, the transistor checked as good during an out-of-circuit check.

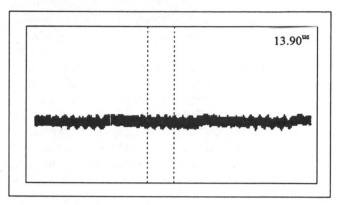

Figure 13-23A   Waveform at Pin 20 of the North-South Pincushion Amplifier IC7501 (*Courtesy of Zenith Data Systems, Zenith Electronics, and Rauland*)

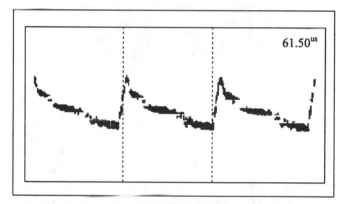

Figure 13-23B   Waveform at Pin 23 of the North-South Pincushion Amplifier IC7501 (*Courtesy of Zenith Data Systems, Zenith Electronics, and Rauland*)

After resoldering the transistor, the technician began checking back through the red video output circuit for the proper supply voltage. He found the correct supply voltage on one side of inductor L6R2, and zero volts on the output side. A further check with a multimeter showed that the coil had opened. After the replacement of the coil, the monitor was restored to its normal operating condition.

➲ **SERVICE CALL**

TOC Monitor Does Not Have Vertical Deflection

The technician found that the monitor returned to his shop had no vertical deflection and displayed only a bright white line across the screen. After locating the vertical output transistor and the vertical deflection IC, the technician performed voltage and resistance checks on the transistor. Everything was correct. The technician then began to concentrate on the integrated circuit.

Since there was no full schematic for the monitor available in the shop, the technician used the label on the IC and a semiconductor reference book to find the pin layout for the device. A check for the proper supply voltages at the IC revealed zero volts instead of the required 24 volts. Tracing along the printed circuit board, the technician found that a burned fusible resistor was preventing the application of the supply voltage.

Given that the resistor was fusible and installed as a protective device, the technician used the ohmmeter function of his multimeter to check the IC for any pos-sible short-to-ground conditions with the monitor pow-ered off. No shorts surfaced. The technician concluded that the IC had an internal short, and replaced both the IC and the fusible resistor. The replacement of those parts restored the vertical deflection of the monitor.

 PROGRESS CHECK

You have completed Objective 7. Troubleshoot computer monitors.

SUMMARY

Chapter thirteen exposed you to two new technology concepts—high-definition television and computer monitors. A preliminary discussion of noninterlaced and interlaced scanning, resolution, and refresh rates provided a foundation for the sections on HDTV and monitors.

HDTV represents a new technology for the repro-duction of clearer pictures and sounds at very high resolutions. It requires a different set of sync signals, and takes advantage of the MPEG video compression and Dolby audio compression formats. The new HDTV standard takes advantage of standards and technologies previously assigned only to computer displays. Be-cause of this convergence, HDTV has many applica-tions, including entertainment, health care, and the simulcasting of broadcast channels.

The second part of the chapter took an in-depth look at computer video adapters and monitors. You learned the terms associated with the computer display tech-nologies, and a short history of the evolution of those technologies. Using a Zenith ZCM-1492 monitor as an example, this chapter explored flat-screen technologies and video, horizontal deflection, vertical deflection, and pincushion circuitry. The operation of these cir-cuits linked with the overview of computer monitor troubleshooting techniques provided at the conclusion of the chapter.

REVIEW QUESTIONS

1. Define the relationship between dot pitch and pixels.

2. An increase in the horizontal scan rate improves picture clarity by:
   a. changing the vertical scan rate
   b. reducing the number of horizontal scan lines squeezed into a vertical frame
   c. increasing the number of horizontal scan lines squeezed into a vertical frame
   d. maintaining a constant number of horizontal scan lines

3. Computer monitors have a bandwidth:
   a. higher than television
   b. lower than television
   c. the same as television

4. What is the difference between interlaced and progressive scanning?

5. True or False   Televisions based on the NTSC standard use interlaced scanning.

6. True or False   Interlaced scanning is desirable because of reduced flicker and visible scan lines.

7. The NTSC broadcast standard uses:
   a. 626 horizontal scanning lines
   b. 525 horizontal scanning lines
   c. 483 scanning lines
   d. does not show horizontal scanning lines

8. The PAL broadcast has a:
   a. 50-Hz frame rate, a color subcarrier frequency of 3.58 MHz, and 626 scanning lines
   b. 50-Hz frame rate, a color subcarrier frequency of 4.43 MHz, and 525 scanning lines
   c. 60-Hz frame rate, a color subcarrier frequency of 4.43 MHz, and 626 scanning lines
   d. 50-Hz frame rate, a color subcarrier frequency of 4.43 MHz, and 626 scanning lines

9. The SECAM broadcast system was developed in:
   a. Japan
   b. the former Eastern Bloc countries
   c. France
   d. England
   e. the United States

10. True or False   The PAL and SECAM systems are directly compatible.

11. What is the Digital HDTV Grand Alliance?

12. Advantages offered by the HDTV system are:
   a. _____;
   b. _____;
   c. _____.

13. The HDTV system has _____ horizontal scan lines and a vertical refresh rate of _____.

14. Compared to the viewing area for an NTSC system, the viewing area for an HDTV system is:
   a. larger
   b. smaller
   c. the same

15. True or False   The HDTV system requires no changes in transmitter or receiver technology.

16. The aspect ratio for the HDTV system describes the ratio of _____ to _____ and is:
   a. 4:3
   b. 16:9
   c. 17.5
   d. 4:2

17. True or False   The HDTV 20-megabit broadband transmission system will allow the simultaneous broadcast of five channels.

18. Along with the entertainment industry, what other industries may have uses for the HDTV system?

19. In the HDTV system, the luminance channel has a bandwidth of:
   a. 4.2 MHz
   b. 15 MHz
   c. 10 MHz
   d. 30 MHz

   Each of the chrominance channels has a bandwidth of:
   a. 4.2 MHz
   b. 15 MHz
   c. 10 MHz
   d. 30 Mhz

20. True or False   The HDTV system offers both interlaced and noninterlaced scanning.
21. True or False   The NTSC and HDTV systems use the same sync signal.
22. The HDTV system has direct compatibility with:
    a. computing systems
    b. the NTSC system
    c. the PAL system
    d. the MUSE system
    e. the SECAM system
23. What advantage is given through 4:2:0 chrominance sampling in HDTV?
24. Why is the use of a packetized data transport structure important in the HDTV system?
25. Each data packet in the HDTV system has:
    a. a 4-byte header and 184-byte information payload
    b. a 4-byte header and 188-byte information payload
    c. an 8-byte header and 184-byte information payload
    d. an 8-byte header and 188-byte information payload
26. In the HDTV system, what is the advantage of vestigial transmission?
27. Briefly describe the MPEG-2 compression standard and the advantages of its use in the HDTV system.
28. Briefly describe the Dolby AC-3 compression standard and its advantages.
29. What is the purpose of video RAM in a computer system?
30. True or False   EGA is a current video adapter card standard.
31. True or False   VGA and super VGA are digital standards.
32. What is the difference between super VGA and VESA?
33. RGB video:
    a. has separate signals for red, blue, and green
    b. mixes the red, blue, and green signals together
    c. uses the I and Q signal method
    d. is a TTL standard
34. The CRT used in the Zenith ZCM-1492 has a
    a. 25% increase in contrast ratio
    b. 70% increase in contrast ratio
    c. 75% increase in contrast ratio
    d. 50% increase in contrast ratio
35. In Figure 13-8, capacitors C5101, C5102, and C5103:
    a. bypass the red, blue, and green analog color signals to ground
    b. couple the red, blue, and green analog color signals to the video output transistors
    c. couple the red, blue, and green analog color signals to the video inputs of IC5101
36. An automatic brightness limiter circuit within IC5101 controls the _____
    of the:
    a. AGC signals
    b. video signals
    c. TTL signals
    d. IC5101 bias
37. How is dc restoration accomplished in the ZCM-1492?

38. In Figure 13-9, C3009, R3007, and R3009 form:
   a. an integrator
   b. a coupling network
   c. a waveshaping network
   d. a filter network

39. What is the purpose of the damping diode in Figure 13-9?

40. In Figure 13-10, the oscillator stage operates from a time constant determined by:
   a. Q2116
   b. IC2101 and Q2101
   c. C2101 and R2102
   d. Q2116 and Q2101

41. Referring again to Figure 13-10, describe the operation of the vertical driver transistor and the Q2107 and Q2108 transistor pair.

42. Describe the operation of the anode voltage regulator.

43. True or False   The FTM monitor requires a more sophisticated pincushion circuit than that in a conventional television.

44. True or False   Computer monitors use many of the same circuit designs as in television receivers.

45. With a symptom of smeared colors and the correct signals at the CRT, suspect the:
   a. monitor input cable
   b. CRT
   c. video amplifiers
   d. cable connector

46. When checking IC401 and finding a good waveform at pins 12, 15, and 18 along with a distorted picture, suspect the:
   a. CRT
   b. video amplifier transistors
   c. IC

47. An incorrect waveform at pin 16 of U303 and no vertical deflection points toward a problem with:
   a. U302
   b. Q307, Q308, Q314, and U302
   c. Q308
   d. U303

48. A good horizontal output transistor and no horizontal deflection suggest:
   a. a defective yoke
   b. a defective horizontal processor IC
   c. a defective flyback transformer
   d. a bad connection at the input cable
   e. an intermittent horizontal output transistor

49. An important waveform to check if a pincushion problem appears in the ZCM-1492 is the:
   a. waveform at pins 3 and 5 of the horizontal processor
   b. vertical ramp waveform at the base of Q7505
   c. pincushion waveform at the emitter of Q7505
   d. sync signal waveform at connector 5A9

50. In the first Service Call, the voltage at the collector of Q6R2 was zero. Why?

51. In the second Service Call, how did the technician check the vertical deflection IC in the TOC monitor?

PART 5

Focusing on the Customer

Now that you have progressed through a wide range of technical information, the last section of this text provides an important reminder. Your success as a technician hinges on more than just technical knowledge. Rather, your success depends on a mix of technical expertise and consideration for the person who owns the product. While technical expertise may come naturally to you, focusing on the customer may be a more difficult assignment. However, the ability both to solve technical problems and to consider customer needs is what will ultimately spell long-term success for you.

Focusing on the customer can take many different forms. As chapter fourteen points out, honesty, ethics, and integrity are key factors in gaining the trust of your customer. Such trust also develops through timely repairs and thorough, understandable explanations of problem causes and solutions. In addition, consideration of customer needs requires consistent billing practices, the efficient use of tools and test equipment, and the continuous upgrading of your knowledge.

If we were to rate all the skills presented in this text, customer consideration would probably carry the most weight. For while technology changes continuously and challenges our ability to adapt, the customer needs remain fairly constant. Every customer expects to receive fair and honest treatment while benefiting from the knowledge that you offer. In everything that you do, you should strive not to disappoint the customer.

CHAPTER
14

Basic Approaches to Service and Professional Development

OBJECTIVES Upon completion of this chapter, you should be able to:

1. Explain the need for good customer, co-worker, and employee relationships.
2. Describe methods for continually upgrading your knowledge.
3. Discuss proper business accounting methods.
4. Explain the need for insurance policies and licenses.
5. Discuss the factors in choosing a parts supplier.
6. Purchase schematics.

KEY TERMS

accounts payable ledger

accounts receivable ledger

casualty policy

Certified Electronics Technician Program

CET examination

Electronics Industries Association/Consumer Electronics Group (EIA/CEG)

ethics

fire and flood basic policy

loss of revenue/loss of use insurance

margin pricing

mark-up pricing

operating statement

performance bond

profit and loss statement

renter's insurance

trust

worker's compensation insurance

zoning regulation

INTRODUCTION

For good reasons, each of the previous chapters has emphasized the technical side of the service industry. In this chapter, however, we will look at some of most troublesome issues for individuals who specialize in a technical profession: management and service. As these issues are explored, you will find that part of managing your career and providing good service for your customers, co-workers, and your company involves:

- Making solid business decisions
- Applying ethics throughout all your actions

- Achieving versatility
- Making wise decisions about choosing parts and service vendors
- Maintaining a focus on your customer needs

It is easy to see that those issues are interrelated, but it may be difficult to understand how they can affect your understanding of technologies. After all, most of us would rather work with "something technical" rather than "something that involves people." The reason for this preference involves the relative ease in measuring and solving technical problems compared to the often intangible challenges posed by customers and peers.

Thus, many factors besides technical knowledge affect your chances for success in the workplace. Two of these factors are especially worth remembering. We live in an extremely competitive world where change occurs almost daily. Individuals who can adjust to changing technologies and changing expectations have a much greater chance of success. Second, and more importantly, people skills should be an indispensable component of your tool kit. All your accomplishments are a direct reflection of the way you treat your customers, co-workers, and employees.

14-1 FINDING SUCCESS THROUGH PEOPLE SKILLS

Several years ago, a service organization began compiling a list of reasons for customer satisfaction and reasons for customer dissatisfaction. As Table 14-1 shows, the reasons for satisfaction fall under the broad categories of caring about the quality of the repair, reliability, understandable advice, responsiveness, and professionalism. The reasons for dissatisfaction include lack of timeliness, unreliable repairs and service, poor planning and inadequate service, unreasonable charges, and lack of communication.

The difference between an excellent job and a poor job is often only a few minutes—and results in return business or referrals through satisfied customers. Indeed, the few minutes required to do an excellent job often eliminate the necessity for redoing the job. Building quality into service also means, if possible, completing the job the first time. Aside from building quality into the service process from the start, many of the initiatives outlined in this section improve communication with customers. Because of this, time-consuming return calls decrease, and customer complaints drop substantially.

Building Ethical Relationships

Ethical relationships hinge on the responsible exercise of trust, professionalism, and equity. In short, the exercise of **trust** means that we act in behalf of our customers, co-workers, and employees, and that we act professionally at all times. When we apply the concept of equity, we find ourselves applying the "golden rule" to our business practices and our relationships with everyone. That is, we treat all co-workers, employees, and customers as we would like to be treated.

Table 14-1   Reasons for Customer Satisfaction and Dissatisfaction

Satisfaction	Dissatisfaction
Professional treatment (customer was treated well)	Repair took too much time
Knowledgeable service personnel	Never did repair the problem
Service representatives communicated and listened well	Poor planning and inadequate service
Product was returned promptly	Service representative disagreed with the customer
Service center maintained direct contacts with the manufacturer of the product	Service representative never returned or called back
Service representative maintained all appointments and arrived on time	Service center had to order parts. The parts order took a long time.
Some part of the service was outstanding	Unreasonable charges for the quality of service
Service representative demonstrated the proper operation of the product	Service center never returned the product

Doing business in an ethical way requires that we maintain consistency, firmness, and honesty whenever we make decisions and provide services. From an overall perspective, the merging of personal and business **ethics** encourages employees to become more sensitive to the implications of their actions. Ethical technicians never replace working parts in an effort to increase profit, and never tell a customer that the parts distributor is the reason for the delay in repairing the product. The following example illustrates how the lack of ethics can adversely affect an individual and an organization.

Several years ago, an employee of a television repair company arrived at a customer site and, after several minutes, realized that he did not have the parts needed to complete the repair. Already behind schedule and not wanting to tell his superior that he had not packed the proper parts, he wrapped aluminum foil around the blown OEM fuse. The technician had previously endured a verbal reprimand for the same type of oversight and did intend to return the next day with the proper fuse. Unfortunately, as often happens, the employee did not go back to the site because of a heavy schedule and personal plans. As you probably already suspect, the original intermittent power supply problem resurfaced, and started a fire in the customer's house. Not surprisingly, the customer sued the television repair company and the employee lost his job. The publicity surrounding the case eventually forced the company to close its doors.

The unreasonable fear of an employee turned an easily preventable situation into a horror story. The story illustrates the negative consequences of unethical actions. The dishonest act of replacing a fuse with a piece of foil showed complete disregard for the technician's customer and employer. Although none of us may make the crucial error of overriding a fuse, we should remember that ethics covers a large number of situations.

Building and Maintaining Communication

Poor communication within an organization detracts from the best uses of resources, may cause duplication of efforts, and may send departments in opposite directions. If the service and sales departments of an electronic retailer do not communicate about major service problems and types of equipment sold, the company may find that customer dissatisfaction coincides with the increasing sales of an electronic product that fails regularly. Yet, because the departments have different priorities and few things in common, departmental managers and the company owners may never perceive the connection between the dissatisfaction and the product.

Looking at communication from another angle, not returning to complete a task involves a fundamental lack of communication that can ruin a business reputation. Many of us have the best intentions when we put off the completion of a job until the next day. Unfortunately, we may not be able to return because of other work, or perhaps going back is simply inconvenient. Rather than communicate with customers and inform them about the delay, we decide that they will simply understand. When a service technician does not return a call or communicate properly with a customer, the lack of communication evolves into an insult and has a long-term affect.

Tracking Service Requests Some repair services implement a scheduling system, customer-contact tracking, and troubleshooting checklists. Whether customer contacts to a repair center occur through phone messages, electronic mail, or face-to-face meetings, an office staff-member translates the customer's contact into a written, numbered work order, which becomes the first stage of the process. With the work order in hand, the office manager places the task into the office work schedule and assigns it to a particular service technician. After scheduling the task, the manager informs the customer about the scheduled day and time, the assignment of the task to the staff member; and that all times are tentative. The last, qualifying statement accounts for unforeseen delays that are a part of the service business.

All this accomplishes several things. Translating the customer's call into a written work order creates a paper trail for the manager and the staff member. After completing the task, staff members ask the customer to sign the work order. For the manager, staff member, and—most importantly—the customer, this signifies the completion of the work order. Placing the task into the office work schedule allows the manager and staff to see the daily progress. If the daily schedule is disrupted by jobs that take longer than anticipated, the service manager has the opportunity to call the customer and reschedule the repair.

Using Management and Scheduling Software Several companies produce service center management software packages that automate almost every part of a business. With the software in place, one can enter any information provided by a customer and link that information to a manufacturer and model number. The resulting data files can then become part of a larger troubleshooting tips file, and can provide an instant reference to a particular customer.

In addition to data entry and address control, the software management packages may also offer a method for tracking and scheduling tasks. Within the tracking information, the system will show job information; the current status of the job; a listing of parts ordered for the particular job; and a schedule of service performed by all technicians. An inventory management program shows which in-stock parts sell, and whether prices and quantities have changed. The data accumulated through the tracking and inventory modules then becomes part of the information written to the customer invoice.

Most service management software packages also generate invoices and progress reports. Combined with operating statements, and accounts payable and receivable statements, those reports can provide an instant method for observing daily business operations. From a business perspective, the use of the service management software also frees the technician's time because of the automatic generation of forms and invoices.

Talking with the Customer   Conversations with the customer are the first steps towards successfully completing a service task. While performing preliminary tests on the product, ask if the customer can list any symptoms that occurred before the actual failure. With modern televisions, simply knowing that the picture disappeared before the sound can ease the problem solving. In addition, always inquire about the history of product. The customer may disclose that the symptoms have appeared earlier; that another technician had attempted to repair the product; or that the receiver had experienced other problems.

After you have completed the repair, ask the customer to observe as you test every possible function of the receiver. As the next section shows, those customer-related tests should cover raster and picture quality, and the ability of the receiver to tune across all channels. This type of communication reassures the customer that the problem is solved and that no other problems have surfaced. After running through the performance tests, ask the customer to agree about the quality of the service and whether any other repairs are needed. The final step in this communications process involves making a follow-up call to the customer one to two weeks after completing the repair.

The Repair Checklist

Building quality into the repair process and satisfying customer needs can be accomplished by establishing routines. When you or your employees begin a repair task, start the task by writing a repair checklist. The progress of a repair can be interrupted by other service calls, the need for parts, or visitors. A checklist allows you to record your progress on a given job through every one of its stages. At the end of the repair, provide a copy of the checklist for your customer. The checklist assures the customer that you have taken a formal, sequential approach to the repair, and establishes a repair record for each item.

The following troubleshooting checklist establishes a routine for service personnel. Often, difficult-to-

diagnose problems seem to send service personnel in circles. Utilized during the repair process, the checklist allows technicians to keep track of their activities. For the customer, the checklist also provides a sense of organization and progress.

☑ PROGRESS CHECK

You have completed Objective 1. Explain the need for good customer, co-worker, and employee relationships.

14-2   GAINING VERSATILITY

Competition drives all of us to become more versatile and to build our skills to the next level. Advances in electronics have pushed service companies to expand from servicing only televisions to becoming total repair centers for VCRs, satellite receiving systems, and computer systems. In addition, the need to sustain competitiveness also pushes us to gain more knowledge about business and organizational practices.

Whether it is technical training or organizational training, only a formalized approach will provide the best results. Attempting to save money and time by providing informal, "in-house" training can sometimes produce good results. However, a formalized, external approach to training will gain more attention and produce greater long-term benefits. Rather than look at the monetary and time costs for training as avoidable and unnecessary spending, consider the money and time as a wise investment.

Methods for Continuing Your Education

Depending on the amount of available funds and time, technicians have a wide variety of opportunities for upgrading their technical and business knowledge. While traditional vocational colleges and universities offer classes in topics ranging from automotive electronics to business management, correspondence and Internet-based courses offer easily-accessed alternatives. The fees for those classes and courses vary with the institution. In addition, most manufacturers and trade associations offer seminars, books, and video tapes. If the service center is part of a manufacturer's warranty support network, the cost of maintaining membership includes a subscription to training and service materials, and a listing of seminars.

The product services department of the **Electronics Industries Association/Consumer Electronics Group**, or EIA/CEG, also offers regional workshops, videotapes, and manuals. Although generic, the manuals provide in-depth coverage of many electronics topics such as basic electronics theory, the use of modern test equipment, and digital/microprocessor applications.

Troubleshooting Checklist

Your Electronic Service Company

Date _____   Job Number _____   Customer Name _____

Safety Checks

1. Leakage _____   2. High Voltage _____   3. X-ray Protection _____

Raster Checks

1. Purity _____   2. Gray Scale _____

Picture Checks

1. Proper Brightness _____   4. Proper Tint _____   7. Vertical Height _____

2. Proper Contrast _____   5. AGC _____   8. Horizontal Width _____

3. Proper Color _____   6. Vertical Linearity _____

Channel Selection Over All Bands

Parts Replaced

_____

_____

I have performed the above tests while working with model _____.

If you require further assistance, please call me at this number (_____) _____.

Technician Signature

_____

The workshops and courses last five days and are offered throughout the year.

The addresses for several trade associations are:

Electronic Industries Association
Consumer Electronics Group
2001 Pennsylvania Avenue, N.W.
Washington, DC 20006-1813
(202) 457-4919

National Association of Retail Dealers of America
10 East 22nd Street
Lombard, IL 60148
(312) 953-8950

National Electronic Servicing Dealers Association
2708 W. Berry Street
Fort Worth, TX 76109
(817) 921-9062

Other sources for training and information are a wide selection of books that cover all facets of electronics, computer-based training courses, seminars produced by test-equipment manufacturers, articles in technical magazines, and the World Wide Web. Many manufacturers, such as Philips and Matsushita, offer computer-based training for sale to all technicians. In addition, several sites on the Web provide technical tips, partial schematics, and detailed information about new products.

The Certified Electronics Technician Program   Developed in 1965, the **Certified Electronics Technician program** gives customers a method for selecting the most qualified service personnel; guarantees the technical competence of a technician; and largely eliminates low-quality service within the technical industry. Any individual with either four years experience as an electronics technician or technical schooling may complete an application form and take the Certified Electronics Technician, or CET, examination. Individuals with less than four years experience or schooling may take the associate level CET examination, which contains the basic portions of the full examination. Once the associate CET has worked in the technical industry for four years, he or she may receive full certification by taking the regular CET examination. A regular examination contains the basic portions as well as special examinations for specific areas of electronics. The examinations are given quarterly at locations around the United States.

CET examinations contain a mix of theoretical and practical questions such as those shown in the following sample questions. The tests cover topics such as electronic devices, circuit theory, television operation, and troubleshooting theory. CET examination study guides and test applications are available from:

International Society of Certified Electronic
   Technicians
2708 West Berry Street
Fort Worth, TX 76109
(817) 921-9101

CET Examination Sample Questions

1. Match the following frequencies with the application:
   - **a.** 455 kHz   1. SCA
   - **b.** 10.7 MHz   2. AM Radio
   - **c.** 67 MHz   3. FM Radio
   - **d.** 4.5 MHz   4. Television Video I-F
   - **e.** 19 kHz   5. FM Stereo Pilot

2. If an amplifier has a gain of 3 dB and an input of 20 mW, the output is:
   - **a.** 60 mW
   - **b.** 40 mW
   - **c.** 23 mW
   - **d.** None of the above

3. A byte is understood to be:
   - **a.** 4 binary bits
   - **b.** 8 binary bits
   - **c.** 32 binary bits
   - **d.** none of the above
   - **e.** all of the above

4. Retrace blanking in an oscilloscope is:
   - **a.** not used
   - **b.** used to blank the vertical retrace
   - **c.** applied to the Z-axis
   - **d.** used only on recurrent sweep

Presenting Yourself

Obviously, one of the important reasons for gaining knowledge and accumulating skills is finding employment and then enjoying your work. Technicians just entering the job market or considering making a job change should complement their technical skills with a well-written cover letter and resume, interviewing skills, and the ability to cooperate with other employees. As you enter the job market, you will also find that employers are searching for individuals who can offer a wide variety of skills. Many studies and surveys have confirmed that knowledgeable workers who have technical skills, the ability to communicate, and confidence to work as team players are in the highest demand.

Cover Letters   One of the most important tools that you can own as you enter the job market is the cover letter. A poorly written cover letter may turn off an employer to the extent that they never bother to read your resume. In brief, the cover letter works as a personal sales tool, a business letter, a letter of introduction, and as a method for generating interest in you and your skills. A cover letter creates an impression as you attempt to obtain a job interview by highlighting skills, accomplishments, positions, and titles that relate to the desired job.

Generally, applicants should limit the length of cover letters to between one and three paragraphs. Cover letters should emphasize these important points:

- Introduction of the applicant to the potential employer.
- An explanation of why the writer is responding to the advertisement.
- A brief description of skills and experience.

In addition, a well-written cover letter should include information that shows that you have knowledge about the company and its direction, and should indicate how you can help the company to achieve its goals. The style of your cover letter expresses your personality, attitude, and your ability to communicate in a written format.

Always proofread your cover letter for typographical errors and grammatical mistakes. Many times, an advertisement will not list a contact person. When this occurs, use "To Whom It May Concern" or "Dear Sir or Madam" as a salutation. A good cover letter is always simple—never use gold borders, fancy stationery, or heavy-duty paper. On the other hand, never send a photocopied letter or resume. As with all correspondence, always make sure that the envelope containing your cover letter and resume has the correct postage and a return address.

Resumes   All professional resume writers provide the following essential tip for anyone writing a resume: keep it simple. Most warn that attention-getting devices such as color, computer graphics, or slogans will prompt most employers to discard the resume before reading it. With the company taking itself and the interview process seriously, you would not want to appear frivolous through an attempt to express personality.

Many employers receive 100 to 400 resumes for every position advertised. To speed the process of selecting applicants, employers quickly glance over resumes, looking for key job titles or skills that match their job requirements. Always prominently display headings such as skills, experience, and education so that an interviewer can pick out the essential pieces of the resume within four to ten seconds. When employers read resumes, they look for problem-solving and strong communication skills. Although this seems generic, every type of job requires these skills. Rather than use "problem solving" and "strong communications skills" as a way of describing your talents and experience, concentrate on specific characteristics that may fall under those headings. Given this concentration, you may be able to list desirable experiences such as project management, customer service, and departmental coordination.

Along with emphasizing your positive characteristics, always address potentially negative attributes as well. However, the challenge with any resume is to downplay those negatives so that each turns into a positive. For example, if you have moved from one short-term job to another, list only the years of your employment rather than the months. Another tactic is listing a variety of odd jobs under one heading.

Interviews More than likely, the least favorite activity for anyone is preparing for and going through a job interview. Many employers have stepped away from the traditional interview process to team interviews or situational interviews. A team interview may involve three or more supervisors and a rapid-fire sequence of questions. Rather than attempting to intimidate the applicant, this process shows the ability of the applicant to think and respond while under pressure. The situational interview places the applicant in a position of applying skills shown on a resume to actual problems or projects. Again, the employer is attempting to find out how the applicant responds to stressful situations.

When interviewing for a position, dress professionally and conservatively. In addition, remember to:

- respond to the objectives and needs of both you and the interviewer
- listen intently to the interviewer
- retain control throughout the interview

When responding to objectives, emphasize your strengths rather than your weaknesses. Instead of stating that you have no experience with a particular circuit or set of test equipment, show that you have experience with a variety of circuits or equipment and that your experience provides flexibility and versatility. Retaining control is defined as being professional, focused, personable, and friendly. Above all, be thorough yet brief when talking, and show enthusiasm for your work.

Common interview question include:

- Tell me about yourself.
- What are your salary expectations?

- Why did you leave your last job?
- How did you get along with your employer and co-workers?
- What are your strengths and weaknesses?
- What are your career goals?
- What is your ideal position?
- Is there anything that you would like to say in closing?

Self-Employment?

In the not too distant past, may electronics technicians were self-employed. Even though the nationwide trend towards self-employment has increased, the number of self-employed who remain in business after four years has decreased. This decrease occurs because of the responsibilities associated with owning a business, the need for a long-term investment of finances and time, and hidden costs of ownership that include insurance and maintenance.

Despite these drawbacks, many individuals continue to consider self-employment as the best option. Most forget that the difference between success and failure rests on a realistic concept of obtaining finances, selling services, and purchasing materials. In addition, many people forget that starting a small business has a dramatic effect on everything else involved with a normal life. A self-employed technician, for example, cannot claim to be his or her own boss. Instead, that person reports directly to his or her customers. Because of this, the financial risks become greater, and a technician may work longer, harder hours. Any mistakes or customer relation problems have a direct link to the financial health of the business.

Several key points are work remembering if you are considering starting your own business:

- Be very realistic about your abilities to manage, communicate, and provide services.
- Prepare a financial safety net that will support a business for at least two years. Generally, a new business will need this or a longer period of time before it changes income into profit.
- Always turn negatives into positives.
- Refrain from working at home. Unless you are very disciplined, distractions at home may reduce your ability to conduct business.
- Along with investing your own money, work with financial institutions to support the needs of the business. The start-up of a small business requires money for purchasing equipment, phone lines, letterhead, business cards, advertising, and other essentials.

Getting Ahead on the Job

Although being visible in the workplace may seem to have the same meaning as bragging, it should carry a different definition as you work through your career. Gaining visibility is a positive and necessary form of communication. Given the busy pace of the workplace, managers rarely have time to find out about everything involved with your duties. As a result, employees sometimes need to remind management about the professional skills that they contribute to the workplace.

Success also depends on your ability to cooperate with other employees who have different interests and talents. In most cases, every employee offers something that the employer needs. When employees with different skills have interoffice battles to determine who has the superior talents, both the employees and the employer lose. The optimal situation occurs when those employees find a common base for cooperation.

☑ PROGRESS CHECK

You have completed Objective 2. Describe methods for continually upgrading your knowledge.

14-3   SUCCEEDING THROUGH BUSINESS SKILLS

Unfortunately, many businesses take a "short-term" attitude toward improvement. Identifying short-term improvements does not require as much effort as that needed for the long-term approach, so this attitude is understandable. Short-term improvements in the service arena may involve goals such as improving timeliness or adding more hours to the work week. Although working harder to achieve short-term goals may seem to be correct, all of us need to consider methods that make our work efficient and effective. Improvement in that context includes innovation in processes, services, skill utilization, and products. Instead of working harder, the organization works smarter.

For a one-person electronics repair business, constant improvement may include scheduling jobs by location. If we couple this idea with the notion of attempting to carry needed parts in the service vehicle, several benefits result. For example, eliminating even small travel expenses should translate into more profits. Even if you can eliminate only a fraction of the "road time," the savings will accumulate.

The other advantages may be more intangible than the more obvious benefits of money and time. As you concentrate on scheduling jobs by location and on carrying a supply of the most needed parts, your service

operation will become more efficient. One consequence of this efficiency may be a reduction in your physical parts inventory. Any reliance on having the necessary parts on hand will increase awareness about the amount of money needed to maintain the inventory. As such, tracking the number and type of parts used for a given time period may become a necessity.

Working with Business Records and Forms

Many individuals engaged in a technical trade shy away from learning about business practices for a number of reasons, including lack of knowledge or experience and lack of interest. However, if we look at the need to gain business sense from an overall perspective, accumulating business skills may be a wise career move. Anyone wanting to progress through today's organizations must have knowledge of not only technical issues but also business practices. Those practices include working with income statements, negotiating with vendors, and handling customer and personnel matters.

Before considering wise business practices, however, you need to build a foundation by learning the following business terms:

- Accounts payable ledger—Statement of amounts owed *by* the business, and the length of time for the obligations.

- Accounts receivable ledger—Statement of amounts owed *to* the business, and the length of time the obligations have existed.

- Assets—Items, such as money or physical property, that hold a value for a business. Assets include parts, test equipment, and vehicles.

- Balance sheet—Statement of net worth at a given point in time. Most business operations produce a balance statement once a month.

- Collateral—Property pledged to a lender to guarantee a loan repayment.

- Daily ledger—A daily listing of all transactions, including gross sales, expenses, and obligations.

- Double-entry—A bookkeeping system in which each transaction is entered on the ledger once as a debit and once as a credit.

- Financial statement—A listing of the assets and liabilities of a person or company.

- Fiscal year—A period of twelve consecutive months established by an individual or company for accounting purposes. A fiscal year may or may not coincide with the calendar year.

- Good will—An intangible asset of a business or professional firm, which consists of relationships with current customers and clients. Usually, business value is assigned to good will during a business negotiation.

- Gross income—Sales volume less the direct costs related to the sale.

- Operating statement—A statement showing the income and expenses of a business for a specific period of time. The income statement lists all sources of revenue (operating income, interest received, and rent) and all expenses (costs of materials, wages, and taxes) with the purpose of showing a profit or a loss. The income statement is also known as a statement of earnings or a profit and loss statement.

- Independent contractor—An individual or business who works for others but exercises control over work hours and working conditions.

- Liabilities—The financial obligations of a business. Liabilities include outstanding bills to be paid, taxes, a loan against a business, and lease obligations.

- Net income—The dollar amount remaining from a given sales period after all expenses and costs for conducting business for the same period are deducted.

- Service charge—An amount assessed against an unpaid account and added to the balance.

- Turn-key expense—The total cost of beginning the operation for each day.

Tables 14-2, 14-3, 14-4, and 14-5 provide examples of an operating statement, a balance sheet, an accounts receivable ledger, and an accounts payable ledger. While you may not need to know everything about accounting to work with technologies, the statements map out the transactions that envelope the technical trades. In addition, they provide an instant picture of profits and losses.

While the operating, or **profit and loss statement** shows whether an operation made or lost money, a balance sheet shows the status of a business at any given point in time. The **operating statement** considers sales volume minus the costs to make those sales develop into a gross profit. Subtracting operating expenses from the gross profit yields the net profit. Because the salary of an employee is not part of the cost of the sale, the only expenses associated with a service call are the materials needed to complete the work.

An **accounts payable ledger** lists items purchased while operating as a business, and the charges levied by

Table 14-2   Operating Statement

Your Electronics Service Business Current Date	
Sales	$ 7,800.00
Cost of Goods Sold	2,300.00
Parts	
Mileage	
Overtime	
Gross Profit	$ 5,500.00
Operating Expenses	
Total Salaries	$ 1,200.00
Rent	800.00
Telephone	100.00
Utilities	150.00
Supplies	1,000.00
Depreciation	550.00
Loan Interest	50.00
Miscellaneous	25.00
Subtotal of Operating Expenses	$ 3,875.00
Net Profit	$ 1,625.00

Table 14-3 Balance Sheet

Your Electronic Service Business
Current Date

Assets		Liabilities	
Cash on Hand	$ 800.00	Accounts	$ 700.00
Cash in Bank	10,500.00	Notes Payable	2,500.00
Accounts Receivable	1,500.00	Taxes Payable	600.00
Inventory	6,750.00	Unpaid Wages	900.00
Equipment	22,250.00		
Insurance	500.00		
Total	42,300.00		$ 4,700.00

Total Equity in Business
$ 37,600.00

Table 14-4 Accounts Payable Ledger

Your Electronic Service Business
Current Date

Name	Total	Current	30 Days	60 Days	120 Days
Quasar	$ 500.00	$ 500.00			
Zenith	250.00	250.00			
Radio Supply	1,750.00	1,000.00	750.00		
Telephone	100.00	100.00			
Utilities	150.00	150.00			
Advertising	1,250.00		1,250.00		
	$ 4,000.00	$ 2,000.00	$ 2,000.00		

Table 14-5 Accounts Receivable Ledger

Your Electronic Service Business
Current Date

Name	Total	Current	30 Days	60 Days	90 Days
Brown, John	$ 300.00	$ 200.00	$ 100.00		
City of Tulsa	2,000.00	1,500.00	500.00		
Green, James	500.00		250.00	150.00	100.00
Jones, Robert	100.00	100.00			
	$ 2,900.00	$ 1,800.00	$ 850.00	$ 150.00	$ 100.00

the supplier. Although most suppliers offer attractive credit terms, always verify the terms of agreement before conducting the transaction. In addition, remember that most suppliers offer discounts for payments made before the agreement date. Even though the standard 2 percent discount may seem insignificant at the time, the accumulation of savings shown at the end of the fiscal year will show as profit.

In our society, offering credit terms to customers has become a norm. The **accounts receivable ledger** records the names of all customers holding charge accounts, and provides a record of the length of credit. While the accounts receivable ledger may seem like a listing of assets, many of the charges shown on it will be uncollectable. When establishing customer credit, always maintain an accurate accounts receivable ledger; customers purchasing items or service on credit terms are taking advantage of your assets.

When working with accounts payable and receivable ledgers, note that both forms have an "aging" characteristic. Along with showing the current status of a payment or an account, the ledgers also show a 30-day, 60-day, and 90-day status report. The aging of the account can provide the first clue about a payment problem. An important difference between accounts payable and accounts receivable ledgers is that an old account receivable has less value than an old account payable. That is, as the account receivable ages, you have a lesser chance of collecting the funds. However, if your account payable with a supplier ages, the supplier begins to add monthly services fees.

Using Ratios to Improve Profitability Wise management practices always involve making good financial decisions. In the consumer electronics business, those decisions cover issues such as materials (parts and service literature); advertising; travel expenses; utility and telephone payments; insurance; licenses; rent; daily office and shop expenses; taxes; and wages. Each of these items either adds to or subtracts from the total gross income.

The regular analysis of business statements can become the basis for sound business decisions. An interpretation of the operating statement can quickly show the development of certain trends and whether the trends fit the objectives of the business. Also, the ongoing analysis of those trends allows everyone to stay on top of changing situations.

To obtain an almost instant picture of expenses versus income, we can apply each item against the gross income as a ratio. For example, the income gained from selling a part becomes part of a ratio that contributes to the gross income, while the expense of buying parts becomes part of another ratio. The use of ratios is particularly handy when comparing different years or time periods against one another, and when looking for trends that may affect your purchasing habits. Table 14-6 shows a balance statement that includes a list of ratios.

When comparing ratios from one year to another, some items—such as annual licensing or subscription fees—will have a higher expense ratio during one month than others. This is because the actual payment for the license or subscription shows during only one month. When considering ratios, however, the actual working cost and worth of the license or subscription spreads over twelve months.

When examining trends, look for ratios that lie outside of the normal ranges, and then determine the reason for the abnormality. In some cases, a dramatic increase in gross income may cause a smaller than normal rent ratio; while the gross income increased, the

Table 14-6   Balance Statement with Ratio Listings

Your Electronics Service Company					
Profit and Loss Statement for October					
Current Date					**Statement Period**
Account #	Account	Month to Date		Year to Date	
		Amount	Ratio	Amount	Ratio
100	Gross Receipts	$ 4,000.00	100.00%	$ 19,895.00	100%
110	Materials	2,700.00	67.00%	10,800.00	54%
120	Telephone	400.00	10.00%	1,200.00	06%
130	Utilities	600.00	15.00%	2,000.00	11%

rent remained the same. In others, the ratio for service income may decrease if your business goals shift toward integrating sales with service.

After discovering the reasons for an abnormal expense ratio, make immediate decisions that will change the ratio and increase profitability. With the sales versus service situation, you may want to find out why the emphasis on sales did not increase income, and if improvements are needed in the service area of the business. Your analysis of the ratio may disclose that technicians lack the test equipment needed to repair modern microprocessor-based products, or that they lack an adequate supply of parts. Or it may show that the costs of sending modules or units to a service center outweigh the costs of completing the work internally.

PROGRESS CHECK

You have completed Objective 3. Discuss proper business accounting methods.

Setting Prices for Materials

When we examined operating statements, we found that the prices charged for components used during a repair play a large role in the ability of a service business to make a profit. Although the costs shown on a parts invoice may reflect the distributor's price for the item, the actual cost of a component includes:

- the cost of shipping the part from the distributor
- the opportunity cost of identifying the correct part, checking the part against the manufacturer's listed number, and ordering the part
- any minimum invoicing cost added by the distributor
- the cost of stocking the part
- the cost of unpacking and handling the part
- costs accumulated because of shrinkage due to misplaced parts
- the costs of replacing the same part (such as a fuse) several times while locating the problem source

Although it may seem logical to factor all these costs into the price of parts because of the need for profitability, that strategy involves the risk of over-charging and being seen as "too high-priced" by your clientele.

The key intangible factor in the pricing equation is the cost of service. For example, stocking a large quantity of many different parts may ensure that a customer will receive quick service. On the other hand, the cost of storing parts on the shelf equates to approximately 25 percent of the value of the total inventory of parts per year. As a result, achieving a reputation for quick

service has an attached cost. The money used to buy the parts remains locked into place until the part sells.

To find the correct balance between profit and fair prices, we can apply two basic pricing strategies, mark-up or margin pricing. **Mark-up pricing** is the least attractive for the business and the most attractive for the customer. With mark-up, a business simply adds a mark-up percentage to 1 as a decimal amount, and then multiplies that value by the net cost of the part. For example, to apply a 40 percent mark-up against a part having a net cost of $12.00, we would multiply 1.4 times $12.00:

$12.00 net cost for part × 1.4 = $16.00 retail price

As a result, the business would accumulate almost five dollars in profit.

With **margin pricing**, we divide the $12.00 cost price by one minus 40 percent or:

$12.00 net cost for part/(1 − 0.40) =
$12.00 net cost for part/0.60 = $7.20

Then we add this value to the cost, for a retail price of $19.20. Using the margin method to establish the retail price provides a greater gross margin, and also sets a limit so that the margin will never exceed 100 percent.

For the small-ticket items, the use of the standard margin percentage provides the balance between profit and fair pricing. When using either method, the net cost always includes shipping and handling costs. However, when we sell large-ticket items such as a CRT, the margin method yields an excessive retail cost. For those items, we apply a smaller margin of profit and still produce more income because of the higher net cost for the item.

Setting Prices for Labor

Despite the emphasis on finding the correct balance for materials costs, almost 80 percent of the profit for a service enterprise arrives in the form of labor. Because of this, a service-oriented business must become proficient at selling service. Obtaining this proficiency requires that both the service center and the customer recognize the value of the service.

In some instances, it is appropriate for technicians to advise customers to buy a new product rather than having the product repaired. Sometimes this free, honest advice establishes a relationship between the customer and the technician that may result in long-term benefits. In other cases, though, the advice to repair rather than replace is good for both the customer and the service center. Here, the customer finds value through information provided by the technician and through the opinion given by an expert.

Aside from actually repairing the product, setting a price for labor is probably the most difficult part of the repair task. As with the charges for parts, many different costs affect labor rates. Those expenses include:

- hiring and paying employees who have the proper skills
- payments for medical insurance
- retirement plans
- management costs
- support costs (secretarial, parts ordering)
- training
- taxes (including social security, Medicare, worker's compensation, and unemployment)
- the cost of reservicing a product for no charge

With all this, the total cost for labor exceeds the wages paid to an employee by almost two times.

The best method for achieving a balance between profitability and fairness for the customer is defining distinct categories for different product types, and then establishing the cost for servicing each product. The categories should fall under headings such as minor labor required, intermediate labor required, and major labor required, and should break the product types into the smallest possible groups. As you can see by looking at Table 14-7, this type of order allows the technician to explain charges to the customer, and matches labor costs to specific items. Generally, service organizations set labor rates on an hourly basis.

When presenting your rates to the customer, always remember that your labor and expertise carries a value. Many technicians have a tendency to back away from charging full labor rates, and may charge for only a small portion of the time needed to repair the item. Having a detailed labor price list defines the guidelines for both the customer and the technician.

Establishing Service Contracts

Many manufacturing and service organizations offer service contracts for sale to customers who have purchased an electronic product. When a customer purchases a service contract, the service company guarantees to perform for no charge any periodic service that falls within the limits of the contract. Most professional service contracts include all parts and service labor charges needed to return a product to its original operating condition following a normal failure, and have separate clauses to compensate for the aging of components. In brief, service problems covered by a contract may be related to either a component failure or the aging of the product.

Customers purchase service contracts for protection and the convenience given by the contract. A service contract offers a fixed expense. Whenever a product covered by a service contract fails, the customer can avoid unexpected charges. It also offers convenience, because the customer does not have to make decisions about which service organization to call. Retail stores prefer service contracts because the point-of-contact for problems changes from the store to the service center.

The benefits for service organizations include the addition of another revenue source and renewed customer loyalty. With service-related income fluctuating, selling and maintaining service contracts provides a closer relationship with the customer and an assured revenue stream. In turn, a service contract makes most consumers feel more comfortable about purchasing the product. That level of comfort covers both the sale and the service of the product.

Insuring Your Property

Insurance policies can cover two facets of a technician's well-being. From the perspective of property, a good insurance policy will cover test equipment, hand tools, the business location and premises, and inventory owned by the technician. This type of policy also protects the livelihood of the technician by providing a cushion against potential disasters. From another perspective of liability, an insurance policy can protect a company or individual technicians from the results of legal actions that may stem from neglect or a wrongful act.

Table 14-7   Setting Labor Rates

Labor Extended Price List—Hourly Charges			
	Minor	Intermediate	Major
Projection Television	$ 45.00	$ 60.00	$ 80.00
13"–20" Television	$ 20.00	$ 35.00	$ 65.00
25"–27" Television	$ 25.00	$ 40.00	$ 70.00

Insurance policies that cover property and livelihood fall under the categories of:

- **Renter's insurance**—covers office equipment, supplies, furniture, and fixtures located within a rented space.

- **Loss of revenue/loss of use insurance**—provides an instant revenue stream for a business that suffers a total loss due to a major catastrophe such as a flood or fire.

- **Fire and flood basic policies**—protect against property losses caused by a fire or flood.

- **Casualty policies**—protect a business from liabilities caused by an on-site accident experienced by either a customer or an employee. A casualty policy for employees usually covers shop-, site-, and travel-related accidents.

- **Worker's compensation insurance**—guarantees compensation to an employee experiencing an on-the-job accident.

Because the value and protection of insurance policies vary from company to company, carefully study the individual parts of each policy and discuss any issues with your insurance agent before signing the policy. Although insurance policies provide protection, the cost of multiple policies sometimes drives companies away from buying the correct policy. In addition, check with local government offices about any regulations concerning insurance, and with manufacturers about insurance qualifications for in-warranty service centers. The combination of cost, necessity, and regulations may affect your insurance policy choices.

Licenses and Regulations

Both large and small service operations should be aware of the license requirements for any type of service offered. Whether considering an operating license or a zoning permit, the business always has the obligation to know about all laws and regulations. In almost all cases, local government offices have jurisdiction over local areas and may require specific licenses. Generally, state and federal departments have minimal requirements. Because of this, license requirements may vary from region to region or from city to city.

Along with license restrictions, local taxes and operation permit regulations also vary. Most communities have local sales taxes that apply only within a city limits. In addition, most communities have some type of **zoning regulation** that limits certain types of industries to different areas of the community, or prevents the establishment of business operations in a residential area. Other communities may require a service business to have a **performance bond** posted for anyone selling a service. Written through an insurance or

bonding company, the performance bond guarantees that a company will provide satisfactory service for all customers.

☑ PROGRESS CHECK

You have completed Objective 4. Explain the need for insurance policies and licenses.

14-4   WORKING WITH PARTS SUPPLIERS

When profit margins become a priority, we often look for sources for cheaper parts. However, we do not want to experience the uncomfortable situation of returning to a customer's house to replace a previously replaced component. As you found during the discussion of accounting methods, most service centers do not charge for the labor time involved with reservicing a product. Because of this factor and the very real need to preserve customer relations, choosing a reputable parts supplier is as important as correctly diagnosing a fault.

Often, only original equipment manufacturer, or OEM, parts meet the tolerance specifications of an electronic product. In chapter twelve, you saw that most manufacturers will shade the portions of a schematic where original parts are required. While most of those portions lie within the low- and high-voltage supply circuits, and the overvoltage and overcurrent protection circuits, many signal circuits will function only with an OEM part.

Even if repairing the electronic device does not require an OEM component, pay close attention to component tolerances. As you know, temperature and frequency can cause component breakdown. Although thousands of transistors, integrated circuits, and diodes exist, each has a range of specifications that match the device to certain applications. Even if shown as a recommended replacement, a generic component offered by a parts supplier may not fit the criteria of the application.

Setting a specific standard for the parts that you use during the repair process also becomes a matter of establishing long-term vendor-buyer relationships. In the service business, this is a sometimes difficult proposition. Some manufacturers and vendors have either gone out of business or have downsized. Warehouses sometimes drop product lines that have a poor sales record. All this requires a careful study of when and how parts and supplies are used, and the realization that the needed part may be available from only one source.

Aside from using quality as a measure of the parts supplier, also take a close look at the vendor's shipping and invoicing policies. The vendor should show the number of shipping locations and how often orders are completed and shipped. In addition, a parts supplier will usually show the amount of elapsed time between receiving and shipping the order, whether they add shipping surcharges, and if they offer several shipping options. Most—but not all—vendors provide a warranty for failed parts and some type of return policy.

When working with parts suppliers, always inquire about payment options such as open accounts and credit. Because of the constant changes in the electronics industry, many vendors will not ship a component unless the order has a minimum dollar amount. Others may attach special charges for handling or for research to the invoice. Having answers to all these questions will ensure that the relationship between you and the vendor will prosper and that you can relay the correct information to your customer.

Sources for replacement parts include manufacturers, national warehouses, independent regional parts suppliers, regional distributors for manufacturers, local electronics stores, or the World Wide Web. Most manufacturers publish lists of authorized parts suppliers, or list the addresses and phone numbers of regional distributors. While the manufacturers may also recommend independent regional parts suppliers, telephone directories for larger cities also usually list the suppliers. Other corporations offer discounted prices on obsolete and hard-to-get electronic parts.

☑ PROGRESS CHECK

You have completed Objective 5. Discuss the factors in choosing a parts supplier.

Locating Schematics and Other Information

Although the quality and amount of information may vary, all manufacturers have schematics and service manuals available for sale. However, some manufacturers may limit the sale of those materials to authorized service centers. While most regional independent parts distributors carry schematics and other reference guides, technicians may obtain detailed schematics and manuals from Eagan Technical Services, Howard W. Sams and Company, and Schematic Solutions, Inc., at the following addresses:

Eagan Technical Services, Inc.
1380 Corporate Center Curve
Suite 107
Eagan, MN 55121
(612) 688-0098

Howard W. Sams & Company
2647 Waterfront Parkway East Drive
Indianapolis, IN 46214
(317) 298-5400

Schematic Solutions, Inc.
11120 Wurzback Road, Suite 206
San Antonio, TX 78230
(512) 696-0404

☑ PROGRESS CHECK

You have completed Objective 6. Purchase schematics.

SUMMARY

Chapter fourteen introduced you to concepts designed to broaden your professional skills. You learned to consider methods for improving interpersonal relationships with customers, co-workers, and employees. Within this context, you found that ethical values are the foundation for professionalism. In addition, you learned about the importance of communication skills in preventing problems from occurring. Those skills ranged from utilizing checklists to simply talking with the customer.

The chapter then presented methods for becoming a more versatile professional. Those methods include formal college and university courses, seminars, trade associations, and the Certified Electronics Technician program. The chapter also provided information about resumes, interviewing, and self-employment issues.

Next you learned the importance of basic business skills, and the usefulness of business forms and reports as organizational and analytical tools. Each of the forms shown in the chapter is designed to complement decision-making. The chapter also examined pricing structures for parts and labor. You found that a balance must exist between achieving an acceptable profit margin and offering fair prices. The business overview concluded by defining service contracts, insurance policies, licenses, and regulations. Finally, the chapter covered issues in choosing a parts supplier and finding service and schematic information. With the information you have acquired, you should have a basic understanding of how purchasing decisions affect the quality of your work and the perception of your customers.

REVIEW QUESTIONS

1. Ethical relationships can be defined within terms such as trust, professionalism, and equity. Briefly explain each of those concepts.

2. Considering the customer and the technician, name three benefits gained by using written work orders.

3. True or False   Service management software allows the data entry of customer contact information and a method for tracking and scheduling tasks.

4. When first talking with the customer about a service problem, you can determine

    _____.

5. What does the use of a repair checklist accomplish?

6. Name three methods for maintaining and upgrading your knowledge.

7. What is the purpose of the CET examination?

8. What is an OEM component?

9. Name three sources for purchasing replacement parts.

10. True or False   You should always define distinct categories for different product types and establish the cost for servicing each product.

11. The reasons for having insurance policies are _____,

    _____, and _____.

12. A renter's insurance policy covers _____.

13. Name three benefits of the sales of professional service contracts.

14. True or False   License regulations are established at the state and federal levels.

15. True or False   An operating statement considers sales volume minus the costs to make those sales develop into a gross profit.

16. True or False   A balance sheet shows the status of a business only at the end of the fiscal year.

17. Anyone wanting to progress through modern service organizations must have knowledge about

    _____ and _____ issues.

18. What is the difference between accounts payable and accounts receivable ledgers?

19. The use of ratios can _____.

20. What unseen factors can increase the cost of a component?

21. What is the difference between margin and mark-up pricing?

22. What are the characteristics of a well-written cover letter?

23. What are the characteristics of a well-written resume?

APPENDIX

Schematics

One of the key purposes for this text is to provide "real-world" instruction for students. With this in mind, the appendix offers complete schematics of circuits studied in chapters seven, nine, and thirteen. As you read through the discussions about the circuits and components, take time to carefully consider each schematic drawing. Familiarity with the schematics will enhance your understanding of the information presented in the text and will improve your troubleshooting ability.

SCHEMATIC 1　**Figure 7-3**　**Schematic Diagram of a Cable-Ready Tuner** **430**

SCHEMATIC 2　**Figure 9-23**　**Schematic Diagram of an MTS/SAP System with dbx Noise Reduction** **432**

SCHEMATIC 3　**Figure 13-11**　**Schematic of the ZCM-1492—East-West Generator/Output Circuit** **434**

SCHEMATIC 4　**Figure 13-12**　**Schematic of the ZCM-1492—North-South Generator/Output Circuit** **436**

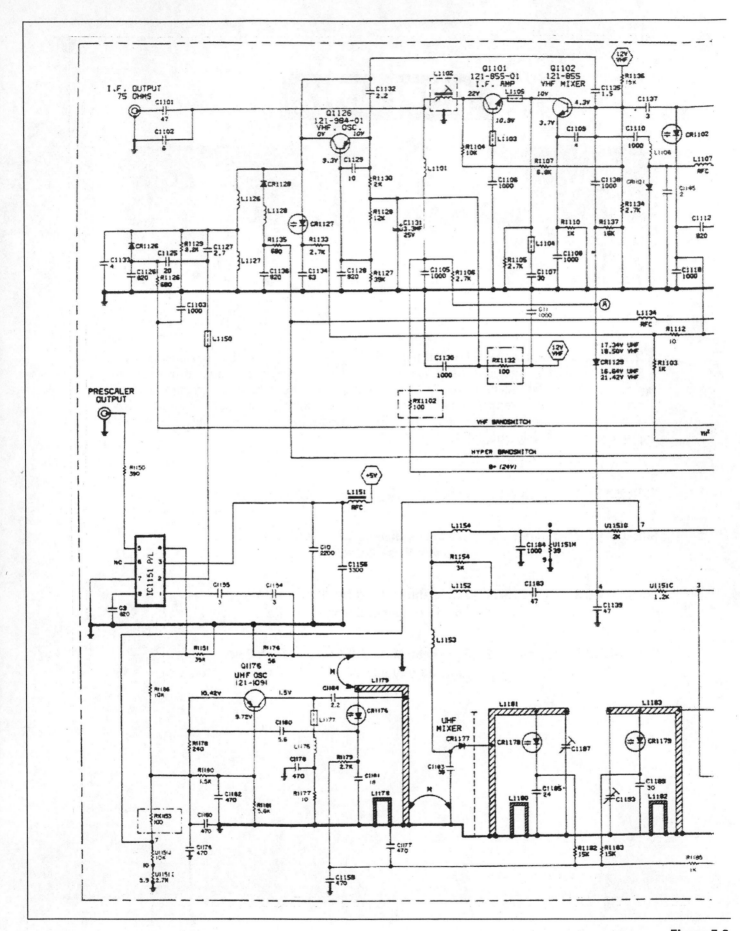

Figure 7-3

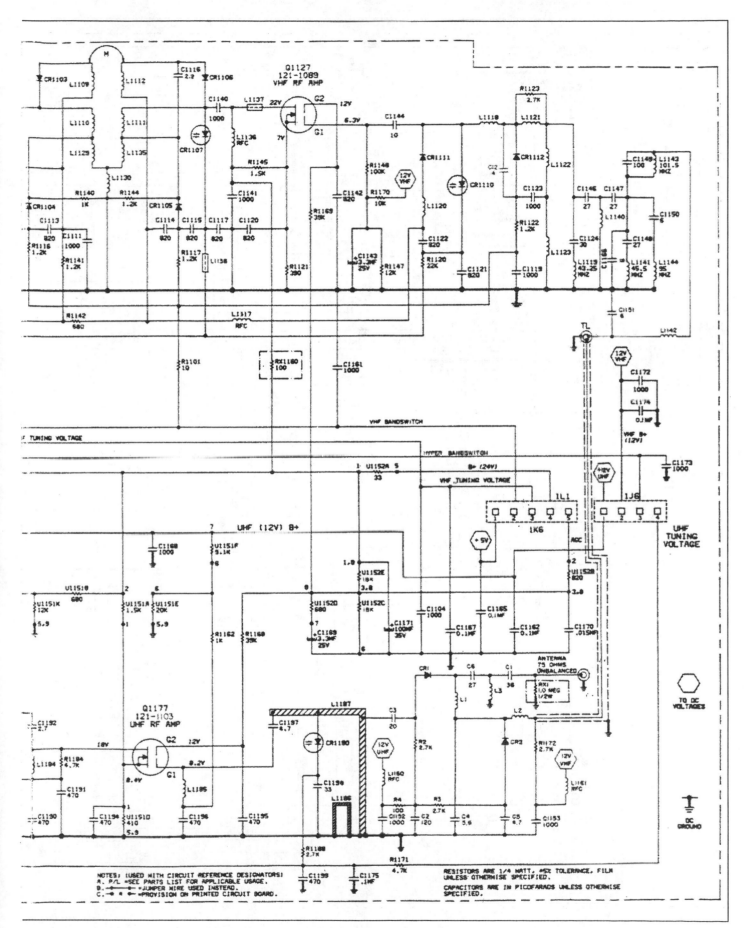

Schematic Diagram of a Cable-Ready Tuner (*Courtesy of Philips Electronics N.V.*)

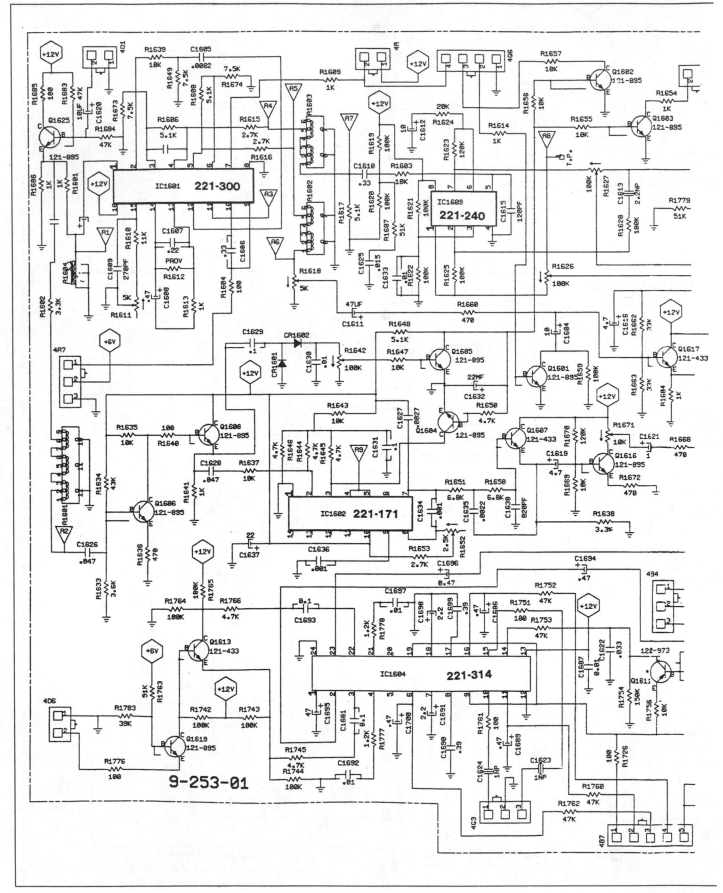

9-253-01

Figure 9-23

432

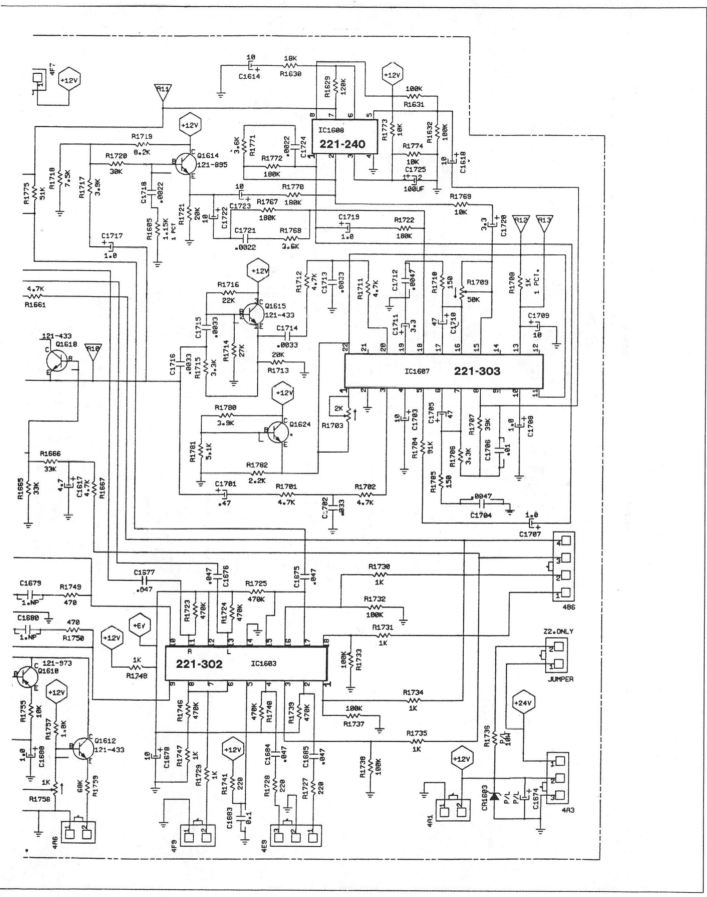

Schematic Diagram of an MTS/SAP System with dbx Noise Reduction (*Courtesy of Zenith Data Systems, Zenith Electronics, and Rauland*)

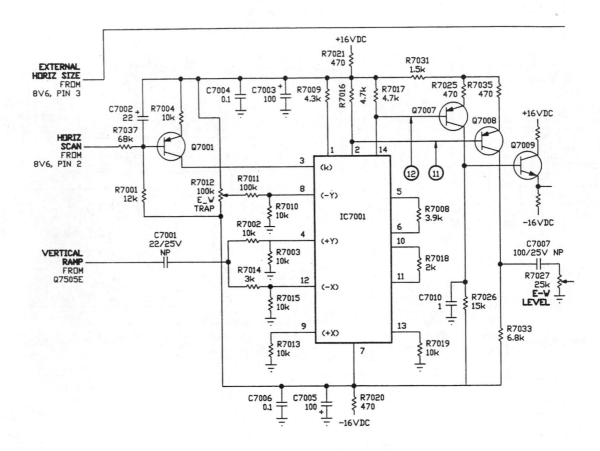

Figure 13-11

434

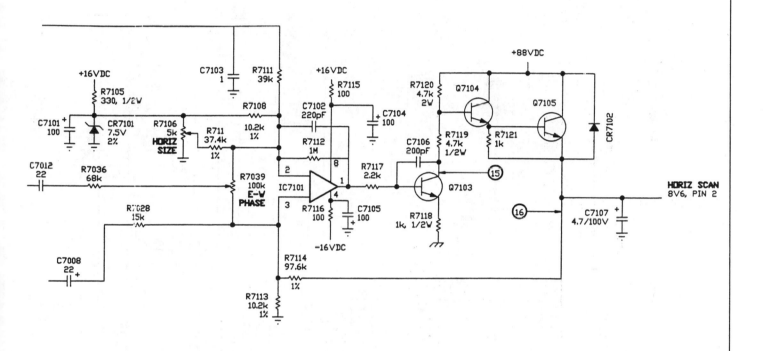

Schematic of the ZCM-1492—East-West Generator/Output Circuit (*Courtesy of Zenith Data Systems, Zenith Electronics, and Rauland*)

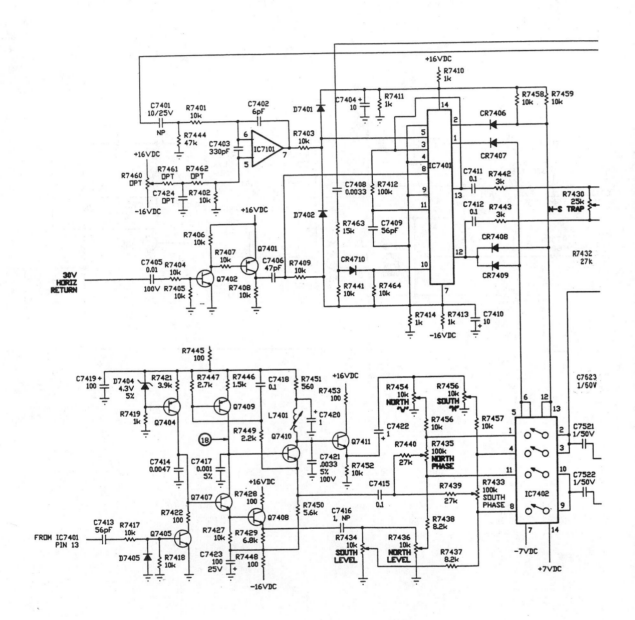

Figure 13-12

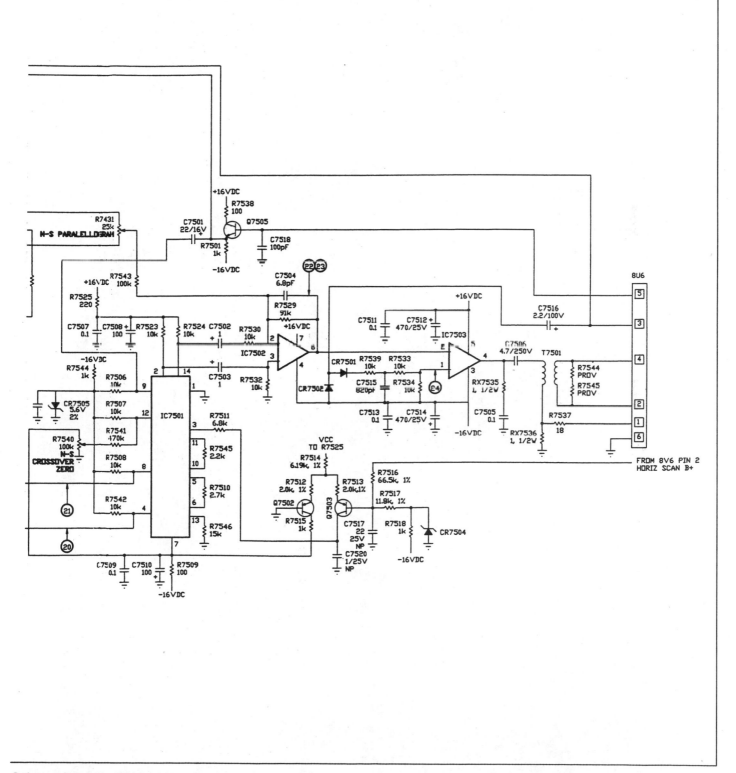

Schematic of the ZCM-1492—North-South Generator/Output Circuit (*Courtesy of Zenith Data Systems, Zenith Electronics, and Rauland*)

GLOSSARY

A

1125/60 standard The resolution standard for HDTV of 1125 scanning lines and a picture refresh rate of 60 Hz.

Absorption trap A wave trap that absorbs energy at a resonant frequency. An absorption trap is a parallel resonant circuit that is inductively coupled to a tank circuit.

Accounts payable ledger A statement of amounts owed by the business, and the length of time for the obligations.

Accounts receivable ledger A statement of amounts owed to the business, and the length of time the obligation has existed.

Adjacent channel A television channel that has a frequency next to the frequency of the desired channel.

Adjacent channel trap In a television tuner, a trap that rejects any 39.75-MHz or 47.25-MHz signals for any selected channel.

Aluminum screen A thin layer of aluminum on the inside surface of the phosphor screen of a CRT. It is transparent to the electron gun but continues to reflect light from the screen.

Amplification An increase in the voltage, current, or power gain of an output signal.

Amplified AGC A circuit that amplifies the AGC voltage for the R-F amplifier of a tuner.

Amplitude modulation (AM) A method of inserting information into a signal in which the amplitude of the carrier changes while its frequency remains constant.

Analog-to-digital conversion (ADC) circuit A circuit that converts an analog signal to a digital number that corresponds to the value of the analog signal.

AND function A Boolean function that says that an output is a 1 only if all the inputs are 1.

Anti-hunt circuit Slows the build-up of the correct voltage in horizontal oscillator circuits, and allows the dc control voltage to shift the oscillator frequency without overcorrecting.

Aperture One of approximately 400,000 holes in a CRT's shadow mask.

Aquadag A conductive coating in a CRT that contacts the screen and provides a path for the electrons to follow from the screen back to the power supply.

Aspect ratio The ratio of television picture width to picture height.

Astable multivibrator A multivibrator that locks into a logic state when triggered by an external signal.

Asynchronous counter Receives input pulses at any time and operates with applications where the inputs may occur at any time.

Attenuation The opposite of gain or a loss in signal strength. With attenuation, the output signal from an electronic circuit has a lower amplitude or signal level than the input signal.

Automatic brightness-control circuit A circuit in a television receiver that monitors the existing light in the room housing the receiver and automatically adjusts both the brightness and the contrast so that the picture will have the correct settings for the current environment.

Automatic chroma control A circuit in a television receiver that provides an even control of gain by monitoring the strength of the received signal, controlling the dc bias of the first chroma amplifier, and controlling the level of color saturation.

Automatic degaussing circuit A circuit that removes the residual effects of a magnetic field on the CRT.

Automatic fine-tuning (AFT) circuit When a frequency change occurs in a tuner oscillator, this circuit generates and feeds either a proportional positive or a negative dc correction voltage back to the tuner local oscillator.

Automatic frequency and phase control A phase-actuated feedback system that consists of a chroma phase detector, a low-pass filter, and a voltage-controlled oscillator. It supplies the proper phase signal to the demodulators and determines the phase of the oscillator signal.

Automatic gain control (AGC) circuit A circuit that controls the gain of a signal amplifier.

Automatic sharpness circuit A circuit consisting of an adaptive feedback loop that uses a high-pass filter and detector to determine the amount of high-frequency content in the picture.

Automatic tint control A circuit that automatically corrects any flesh-tone problems that may occur in a received picture by phase-shifting the incoming chroma signal by a small percentage.

B

Bandpass filter Attenuates any frequency outside of established band limits.

Bandwidth A frequency range over which the amplifier has relatively constant gain.

Base In a CRT, contains connectors that allow the attachment of external electronics to the electrodes.

Bias The no-signal dc operating voltage or current between two of the elements of a semiconductor device.

Binary-coded decimal A counting system that uses 4 bits to represent each digit of a decimal number.

Binary numbering system Allows only two values, 0 and 1.

Bistable multivibrator A multivibrator configured so that an input pulse will cause the gates to move from one logic state to the other; a flip-flop.

Bit A binary position in a binary number system.

Black level A dc reference component in the luminance signal.

Blanking pulse In the composite video signal, a pulse with a greater amplitude than any of the picture signals that rises above the designated blanking level.

Boolean algebra A mathematics system designed to test logical statements and show if the result of those statements is true or false.

Bridged-T trap A wave trap that blocks all frequencies that fall within its bandwidth. Also called a notch trap.

Brightness control Allows the customer to change the bias of the CRT and adjust the picture from black to white display.

Brightness-limiter circuit A circuit that monitors changes in the beam current and changes the bias of the CRT to prevent the change from occurring.

Brightness reference level Provides a basis for adjusting the brightness controls. It may range from the signal level that corresponds with a maximum white raster, the signal level that produces a black raster, or a level that corresponds with a definite shade of gray.

Buffer An integrated circuit that establishes isolation, fault-protection, the simultaneous distribution of gate output signals, or the accommodation of different voltage levels throughout a digital system. It also reduces the electrical load on the output of an IC while providing a known load, and provides additional electrical drive power on signal lines.

Burst amplifier The second color-processing stage, which selects and amplifies the color burst signal from the chroma subcarrier signal.

Burst gate See burst amplifier.

Burst separator See burst amplifier.

Byte A group of 8 bits.

C

Cable-ready tuner A tuner that selectively tunes to the channels included in both the VHF and UHF bands as well as the cable mid-, super-, and hyperband channels.

Capacitance meter Checks the value of capacitance.

Cascode mixer Provides good mixer stability at high frequencies, high input impedance, and low overall capacitance.

Casualty policy Insurance policy that protects a business from liabilities caused by an on-site accident experienced by either a customer or an employee.

Cathode A conducting element in a CRT.

Certified Electronics Technician program A program that administers tests for the certification of electronic technicians.

CET Examination An examination offered through the Certified Electronics Technician program that covers topics such as electronic devices, circuit theory, television operation, and troubleshooting theory.

Charge-coupled device A semiconductor that operates like a capacitor, storing a given amount of charge. In television applications, a CCD stores video information.

Chroma phase detector A phase-locked loop that controls the frequency and phase of a locally generated reference signal so that it matches the frequency and phase of the subcarrier.

Chrominance The signal containing the color and tint information for a received color picture.

Clamping circuit Used in luminance signal-processing circuits to hold the sync tips of the composite video signal to a fixed voltage level.

CMOS logic A type of digital logic integrated circuit.

Color bandpass amplifier An amplifier stage that separates the color subcarrier and burst signals from the video signal, and amplifies the 3.58-MHz subcarrier sidebands.

Color bar/Cross-hatch generator Used for television receiver set-up adjustments, it supplies color bars for the primary and complementary colors.

Color burst signal A signal that is a reference for the control of hue.

Color control A customer control used for adjusting the saturation of the received picture.

Color-difference amplifiers Amplifies the R − Y, G − Y, and B − Y signals to the level needed to drive the CRT.

Color-difference drive matrixing A matrixing technique in which the R − Y, B − Y, and G − Y signals travel to the CRT control grids, while the luminance signal travels to the red, blue, and green CRT cathodes. Matrixing of the red, blue, and green colors occurs within the CRT.

Color graphics adapter (CGA) An early video monitor display consisting of sixteen-colors of text or graphics.

Color killer A circuit that interfaces with the last bandpass amplifier and checks for the presence of the color burst signal.

Color oscillator A crystal oscillator tuned to 3.579545 MHz that supplies the reference signal for demodulating the color signal.

Colpitts oscillator An oscillator configured to provide oscillations through regenerative feedback. A Colpitts oscillator produces a 180-degree phase shift.

Comb filter Separates the chroma and luminance components from the composite video signal by delaying the composite video signal one horizontal scan period, or approximately 64 microseconds, and then either adding or subtracting the undelayed composite video signal.

Combinational logic The connecting of gates to provide a more complex logic function.

Comparator A digital circuit that detects the output from registers and compares either the logic states or the polarity of the signal voltages.

Complementary colors A color that produces white light when added to a primary color.

Contrast The difference between the light and dark areas of a scene.

Contrast control A customer control used for controlling the amount of gain at the CRT of a television receiver, and for allowing the adjustment of the picture contrast.

Control grid In a CRT, controls the number of electrons in the beam and has a negative potential with respect to the cathode. Also called grid one.

Corrector lens A lens mounted around the neck of each CRT in a projection television that reduces possible optical irregularities around the edges of the projected image.

Counter A digital circuit that counts input pulses as the pulses arrive and then passes the total to another circuit.

Coupling The connecting of one or more amplifier stages through various circuit configurations.

Critical coupling In transformer coupling, a condition where a frequency response curve broadens until the secondary voltage of a transformer reaches its maximum value.

Cross-over distortion Occurs when neither transistor in a Class B amplifier has any forward bias, and the output voltage of the amplifier circuit equals zero.

Cross-over network A component that splits the audio output signal into separate low- and high-frequency signals.

CRT cut-off Establishes a black screen in a television receiver CRT circuit.

D

Damper diode Prevents unwanted oscillations and provides a high-voltage pulse through the operation of the output stage and the horizontal output transformer.

Darlington stage A special type of emitter follower that uses two transistors to increase the overall current gain and amplifier input impedance.

dc restoration The restoration of a dc reference voltage in the video processing circuits of a television receiver.

Decoder   Converts a combination of digital inputs into another combination of digital circuits.

De-emphasis   Occurs after the modulation of the sound I-F signal; the higher-frequency noise voltages are reduced, and a desired signal-to-noise ratio is established.

Deflection angle   A standard used to measure CRTs; the size of the raster produced on the face of the tube is proportional to the angle of the electron beam deflection and the distance between the tube face and the point of beam deflection.

Degaussing   To remove the induced magnetic field in a CRT and build a local magnetic field to offset the effects of the earth's magnetic field.

Delay elements   Establishes a time delay between the input and output signals.

Delayed AGC   A circuit that senses increased R-F signal levels and reduces the gain of the I-F amplifiers in proportion to the R-F signal level increases.

Delta-gun configuration   A type of configuration of the three electron guns in a CRT in which the guns form an equilateral triangle, or delta.

Demodulation   The recovery of intelligence from a carrier signal.

Demodulator probe   Used with an oscilloscope, it allows the observation of modulated R-F signals by demodulating the signal before its application to the vertical input terminals.

Demultiplexer   Takes signals from a single line and allows the direction of the signals to one of several output lines.

Differential peak detector   Compares the peak voltages of two reference signals and uses the difference to produce a demodulated FM signal.

Differentiator   A circuit at the input to the horizontal oscillator stage that prevents vertical sync pulses from entering the horizontal circuit and triggering the oscillator.

Digital HDTV Grand Alliance   An HDTV design and standards team consisting of AT&T and Zenith Electronics; the General Instrument Corporation and the Massachusetts Institute of Technology; and Philips Consumer Electronics.

Digital storage oscilloscope   Allows the storage of signals into digital memory and the recalling of the signals at a later time. This capability allows simultaneous viewing and comparing of stored and real-time signals.

Digital-to-analog conversion (DAC) circuit   Produces a dc output voltage that corresponds to a binary code, and converts digital properties to analog voltages.

Direct-coupled doubler   A circuit that adds the output voltages given by two half-wave rectifier circuits.

Discriminator   In television I-F circuits, converts the 45.75-MHz ac signal found at the output of the I-F amplifier to a dc control voltage.

Divider   Divides input pulses by an output pulse.

Dolby AC-3   The audio-compression standard for HDTV.

Dot pitch   The distance between the center points of adjacent horizontal pixels on the CRT screen.

Drive   The capability of the amplifier to vary the light output of the tube by changing the bias of the tube over a wide range of conditions.

Dynamic convergence   Adjustments that control the convergence of the electron beams to the outer edges of the CRT screen.

E

Electromagnetic waves   Magnetic and electric fields placed at right angles to each other and at right angles to their direction of travel. The magnetic and electric fields vary continually in intensity and periodically in direction at any given point.

Electron gun   A component in a tricolor CRT that forms and directs a narrow beam of electrons toward the screen, and accelerates them to the high speeds needed to cause light emission on the screen phosphors.

Electronic Industries Association/Consumer Electronics Group (EIA/CEG)   One of the world's leading electronics service and trade associations.

Electronic system   A group of electronic components interconnected to perform a function or group of related functions.

Electrostatic focus   Occurs when the narrow beam of electrons in a CRT travels through two electron lenses.

Encoder   Converts BCD data into a multiple-line data set or the multiple lines into a BCD format.

Envelope   The large, bell-shaped glass bulb of a CRT.

Equalizing pulses   A type of sync pulse that occurs directly before and after the vertical pulses. Equalizing pulses determine the exact location of the scanning lines of a field in relation to the lines found in the preceding field.

Ethics   The maintenance of consistency, firmness, and honesty whenever decisions are made and services are provided.

Evacuation   A process that establishes a vacuum inside the glass envelope of a CRT.

Extended graphics adapter (EGA)   An early video monitor display consisting of sixty-four colors.

F

Feedback   A signal that travels from the output signal to the input of the amplifier.

Feedback bias   The feeding of part of the output signal back to the input as a 180-degree out-of-phase, or degenerative, signal.

Figure of merit   A measure of the efficiency of an amplifier.

Fire and flood basic policy   Insurance policy protects against property losses caused by a fire or flood.

Fixed bias   A type of bias in which a dc source external to the amplifier circuit supplies a very stable bias voltage.

Flip-flops   A cross-connection of two gates that acts like a 1-bit memory element and generally consists of NOT and NAND gates; a bistable multivibrator.

Flyback   A horizontal output transformer.

Focus adjustment   Ensures that the electron beam in a CRT will retain a narrowly defined shape.

Focus grid   A component of an electron gun that combines with grid two to form an electrostatic lens. With a voltage of several kilovolts at the focus grid, the lens combination forces electrons to stay on paths that come to a point at the phosphor-coated screen. Also called grid three.

Forward AGC   Automatic gain control in which the AGC circuit begins with a no-signal bias which produces the maximum gain of the amplifier.

Foster-Seeley discriminator   A type of sound-detector circuit that uses frequency variations in the I-F carrier to cause corresponding phase shifts.

Frequency   The number of waves that passes a point each second, and the rate of polarity change seen in those waves.

Frequency division   Involves producing an output signal that has a fractional relationship, such as one-half, one-third, or one-tenth, to the input signal.

Frequency interleaving   The interlacing of odd and even harmonic components of two different signals to minimize interference between the signals. This allows the transmission of the chrominance signal within the same 6-MHz channel as the luminance and sound signals.

Frequency modulation (FM)   A method of inserting information into a carrier wave in which the frequency of the carrier changes while its amplitude remains constant.

Frequency multiplication   Occurs when 1) a time-varying circuit introduces harmonics at the output along with the fundamental frequency; and 2) a resonant circuit tunes to the desired output frequency.

Frequency response curve   Shows the amount of gain found at given frequencies.

Frequency synthesis   A method of digitally generating a single desired, highly accurate, sinusoidal frequency from the range of a highly stable master reference oscillator.

Front-projection system   A projection television that allows the image to be viewed from the same side of the screen as that used for the projection.

G

Gain   The ratio of input signal voltage, current, or power to the output signal voltage, current, or power of an amplifier.

Gates   An integrated circuit that performs fundamental Boolean logic functions and combinations of the functions. A gate "sees" the logic values at its inputs and produces the output that corresponds with the correct logic function.

Gated AGC   A circuit that compares the video sync tip level with a reference voltage derived from a stage that operates only when a sync signal is transmitted.

Gated pulse   A reference pulse for a gated AGC system.

Glass faceplate   The front of a CRT, which either consists of a laminate of thick glass layers, or has a prestressed steel tension band mounted around it.

Graticule   The grid-like display area of an oscilloscope.

Gray-scale adjustment   Provides a correctly balanced picture through red, blue, and green screen controls.

Grid four   A component of an electron gun that is part of the high-voltage connection.

Grid one   See control grid.

Grid three See focus grid.

Grid two A component of an electron gun that has a positive potential with respect to the cathode, and prevents any interaction between the control grid and anode electric fields.

H

Hercules graphics adapter (HGA) An early video monitor display consisting of monochrome, high-resolution text.

Heterodyne To combine two output frequencies in a mixer circuit.

High-definition television (HDTV) The new standard for television transmission and reception; it doubles the luminance definition and displays four times as many pixels as the older NTSC standard.

Horizontal AFC stage Compares the frequency and phase of feedback pulses taken from the horizontal output stage with horizontal sync pulses arriving from the differentiator.

Horizontal driver Reshapes the rectangular pulse taken from the oscillator output to provide the waveshape needed to produce a horizontal output signal.

Horizontal foldover One portion of the television picture folds back over one edge of the display.

Horizontal hold control Adjusts the free-running frequency of an oscillator.

Horizontal output transistor Part of the horizontal deflection system of a television; develops a sawtooth current through the deflection yoke and produces a high-voltage pulse, which is rectified and used as the accelerating dc voltage for the CRT.

Hue The color of an object, which is represented as an angular measurement.

Hue control Determines the phase of the color burst.

Hyperband channels Range from AA to BBB and cover frequencies from 300 to 776 MHz.

I

In-line configuration A type of configuration of the three electron guns in which the guns are aligned along one horizontal plane whose width equals the diameter of the tube neck.

In-phase signal A color-information-carrying signal used to modulate the color subcarrier. No phase shift occurs with modulation by the in-phase signal.

Integrated high-voltage transformer Supplies the source voltages and high voltages to a television receiver.

Integrator circuit Precedes a vertical oscillator circuit in a television receiver and prevents horizontal sync pulses from entering the vertical oscillator circuit.

Intercarrier method A method of processing I-F signals in which one strip of amplifiers amplifies both the video and sound I-F signals.

Interlaced scanning A process in which electron guns draw only half the horizontal lines with each pass.

Intermediate frequency Frequency with a much lower value than radio frequencies, produced through heterodyning. An intermediate frequency equals the difference between the oscillator frequency and the frequency of the corresponding R-F channel.

Iso-hot chassis Combines cold and hot grounding systems through the connection of one portion of the receiver to earth ground and the other portion to the ac line ground.

Isolation transformer Provides the necessary isolation of the cold chassis ground and the ac power line hot ground through separate, isolated primary and secondary windings.

K

Keyed AGC See gated AGC.

Keyed amplifier See burst amplifier.

Keyed pulse See gated pulse.

Keyed rainbow generators A tool that contains an oscillator tuned to 3.563795 MHz (3.58 MHz minus 15,750 Hz), or one scanning line below the color-carrier frequency.

Kick-start circuit In scan-derived power supplies, supplies a small amount of voltage to the horizontal oscillator after the receiver is turned on.

L

Latch To set desired conditions in a digital circuit.

Light valve An optical valve in a particular type of projection television that modulates a high-intensity, fixed light source.

Line regulation Indicates the amount of change in output voltage that can occur per unit change in input voltage.

Load regulation Indicates the amount of change in output voltage that can occur per unit change in load current.

Low electron gun emission A common CRT fault that results from the aging of the CRT; the symptom is a dim raster or a raster that lacks one of the primary colors.

Low-level RGB matrixing and CRT drive A matrixing technique in which the RGB signals matrix at a level of only a few volts. The CRT drive circuits then amplify the signals to the 100- to 200-Vdc level needed to drive the CRT.

Low-voltage focus Controls the electron beam in a CRT by using a low voltage, ranging from 0 to 400 Vdc, from the low-voltage power supply.

Luminance signal A signal, consisting of proportional units of red, green, and blue voltages, that contains the brightness information for the picture.

Luminescence The absorption of energy and the emission of light without the absorbing material becoming hot enough for incandescence.

M

Magnetic focus Controls the electron beam in a CRT by utilizing the magnetic fields developed through the horizontal and vertical yoke deflection coils placed around the picture tube neck.

Margin pricing A pricing strategy for merchandise and service in which the cost price is divided by 1 minus 40 percent.

Mark-up pricing A pricing strategy for merchandise and service in which a mark-up percentage is added as a decimal amount to 1, and then that value is multiplied by the net cost.

Matrix circuits Combines the demodulated color-difference signals with the luminance signal by adding proportional amounts of input voltages to form new combinations of an output voltage.

Metal-oxide varistor (MOV) Provides protection against severe surges; located in the ac line circuits.

Microprocessor A complex integrated circuit that provides the functions of a full-scale computer and contains registers, memory, and buffers. It executes instructions through a precise method and follows a sequence of instructions.

Midband channels Range from A to I and include frequencies from 120 to 174 MHz.

Miller effect A rule stating that the value of an input capacitance will increase with the gain of the stage.

Mixer stage Converts the selected VHF television channel frequency into the lower intermediate-frequency signal.

Modulation The encoding of a carrier wave with another signal or signals that represent some type of intelligence.

Monaural sound A sound system that has only one output channel for the audio signal.

Monochrome display adapter (MDA) An early video monitor display that allowed only monochrome graphics.

Monostable multivibrator A multivibrator that has one stable condition. When excited by an external signal, it will go to its unstable state and then automatically return to its stable state.

MPEG-2 The video-compression standard for HDTV.

Multichannel sound A sound system that relies on the broadcast of a composite audio signal and the splitting of that signal into left and right sound components.

Multiplexer Allows a user or a circuit to control the combining of signals from logic gates. It receives a large cluster of input lines and allows only the logic state of one of the 1s through to the processor.

Multisync A video display standard that allows the monitor to adjust its refresh rate to the display adapter.

Multivibrator A type of oscillator built around two active devices; it may have square waves or abrupt changes between logic states at the output.

Mutual coupling impedance A property of coupled filters that provides the maximum amount of energy transfer.

N

NAND function Boolean operation that adds a NOT function to the output of an AND function. The output of a NAND is a 0 only if all the inputs are 1; otherwise, the output is 0.

National Television Systems Committee (NTSC) Sets transmission and reception standards for television. That standard utilizes 525 horizontal scanning lines, interlaced scanning, the transmission of separate luminance and chrominance signals, and relies on a 60-Hz frame rate.

Noise   Any type of disturbance other than the desired signal.

Noninterlaced scanning   A process in which every line of information on the display is scanned by the electron gun at each pass across the panel. Also called progressive scanning.

NOR function   A Boolean function that adds a NOT function to the output of an OR function. The output of a NOR is a 0 if any one or more of the inputs is a 1; if all the inputs are 0, then the output is 1.

Notch trap   See bridged-T trap.

NOT function   A Boolean function that states that the output is the opposite of the input—a 0 is converted to a 1 and a 1 to a 0.

Novabeam   Used in some projection television systems, a CRT with a large aperture and low surface reflectance. With this design, a relatively small, metal-backed target screen and magnetic focusing provide the small dot resolution needed to produce a viewable picture.

O

Operating statement   Shows whether an operation made or lost money. Also called a profit and loss statement.

Optoisolator   A combination of an LED and a photodiode; it establishes an isolation barrier between low-voltage secondary outputs and the ac line.

OR function   A Boolean function that may have two or more inputs. The output of an OR is a 1 if any of the inputs are a 1, and a 0 if all the inputs are 0.

Oscillator   An amplifier with a feedback circuit connected to the output of the amplifier. With the feedback circuit feeding back a portion of the output to the input of the amplifier, the amplifier begins to produce its own input.

Oscilloscope   A measuring device that provides a visual representation of electrical signals. It is used to check the display of a waveform for variations in amplitude and time; to determine phase relationships between voltages and currents; or to check the frequency response of a circuit.

Overcoupling   In transformer coupling, a condition where an increase in coupling spreads the response humps of a frequency response curve further apart.

P

Parasitic oscillations   High-frequency oscillation that occurs within an I-F amplifier.

Performance bond   Guarantees that a company will provide satisfactory service for all customers.

Phase   A description of when signal repetitions occur in time.

Phase control regulator   A regulator that uses a silicon-controlled rectifier rather than a series-pass transistor.

Phase-locked loop (PLL)   Consists of a voltage-controlled oscillator, prescaler and divider circuits, a comparator, and a quartz crystal. The PLL receives an input signal and then compares that signal with the feedback of an internal clock signal generated by the VCO.

Phosphorescence   The persistence of a phosphor.

Pincushion   A condition where the raster of a CRT display has a distorted shape.

Pincushion circuits   Establishes the correct shape of the rectangular picture on a CRT display.

Pixel   A unit consisting of three dots: one each of red, blue, and green.

Pre-emphasis   Improves the signal-to-noise ratio of transmitted high frequencies by boosting high frequencies and attenuating low frequencies.

Prescaler   Offers precise division of an input signal by variables preset by a mode selection.

Primary colors   A color that does not result from the mixing of other colors.

Profit and loss statement   See operating statement.

Progressive scanning   See noninterlaced scanning.

Projection television   Projects the picture onto a large viewing screen through an optical assembly, an imaging source, and an electronics system.

Pulse   A fast change from the reference level of a voltage or a current to a temporary level, and then an equally fast change back to the original level.

Pulse-rate modulation   Varies the frequency of the dc pulses.

Pulse-width modulation   Varies the duty cycle of a dc voltage.

Purity magnet   An assembly mounted on the CRT neck consisting of two magnetic rings; it adjusts the three electron beams for the purpose of obtaining pure primary colors.

Q

Quadrature detector   A sound-detector circuit that measures the phase shift across a reactive circuit as the carrier frequency varies.

Quadrature signal   A color-information-carrying signal that modulates the color subcarrier, causing a phase shift.

Quasi-parallel sound method   A method for transferring the sound I-F signal to the sound signal-processing section; it uses the best portions from the intercarrier and split-sound carrier methods.

R

Random-access memory (RAM)   A device that allows changes to occur once information is written to memory; it does not retain its contents if the power is removed.

Ratio detector   A sound-detector circuit that uses diodes to compare the phase relationship between two applied signals.

Read-only memory (ROM)   A device that does not allow changes to occur once information is written to the memory; it retains its contents with or without applied power.

Rear-projection system   A projection television in which the image is viewed from the opposite side of that used for the projection.

Rectification   The conversion of the required ac voltage value to a pulsating dc voltage, which may have either a positive or negative polarity.

Refractive system   A projection television that takes advantage of the magnification produced by convex lenses to reproduce a picture on the viewing screen.

Refresh rate   The rate at which a screen image is redrawn; shows how many frames are scanned per second; and is the vertical scanning rate.

Register   A digital circuit that temporarily stores 2 to 16 bits of information.

Regulation   The maximum change in a regulator output voltage that can occur when the input voltage and load current vary over rated ranges.

Renter's insurance   Insurance policy that covers office equipment, supplies, furniture, and fixtures located within a rented space.

Resonance   Occurs when a specific frequency causes the inductances and capacitances in either a series or parallel ac circuit to exactly oppose one another, with a single particular frequency emerging as the resonant frequency.

Reverse AGC   Decreases the forward bias of the amplifier circuit.

R-F amplifier   Provides both selectivity and amplification while increasing the voltage level of an I-F signal applied to its input.

RGB video   A type of video signal composed of three separate signals for red, blue, and green.

Ripple voltage   The varying voltage across a filter capacitor.

S

Saturation   The degree of white in a color, represented through amplitude measurements.

Sawtooth voltage   Results from the relatively slow charging of a capacitor through a large resistance and then the rapid discharging of the capacitor through a small resistance.

Scan-derived power supply   A derivative of a switched-mode power supply, it operates at the horizontal frequency and supplies much higher voltages and currents; used in modern television receivers.

Scan-velocity modulation   The selective modulation of the constant horizontal scanning velocity of the CRT electron beam.

Schmidt optical system   A projection television that uses a spherical mirror to fold the reproduced image back on itself.

Schmitt trigger   A bistable element built around a multivibrator circuit that senses voltage levels and shapes of signals.

SCR crowbar circuit   Protects a voltage-sensitive load such as the shunt regulator from excessive increases in the dc power supply voltages; it consists of a zener diode, a gate resistor, and an SCR.

Screen grid   See grid two.

Selectivity   A measure of the capability of a tuned section to discriminate against unwanted frequencies.

Separate audio programming   Provides a bilingual alternative to the standard English-language audio reproduction.

Sequential logic   The logic used during the synchronization of the input pulses at the gates. In a

sequential system, a clock oscillator outputs a continuous string of pulses that go to all parts of the system.

Series-feedback regulator   Uses a comparator as an error detector and improves the line and load regulation characteristics seen with traditional series-pass regulators.

Series-pass regulator   Uses a power transistor to regulate the load voltage and a voltage reference circuit to sense any source or load variations.

Series regulator   A regulator in series with the load.

Shadow mask   Properly aimed electron beams strike specific perforations on a thin sheet of metal mounted inside the CRT that keeps the electron beams separate before they contact the phosphor screen.

Sharpness control   Enhances detail by adjusting the picture so that a clearly defined line separates light and dark areas.

Shift register   Performs multiplication and division or converts binary pulses from a serial to a parallel format.

Shunt-feedback regulator   Uses an error detector to control the conduction of the shunt transistor.

Shunt regulator   A regulator in parallel with the load.

Signal   A voltage or current with deliberately induced, time-varying characteristics.

Sine wave Sweep Oscillator   Uses pulse-generating circuits that produce oscillations; it usually takes the form of a Hartley oscillator.

Sound I-F signal   Contains modulated audio information.

Source   A device that develops a voltage or a combination of voltages.

Split-sound method   A method for transferring the sound I-F signal to the sound signal-processing stages that takes the 41.25-MHz sound I-F carrier from the output of the mixer.

Split-supply Class AB amplifier   An amplifier in which the two power supply connections have equal but opposite polarities. By using matched power supplies, each amplifier drops its own supply voltage.

Stage   Each active element and its associated passive elements in an electronic system.

Stagger tuning   An arrangement in which a single-tuned circuit works as the common coupling impedance between transistor amplifier stages.

Start-up power supply   In a scan-derived power supply, supplies a small amount of voltage or

current to the horizontal oscillator so that the oscillator can energize and drive the output stage.

Static convergence   Adjusts the electron beams to the center of the CRT screen.

Stereo sound   A sound system that uses separate sound channels to amplify left and right sound signals.

Storage register   Temporarily stores 2 to 16 bits so that the information represented by the bits is used as a group.

Subcarrier reference system   The part of a luminance/chrominance processor that converts the synchronizing bursts obtained from the burst gate into a continuous carrier with a stable frequency and phase.

Superband channels   Range from J to W and include frequencies from 216 to 300 MHz.

Superheterodyne receiver   A receiver that operates with a fixed intermediate frequency.

Surface acoustic wave (SAW) filter   A transducer found at the beginning of television I-F amplifier circuits; it establishes the intermediate-frequency bandwidth and provides waveshaping.

Switch   A device that opens and closes the path for electrons.

Switched-mode power supply (SMPS)   Consists of a full-wave rectifier circuit connected directly to the line, a high-frequency transformer, a power transistor, and a pulse generator.

Switching element   Causes the logic state found at the output of a device to switch to the opposite condition.

Synchronous counter   A counter that is accompanied by a system clock that counts pulses only if they are synchronized with the system clock.

Sync separator stage   Eliminates the video and blanking signals while amplifying only the horizontal sync, vertical sync, and equalizing pulses.

Sync signal   A signal that controls the oscillation frequencies of the sawtooth current generators.

Synthesized sound   A sound system that uses a monaural sound signal to fool the brain into thinking that a left and right channels exist.

T

Thermal runaway   A condition in which a higher than normal internal current causes a transistor to overheat, which, in turn, breaks down the internal resistance of the device.

Tint See hue.

Tint control See hue control.

Transistor tester Used for the in-circuit or out-of-circuit testing of transistors, diodes, and field-effect transistors.

Transistor-transistor logic (TTL) video An early video signal input standard for video display monitors in which two bipolar transistors worked together to control the logic level of a signal.

Trickle-start circuit In scan-derived power supplies, uses a large resistor connected in series with either a diode or regulator transistor and in between the unregulated positive voltage obtained at the rectifier circuit and the horizontal oscillator.

Triggered sweep An oscilloscope test that displays complicated low- or high-frequency waveforms in the same position.

Trinitron A tube produced by the Sony Corporation that features three cathodes and one electron gun that contains all three electrodes.

Trust Acting in behalf of one's customers, co-workers, and employees.

Tuner oscillator Produces a sine-wave frequency with a much higher level than the incoming R-F signal.

U

UHF tuner Tunes to the channels within the ultra high frequency band (470 to 890 MHz).

Ultor The anode button located at the top center of the CRT.

Unipotential focus See low-voltage focus.

V

Varactor diode A P-N junction diode specifically designed to have a variable capacitance when its reverse-bias voltage varies.

Variac A variable ac transformer; it allows the varying of the ac line voltage.

Vector Defines both quantity and direction. When working with electrical quantities, phase is the angular measure while amplitude corresponds with the size of the vector.

Vertical blanking pulse Part of the vertical sync pulse, the vertical blanking pulse—a long duration rectangular wave—carries sync and equalizing pulses on top of the waveform.

Vertical foldover The television picture overlaps at either the top or the bottom and distorts another part of the picture.

Vertical linearity Adjusts the linearity of the sawtooth ramp waveform; it is set to eliminate the overcrowding or overspreading of the scanning lines at either the top or the bottom of the picture.

Vertical sweep oscillator Includes a pulse generator, discharge transistor, and a sawtooth-forming RC circuit.

Vertical sync pulse A pulse consisting of six serrated pulses that occur at the end of the field scan.

Vestigial transmission The suppression through filtering of a major portion of the lower video sideband, which allows the entire sound and picture signal to fit within the 6-MHz bandwidth.

VHF tuner Tunes to any of the 6-MHz-wide channels within the very high frequency bands (54 to 88 MHz and 174 to 216 MHz).

Video amplifier Increases gain, attenuates specific signals, shapes other signals, clamps voltages to a specified level, and ensures the proper amplification of the video signal.

Video detector Demodulates the video I-F signal and recovers the composite video signal.

Video graphics array (VGA) An analog signal-transmission standard for video display monitors that offers a maximum resolution of 640×480 pixels.

Video I-F signal Contains picture information.

Voltage-controlled oscillator (VC) An oscillator with a square-wave output frequency that is inversely proportional to the input voltage.

Voltage divider A voltage divider provides more than one dc output voltage from the same power supply.

Voltage-divider bias Two resistors form a voltage-divider circuit within the supply of a transistor circuit.

Voltage multiplier Provides a dc output that is the multiple of the peak input voltage to the power supply circuit.

W

Wavelength The distance from any given point or condition in one electromagnetic wave to the corresponding point of the next wave.

Wave trap A circuit that rejects, absorbs, or provides a path to ground for a particular signal.

Worker's compensation insurance   Guarantees compensation to an employee experiencing an on-the-job accident.

X

XOR function   A Boolean function that states that if either input, but not both, is a 1, the output will be a 1; if both inputs are a 1 or both are a 0, then the output will be 0.

Z

Zero bias   Establishes a white screen on a CRT.

Zoning regulation   Limits certain types of industries to different areas of the community, or prevents the establishment of business operations in a residential area.

INDEX

Absorption traps, 223–224
Ac. *See* Alternating current (ac)
ACC. *See* Automatic chroma control (ACC)
Accounts payable ledger, 421–423
Accounts receivable ledger, 422, 423
Active filters, 55–57, 272, 274
Adaptors, 339, 386–392
ADC. *See* Analog-to-digital conversion (ADC) circuits
Add-on sound card, 386
Adders, 35, 36
Adjacent channel traps, 223–224, 242
Adjacent channels
  sound signals and, 253
  video signals and, 222–225, 243–244
Adjustable voltage regulator, 62
Advanced Television Service (ATS), 384
Advisory Committee, 384
AFC circuits, 119, 122, 123–129, 206.
  See also Horizontal AFC circuits
AFPC. *See* Automatic frequency and phase control (AFPC)
AFT. *See* Automatic fine-tuning (AFT) circuits
AGC. *See* Automatic gain control (AGC) circuits
AGC gating, 138
Alternating current (ac), 5, 7
  conversion of/to, 50, 121
  inductors and, 11
  for radio waves vs. power lines, 15
  transformers and, 50
  video troubleshooting and, 305
Aluminum screen, 359
AM. *See* Amplitude modulation (AM)
Ampere, 5, 7
Amplification, 4, 5, 11–12, 13–15, 193
  receiver operation and, 18
  tuners and, 197
Amplified AGC, 231
Amplifiers
  audio vs. video, 15. *See also* Audio amplifiers; Video amplifiers
  burst/keyed, 329, 344
  cascaded, 226
  cascode, 197, 338–339, 398
  chroma, 313–350
  circuit configuration of, 14

classifications of, 13–14, 123. *See also entries beginning* Class
color bandpass (BPA), 326–329, 340, 341, 343
color-difference, 335, 339
complementary-symmetry, 262, 399
coupling of, 14
deflection and, 116
differential, 260
digital, 32, 38
efficiency of, 13, 14, 123
feedback and, 13
frequency ranges of, 15
horizontal deflection and, 121
I-F. *See* I-F amplifiers
oscillators and, 15, 121
overheating of, 14
R-F, 193–195, 209, 210
summing, 325, 326
sync separation and, 119
troubleshooting SMPS and, 102
vertical output, 159–160
video. *See* Video amplifiers
Amplitude, 6–7, 22
  color burst and, 318
  comb filter circuits and, 321–326
  demodulation and, 20
  feedback and, 13
  oscillation and, 15
  phase and, 8
  troubleshooting SMPS and, 102
  vector and, 315
Amplitude modulation (AM), 9, 16–17, 19
Analog circuits, 5, 28, 34
Analog-to-digital conversion (ADC) circuits, 37
Analog signals, computer monitor and, 391, 392
AND function, 29, 33, 34
  conversion and, 37
  counters and, 36
  integrated circuits and, 38
  troubleshooting and, 42
Animations, 382
Anode button, 359
Anodes, 64–65, 91, 93–94, 105, 357, 359
  auxiliary power supply and, 93

discharging, 377
electric shock and, 71
measuring voltage at CRT, 377
sync separation and, 116–117
Antennas, 9, 15
  of radio receivers, 18
  keyed rainbow generator and VHF, 342
  transformerless power supplies and, 52
Anti-hunt horizontal AFC circuits, 127–129
APC system, 135–136, 139–141
Apertures, 360
Apple computer monitor, 405
Aquadag, 359
Arithmetic logic unit, 41
Armstrong oscillator, 162, 196
Aspect ratio, 384
Astable multivibrator, 32, 42
  computer monitor and, 400–401
  diagram of, 206
Astable PNP multivibrator circuits, 90
Asynchronous connection, 33
Asynchronous counter, 35–36
AT&T, 384
Atmospheric noise, 9
ATS. *See* Advanced Television Service (ATS)
Attenuation, 4, 12, 252–253
  computer monitor and, 395
  intermediate frequency and, 223, 239, 242
  sound troubleshooting and, 277
Attribute controller, 391
Audio amplifiers, 15, 260–266. *See also* Sound
Audio crosstalk, 9, 242, 322
Audio-frequency signals, 16, 17
  computer monitor, 386
  IC and, 203
  intermediate frequency and, 219
  microprocessors and, 41
Audio frequency signal generators, 277
Audio switching circuits, 272–273, 275
Audio system
  auxiliary power supply and, 93
  HDTV, 386
  linear power supplies for, 84
Automatic brightness control circuit, 299
Automatic brightness limiter circuit, 299

Automatic chroma control (ACC), 326, 328, 331

Automatic CRT tracking, 337–338

Automatic degaussing circuit, 361, 363–364

Automatic fine-tuning (AFT) circuits, 220–221
  troubleshooting, 242, 244

Automatic frequency and phase control (AFPC), 329, 330–331, 341

Automatic frequency control system, 99, 116, 123

Automatic gain control (AGC) circuits, 194, 218, 221, 229–232
  diagram of, 229
  forward vs. reverse, 229
  R-F and I-F, 230–232
  troubleshooting, 242, 243, 245
  video amplifiers and, 300, 307

Automatic sharpness control, 293

Automatic tint control, 329

Auxiliary power supplies, 50, 74
  flyback and, 81, 92–93
  SMPS and, 104
  troubleshooting SMPS and, 104, 105

B+ voltage, 60, 61, 62, 74, 75–76
  damper diode and, 123
  troubleshooting SMPS and, 103

B – Y. See Blue-minus-luminance (B – Y) signal

Balance sheet, 421, 422, 423

Balanced modulator, 401

Bandpass amplifer (BPA), 326–329, 340, 341, 343

Bandpass filters, 18, 223–225

Bandwidth, 12, 382
  amplifiers and, 14
  chrominance and, 315, 319, 339
  coupling and, 227–228
  HDTV and, 383–386
  and interlaced vs. noninterlaced scanning, 382–386
  intermediate frequency and, 219–225
  luminance and, 315, 319, 321
  modulation and, 16
  tuners and, 192

Barrel effect, 370

Base, 356, 357, 360

Base pulse, 135

Base waveform, 134, 162, 164

BCD. See Binary-coded decimal (BCD)

"Beating," 17

Bias/biasing, 32, 63–64
  forward and reverse, 52, 53, 86, 120.
    See also Forward bias; Reverse bias
  horizontal deflection and, 135
  sync separation and, 117–118
  troubleshooting and, 74

tuners and, 192–193
video amplifiers and, 295–298
zero, 86, 288

Binary counters, 33, 35, 37

Binary number system, 30–31, 199–200
  ADC circuits and, 37
  dividers and, 36
  registers and, 36

Binary pulses, 37, 39

Binary-coded decimal (BCD), 31, 37
  tuners and, 199–200

Bipolar-junction transistors, 15, 84, 85, 86, 100

Bistable multivibrator, 32, 34

Bits, 31

Black level, 295, 338

Black lines/bars, 74

Black mask, 356

Black portion, 242

Black screen, 288, 293, 296

Blacker-than-black edge, 326

Blank level, 338

Blanking pulses, 295
  color burst signal and, 328
  computer monitor and, 399

Blanking signals, 116, 152, 201, 239–240, 242
  computer monitor and, 391
  video amplifiers and, 300–301

Bleeder network, 91

Blocking oscillators, 121, 122, 129
  vertical, 160–162, 174–176

Blue-minus-luminance (B – Y) signal, 331–335

Boolean algebra, 29–30, 41

BPA. See Color bandpass amplifier (BPA)

Brass cores, 11

Bridged-T traps, 223–224

Brightness control, 298–299, 307, 369
  CRT troubleshooting and, 377

Brightness-limiter circuits, 299

Brightness reference level, 295

Buffer transistors, 201

Buffers, 16, 34–35
  computer monitor, 388
  luminance/chrominance and, 326
  microprocessors and, 40, 41
  noninverting vs. inverting, 35
  in phase detector, 40
  registers compared to, 35, 36

Burst amplifier, 329, 344

Burst gate, 138, 329

Burst separator, 329

Business ratios, 423–424

Business records and forms, 420–421

Business skills, 420–426

Bypass capacitors, 342

Bypass paths, 85

Bytes, 31

Cable, video input, 386

Cable-ready tuners, 192

Capacitance, 10, 11, 54
  amplification and, 14
  CRT troubleshooting and, 376–377
  intermediate frequency and, 227
  in transistor switches, 85
  tuners and, 208

Capacitance coupling, 14

Capacitance meters, 245

Capacitor filters, 20, 54–57, 61, 74
  common problems with, 74, 75, 76, 78
  discharging, 100
  SMPS and, 85, 100, 104, 108

Capacitor plates, 10

Capacitor tester, 101–102

Capacitors, 10, 11, 20, 50
  bypass, 342
  coupling, 226, 342
  differentiators and, 121–122
  high-voltage transformer and, 91
  integrated circuits and, 38
  leaky. See Leaky capacitors
  neutralizing, 194
  open, 308
  scan-derived power supply and, 89, 90
  SMPS and, 85
  speed-up, 128
  troubleshooting chrominance and, 342, 345
  vertical deflection and, 155

Carrier signal/wave, 16–17, 19, 192
  demodulation and, 116
  intermediate frequency and, 219

Cascaded amplifiers, 14, 226

Cascaded flip-flops, 33

Cascode amplifiers, 338–339, 398

Cascode mixer, 197

Casualty policies, 426

Cathode drive, 241

Cathode-to-heater short circuits, 377

Cathode-ray tubes (CRTs), 13, 47, 89, 91, 93–94, 355–380
  auxiliary power supply and, 93, 105
  common problems with, 377
  composite video signal phase and, 241–242
  computer monitor, 390
  deflection and, 121. See also Deflection
  deflection angles for, 359–360
  diagrams of, 357–358
  display adjustments of, 361–369
  electron guns in. See Electron guns
  handling, 377
  high-intensity, 370
  horizontal deflection and, 116
  laser-screen projection, 375, 376
  monochrome. See Monochrome picture
  Novabeam, 370, 371
  projection television and, 369–376
  refractive system, 371–372

RGB driver and, 336–338
scan-velocity moduation and, 293–294
sync separation and, 116
troubleshooting, 376–378
unitized guns in, 336
video signal gain and, 294–301
for Zenith computer monitor, 392–401
See also Picture tube
Cathodes, 64–65
chrominance and, 337–338
defined, 357
sync separation and, 117
CCD. *See* Charge-coupled devices (CCD)
Center-tapped transformer, 53
Central processing units (CPUs), 391–392
Certified Electronics Technician (CET) program, 417–418
CET examinations, 418
CGA. *See* Color graphics adaptor (CGA)
Channel display, microprocessors and, 41
Channel selection, 38, 41, 199, 201
favorite, 203–204, 205
Channels, 192, 209
adjacent. *See* Adjacent channels
HDTV, 386
improper tuning of, 212–213, 238
intermediate frequency and, 222–223
stereo sound and, 279
Character-generator circuits, 387
Charge-coupled devices (CCD), 323–326
"Charge pump," 40
Chopper-transformer driver transistor, 107
Chroma amplifiers, 313–350
Chroma/color subcarrier, 321, 326, 383
Chroma paths, diagram of, 287, 320
Chroma phase detector, 329
Chroma signals, 320, 331–334
combed filter and, 321–326
diagram of demodulation of, 333
vertical detail and, 325
Chrominance/luminance IC, 303–304, 323, 326–335
troubleshooting, 341–342
Chrominance signals, 218–219, 220, 238–239, 313–315
auxiliary power supply and, 93
components of, 314–315
frequency interleaving and, 321
frequency synthesis and, 38
HDTV and, 383, 386
horizontal deflection and, 116
in-phase (I), 314–315, 317, 318, 319
iso-hot grounding system and, 88
luminance vs., 315
on-screen display and, 201

quadrature (Q), 314–315
troubleshooting and, 308, 339–345
Circuit breakers, 66, 74
Circuit configuration
amplifiers and, 14
oscillators and, 15
SMPS and, 87
Circuit Q, 198–199
Circuits
amplifiers and, 14
complete, 5
functions of, 11–13
quality of (Q), 198–199, 223
R-F signals and, 10, 11
time-varying, 38
See also specific types
Clamping circuit, 297–298
Clapp oscillators, 196
Class A amplifiers, 13, 14, 123, 193, 265, 277, 326
Class AB amplifiers, 13, 14, 123, 262, 263–265
troubleshooting, 277, 278
Class B amplifiers, 13, 14, 123, 167, 263, 265, 277
Class C amplifiers, 13, 14, 123, 265
Cleaning supplies, 21
Clear condition, 33
Clip level, 117, 130
Clip-on adaptors, 209
Clipping, symmetrical, 280
Clock oscillators, 32, 33
Clock pulses, 35, 36
Clock signals, 39, 41
ADC circuits and, 37
computer monitor, 388
PLL and, 40
Closed-switch equivalent circuit, 155
CMOS logic, 38
Coils, 11, 14, 50, 110
Cold chassis ground, 51, 53
electric shock and, 100
iso-hot chassis and, 87–88
isolation transformer and, 71
Collector feedback, 192
Collector pulse, 135
Collector supply, 15, 59, 73, 85
computer monitor and, 399
intermediate frequency and, 227
Collector-supply bias circuits, 63
Collector waveform, 106, 134, 162, 164
Color, 296
common problems with, 74
complementary. *See* Complementary colors
enhanced, 387, 390
fading in and out of, 377–378
primary. *See* Primary colors
problems with, 242, 341

in television vs. computer monitor, 386
troubleshooting and, 339–345
See also entries beginning Chroma; Chrominance
Color bandpass amplifier (BPA), 326–329, 340, 341, 343
Color-bar generators, 339–341, 364–365
Color burst signal, 218, 318–319, 326, 328–331
computer monitor and, 388
troubleshooting, 340, 344
Color control, 328
color-bar generator and, 340
gray scale and, 369
Color demodulators, 331–335
Color-difference amplifiers, 335, 339
Color-difference matrixing, 334–335
Color electron guns, 357–359. *See also* Electron guns
Color graphics adaptor (CGA), 387, 390, 391, 398
Color information, demodulation and, 20
Color killer, 328–329, 331, 343
Color matrix circuits, 334–335
Color monitors, 390–392. *See also* Monitors
Color oscillator, 331, 341, 342
Color picture tube, 360. *See also* Picture tube
Color sidebands, 218–219
Color sync signals, 218
Color television, 46, 47–183
diagram of, 388
monochrome vs., 81
See also Television
Color wheel, 319
Colpitts oscillator, 162, 195, 196
Comb filter circuits, 321–326, 342, 343
Combinational logic, 31, 33
Common-base configuration, 14
Common-collector configuration, 14
Common-emitter configuration, 14, 85
Common-mode interference, 9
Comparators, 35, 36, 37
frequency synthesis and, 38
horizontal AFC circuits and, 124
in PLL, 40
regulation and, 59, 61
tuners and, 207–208
Comparison, 35–36
Complementary colors, 314–315, 319
color-bar generator and, 339–341
See also Secondary colors
Complementary-symmetry amplifier, 262, 399
Complementary transistors, 262
Complete circuits, 5
Composite video signal, 286, 295–296
chrominance and, 315
RGB video vs., 392

separating chrominance and, 319–326
separating luminance and, 322–326
TTL vs., 392
Computer video adaptors, 387–392. *See
also* Adaptors
Computers
embedded controller in, 41
integrated circuits in, 38
microprocessors and, 40. *See also*
Microprocessors
monitors for. *See* Monitors
power supply for, 84
Conductance, 5, 10, 58
Conducted interference, 9
Conductors, capacitance and, 11, 54
Connecting leads, 10
Connections, bad/open, 20, 22, 23, 74
Connector pin-out, 390–391, 392
Constant amplitude carrier wave
(CW), 16
Continuity testing, 74
Contrast, 356, 299
Contrast control, 299–300, 307
CRT troubleshooting and, 377
Control, 34–35, 38
customer preferences and, 200–201
horizontal AFC circuits and, 123–129
microprocessors and, 38, 40, 41,
200–209
SMPS and, 85
troubleshooting and, 74, 76
Control grid, 358
Controllers, embedded, 41
Convergence, 361–362, 363, 364–367
Convergence yoke, 363, 365
Conversion
of ac to dc voltage, 50
of analog to digital, 34–35, 37
of dc to ac voltage, 121
of digital to analog, 391
of rectified line voltage into
unregulated power supply, 89
of serial to parallel, 36
Converters, 37
Copper cores, 11
Copper wires, 5
Corrector lens, 370
Corrosion, 20
Cosmic noise, 9
Cosmic waves, 8
Countdown circuits, 121, 135, 167–169
Counters, 35–36
electromotive, 10
flip-flops as, 33, 35
frequency synthesis and, 38
in PLL, 40
troubleshooting and, 42
tuners and, 209–210
See also Binary counters
Couplers, SMPS and, 98
Coupling, 12, 123, 226–228

amplifier, 14
capacitor, 342
direct. *See entries beginning
Direct-coupled; Direct coupling*
interstage, 32
CPUs. *See* Central processing units
(CPUs)
Cross-hatch generators, 339–341
Cross-over audio network, 272, 274
Cross-over distortion, 263
Cross-over point, 368, 401
Crosstalk, 9, 242, 322
CRT base, 356, 357, 360
CRT burn, 375
CRT circuits, 5, 64, 75, 76. *See also*
Cathode-ray tubes (CRTs)
CRT controller, 387, 391
CRT cutoff, 288
CRT drive, low-level RGB matrixing
and, 334–335, 336
CRT tracking, 337–338
CRTs. *See* Cathode-ray tubes (CRTs)
Crystal, in PLL, 40
Crystal oscillators, 39
Crystalline screen, 375
Current, 5, 6
alternating. *See* Alternating current (ac)
amplification and, 13
direct. *See* Direct current (dc)
excessive, 75–76, 375. *See also*
Overcurrent protection
measuring, 71, 72, 73, 74
oscillation and, 15
troubleshooting and, 70
Current gain, 13, 14, 228–229
Current-limiting circuits, 66, 110
Customer preferences, 200–204
Customer service, 411–428
communication and, 415–416
software relating to, 415–416
trust and, 411
Cutoff, 15, 85, 86
CRT, 288, 296
rectangular wave and, 122
CW. *See* Constant amplitude carrier
wave (CW)
Cycle time, 7, 122

D-type flip-flop, 33
DAC. *See* Digital-to-analog (DAC)
circuits
Damper diode, 99, 121, 131
computer monitor and, 398
defined, 123
shorted, 140
Damping, 15
Darlington Pair/Stage, 59, 119, 264
David Sarnoff Research Center, 384
Dbx compressor circuit, diagram of, 272

Dbx expander circuit, diagram of, 273
Dc. *See* Direct current (dc)
Dc restoration, 297–298, 395, 398
Decibels, 12
Decimal display, 37
Decoders, 37
frequency synthesis and, 38
microprocessors and, 41
De-emphasis, 255–256, 258
Deflection, 5, 38, 47
amplifiers and, 116
auxiliary power supply and, 93
computer monitor troubleshooting
and, 403
focus and, 367–368
horizontal. *See* Horizontal deflection
vertical. *See* Vertical deflection
Deflection angle, 359–360, 367–368
Deflection yoke, 47, 363, 364, 367
cross section of, 158
Degaussing circuit, 361, 362–364
Degaussing coils, 110
Degaussing, defined, 363
Delay element, 32
Delay line, 320, 322–326
Delay time, 15, 16, 32, 33
rectangular wave and, 122
transistor switches and, 85
troubleshooting and, 42
See also Propagation delay
Delayed AGC, 231–232
Delta-gun configuration, 357–358, 361, 362
Demodulation, 12, 16, 19–20
chrominance and, 317, 319, 320,
331–335, 341–345
I-Q signal, 333–334
intermediate frequency and, 221,
239–241
sound, 256–257, 261
sync separation and, 116
synchronous, 332, 333–334
troubleshooting color, 341–345
typical AM, 17
X/Y, 333, 334
Demodulator probe, 102
Demultiplexer, 34
Detail, 325, 377
Detectors, 19, 238–242, 244, 255, 257–260
chroma phase, 329
differential peak, 259
quadrature, 258–259
ratio, 257–258
See also Phase detectors
Deviation, 17
Dielectric, 10, 54, 360
Differential amplifiers, 260
Differential peak detector, 259
Differentiator circuits, 121–122, 123
computer monitor and, 398–399
Digital-to-analog (DAC) circuits, 37
Digital-to-analog converter, 391

Digital circuits, 5, 28–45
  amplifiers in, 32
  Boolean algebra and, 29–30, 41
  conversion of/to, 34–35, 37, 391
  gates and, 31, 32, 34
  manipulating frequencies with, 38–40
  troubleshooting of, 42–43
Digital HDTV Grand Alliance, 384, 385
Digital number systems, 30–31
Digital signals, computer monitor and,
    391, 392
Digital storage oscilloscope (DSO),
    138–139
Digital technology, HDTV and, 386
Digital video disks, 392
Diode detector circuits, 19–20
Diode sync-separator circuits, 116–118
Diodes, 38, 39, 52–54, 70, 74, 100
  chrominance and, 337–338
  SMPS and, 85
  sound signals and, 256
  troubleshooting SMPS and, 108, 110
  varactor, 39
  zener. See Zener diode
DIPS. See Dual in-line packages (DIPS)
Direct coupling, 12, 14, 289, 296–297
Direct-coupled doubler circuits, 64–65
Direct-coupled half-wave rectifier
    circuits, 91
Direct-coupled voltage doubler, 50
Direct current (dc), 5, 6
  continuity testing and, 74
  conversion of/to, 50, 121
  inductors and, 11
  peak value and, 7
  video signals and, 296
Direct-view color picture tube
  electrostatic focus in, 367, 368
  projection vs., 369
  shadow mask in, 360
Direct waves, 9, 10
Discriminators, 221, 258
Disk drives, 41
Display, 361–376, 386–392
  adjustments for, 361–369
  customer preferences in, 200–204
  decimal, 37
  light-emitting. See Light-emitting
      display (LED)
  liquid-crystal, 370
  menu or on-screen, 201, 202
  microprocessors and, 41
Display memory, 391
Dividers, 35, 36
  frequency synthesis and, 38, 39
  in PLL, 40
  tuners and, 207–208
  See also Voltage dividers
Documentation, 21, 22
Dolby AC-3 five-channel sound
    system, 386

Dot pitch, 382, 386
Doubler circuits, 64–65, 69–70
  SMPS and, 85, 86, 104
Drain, 85, 86
Drive, defined, 288
Drive power, buffers and, 35
Drivers
  horizontal, 116, 121, 122, 130–134,
      135
  vertical, 162, 164–166
DSO. See Digital storage oscilloscope
    (DSO)
Dual-gate MOSFET, 194–195
Dual in-line packages (DIPS), 209
Dual-trace oscilloscope, 306
Dual-tracking regulator, 62
Dummy loads, 101, 105, 108
Dust, 105
Duty cycle, 86, 87, 90
Dynamic convergence, 339, 361–362,
    365–367
Dynamic troubleshooting, 4, 21–22, 72

Eagan Technical Services, Inc., 427
East-West waveform generation/output,
    400
Education, continuing, 416–418
EGA. See Extended graphics adaptor
    (EGA)
EHF band, 9, 10
EIA/CEG. See Electronics Industries
    Association/Consumer Electronics
    Group (EIA/CEG)
8514/A standard, 391
Einzel lens, 358
Electric fields, 6–9, 356, 367, 368
Electrical shock, 20, 52, 71
  CRT troubleshooting and, 377
  SMPS and, 100
Electromagnetic induction, 50
Electromagnetic interference (EMI), 9
Electromagnetic spectrum, 189
Electromagnetic waves, 6–9
Electromotive force (emf), 5
  counter, 10
Electron beam, 47, 356–357, 359
  computer monitor and, 390, 391
  convergence and, 364–369
  deflection of, 121. See also Deflection
  display adjustments and, 361–369
  flyback and, 91
  focusing, 367–369
  light valves and, 375
  purity magnets and, 364
  shadow mask and, 360–361, 362
  sync separation and, 116
Electron flow, 5, 54–55
  intermediate frequency and, 227
Electron guns, 91, 356, 357–359
  gray scale and, 369

  grids for, 358–359
  light valves and, 375
  in monochrome vs. color picture tube,
      360
  refractive systems and, 371
  troubleshooting and, 377
  unitized, 336
Electronic bandswitching, 199, 208–209
Electronic counters, 36. See also Counters
Electronic systems
  basic parts of, 4–5
  fundamental concepts of, 4–23
  power supply for, 84. See also Power
      supplies
Electronic tuner control, 199
Electronics Industries Association/
    Consumer Electronics Group
    (EIA/CEG), 416–417
Electrostatic focus, 367, 368
Electrostatic induction, 9
Emerson television, 144
Emf. See Electromotive force (emf)
EMI. See Electromagnetic interference
    (EMI)
Emitter feedback, 192
Emitter follower, 119, 135, 395
Emitter pulse, 135
Emitter supply, 15, 59, 73, 85
  intermediate frequency and, 227
Enable function, 31, 37
Encoders, 37, 38
Energy, 6–9
  and circuit quality (Q), 198–199,
      223, 227
  resistance and, 11
Enhanced color, 387, 390
Envelope, 356–357, 359–361
Equalizing sync pulses, 116, 152
Error detectors/amplifiers, 58, 60
Error latch, 98
Error voltage, 331
Ethics, 413, 414–416
European HDTV, 383–384
Evacuation, 359
Extended graphics adaptor (EGA), 387,
    389, 390–391, 398

Fall time, 15, 16, 42
  rectangular wave and, 122
  transistor switches and, 85
Farad, 11
Fault protection, 34, 35
Favorite station selection, 203–204, 205
FCC, 384
Feedback, 4, 12, 13
  AFC stage and, 122
  chrominance and, 330–331
  degenerative vs. regenerative, 64, 196
  horizontal AFC circuits and, 124
  neutralization of, 194

oscillation and, 15, 121
PLL and, 40
SMPS and, 85, 87, 98
sync separation and, 119
tuners and, 192–193
video amplifiers and, 289
Feedback bias, 64
FETs. *See* Field-effect transistors (FETs)
Field-effect transistors (FETs), 15–16,
    85–86, 194, 195
    computer monitor and, 398
Figure of merit, 262
Filter choke, 55, 56
Filter driver transistor, 55–57
Filter network, 272
Filter resistors, 56
Filters, 5, 6, 50, 54–57
    active, 55–57, 272, 274
    bandpass, 223–225
    capacitor. *See* Capacitor filters
    comb, 321–326, 342, 343
    differentiators as, 121
    frequency synthesizer and, 39
    inductive, 55
    in PLL, 40
    SAW, 223, 234–237, 244
    SMPS and, 85
    vestigial transmission and, 219–220
Fire and flood basic policies, 426
Fixed bias, 63
Fixed-negative voltage regulator, 62
Fixed-positive voltage regulator, 62
Flicker, 382, 383, 389
Flip-flops, 32–33, 137
    counters and, 33, 35
    dividers and, 36
    in phase detector, 40
    registers and, 36
    troubleshooting and, 42–43
Flux lines, 10, 365
Flyback, 47, 89, 91–93
    auxiliary power supplies and,
        92–93
    computer monitor and, 398
    convergence and, 366
    horizontal deflection and, 121
    troubleshooting SMPS and, 104
FM. *See* Frequency modulation (FM)
Focus
    computer monitor and, 399
    low-voltage or unipotential, 368
Focus adjustments, 367–369
Focus circuits, 5
Focus grid, 358
Focus voltage, 91
Forward AGC, 229
Forward bias, 52, 53, 86, 120, 229
    sound signals and, 256
Foster-Seeley discriminator, 258
Free-running multivibrator, 32
French HDTV, 383–384

Frequency, 6–9
    amplifier classification and, 15
    center, 17, 18
    chrominance, 321. *See also*
        Chrominance signals
    computer monitor and, 389
    de-emphasis network and, 256
    digital circuits to manipulate, 38–40
    fundamental, 8, 38, 275, 277, 308,
        317
    horizontal AFC circuits and, 124
    I-F. *See* Intermediate frequency (I-F)
    luminance, 321. *See also* Luminance
        signals
    PLL and, 40
    pre-emphasis network and, 254–255
    reduced, 172
    resonant, 11. *See also* Resonance
    scan-derived power supply and, 90
    switching, 86
    video amplifiers and, 289–293
    video troubleshooting and
        measuring, 306
Frequency bands, 4
Frequency compensation, 220, 292–293
Frequency discriminator, 20
Frequency division, 38
Frequency interleaving, 320–321
Frequency locking, 40, 99
Frequency modulation (FM), 10, 16,
    17, 19
Frequency multiplication, 12, 38, 39
Frequency response curve, 193
    bridged-T trap, 224
    chroma bandpass amplifer circuit
        and, 328
    compensated vs. uncompensated, 220
    coupling and, 227–228
    flat, 253
    intermediate, 219–222, 233
    stereo sound, 279
    video amplifier, 288–292
Frequency shift keying, 17
Frequency spectrum, 8–9
    comb filtering and, 323
    radio (R-F), 9–10
Frequency synthesis, 12, 38–40
    common problems with, 212–213
    indirect, 39
    PLL as alternative to, 40
    troubleshooting, 212–213
    tuners and, 199, 204, 207, 209,
        212–213
Frequency tripler circuit. *See* Tripler
    circuit
Fresnel lens, 372
Front-projection system, 370
Full-wave bridge rectifier, 50, 53–54, 68
    SMPS and, 86
Full-wave rectifiers, 52–53, 67–68
    filter capacitor and, 55

SMPS and, 84
Fundamental frequency, 8, 38, 275, 277,
    308, 317
Fuses, 65, 66
    blown, 52, 66, 70, 74, 76, 103, 107

G – Y. *See* Green-minus-luminance (G – Y)
    signal
G1, G2, etc. *See* Grid one (G1), Grid two
    (G2), *etc.*
Gain, 4, 12
    amplifier coupling and, 14
    in digital systems, 32
    feedback and, 13
    intermediate frequency and, 218, 221,
        225–226, 228–232
    op-amp and, 119
    in superheterodyne receivers, 19
    troubleshooting SMPS and, 102
    tuners and, 197
    video amplifiers and, 294–301
    See also Power gain; Voltage gain
Gate resistor, in SCR crowbar circuits, 67
Gated AGC circuit, 230–231
Gated pulse, 230–231
Gates, 31, 34, 85, 86
    counters and, 35
    dividers and, 36
    flip-flops and, 32–34, 137
    troubleshooting and, 42
Gates differential amplifier, 136
General Instrument Corporation, 384
Glass envelope, 356–357, 359–361
Glass faceplate, 359
Government regulations, 94, 133, 426
    HDTV and, 384
Grand Alliance, 384, 385
Graphics controller, 391
Graticule, 102
Gray scale, 337–338, 341, 362,
    368–369, 371
    troubleshooting and, 377
Gray-scale adjustments, 368–369, 377
Green-minus-luminance (G – Y) signal,
    331–335
Grid, 296, 357–359
Grid-to-cathode short circuits, 377
Grid drive, 241
Grid one (G1), 358
Grid two (G2), 93, 358, 368
Grid three (G3), 358, 368
Grid four (G4), 359, 368
Ground waves, 9
Grounding, 50–51, 70, 71
    CRT troubleshooting and, 377
    diagram of, 98
    for modern televisions, 88
    op-amp, 266
    SMPS and, 98–99, 106

troubleshooting SMPS and, 106
See also Cold chassis ground; Hot
chassis ground

H. *See* Henry (H)
Half-wave rectifier, 50, 52, 54, 56
direct-coupled, 91
filter capacitor and, 55
filter choke and, 55
SCR as, 61
Harmonics, 8, 38
chrominance, 317
high-voltage transformer and, 91
I-F, 221
luminance, 321
sound troubleshooting and, 275, 277
Hartley oscillator, 129, 136, 162, 196
HDTV. *See* High-definition television
(HDTV)
Heat sink, 59, 66, 106
Henry (H), 11
Hercules graphics adaptor (HGA), 386,
389–390, 391
Hertz (Hz), 6, 305
Heterodyning, 12, 17–18, 19, 238–239
Hexadecimal seven-segment display, 37
HF band, 9, 10
HGA. *See* Hercules graphics adaptor
(HGA)
High-definition television (HDTV), 369,
375, 381–386
1125/60 standard of, 384
sync waveform of, 385–386
High-voltage circuitry, computer
monitor and, 399–400
High-voltage focus, 368, 399
High-voltage shutdown, 74
Hold-up time, 67
Horizontal AFC circuits, 123–129, 134
anti-hunt, 127–129
diagram of, 124
troubleshooting, 139–141, 142, 143
Horizontal AFC equivalent circuits,
125–127, 140
Horizontal AFC stage, 122
Horizontal deflection, 28, 47, 94
classifications of, 129
complete systems of, 134–135
computer monitor and, 398, 403, 404
diagram of, 121
IC-based scan systems of, 135–137
retrace time for, 130
sync signals and, 115–150
troubleshooting, 137–147
Horizontal driver, 121, 122,
130–134, 135
Horizontal dynamic convergence,
365–367
Horizontal foldover, 141
Horizontal frequency, 124, 139–140

Horizontal hold control, 122, 142,
145, 147
Horizontal oscillators, 116, 119–122,
129–130
computer monitor and, 390, 398
horizontal AFC circuits and, 123, 134
noise spike and, 120
overvoltage protection and, 94
scan-derived power supply and, 89
SMPS and, 98, 103
vertical oscillators combined
with, 129
Horizontal output, 121, 123,
130–134, 135
Horizontal output stage, 116
Horizontal output transistors, 131–133,
135, 140
Horizontal pull, 74, 75
Horizontal scanning, 91, 123
chrominance and, 321
computer monitor, 389, 390
IC-based, 135–137
interlaced vs. noninterlaced,
382–386, 391
luminance and, 321, 322
SVM and, 293
Horizontal sync pulses, 116
color burst vs., 318
luminance and, 321
Horizontal sync signals
computer monitor and, 390, 398
and interlaced vs. noninterlaced
scanning, 382, 383
Horizontal waveforms, 106
Hot chassis ground, 51–52, 53, 64
electric shock and, 71, 72, 100
iso-, 87–88, 106
Howard W. Sams & Company, 427
Hue, 315–319, 329
troubleshooting and, 339, 341
Hue control, 329, 331
Hum, 9
Hum bars, 74, 75
Hybrid regulators, 58
Hyperband channels, 192, 222
Hz. *See* Hertz (Hz)

IBM, 391
IC audio amplifiers, 265–266
troubleshooting, 278–279
ICs. *See* Integrated circuits (ICs)
I-F. *See* Intermediate frequency (I-F)
I-F amplifier stage gain, 225–232
I-F amplifiers, 225–226
color-bar generator and, 340
sound, 255–257
troubleshooting, 242, 243, 245
video amplifiers and, 300
I-F carrier, 18–19
I-F response curve, 219

I-F/AFT/AGC circuit
diagrams of, 236–239
troubleshooting, 244
IGBT. *See* Insulated gate bipolar transistor
(IGBT)
IHVT. *See* Integrated high-voltage
transformer (IHVT)
I-Q signal-demodulation system, 333–334
Image viewers, 392
Impedance, 11
amplifier circuit configuration and, 14
op-amp and, 119
troubleshooting and, 22
Impedance coupling, 12, 14, 226
mutual, 228
Impedance-matching network, 272
Impulse noise, 9
In-line configuration, 358, 361, 362
In-line projection system, 372
In-phase (I) signals, 6, 12, 192–193,
314–315, 317–319
luminance, 322–324
Incidental-carrier phase modulation,
252–253
Indicator lamp, 74
Inductance, 10, 11, 54
filtering and, 54
high-voltage transformer and, 91
intermediate frequency and, 227
Induction, 9, 50
Inductive filters, 55
Inductive reactance, 10, 11, 14
efficiency of, 11
R-F signals and, 11
Information, locating, 127
Infrared light, 6
Infrared remote transmitter, 203
Inhibit function, 31
Input, 4
for analog vs. digital circuits, 5, 34
feedback and, 13
integrated circuits and, 38
microprocessors and, 40, 41
multivibrators and, 32
oscillation and, 15
switched-mode regulators and, 86–87
Insulated gate bipolar transistor (IGBT), 85,
86
Insulation, 50, 377
Insulators, 10, 11
Insurance, 425–426
Integrated circuits (ICs), 31, 37–38
automatic CRT tracking and, 337–338
buffers and, 35
chrominance and. *See* Chrominance/
luminance IC
clip-on adaptors for testing, 209
computer monitor and, 386, 405–406
counters and, 36
dividers and, 36
flip-flops and, 32, 33

frequency synthesizer and, 39
horizontal deflection and, 129,
135–137, 143–144
intermediate frequency and, 221,
234–241
monaural sound and, 266–268
multichannel sound and, 271–272
op-amp and, 119
SMPS and, 84–85, 98, 99
sync processor and, 119, 137
testing, 209
tuners and, 203–204
vertical oscillators and, 123
vertical scanning and, 166–171
See also Microprocessors
Integrated flyback transformer, 92. *See
also* Flyback
Integrated four-terminal regulators, 75
Integrated high-voltage transformer
(IHVT), 92, 99
troubleshooting SMPS and, 105, 106
Integrated rectifiers, 54
Integrated three-terminal regulators,
58, 62
Integrating network, 153
Integrator circuit, 152–153
Intercarrier sound method, 218, 252
Interference
HDTV and, 386
intermediate frequency and, 222–225,
243–244
luminance and, 321
sound signals and, 252, 253
tuners and, 209, 222
See also Noise; Radio-frequency
interference (RFI)
Interlaced scanning, 382–386, 391
Intermediate frequency (I-F), 17–19
auxiliary power supply and, 92
chrominance and, 326
extraction points of sound,
238–242, 255
extraction points of video,
238–242, 255
processing, 218–222, 255
sound, 253, 255–256
stagger-tuned, 234, 235, 244
sync signals and, 116
troubleshooting, 242–246
tuners and, 209, 211
video, 238–242
See also entries beginning I-F
Intermittent power cycling, 103,
104–105, 108
Internal arcing, 377
International Society of Certified
Electronic Technicians, 418
Interstage coupling, 32
Inventory management, 416
Inverters, 16, 40

Inverting switched-mode regulator,
94–95
Iron cores, 11
Iso-hot chassis, 87–88, 106
Isolation, 34, 35, 50, 51, 98
SMPS and, 85, 87
Isolation transformer, 70, 71, 106

Japan Broadcasting Corporation (NHK),
384
JFET, 398
J-K flip-flop, 33

Keyboards
BCD system and, 31, 200
microprocessors and, 41, 200
Keyed AGC circuit, 230–231
Keyed amplifier, 329
Keyed pulse, 230–231
Keyed rainbow generators, 339–341, 342
Keypad, 201
Keystone effect, 370
Keystone generators, 370
Kick-start circuits, 89, 90
Kloss Novabeam CRT, 370, 371

Labor, setting prices for, 424–425
Lambda, 6
Large-screen televisions, 53
Laser printers, 41
Laser-screen projection CRT, 375, 376
Latching, 36, 62, 98
LC filters, 55
LC network, 259
LC resonant circuit, 271
LC tank circuits, 38
Leakage inductance, 91
Leakage test, 70, 71
Leaky capacitors, 20
chrominance and, 341, 345
filter capacitor and, 74, 78
"pie crust" distortion and, 140
power supply and, 70
Leaky components, 107, 108, 140, 277
chrominance and, 342, 345
power supply and, 74, 75, 76
LED. *See* Light-emitting display (LED)
LF band, 9, 10
Licenses, 426
Light, 6
Light valves, 370, 375, 376
Light-emitting display (LED), 100,
200, 201
BCD system and, 31, 200
diagram of, 202
optoisolator and, 87
sound signals and, 256
SMPS and, 108
troubleshooting and, 42, 108, 211–212

Line isolation, 87
Line regulation, 57
Line voltage, transformers and ac, 50–51
Linear power supplies, 48–81
common problems with, 74–75
drawbacks of, 67
efficiency of, 84
transformerless, 51–52
troubleshooting SMPS, 105
Linear ratio, 12
Linear regulators, 58, 61
Liquid-cooled projection tubes, 372
Liquid-crystal display, 370
Load, 5, 13
amplifiers and, 14
buffers and, 35
dummy, 101, 105, 108
isolation between ac line voltage and, 50
SCR and, 62
zener diode regulator and, 58
Load regulation, 57, 58
Load resistors, open, 308
Logic, sequential, 32–33
Logic analyzers, 42
Logic clip, 42
Logic elements, schematic symbols for, 31
Logic probe, 42
Logic states, 32, 34, 35
integrated circuits and, 37–38
Logical statements, 29
Logical troubleshooting, 20–21, 70, 102,
213, 305
Loop
AFPC, 330–331
feedback. *See* Feedback
phase-locked. *See* Phase-locked loop
(PPL)
SMPS and, 98
Loss of revenue/use insurance, 426
Low electron gun emission, 377
Low-level RGB matrixing and CRT drive,
334–335, 336
Low-voltage focus, 368
Low-voltage regulator problems, 74–75
L-R audio signals, 270–271
Luminance signals, 285–312
applications of, 301–304
auxiliary power supply and, 93
bandwidth and, 319
chroma signals and, 331–335
chrominance vs., 315
dc component of, 296, 297
delay line and, 320
frequency interleaving and, 321
frequency synthesis and, 38
HDTV and, 383, 384, 386
horizontal deflection and, 116
IC and. *See* Chrominance/luminance IC
in-phase, 322–324
intermediate frequency and, 219
iso-hot grounding system and, 88

light valves and, 375
on-screen display and, 201
out-of-phase, 322–324
troubleshooting and, 308, 339–345
Luminance signal inverter, 323–324
Luminescence, 356

Magnavox television, 76, 77, 180–182
Magnetic fields, 6–9, 10, 50, 356
degaussing circuit and, 361, 362–364
focusing with, 367–368
induction and, 54
SMPS and, 85
Magnetic flux, 116, 121, 365
Magnetic focus, 368
Magnetic induction, 9, 50
Management/scheduling software, 415–416
Margin pricing, 424
Mark-up pricing, 424
Massachusetts Institute of Technology, 384
Materials, setting prices for, 424
Mathematical functions, 36
Matrix circuits, 334–335, 386
Matsushita, 417
MDA. *See* Monochrome display adaptor (MDA)
Medium-scale integration, 31
Memory
display, 391
flip-flops and permanent, 32, 33
microprocessors and, 40, 41
multivibrators and, 32
RAM vs. ROM, 200. *See also* Random-access memory (RAM); Read-only memory (ROM)
registers and, 36
tuners and, 200, 203
Memory IC, 204
Menu display, 201, 202
Metal-oxide resistors, 85
Metal-oxide semiconductor field-effect transistors (MOSFETs), 85–86
computer monitor and, 398
dual-gate, 194–195
as R-F amplifiers, 194–195, 210, 245
troubleshooting SMPS and, 102–103, 104
Metal-oxide silicon FETs, 15
Metal-oxide varistors (MOVs), 85, 86, 110
MF band, 9, 10
MH. *See* MilliHenry (mH)
Mhos, 5
Microcomputer, 386, 389
Microcontrollers, 41
MicroFarad, 11
MicroHenry (uH), 11
Microprocessors, 40–42, 199–209

ADC circuits and, 37
counters and, 36
in CPUs, 391–392
dedicated, 41, 192, 199
integrated circuits and, 38
J-K flip-flops and, 33
registers and, 36
troubleshooting, 211–213
Microwaves, 10
Midband channels, 192
Miller effect, 291
MilliFarad, 11
MilliHenry (mH), 11
Mitsubishi television, 179–180
Mixer, 17–18, 19
cascode, 197
frequency synthesizer and, 39
incidental carrier phase modulation and, 252
tuners and, 196–197, 209, 211
Modems, embedded controller in, 41
Modern television, 84
color killer and, 329
countdown circuits in, 121, 135
faceplates of, 360
flyback for, 91
grounding for, 88
horizontal hold control in, 122
pincushion and, 367
power supply diagram for, 91
shutdown of, 105–106
Modulation, 12, 16–17
chrominance and, 314–315, 317, 318
computer monitor and, 401
incidental-carrier phase, 252–253
intermediate frequency and, 219
over-, 252
phase, 16, 17
pulse, 17, 86, 87
scan-velocity, 293–294
single-sideband. *See* Single-sideband modulation
Modulation envelope, 16, 17, 19, 20
Modulators, balanced, 401
Monaural sound systems, 266–273
Monitors, 84, 381, 382, 386–409
adaptors for, 339, 387–392
cascode video output circuits in, 339
circuits for, 395–401
display adaptors for, 386–392
newer, 387
older/early, 382, 388–389, 390
RGB, 390, 391
RGBI, 390
SMPS and, 108, 110
standards for, 391
text-only, 389–390
troubleshooting, 108, 110, 401–406
viewing distance for, 389
Zenith, 392–401

Monochrome display adaptor (MDA), 387, 389–390, 391
Monochrome picture, 81, 253, 320
base tube layout for, 360
beam current for, 357
chrominance disabling for, 319
color control and, 328, 341
computer monitor, 386, 387
gray scale and, 369
low-voltage focus for, 368
Monostable multivibrator, 32, 42
MOSFETs. *See* Metal-oxide semiconductor field-effect transistors (MOSFETs)
Moving Picture Experts Group, 386
MOVs. *See* Metal-oxide varistors (MOVs)
MPEG-2 video, 386
MTS. *See* Multichannel television sound (MTS)
Multichannel television sound (MTS), 268, 270–272
Multimeter, 22, 70, 71, 72–74
collector voltage and, 140
computer monitor troubleshooting and, 406
example of use of, 76
ohmmeter function of, 74
purchasing, 72
SMPS and, 101–102, 106
tuners and, 212
Multiple display traces, 174
Multiple-line data set, 37
Multiplexer, 34, 388
Multisync displays, 386, 388–392
Multivibrator-based kick-start circuits, 89
Multivibrators, 32
astable. *See* Astable multivibrators
horizontal oscillators and, 121, 122, 142
troubleshooting and, 42–43
tuners and, 205–206
vertical, 162–164, 176–179
MUSE HDTV interlaced system, 384
Mutual impedance coupling, 228

NAND function, 30, 32, 36, 137
troubleshooting and, 42
National Association of Retail Dealers of America, 417
National Electronic Servicing Dealers Association, 417
National Television Systems Committee (NTSC), 339, 341, 382, 383–386
Negative value, 5, 7
Network
AFPC, 329
cross-over audio, 272, 274
de-emphasis, 255–256, 258
dual time-constant, 120
filter, 272
impedance-matching, 272
integrating, 153

LC, 259
pre-emphasis, 254–255
RC. *See* RC network
series RC, 121
sine wave stabilizing, 122
sync separation and, 118
Neutralization, 14, 194, 226–227
NHK. *See* Japan Broadcasting
Corporation (NHK)
Noise, 6, 8–9, 28
AGC and, 230
amplifier coupling and, 14
buffers and, 35
diode sync separator and, 118
flip-flops and, 33
intermediate frequency and, 227,
242, 243
oscillators and, 123
PLL and, 40
R-F amplifiers and, 193
SMPS and, 86
squealing, 103
troubleshooting SMPS and, 103
types of, 9
Noise-reduction circuits, 119–120
Noninterlaced scanning, 383–386, 391
NOR function, 30, 42
North-South waveform generation/
output, 400, 401
NOT function, 30, 32, 33, 137
Notched traps, 223–224
Novabeam CRT, 370, 371
NPN transistors, 167, 227, 262, 337
NTSC. *See* National Television Systems
Committee (NTSC)

OEM. *See* Original equipment
manufacturer (OEM)
"Off," 5
Ohm's Law, 73
Ohmmeter, 74, 406
Ohms, 5, 11
Older computer monitors, 382,
388–389, 390
Older televisions, 64, 65, 66, 91, 122,
142, 326
"On," 5
One-shot multivibrator, 32
1125/60 standard, 384
Op-amp. *See* Operational amplifier
(op-amp)
Open capacitors, 70
Open components, 74, 76
square waves and, 308
Open/bad connections, 20, 22, 23, 74
Open resistors, 20, 70, 76, 103
Open-switch equivalent circuit, 155
Open transistors, 140
Operating speed, integrated circuits
and, 38

Operating statement, 421
Operational amplifier (op-amp), 15, 16,
61, 119
as audio amplifier, 265–266
Optical system, 370–375
Optocouplers, 85, 98
Optoisolators, 85, 87, 98
OR function, 29–30, 34
integrated circuits and, 38
troubleshooting and, 42
Original equipment manufacturer
(OEM), 426
Oscillation, 4, 11–12, 15–16
amplifier coupling and, 14
frequency synthesis and, 38, 39
parasitic, 227
Oscillators, 15, 121
Armstrong, 162, 196
blocking, 121, 122, 129
Clapp, 196
clock, 32, 33
color, 331, 341, 342
Colpitts, 162, 195, 196
computer monitor and, 398–399
counters and, 36
crystal, 39
crystal-controlled, 196, 210
diagram of, 15
in digital systems, 32–34
feedback and, 13
flip-flops and, 32
Hartley, 129, 136, 162, 196
heterodyning and, 18, 19
receiver operation and, 18–19
reference, 38, 39
relaxation, 136
sawtooth wave, 142
sine wave. *See* Sine wave oscillators
sweep voltage and, 121
troubleshooting SMPS and, 104
tuner, 39, 195–196, 209, 210–211
types of, 162, 196
vertical, 390
vertical sweep, 157
VCO. *See* Voltage-controlled
oscillator (VCO)
Wien-Bridge, 162, 196
Oscilloscope, 6, 22, 42, 70, 71
chrominance and, 340, 342
color-bar generator and, 340
dual-trace, 306
example of use of, 76
front panel of, 101
horizontal deflection and,
138–139, 140
rectangular waveforms and, 122
SMPS and, 101–102, 106
stereo sound and, 279–280
tuners and, 211
vertical deflection and, 173–174
video troubleshooting and, 304–307

Out-of-phase signals, 6, 12, 14, 57,
192–193, 317
luminance, 322–324
Output, 4
of analog vs. digital circuits, 5, 34
feedback and, 13
integrated circuits and, 38
multivibrators and, 32
oscillation and, 15
regulated, 5
slew rate and, 16
SMPS and, 103
switched-mode regulators and, 86–87
Overcoupling, 228
Overcurrent protection, 65–66
horizontal deflection and, 133–134
SMPS and, 85, 94, 105–106
Overload, 74
SMPS and, 100, 106
video, 242, 243
Overmodulation, 252
Overvoltage protection, 65, 67
horizontal deflection and, 133–134
SMPS and, 85, 93–94, 105–106

PAL. *See* Phase Alternate Lines (PAL)
Parabolic generators, 370
Parallel data, registers and, 36
Parallel-resonant circuits, 11
Parallel-resonant traps, 223–224
Parallel switched-mode regulator, 95
Parasitic oscillations, 227
Parts substitution, 22–23
Parts suppliers, 426–427
Pass-resistor regulators, 58
PBS, 386
Peak inverse voltage, 54
Peak value, 7
Peaking, 325–326
differential peak detector and, 259
shunt vs. series, 292–293. *See also* Shunt
peaking
video amplifiers and, 289, 292–293
video troubleshooting and, 306
Pedestal level, 119
People skills, 414–416
Performance bond, 426
Persistence of vision, 47, 356
PF. *See* PicoFarad (pF)
Phase, 6, 12
amplitude and, 8
feedback and, 13
horizontal AFC circuit and, 124, 142
oscillation and, 15
tuners and, 192–193
vector and, 315
video troubleshooting and, 306–307
Phase Alternate Lines (PAL), 383, 384
Phase angles, 315–317, 318, 331
troubleshooting and, 339

Phase changes, horizontal oscillators and, 122
Phase-control regulators, 58, 60–61
Phase detectors, 40, 142
Phase-locked loop (PLL), 38, 39, 40, 124, 136
  chrominance and, 329, 341
  intermediate frequency and, 221
  tuners and, 206–207, 208
Phase modulation, 16, 17
  demodulation and, 20
  keyed rainbow generators and, 339–340
Phase reference, 318–319
Phase reversal, 4, 15
Phase shift, 12, 16, 259
  chrominance and, 317
  hue and, 329
  tuners and, 193
Philips Consumer Electronics, 384, 417
Phosphorescence, 356
Phosphors, 356, 357, 359
  interlaced scanning and, 382
  purity and, 364
  shadow mask and, 360–361, 362
Photodiodes, 87
Pi filter, 104
PicoFarad (pF), 11
Picture
  aspect ratio of, 384
  breakup of, 142, 144
  color. See entries beginning Chroma; Chrominance; Color
  color problems with, 341, 343–344
  common problems with, 74, 75–76
  display adjustments and, 361–369
  herringbone pattern in, 209, 252
  horizontal frequency of, 124, 140
  horizontal line in, 105
  horizontal pull of, 74, 75
  horizontal sync of, 122
  hourglass-like, 76
  hum bars in, 74, 75
  monochrome. See Monochrome picture
  multiple images in, 142
  sync signals and, 116
  tables listing problems with, 242–244
Picture tube, 93–94, 356
  base layout for monochrome and color, 360
  discharging, 377
  high-intensity, 370
  measuring, 360
  troubleshooting, 376–378
  video camera, 367
  See also Cathode-ray tubes (CRTs)
"Pie crust" distortion, 140
Piezoelectric effect, 196
Pincushion, 367, 370, 400–401

computer monitor troubleshooting and, 403, 404–405
Pixels, 382, 384, 389, 391
Plastic leaded chip carriers (PLCCs), 209
Plastic quad flat packs (PQFPs), 209
Platinum wires, 5
PLCCs. See Plastic leaded chip carriers (PLCCs)
PLL. See Phase-locked loop (PLL)
PLL detectors, 259–260
P-N junctions, 38, 52, 198, 256
PNP transistors, 135, 167, 227, 262, 337
  astable, 90
Point-to-point transmission, 9
Polarity, 5, 6
  comparison and, 35
  feedback and, 13
  oscillation and, 15
  rectification and, 52–54
  regulators and, 62
Polarized line plugs, 52
Portable electronic devices, 84
Positive value, 5
Potentiometer, 91, 122, 366
  color control, 328
Power amplifiers, 13, 123, 262–266. See also Amplifiers
Power cycling problems, 103, 104–105
Power gain
  amplifiers and, 13, 14
  current gain vs., 229
Power lines
  electric shock and, 71
  radio waves vs., 15
Power supplies, 4, 47, 50–81
  auxiliary, 50, 74
  computer monitor and, 399, 402
  horizontal deflection and, 115–150
  integrated circuits and, 38
  linear. See Linear power supplies
  scan-derived, 88–94, 99
  scan-rectified, 93–94
  start-up, 89–90
  switched-mode. See Switched-mode power supply (SMPS)
  sync signals and, 115–186
  transformerless, 50, 51–52
  troubleshooting, 70–78
  vertical deflection and, 151–186
  See also Source voltages
Power transformers, 50–51. See also Transformers
Power transistors
  filtering and, 54, 55–57
  regulators and, 59
  SMPS and, 84, 85
  troubleshooting and, 74, 76
  See also Transistors
PQFPs. See Plastic quad flat packs (PQFPs)

Pre-emphasis network, 254–255
Preamplifiers, 84, 200
  remote, 203, 204
Prescalers, 35, 36, 212
  frequency synthesis and, 38, 39
  in PLL, 40
Preset condition, 33
Pricing, margin vs. mark-up, 424
Primary colors, 314–315, 319
  color-bar generators and, 339–341
  computer monitor and, 390
  convergence and, 364–369
  CRT troubleshooting and, 377
  delay lines and, 320
  electron guns and, 356
  purity magnets and, 364
  shadow mask and, 360, 362
Printers, 41
Probe, 42, 100, 102, 103, 106
  CRT troubleshooting and, 377
  demodulator, 102
Process, 4
Processing signals, 189
Professional development
  continuing education in, 416–418
  cover letters and, 418
  interviews and, 419
  parts suppliers and, 426–427
  resumes and, 418–419
  self-employment and, 419
  versatility and, 416–420
Profit and loss statement, 421
Progressive scanning, 383, 386
Projection television, 355–380
  disadvantages of, 369
  display methods for, 369–376
  electrostatic focus in, 367, 368
  front-projection vs. rear-projection, 370, 372
  shadow mask and, 360
  Sylvania, 95–98
  Zenith, 372–375
Propagation delay, 31, 85
Pulse-amplitude modulation, 17
Pulse generators, SMPS and, 84
Pulse modulation, 17, 86, 87
Pulse-pulse modulation, 17
Pulse-rate modulation, 86
Pulse train, 86
  troubleshooting and, 42, 43
Pulse width, 7, 42, 61, 87
  driver stage and, 122
Pulse-width modulation, 17, 86, 87
Pulse-width regulator, 98
  troubleshooting SMPS and, 107, 108
Pulses, 7, 32, 33, 42, 86
  binary, 37, 39
  blanking, 295. See also Blanking signals
  clock, 35, 36
  conversion and, 37
  high-voltage transformer and, 91

horizontal oscillator and, 122
IHVT and, 92
measuring, 287
sampling, 338
scan-derived power supplies and, 88–89
sine wave oscillators and, 130
sync, 116–120, 125, 152. *See also* Sync signals
vertical deflection and, 156–157
Purity, 339, 361, 364
Purity magnet, 363
Pushbutton device, 201

Q. *See* Quality (Q) of inductor
Q signal. *See* Quadrature (Q) signals
Quadrature (Q) signals, 314–315, 317, 318, 331
Quadrature detectors, 258–259
Quadrupler circuits, 64, 65
Quality (Q) of inductor, 11, 198–199, 223, 227
Quartz crystal, in PLL, 40
Quartz crystal oscillators, 196, 210
Quasar television, 144–147
Quasi-parallel sound method, 252–253
Quasi-split-sound system, 253

R – Y. *See* Red-minus-luminance (R – Y) signal
Radiated interference, 9
Radio
    superheterodyne receivers in, 19
    troubleshooting, 21–22
Radio bands, 9–10
Radio broadcasts
    AM, 9, 16–17, 19
    amateur, 9, 16
    FM, 10, 16, 17, 19
    foreign, 9
Radio-frequency interference (RFI), 9, 86
Radio-frequency (R-F) signals, 9–10
    AGC, 245. *See also* Automatic gain control (AGC)
    amplifers and, 15
    chrominance and, 317
    heterodyning and, 17–18, 19
    modulation and, 16–17
    receiver operation and, 18–19
    troubleshooting SMPS and, 102
    tuners and, 192, 209, 210
Radio waves, 6
    power lines vs., 15
    range of, 8–9
    types of, 9
"Rainbow generator," 339
RAM. *See* Random-access memory (RAM)

Ramp waves, 6, 7, 37
Random noise, 9
Random-access memory (RAM), 200, 203, 387, 391
    display memory and, 391
    video (VRAM), 387, 392
Raster, 74, 76
    computer monitor, 389, 398, 405
    horizontal output stage and, 123, 140
    keystone, 141
    purity and, 364
    sync signals and, 116
    troubleshooting SMPS, 105–106, 107, 144
    tuners and, 209
    video amplifiers and, 307
Ratio detectors, 257–258
Rauland CRT, 392–395
RC coupling, 123
RC-coupled transistor amplifier, 290–291
RC filters, 55, 152
RC network, 398
    series, 121
    vertical deflection and, 153–157
RCA television, 76, 78, 89, 108, 109, 303, 345
Reactance, 10, 11
Reaction scanning, 123
Read-only memory (ROM), 200, 387
Real-time analysis, 42
Rear-projection system, 370, 372
Receiver operation, 18–19
    sequence scanning and, 47
    troubleshooting, 21–22, 70
    volume control and, 203
Receiver stages, 46
Receivers
    flyback circuit for, 95. *See also* Flyback
    large-screen television, 53
    linear regulators for television, 58
    locking problem with, 245
    scan-derived power supply and, 90
    small-screen television, 52, 64
    television, 52, 53, 64, 75
    transformerless power supplies for, 50, 51–52
    tuners and, 209
Rectangular waves, 6, 7, 15, 122
    horizontal frequency and, 140
Rectification, 5, 50, 52–54
    filtering and, 54–57
    IHVT and, 92
    scan-derived power supply and, 89
    SMPS and, 85, 86, 106
Rectifier diodes, 240–241
Rectifiers, 50
    chrominance and, 342
    faulty, 20
    integrated, 54

shorted, 108, 110
SMPS and, 84
troubleshooting and, 70, 74, 342
Red, green, blue projection tubes, 372
Red, green, blue rasters, purity and, 364
Red, green, blue signals
    computer monitor and, 390–391
    luminance, 335, 336, 341
    See also entries beginning RGB
Red-minus-luminance (R – Y) signal, 331–335, 342
Reference oscillator, 38, 39
Reference signals, 12, 37, 39, 40
Refractive system, 370, 371–372
Refresh, 382, 383, 384–385
    CGA and, 390
    vertical, 389, 390
Registers, 35, 36, 40, 41
Regulation, 5–6, 50, 57–64
    high-voltage transformers and, 91
    load vs. line, 57
    scan-derived power supply and, 89
    SMPS and, 85, 106
Regulations. *See* Government regulations
Regulators, 50
    chrominance and, 342
    common problems with low-voltage, 74
    efficiency of, 59
    hybrid, 58
    linear, 58, 61
    overcurrent protection and, 66
    series, 57, 58, 66
    shunt, 57–58, 67
    SMPS and, 98, 103
    troubleshooting and, 74–75, 342
Relaxation oscillators, 136
Reliability, 19, 38
Remote control device, 41, 84, 201, 203, 204, 211–212
Renter's insurance, 426
Repair checklist, 416
Reset condition, 33
Resistance, 5, 10
    continuity and, 74
    inductors and, 11
    measuring, 73, 74
    regulators and, 58–59
    transistor switches and, 85
    troubleshooting and, 70, 76
    video amplifiers and, 289
Resistance-capacitance coupling, 12, 14
Resistance-capacitance RC-coupled transistor amplifier, 290–291
Resistors, 5, 50
Resume, 418–419
    amplifiers and, 123
    computer monitor and, 395, 406
    differentiators and, 122
    filter, 56
    metal oxide, 85
    open, 20, 70, 103

overcurrent protection and, 66
power dissipation and, 123
R-F signals and, 10
scan-derived power supply and, 90
SMPS and, 85
troubleshooting and, 108, 342, 406
Resolution, horizontal vs. vertical, 389
Resonance, 10, 11
high-voltage transformer and, 91
tuners and, 196
Resonant circuits, 10, 11, 38
Retrace, 121, 130, 338
CRT troubleshooting and, 377
interlaced scanning and, 382
Reverse AGC, 229
Reverse bias, 52, 53, 56, 120
maximum, 54
sound signals and, 256
R-F. See Radio-frequency (R-F) signals
R-F amplifiers, 193–195, 209, 210
video amplifiers and, 300
R-F signal generators, 277
RFI. See Radio-frequency interference
(RFI)
RGB. See entries beginning Red, green,
blue
RGB driver, 334–339
RGB monitors, 390, 391
RGB video, 392
RGBI monitors, 390
Ringing, 242, 243
Ripple voltage, 55–57, 61, 62
SMPS and, 103
Rise time, 15, 16, 34, 42
rectangular wave and, 122
transistor switches and, 85
video troubleshooting and, 305
Rms. See Root mean square (rms)
ROM. See Read-only memory (ROM)
Root mean square (rms), 7

Safety, 4, 20
CRT troubleshooting and, 376–377
projection television and, 372, 375,
376–377
SMPS and, 100
with transformerless power
supplies, 52
troubleshooting and, 70, 71
tuners and, 209
X rays and. See X rays
Safety capacitors, 140
Safety-sensitive components, 71
Sampling pulse, 338
Sampling rate, 37
Samsung television, 108
Sandcastle signal, 329, 330, 342
SAP. See Separate audio programming
(SAP)
Satellite transmission, 9, 41, 209

Saturation, 15, 85, 86
chrominance troubleshooting and,
339, 340
color, 315–319, 320, 328, 329
rectangular wave and, 122
SAW filter. See Surface acoustical wave
(SAW) filter
Sawtooth voltage, 124
Sawtooth wave oscillators, 142
Sawtooth waves, 6, 7, 15, 116
horizontal AFC circuits and, 124–127
horizontal frequency and, 140
horizontal output transistors and,
131–133
vertical deflection and, 155–157, 162
Scan-derived power supplies, 88–94, 99
high-voltage transformer and, 92
horizontal deflection and, 137–138
troubleshooting SMPS and, 102–103
Scan-rectified power supply, 93–94
Scan-velocity modulation (SVM),
293–294
Schematic Solutions, Inc., 427
Schematics, locating, 427
Schmidt optical system, 370
Schmitt trigger, 32, 34, 43
SCR. See Silicon-controlled rectifier
(SCR)
SCR crowbar circuits, 67
Screen, 356, 357
aluminizing of, 359
blanking of, 339
grid for, 358
SECAM. See Sequential Coleur Avec
Memoire (SECAM)
Secondary colors, 377. See also
Complementary colors
Selecting channels, 38, 41
Selecting operating frequencies, 38
Selectivity, 18, 19, 193
intermediate frequency and,
222–223, 242
Self-employment, 419
Semiconductor material, 37, 38, 39
Semiconductor testor, 101–102
Semiconductors, 5, 14
countdown circuits and, 121
integrated rectifiers and, 54
peak inverse voltage and, 54
transformers and, 50
Sensitivity, 18, 19
Separate audio programming (SAP),
270–272
Sequence scanning, 47
Sequencer, 391
Sequential Coleur Avec Memoire
(SECAM), 383–384
Sequential logic, 32–33
Serial data, registers and, 36
Serializer, 391
Series circuits, 63, 66

Series-feedback regulators, 60
Series peaking, 292–293
Series-pass regulators, 58–59, 61, 66,
68–69
troubleshooting and, 74, 75
Series RC network, 121
Series regulators, 57, 58
Series-resonant circuits, 10
Series-resonant traps, 223–224
Service calls, 245
chrominance, 345
CRT, 377–378
HDTV, 405–406
luminance, 308–309
power supply, 76–78
SMPS, 107–110
sound, 280
sync signals, 144–147, 179–180
Service contracts, 425. See also Customer
service
Service manuals, 72
Set condition, 33
Seven-segment display, 37
Shadow mask, 356, 360–361, 362
in Zenith computer monitor, 393, 395
Sharp television, 75, 76
Sharpness control, 293
SHF band, 10
Shift register, 36, 387
Shunt capacitors, 62
Shunt-feedback regulators, 58, 59
Shunt peaking, 292–293, 339
Shunt regulators, 57–58, 67, 84
Shutdown, 105–106, 140–141, 399–400
Shutdown circuits, 372, 375
Sidebands, 192
chrominance and, 315, 320
color, 218–219
vestigial transmission and, 219–220
Siemens, 5
Signal generator, 304
Signal injection, 21–22
horizontal AFC and, 142
intermediate frequency and, 245
sound troubleshooting and, 273–274
troubleshooting and, 43, 102
Signal-to-noise ratio, 8, 230
comb filter and, 322
Signal paths, 4, 46
Signal processing, 41
Signal tracing, 21, 22, 102
chrominance troubleshooting and, 339
Signal voltages, 4, 6
Signals, 6
clock. See Clock signals
high vs. low, 28, 30–31, 200
in-phase. See In-phase (I) signals
manipulating, 16–20
out-of-phase, 6, 12, 14, 57, 192–193, 317
processing. See Processing signals
quadrature. See Quadrature (Q) signals

radio-frequency. *See* Radio-frequency (R-F) signals
reference, 12, 37, 39, 40
reshaping, 34
separation of, on frequency response curve, 193
sync. *See* Sync signals
test, 21–22
two-way flow of, 35
Silicon-controlled rectifier (SCR), 58, 60–62
  auxiliary power supplies and, 92
  SMPS and, 84, 85, 86, 104
  sweep voltage and, 121
  troubleshooting SMPS and, 104
Simultaneous distribution, 34–35
Sine wave. *See* Sinusoidal (sine) wave
Sine wave oscillators, 121, 129–130, 134–135, 142
  schematic drawing of, 142
  troubleshooting, 142–143
Sine wave stabilizing network, 122
Single-sideband modulation, 16, 17, 40
Single-sideband transmission (SSB), 17
Sinusoidal (sine) waves, 6–7, 15, 38, 193
  Schmitt trigger and, 34
  sound troubleshooting and, 277
  tuner oscillators and, 195
  vector angles and, 315–317
Skin effect, 10, 11
Sky waves, 9
Slew rate, 16
Slice level, 119
Small outline integrated circuits (SOICs), 209
Small-scale integration, 31
Small-screen television receivers, 52, 64
Smearing, 242, 243
SMPS. *See* Switched-mode power supplies (SMPS)
SOICs. *See* Small outline integrated circuits (SOICs)
Soldering, 20, 21, 103, 378
  computer monitor and, 405
Sony Corporation, 358, 361. *See also* Trinitron tubes
Sony monitors, 402
Sound, 251–284
  flub-flub-flub, 104
  HDTV, 386
  monaural, 266–273
  multichannel, 268, 270–272
  problems with, 107, 140, 209
  for projection television, 370
  stereo, 268–272
  synthesized, 268, 270
  problems with, 74, 76, 105, 144, 242–244
  "ticking (tic-tic-tic)," 107
  troubleshooting, 273–280
  tweet-tweet-tweet, 104

Sound carriers, 238–242
  diagram of extraction points of, 240
  MTS signals vs. standard, 271
Sound detectors, 255, 257–260
Sound intermediate frequency, 218, 238–242
Source, 5, 85, 86
Source voltages, 5–6
Space width, 7, 122
Speakers, 272
Speed-up capacitor, 128
Split-sound method, 218, 252, 253
Split-supply Class AB amplifiers, 264–265
Square waves, 7, 15, 90, 122
  in digital systems, 32, 34
  Schmitt triggers and, 34
  sound troubleshooting and, 274–275, 277
  SMPS and, 98, 106
  VCO and, 39
  video troubleshooting and, 305, 307, 308
S-R flip-flop, 33
SSB. *See* Single-sideband transmission (SSB)
Stability, 15, 85
  in digital systems, 32, 39
  feedback bias for, 64
  oscillators and, 122, 196
Stages, 11, 12, 13, 46
  amplifier, 193–195
  cascaded amplifier, 14
  of horizontal deflection, 116
  interstage coupling and, 32
  mixer, 196–197
  oscillator, 195–196
  troubleshooting SMPS and, 102
Stagger tuning, 234, 235, 244
Start-up voltage, 89–90, 98
  troubleshooting SMPS and, 103, 106
Static convergence, 361–362, 365
Static troubleshooting, 4, 21–22, 72
Step-down switched-mode regulator, 94–95
Step-up switched-mode regulator, 94–95
Stereo sound systems, 268–272, 279–280
  embedded controller in, 41
Storage register, 36
Storage time, 16, 42, 85
  rectangular wave and, 122
Subcarrier reference system, 330–331
Subharmonics, 8, 38, 39
Summing amplifier, 325, 326
Super VGA (SVGA), 387, 388–392
Superband channels, 192, 222
Superheterodyne receivers, 19
Supply voltage, chrominance and, 341–342

Surface acoustical wave (SAW) filter, 223, 234–237, 244
  sound signals and, 253
Surround sound, HDTV, 386
SVGA. *See* Super VGA (SVGA)
SVM. *See* Scan-velocity modulation (SVM)
Sweep circuits, 84
Sweep rate, 173–174
Sweep voltage, 121, 141
Switched-mode power supplies (SMPS), 83–113
  auxiliary power supplies and, 92–93
  diagrams of, 96–97, 107, 109
  electric shock and, 100
  overvoltage protection and, 93–94
  in projection television, 372, 374
  scan-derived and, 88
  testing, 101–105
  troubleshooting SMPS, 99–110
  typical problems with, 103–105
Switched-mode regulators, 86
Switches/switching, 4, 5, 11–12, 15–16
  audio, 272–273, 275
  conversion and, 37
  BCD system and, 31, 200
  horizontal driver and, 122
  intermediate frequency and, 221
  latching, 62, 85
  microprocessors and, 41, 200
  overload and, 100
  in PLL, 40
  transistor, 85
  troubleshooting and, 42
Switching element, 32
Sylvania television
  luminance/chrominance for, 303
  no video and dim raster on, 342, 345
  no video on, 308–309
  troubleshooting for no front panel display in, 76
Sylvania television diagrams, 345
  luminance circuit, 302
  projection television, 95–99
  SMPS, 107
  sound processing system, 269
  waveforms, 106
Symbols, 31, 51
Symmetrical clipping, 280
Symptom-function analysis, 21
Sync circuits, video amplifiers and, 300
Sync paths, diagram of, 287, 320
Sync processor, 98–99
Sync processor IC, diagrams of, 119, 137
Sync pulses, color, 318–319, 329, 341
Sync selector switch, 102
Sync separation, 116–120, 154
  troubleshooting and, 139, 142
Sync-separator stage, 116
Sync signals, 115–186
  chrominance and, 319–320
  color, 218

computer monitor and, 390–391, 398
defined, 116
HDTV, 385–386
horizontal deflection and, 115–150
intermediate frequency and, 219
troubleshooting and, 137–147
tuners and, 192
vertical deflection and, 151–186
Synchronization, 32, 35, 116, 151
    chrominance and, 319
    computer monitor, 387, 389
    intermediate frequency and, 239–240
    oscillators and, 123
    PLL and, 40
    vertical deflection and, 174
Synchronizing signals, 116
Synchronous counter, 36
Synchronous demodulator, 221, 241,
    332, 333–334
    diagram of, 261
Synchronous detectors, 259–260,
    332–333
Synthesized sound, 268, 270
Synthesized stereo circuits, 268, 270

Teleconferencing, 384
Television
    AM and FM in, 16, 19
    amplifiers in, 15
    common problems with power
        supplies for, 75–78
    complex problem with, 238
    diagram of, 4
    direct-view, 360, 367, 368, 369
    electric shock and. See Electric shock
    embedded controller in, 41
    frequency synthesis and, 38
    front panel display of, 76
    high-definition. See High-definition
        television (HDTV)
    large-screen, 53
    linear regulators for, 58
    microprocessors in, 40, 41
    modern. See Modern television
    older, 64, 65, 66, 91, 122, 142, 326
    projection vs. direct-view, 369. See
        also Direct-view color picture tube;
        Projection television
    receiver stages and signal paths
        for, 46
    repair of, 21. See also Trouble-
        shooting
    satellite. See Satellite transmission
    small-screen, 52, 64
    superheterodyne receivers in, 19
    troubleshooting, 20, 21. See also
        Troubleshooting
    tuners of, 209. See also Tuners
    UHF band and, 10
    VHF band and, 9

See also Color television
Temperature, 20, 59, 72
Thermal noise, 9
Thermal runaway, 263
Thermistors, 85
Thomson Consumer Electronics, 384
3–D graphics, 392
Thyristor, 62
Time
    delay. See Delay time
    switching and, 15–16
    transient recovery, 67
    transit, 227
    video troubleshooting and, 305–307
Time domain, 6
Time-varying circuits, 38
Timing, 35–36
Tint, 315–319, 329
Tint control, 329
    color-bar generator and, 340
TOC computer monitor, 405–406
Toggle (T-type) flip-flop, 33
Tools, 21
Trace, 121, 123–127
    horizontal AFC and, 142
    interlaced scanning and, 382
    multiple, 174
Tracking, 337–338
    counters for, 36
    gray-scale, 369
Trade associations, 417
Train, 7. See also Pulse train
Transducers, 237
Transformer coupling, 14, 123
Transformerless systems, 50, 51–52
Transformers, 5, 50–51
    center-tapped, 53
    filters and, 54–55
    flyback. See Flyback
    high-voltage, 91–92
    intermediate frequency and, 227–228
    isolation, 70, 71, 106
    in older televisions, 64, 65
    scan-derived power supply and,
        89, 90
    SMPS, 84, 85, 87, 98–99, 106
    variable ac, 101–102
Transient recovery time, 67
Transistor audio amplifiers, 262–265
    troubleshooting, 277–278
Transistor noise switch circuits, 120
Transistor switches, 85
Transistor sync-separator circuits,
    118, 154
Transistor tester, 245
Transistor-transistor logic (TTL),
    38, 392
Transistor video amplifiers, 289–292
Transistors, 37, 38, 39
    amplifiers and, 123
    buffer, 201

horizontal deflection and, 129
horizontal output, 131–133, 135, 140
intermediate frequency and, 227
leaky, 107, 277
monaural sound and, 266–268
open, 140
overcurrent protection and, 66
overvoltage protection and, 67
power dissipation and, 123
scan-derived power supply and, 90
as series-pass regulators, 58
shorted, 20, 74, 75, 104, 140, 277–278
SMPS and, 85, 98–99
sync separation and, 154
troubleshooting and, 74–75
vertical output, 162
Transit time, 227
Trapezoidal waveform, 158–160, 307
Traps, 223–225, 239, 242
    sound signals and, 253
Triac, 85, 86
Triangular waves, 6, 7, 15
Trickle-start voltage, 89, 90
Triggered sweep, 174
Trinitron tubes, 358, 361, 368
Tripler circuit, 38, 64, 65, 91, 92, 324–325
Troubleshooting, 4
    checklist for, 20–21, 417
    chrominance/luminance IC, 341–342
    color bandpass amplifier circuits, 340,
        341, 343
    color burst amplifier circuits, 344
    color demodulators, 341–345
    color killer circuits, 343
    color-processing circuits, 339–345
    comb filter circuits, 342, 343
    computer monitor, 401–406
    digital circuits, 42–43
    hand tools for, 21
    horizontal deflection, 137–147
    IC audio amplifiers, 278–279
    intermediate frequency, 242–246
    logical, 20–21, 70, 102, 213, 305
    luminance/chrominance processors, 308
    measurements in, 71–73, 74
    microprocessor controls, 211–213
    picture tubes, 376–378
    power supply circuits, 70–78
    SMPS, 88–110
    sound circuits, 273–280
    static vs. dynamic, 4, 21–22, 72
    stereo amplifiers, 279–280
    sync circuits, 137–147
    transistor amplifer circuits, 277–278
    tuner systems, 209–213
    vertical deflection, 171–182
    video, 20–23
    video amplifiers, 304–309
    video output circuits, 344
Truth tables, 29, 33, 42
TTL. See Transistor-transistor logic (TTL)

Tuner mixer stage, 197
Tuner oscillator, 39
Tuners, 28, 38, 40, 41, 191–215
  auxiliary power supply and, 92
  bandswitching circuit and, 199,
    208–209
  color-bar generator and, 340
  frequency synthesis and, 199, 204,
    207, 209, 212–213
  intermediate frequency and, 234
  microprocessors and, 199–209
  for projection television, 370
  sound signals and, 253
  troubleshooting, 209–213
Tuning
  improper, 212–213, 238
  stagger, 234, 235, 244
Turn-off protection, 65–66
Turn-off time, 16, 85
"Turn-off" winding, 60
Turn-on time, 15, 61, 84
"Turn-on" point, 61
Tweeter, 272

UH. *See* MicroHenry (uH)
UHF band, 10, 192–199, 222
  dividers and, 208
  improper tuning of, 212–213
  varactor diodes and, 198
UJT. *See* Unijunction transistor (UJT)
Ultor, 359
Ultraviolet light, 6
Unijunction transistor (UJT), 61
Unipotential focus, 368

Vacuum tube, 356
Varactor diode, 39, 198–199, 208
Variable ac transformers, 101–102
Variac, 101–102, 105, 106, 108
Varistors, 85, 86
VCO. *See* Voltage-controlled oscillators
  (VCO)
VCR, 87
  counters and, 209
  embedded controller in, 41
  encoding and decoding for, 37
  power supply for, 84, 87
  tuners of, 209
Vector, defined, 315
Vector addition/subtraction, 316–319
Vector angles, 315–317
  comparison of I and Q signal, 318
Velocity, 6
Vertical blanking pulse, 152
Vertical centering, 166
Vertical deflection, 28, 47, 93, 116
  blocking oscillator and, 160–162,
    174–176

computer monitor and, 398–399, 403,
  405–406
diagram of, 157
multivibrators and, 162–164, 176–179
retrace time for, 130
sync signals and, 151–186
troubleshooting, 171–182
Vertical detail, 325–326
Vertical drive, 119
Vertical drivers, 162, 164–166
Vertical dynamic convergence, 365–367
Vertical foldover, 171–173
Vertical hold control, 157, 176
Vertical integrator, 119, 156
Vertical linearity controls, 158–159, 172
Vertical oscillators, 116, 119, 123,
    155–157, 160–164
  computer monitor and, 390
  horizontal oscillators combined
    with, 129
  noise spike and, 120
Vertical output amplifiers, 159–160,
    164–166
Vertical output transistors, 162
Vertical refresh, 389, 390
Vertical scanning, 382
  computer monitor, 389
  IC-based, 166–171
  troubleshooting, 179
Vertical sensitivity, 174
Vertical size control, 158, 172
Vertical sweep oscillators, 157
Vertical-sweep shutdown circuit,
    372–375
Vertical sync pulses, 116, 121, 152
Vertical sync signals, computer monitor
    and, 398, 404
Very large-scale integration (VLSI), 31
VESA. *See* Video Electronics Standards
    Association (VESA)
VESA SVGA, 391
Vestigial sideband compensation,
    219–220
Vestigial sideband system, 386
VGA. *See* Video graphics array (VGA)
VHF band, 9, 10, 192–199, 222
  dividers and, 208
  improper tuning of, 212–213
  keyed rainbow generator and, 342
  mixer stage and, 196
  varactor diodes and, 198
Vibration, 20
Video accelerators, 391–392
Video amplifier stage frequency
    response, 288–292
Video amplifiers, 15, 285–312
  chrominance and, 320
  color-bar generator and, 340
  computer monitor and, 395, 398
  diagram of, 300
  direct coupling and, 289, 296–297

high frequencies and, 291–292
low frequencies and, 289
medium frequencies and, 289–291
simplified drawing of, 289
troubleshooting, 304–309
Video camera picture tubes, 367
Video cards, 41, 386
Video carriers, 238–242
  diagram of extraction points of, 240
Video compression system, 386
Video conferencing, 392
Video detectors, 116, 238–242, 244
  color-bar generator and, 340
Video Electronics Standards Association
    (VESA), 391
Video games, 87, 392
Video graphics, 382
Video graphics array (VGA), 386, 388–392
  Super (SVGA), 387, 388–392
  ZCM-1492, 392–401, 403–406
Video intermediate frequency, 218
Video monitors. *See* Monitors
Video output amplifier circuits, 336–339,
    344. *See also* Video amplifiers
Video output stage, 398
Video output transistors, 377
Video paths, diagram of, 287, 320
Video RAM (VRAM), 387, 392
Video receivers
  HDTV and, 384
  safety with, 71
Video signal gain, 294–301
Video signals
  black portion of, 242
  blocking dc component of, 296
  composite, 153, 240, 241–242, 286,
    295–296
  computer monitor, 390–392, 395
  demodulating I-F, 239–241
  intermediate frequency and, 219, 238–242
  luminance and, 321. *See also* Luminance
    signals
  sound signals and, 253
  sync separation and, 116–120
Video systems, 218
  AGC in, 229
  auxiliary power supply and, 93
  HDTV, 386
  microprocessors in, 41
  transformerless power supplies for, 50
VLF band, 10
VLSI. *See* Very large-scale integration (VLSI)
Voltage, 5–6
  bias, 63–64. *See also* Bias/biasing
  de-emphasis network and, 256
  excessive, 70, 75–76. *See also*
    Overvoltage protection
  measuring, 71–73, 74
  oscillation and, 15
  pre-emphasis network and, 254–255
  ripple. *See* Ripple voltage

sawtooth, 124
scan, 88–89
sine waves and, 6–7
sweep, 121, 141
transformers and, 50–51
yoke, 91, 99
zener diode regulator and, 58
Voltage amplifiers, 13
Voltage dividers, 62–63, 74
   SMPS and, 85, 108
   sync separation and, 119
   See also Dividers
Voltage multipliers, 64–65
Voltage regulators. *See* Regulators
Voltage spikes/surges, 86, 100, 104
Voltage supplies, 5–6
Voltage-controlled oscillators (VCO),
      38, 39–40, 136, 138
   AFPC circuit and, 331
   diagram of, 206
   multivibrators and, 205–206
   PLL and, 206
   tuners and, 205–206
Voltage-divider bias circuits, 63–64
Voltage-regulating transformers, 51. *See*
      also Regulators; Transformers
Volts, 5, 7
Volume control, 203, 205, 212
VRAM. *See* Video RAM (VRAM)

Wave traps, 223–225
Waveforms, 70
   of auxiliary power supply, 93

Class AB operating, 263
composite video signal, 153, 240, 241
computer monitor and, 398, 400–405
cross-over distortion, 263
East-West, 400
filters and, 55, 56
for full-wave rectifier, 53
HDTV sync, 385–386
horizontal deflection, 133, 134–135
horizontal output equivalent circuit,
      132, 140
horizontal oscillators and, 122
negative-polarity composite
      video, 240
North-South, 400, 401
oscillation and, 15
oscilloscope and, 174, 307
parabolic, 366–367
positive-polarity composite video, 241
pulse and, 7
at R – Y demodulator, 342
of red, green, blue luminance
      signals, 336
sine wave oscillators and, 130
at 3.58-MHz crystal, 342
troubleshooting SMPS and, 102, 106
types of, 6–7
vertical deflection, 157–160, 162,
      174–176
vertical integrator, 156
Wavelength, 6–9
White, luminance signal and, 323
White noise, 9

White screen, 288, 293
   gray scale and, 369
Whiter-than-white edge, 325–326
Wien-Bridge oscillator, 162, 196
Wires, 5, 9
Woofer, 272
   sub-, 386
Worker's compensation insurance, 426
World Wide Web, 417

X rays, 6, 93–94, 105
   CRT troubleshooting and, 377
   evacuation and, 359
   horizontal deflection and, 133, 141
XGA standard, 391
XOR function, 30
X/Y demodulator system, 333, 334

Yoke coils, 121, 135, 371–372
Yoke voltage, 91, 99

Zener diodes, 57–58, 59, 61
   leaky, 342
   in SCR crowbar circuits, 67
   sound signals and, 256
   troubleshooting and, 75, 76
Zenith computer monitor (ZCM-1492),
      392–401
Zenith Electronics, 384
Zenith television, 303, 372–375
Zero bias, 288
Zoning regulation, 426